SCIENCE IN THE TWENTIETH CENTURY AND BEYOND

SCIENCE IN THE TWENTIETH CENTURY AND BEYOND

JON AGAR

polity

For Kathryn, Hal and Max

CONTENTS

CONTENTS

ACKNOWLEDGEMENTS

This book has been on my mind for over a decade. It has benefited immensely from my efforts to explain the subject to undergraduates at Manchester, Cambridge, Harvard and University College London. I also thank the staff at those institutions. There are many individuals who have helped. I would like particularly to thank John Pickstone and Peter Bowler, the anonymous referees, and Andrea Drugan at Polity Press.

— 1 —

INTRODUCTION

This is a history of science in the twentieth and twenty-first centuries. There are weaknesses as well as strengths in such a project. By necessity, this history is a work of synthesis. It draws extensively on a wealth of secondary literature, to which I am indebted, but which has addressed twentieth- and twenty-first-century science in an uneven fashion. Where good secondary literature is scarce I have been dependent on received histories to an uncomfortable extent. Furthermore, to borrow Kuhnian terminology, the 'normal' strategy of recent history of science has been to take a received account and to show, through documentary and other methods, that an alternative narrative is more plausible. This is good historiographical practice, and where it works it has led to the recognition of missing actors, themes and events as well as the questioning of assumptions and the encouragement of scepticism towards the uses of history in the support of ingrained power. But, just as in normal science, the projects tend to be devised to be at the scale of the doable, which leads to a narrowing in scope. This tendency can lead to the generation of accounts that are mere tweaks of received stories.

On the positive side, I have found that forcing myself to think on the scale of all science in the twentieth century has made the asking of certain research questions unavoidable, in ways I feel that the traditional focused, case-study-led history of science has avoided. I was determined, for example, not to break the subject down into disciplinary strands, since that would replicate existing histories. Instead I have been forced to look out for patterns that were held in common across scientific projects in the twentieth century, across disciplines, across nations. That is not to say that the national stories drop away. Indeed, comparative assessments of national science become

1

particularly important, and again lead to new questions being asked. I think the most important of these, for a history of twentieth-century science, concerns why the United States rose to become the predominant scientific superpower. I will return to this question later.

But, first, a prior question. This is a history of science in the twentieth century, and beyond. But what do I mean by 'science'? The guidance offered by historian Robert Proctor that 'science is what scientists do' provides a good working definition, although it does downplay activities such as science teaching.[1] I have tried to cover the range of activities that people who have been granted the role 'scientist' have pursued. Physical and life sciences are squarely within the scope of this book. In an extended treatment, social scientists would be present when they have considered their work to be within the sciences. The boundaries are both fuzzy and interesting – fuzzy in the sense that the category of 'scientist' can be contested, and different cultures have different senses, some broad, some narrow, of what counts as 'science'. (It is a cliché at this point to contrast the broad German notion of *Wissenschaft* with the narrower English sense of 'science'.) They are interesting because the label of science was a prized cultural attribute in the twentieth century, and the controversies sparked by efforts to police the boundary of the term, to separate insiders from outsiders, what sociologists of science call 'boundary work', are revelatory.[2]

I propose, however, a more substantial model of 'what scientists did' in the twentieth century. As a step towards this model, imagine that it is night, and you are in an airliner which has just taken off and is banking over a city at night. What you see is a glittering set of lights. You are impressed – it is a beautiful, sublime sight. After you rise above the clouds you see nothing but darkness. The flight is a long one. By the time you descend it is very early morning. On the approach to your destination you see a second city. This time you can see not only the lights but also the roads, the buildings, the parks and the factories. Even this early in the morning, there is a bustling sense of activity, of a world going to work. The pattern of lights now makes sense.

I think these two images – of a city at night and a city at dawn – provide a metaphor for how we can better make sense of science in the twentieth century. At first what we see is a dazzling and awesome array of isolated lights – quantum theory here, the sequencing of the human genome there, the detonation of atomic bombs in the middle, famous experiments, celebrated scientists, revolutionary theories. This image of history of science leads ultimately to 'timeline histories'

2

which can be found on the web: history as isolated, bright moments. What we don't see is why the lights of science form the pattern they do.

Working worlds

I argue that sciences solve the problems of 'working worlds'. Working worlds are arenas of human projects that generate problems. Our lives, as were those of our ancestors, have been organized by our orientation towards working worlds. Working worlds can be distinguished and described, although they also overlap considerably. One set of working worlds are given structure and identity by the projects to build technological systems – there are working worlds of transport, electrical power and light, communication, agriculture, computer systems of various scales and types. The preparation, mobilization and maintenance of fighting forces form another working world of sometimes overwhelming importance for twentieth-century science. The two other contenders for the status of most significant working world for science have been civil administration and the maintenance of the human body, in sickness and in health. It is my contention that, just as we can make sense of a pattern of lights once we can see them as structured by the working movement of a city at dawn, so we can make sense of modern science once we see it as structured by working worlds. It is a historian of science's task to reveal these ties and to describe these relations to working worlds.

One of my motivations for talking of 'working worlds' was the feeling that the use of the metaphor of 'context' by historians of science had become a cliché. Routinely we speak of understanding science in its social, cultural or political context. I think it is a cliché in George Orwell's sense of clichés as worn ways of writing that were once alive and are now dead; a clichéd metaphor was a metaphor that was once startling and thought provoking and which now passes unnoticed.[3] 'Context' originally made us think of 'text' and 'context', which in turn invited all kinds of associated questions of interpretation. While the term 'working worlds' is not a like-for-like replacement of 'context', I have deliberately chosen to avoid the word 'context', as far as possible, in order to force myself to reconsider science's place.

I will go further. Sciences solve the problems of working worlds in a distinctive way. Working worlds are far too complex to have their problems solved directly. Instead there are, typically, a series of steps.

First, a problem has to be recognized as such – problematization is not a given but an achievement.[4] Second, a manageable, manipulable, abstracted representative of this problem has to be made. Across disciplines, across the decades, science has sought such representatives – a quick list might include mouse models that are representatives of cancerous human bodies, census data as representative of national population to be governed, a computer-based General Circulation Model as a representative of the world's climate, or the abstraction from the messy reality of Amazonian soil of data that can said to represent the Amazonian rainforest many thousands of miles away.[5] These abstractions have been well named, given their origin in working worlds, 'microworlds'. They are of a scale that can be manipulated in controlled manner, measured, compared and moved.[6] The representatives are also, notice, human-made: science, even when talking about the natural world, talks about artificial worlds. Irving Langmuir's study of electron emission and light bulbs for General Electric, an exemplary early twentieth-century piece of research, has been aptly called the 'natural study of the artificial'.[7]

A special case of this abstractive effort concerns phenomena: natural effects that are deemed important and interesting but are difficult to manipulate in their wild state. The special place of laboratories, especially in the sciences since the mid-nineteenth century, can be explained because they have been places where 'phenomenal microworlds' are made.[8] Sociological documentaries of science support this account of the power of the laboratory.[9] But scientists, too, have offered strikingly similar descriptions. 'What is a scientific laboratory?' asked Ivan Pavlov in 1918. 'It is a small world, a small corner of reality. And in this small corner man labours with his mind at the task of . . . knowing this reality in order correctly to predict what will happen . . . to even direct this reality according to his discretion, to command it, if this is within our technical means.'[10]

Once the model of the working world problem is in place, its use can move in two directions. First, it can become subject to one of the stock of developed techniques for manipulating, analysing and comparing. Second, the conclusions have to be moved back into the working world. (Like problematization, this is not a given: 'solutionization' is also an achievement.) Because we are now dealing with simplified versions, there are often commonalities of structures and features, which in turn lead to commonalities of techniques. These similarities are the reason, for example, for the ubiquity of statistics in modern science. The discipline of statistics is, at least partly, a meta-discipline; it is a science of science. But neither the particular

4

abstraction made nor its meaning are isomorphically determined by the working world (borrowing the precise mathematical sense of isomorphism – of a one-to-one relationship between structure of working world problem and model or meaning). Different representatives can be made, and they can be made sense of in different ways. This leads us to theory building, rival data collection, and experiment – the stuff, of course, of science. Also this is where the agonistic social character of science matters: the active encouragement of challenge, scepticism and criticism – between possessors of representative models.[11] The fact that the scientists are dealing with abstractions, and that the social organizations for framing challenge are distinguishable (although not separate) from others, encourages a sense of science's autonomy, jealously guarded. But, crucially, there is an unbreakable thread that passes back to the working world – which may be ignored, often forgotten, but is never absent. My suspicion is that no meaningful science has been generated that cannot be identified with a working world origin.

I am arguing that science is the making, manipulation and contest of abstracted, simplified representatives of working world problems. Part of the outcome of this work was bodies of knowledge, but science should not be solely or even primarily identified with these. As philosopher John Dewey reminded his audience in his widely circulated essay 'Method in science teaching', science was 'primarily the method of intelligence at work in observation, in inquiry and experimental testing; that, fundamentally, what science means and stands for is simply the best ways yet found out by which human intelligence can do the work it should do' – solving working world problems.[12] I should also clarify what I am not arguing. There have been attempts in recent decades to find a classification more helpful, and more realistic in its description of science to the world, than 'basic science' and 'applied science'. Gibbons and colleagues offered a new mode of production of knowledge ('mode 2'), in which 'knowledge produced in the context of application' has supposedly become more prominent since the Second World War. Historians (Shinn, Godin) have torn down the division, as I discuss in chapter 18.[13] Motivated by a similar prompt to find policy-useful language, Donald Stokes added to the simplistic basic–applied dyad another axis and argued that 'use-inspired basic research' ('Pasteur's quadrant') was most strategically important.[14] I turn the notion of 'applied science' upside down. I am not arguing that human health, efficient administration, weapons or industry are merely (if at all) applied science. Rather, science is applied world.

The possession of skills to manipulate representatives of working world problems granted scientists considerable authority and power in the modern world. The powerful intervention back into the world, justified by this possession, was neither straightforward nor uncontested. Such interventions occurred at different scales. The possession of demographic representations – such as census data – worked at national and municipal scales and reformed power relations at these levels. 'Yet today', for example, wrote Don K. Price in the middle of the century, 'the most significant redistribution of political power in America is accomplished by the clerks of the Bureau of Census.'[15] The intervention could be as local as the enforcement of an individual's dietary regime. Or the ambitions could be global in scope. The X-ray crystallographer and would-be planner J. D. Bernal wrote in 1929, for example, of his hopes that, soon, the whole world would be 'transformed into a human zoo, a zoo so intelligently managed that its inhabitants are not aware that they are there merely for the purposes of observation and experiment'.[16]

Overview

If the existence of working worlds, and their crucial relationship to the sciences, is my primary finding, what are the other headlines from this history of twentieth-century science and beyond? I identify four: the extraordinary and unambiguous importance of the working world of warfare in shaping the sciences, the emergence of the United States as the leading scientific power, the missing stories, and the swing from physical to life sciences in the second half of the twentieth century. Each of these four themes is to be found in the following history. I will also address them again directly in the conclusion. However, this history is written chronologically, and the following is a brief synopsis of the contents.

Send out the clones

The first question we might ask is what the relationships were between the nineteenth and twentieth centuries. Why should the twentieth century be like the nineteenth century? Did like beget like? Or, if not, where have the centuries differed, then why? Of course 1900 is an arbitrary date. But picture in your mind another guiding image: the river of history, temporarily frozen. If we slice this river at 1900, then we can ask what was being carried forward from the nineteenth into the twentieth century. The great distinctive achievement of the nineteenth century was the invention of methods of exporting similarity. Similar things were created by the methods of mass production. Similar ways of conducting research were invented in Paris – think of the modern hospital and the museum – and in the German states – think of the research universities and the research-based industries. These models were exported across the globe. Similar scientists were made at institutional innovations such as the teaching laboratory, the research school and the professional society. They, too, exported similarity as the similar scientists travelled. Laboratories were places where entities of interest to working worlds, such as pathological bacteria, could be made visible and manipulable – or, in the case of standard units, stable and exportable. This mass export of similarity – which makes the river of our image a vast floodtide – made the continuity between the nineteenth and twentieth centuries a social achievement.

Science 1900

Chapter 2, 'New Physics', begins the story of science as it emerged from the nineteenth into the twentieth century. I survey the remarkable decades from the 1890s to the 1910s, which witnessed the observation of new phenomena of rays and corpuscles, the articulation and challenge of new theories, and the development of new instruments and experimental procedures in physics. Topics include Röntgen's X-rays, Curie's radioactivity, Planck and Einstein's quantum mechanics, and Einstein's special and general theories of relativity. Drawing on the recent insights of historians of physics, I relate these extraordinary developments to the working worlds of late nineteenth- and early twentieth-century industry.

Chapter 3, 'New Sciences of Life', begins with another turn-of-the-century event present in any simplistic timeline history of science – the rediscovery of Mendel's theory of inheritance – and summarizes how historians have shown that the 'rediscovery' was constructed as part of disciplinary manoeuvres by interested scientists. The working world of relevance here was that of 'breeding' in two closely intertwined senses: of eugenic good breeding and in practical agricultural improvement. The power of science, I have argued above, lies in its ability to abstract and manipulate representations relevant to working world problems. At this abstracted level a science of genetics, typified by the work of Morgan's fruit fly research school, was articulated. Biochemistry and plant physiology were also sciences of these working worlds.

7

Chapter 4, 'New Sciences of the Self', looks at the sciences of development, including anthropology and child study, but particularly psychology and psychoanalysis. The Freud I describe is a scientist by training. The working worlds of human science were the administration of institutions – the classic Foucauldian sites of the asylum, the school and the army. The representatives of working world problems included the Pavlovian dog and the behaviourist rat. I also survey the early twentieth-century disputes in immunology as the science emerged from nineteenth-century bacteriology and offered solutions, if perhaps not 'magic bullets', that could move out of laboratories to address the problems identified by the working worlds struggling with infectious disease.

Science of a world of conflict

Part II surveys the sciences of a world torn by conflict. Chapter 5 examines the sciences called forth by the working world of military mobilization during the First World War. The career of Fritz Haber demonstrates how the skills and knowledge developed to provide mass nutrition could be so easily transferred to the provision of new methods of mass killing. The mass destruction offered by the machine gun also illustrates how a world geared to the export of similarity was structurally prepared for mass death. The cult of the individual was encouraged by the contrast it drew with the realities of organization and mobilization. Henry Gwyn Jefferys Moseley is my case study. The mobilization of civil scientists, complementing an already deep commitment to organization for war, is traced in Britain, the United States and Germany. The success of American psychology shows how a discipline could benefit from the ratchet of war.

Chapter 6 surveys how Germany, the leading pre-war scientific nation, responded to the widespread sense of cultural, political and social crisis in the post-war decades. Paul Forman called historians' attention to the fact that crisis talk seemed to inflect physical theory, including one of the most profound intellectual developments of the twentieth century, the recognition of the depths and strangeness of quantum theory in the 1920s. Forman's claim for the causal effect of the intellectual milieu has proved controversial but also productive. Among other responses to crisis were the later shaping of Gestalt theory in psychology, parasitology, embryology, genetics and, with the Vienna Circle's project to establish the 'unity of science', philosophy.

The working world of imperial administration continued to be an important one for science. In chapter 7, 'Science and Imperial Order', I examine the contributions of preventative and tropical medicine and ecology as sciences that could address imperial and colonial working world problems. The notion of an 'ecosystem', for example, was proposed as an intervention in a dispute between rival British and South African imperial visions. Likewise, food chains and models of population fluctuations were of great interest to those who wanted to manage diverse parts of the world to maximize the extraction of resources. Conservation has always been a technocratic project.

The recycling of industrial wealth as philanthropy is of crucial importance to understanding science in the United States in the twentieth century. Chapter 8, 'Expanding Universes: Private Wealth and American Science', focuses on the tapping of the fortunes of Carnegie and Rockefeller by scientist-entrepreneurs. George Ellery Hale persuaded philanthropists to build new peaks of science, not least great reflecting telescopes that shifted leadership in optical astronomy from Europe to the United States and made Edwin Hubble's observations possible. Second, I emphasize the importance of private science, not least in the under-appreciated significance of petroleum science – even in such matters as the fate of Wegener's theory of continental drift. Finally, I argue that the working world of industry generated its own sciences – sciences that took the working world as its subject. Scientific management is perhaps the best-known example.

Chapter 9, 'Revolutions and Materialism', visits the extraordinarily fraught position of the sciences in the Soviet Union as communism emerged from civil breakdown, struck out in new directions in the 1920s, and was bloodied by Stalinism from the end of the decade. Scholarship from before and after the end of the Cold War is brought together. The Soviet Union is fascinating for historians of science, not least because the state philosophy was a philosophy of science: dialectical materialism. What 'materialism' might be made to mean had sometimes lethal implications for physicists interested in quantum mechanics or relativity, or for biologists interested in genetics. Just as the working worlds of private industry shaped American science, so the demands from the working world of Soviet administration conjured new directions and made certain research questions of overriding importance. I discuss the distinctive work of the psychologist Vygotsky, the biologist Oparin, the biogeochemist Vernadsky, the plant geneticist Vavilov, and the agronomist Lysenko. Finally, I raise, as a case of both continuity and contrast, a second revolutionary and materialist project: the extension of the range of Darwinian explanation that carries the label the 'Evolutionary Synthesis'.

9

Chapter 10, 'Nazi Science', explores the sciences in Germany after the seizure of power by Hitler's National Socialist Party. In common with the conclusions of our best historians of Nazi science, I reject the simple but consoling argument that Nazi power was purely destructive of science. The destruction of careers and lives for some created opportunities for others. And, among those who seized the opportunity to promote their sciences, not all were self-serving ideologues, such as the proponents of 'Aryan physics'. Some sciences were fundamental to Nazism: 'National Socialism without a scientific knowledge of genetics', one German said, 'is like a house without an important part of its foundation.'[17]

Chapter 11, 'Scaling Up, Scaling Down', surveys the increasing scale of instrumentation in the 1920s and 1930s and asks where the skills and ambitions of scaled-up science came from. The sciences include cyclotron physics, macromolecular chemistry and molecular biology. While we will visit sites as diverse as Cambridge's Cavendish laboratory, Eli Lilly's factories and the workshops of IG Farben, the geographical focus is distinctively Californian. The scientific actors are Ernest Lawrence, Linus Pauling and Robert Oppenheimer, who appears on the stage for the first time.

The Second World War and Cold War

Part III traces the consequences of this scaling up for a world in conflict in the Second World War and the Cold War. Scaling up is the common feature of the stories of penicillin, radar and the atomic bomb, as well as of the less well-known development of biological weapons, traced in chapter 12. The sciences were placed on a permanent war footing in the Cold War. I treat the subject across two chapters. Chapters 13 to 16 examine the Cold War reconstruction of the sciences in broad thematic terms: the sustained increase in funding, the expansion of rival atomic programmes, the consequences of secrecy and national security, and the formation of Cold War problems as a working world for post-war sciences. The Cold War framed competition and collaboration in space research and geophysics, as illustrated, in particular, by the International Geophysical Year of 1957–8. I examine what I call the 'Cold War sciences': human genetic health, systems ecology, fundamental particle physics, cosmology (the last two share the 'standard model', another highpoint of twentieth-century science), radio astronomy, cybernetics, and digital electronic computing, information theory and post-war molecular biology. Many of these shared a distinctive language of information and codes, a commonality that makes sense only when the shared working world of the Cold War is brought into their historical analysis.

Sciences in our world

Part IV offers an account of transition. While the Cold War remained a frame, freezing some relations, others were in movement as working worlds structured by late capitalism came to the fore. Markets were deregulated, entrepreneurial activity was increasingly valorized, and a preference for networks of entities reconfigurable in the name of profit was pronounced. Chapter 17 asks whether the long 1960s, a period lasting from the mid-1950s to the mid-1970s, is a useful category for historians of science. I support arguments that identify a dynamic that led to a proliferation of experts, which could increasingly be heard offering conflicting testimonies in the public sphere. The existence of conflicting public expert testimony led to the question being asked (a revival from the 1930s) of what were the grounds for belief in knowledge claims. Social movements provided the institutional support for the questioning of authority. My examples include pesticide chemistry and psychopharmacology. Meanwhile what I call 'neo-catastrophism' flourished in population science, paleontology, evolutionary biology, astronomy and climate science.

Chapter 18 examines the capitalization of the sciences. Trends towards the re-engineering of life in the name of visions of new economies, vast increases in biomedical funding, and the continuing and deepening informatization of the sciences were punctuated by events such as the end of the Cold War and the emergence of new diseases such as AIDS. Networks of various kinds provided the infrastructure and the subject of sciences. Giant programmes, such as the Human Genome Project, one of many organism sequencing projects, were both organized as networks and revealed networks. Chapter 19 begins by accounting for how one of the ubiquitous networks of twenty-first-century life, the World Wide Web, was prompted by the managerial need to coordinate the documents of networks of physicists. Networks of science reporting had diversified in the last third of the twentieth century. Scientific disciplines were reconfigurable, as the vogue for interdisciplinarity encouraged. Nanoscience provides an example. Finally, in chapter 20, I offer two surveys. One looks backwards and analyses the four major thematic findings of this book. The other looks forward, reviewing the sciences of the twenty-first century.

Part I

Science after 1900

— 2 —

NEW PHYSICS

Waves, rays and radioactivities

Around 1900, physicists were confronted with a bewildering array of new phenomena. Understanding these waves, rays and forms of radioactivity would transform physics. However, they did not spring from nowhere. Crucially, the new physics was a response to the working worlds of nineteenth-century industry and global networks of communication.

From the mid-nineteenth century, in Britain and Germany, but especially Germany, physics laboratories, within universities dedicated to research, emphasized the value of precision measurement of physical quantities. Very little of this work was what in the twentieth century would be considered theoretical physics. Instead it was intimately connected with the industrial and commercial projects of the day. Under a gathering second industrial revolution, science-based industries exploited electrical phenomena and new chemistry. The German synthetic dye industry expanded hand in glove with organic chemistry. In William Thomson's physics laboratory at Glasgow University, the design of instruments capable of precision measurement responded to the projecting of submarine telegraph cables that spanned the world. In Germany, Britain and the United States, measurement of ohms, amperes and volts was essential to new electrical systems of electrical power and electric light. Science, industry, economy, and national and international competition were ever more intertwined.

The laboratories isolated and reproduced, controlled and manipulated phenomena that outside, 'in the real world', were compounded or transient, undomesticated or invisible. These abilities, which could

15

only be achieved with considerable discipline and skill, were the sources of the power of the laboratory, and they were what made laboratories highly valued, even essential, by the second half of the nineteenth century. If the skills and laboratory discipline could be replicated – possible only once teaching laboratories and research schools had inculcated the often highly tacit knowledge in students who could travel, and once metrological networks had made standard scientific units mobile – then there existed one of the necessary conditions for the worldwide replication of new physical phenomena. In addition, the globalization of the world economy in the nineteenth century confirmed and expanded trade routes whereby people and materials could move around the world.

For example, the development of electromagnetic theory was intimately connected to changing industrial practices and concerns. James Clerk Maxwell, a young Cambridge graduate in the mid-nineteenth century, had written to William Thomson for advice on what paper of Faraday's to read; the advice was direct: read Faraday on retardation, the great bane of telegraph transmission.[1] Nor were industrial problems merely prompts for kick-starting theoretical investigations. Maxwell's mathematical version of Faraday's fields required much fine-tuning of constants, such as the ratio of electromagnetic and electrostatic units, which led Maxwell into the practical, and industrial, science of metrology. Even when he retreated from academia to his Scottish estate to write his magnum opus, *Treatise on Electricity and Magnetism*, published in 1873, Maxwell would record the relevance his theory had to the working world: the 'important applications of electromagnetism to telegraphy have . . . reacted on pure science by giving commercial value to accurate electrical measurements, and by affording to electricians the use of apparatus on a scale which greatly transcends that of any ordinary laboratory. The consequences of this demand for electrical knowledge, and of the experimental opportunities for acquiring it, have already been very great.'[2]

James Clerk Maxwell is a name known to generations of physics students primarily for 'Maxwell's equations', which express in mathematical language the relationships between the changing quantities of electric and magnetic fields. Yet the form in which they are universally taught was not that given by Maxwell in 1873, but one by a group historians label the 'Maxwellians': Oliver Lodge, George F. FitzGerald, to some extent Heinrich Hertz and, most centrally, Oliver Heaviside.[3] Heaviside had been so successful as a telegraph cable engineer that he had retired at the age of twenty-four to devote his life to electrical theory.[4] It was Heaviside who wrote out Maxwell's equations in their familiar and compact four-line form. Just as importantly, the Maxwellians in the 1880s interpreted the equations as permitting wave solutions and deduced some properties of these 'electromagnetic waves'. Lodge, in his laboratory in Liverpool, began a series of experiments, discharging currents from Leiden jars. In Germany, Hermann von Helmholtz, who had identified the wave solutions, urged two junior physicists, Heinrich Hertz and Wilhelm Conrad Röntgen, to test Maxwellian theory. Hertz was successful, producing in 1887 at the University of Bonn electric spark discharges that indicated that an electromagnetic wave had propagated across space.[5] He was using old instruments (Riess coils, used for producing impressive sparks) for new purposes: the production, control and manipulation of electromagnetic waves. 'Electromagnetic waves had existed before their artificial production', notes historian Sungook Hong, 'but with Hertz, these waves became the subject and the instrument of research in physicists' laboratories.'[6] The waves could be made to reappear in other laboratories, and replication of Hertz's success encouraged continental physicists to take up Maxwell's theory.[7] Indeed the export of similar laboratory spaces had spread to such a degree that, before 1895, J. Chandra Bose could not only re-create Hertzian waves in his lab at Presidency College, Calcutta, he could also generate them at much shorter wavelengths. Indeed, Bose then turned the town hall of Calcutta into an extension of his laboratory practices with a spectacular public demonstration, using Hertzian waves to ignite gunpowder and trigger the ringing of a bell.

Another episode that illustrates this process of discovery and replication was Wilhelm Röntgen's extraordinary revelation of X-rays. Röntgen was in many ways a typical product of the German-model research university. He had a PhD from Zurich (in German-speaking Switzerland) and worked as a physics professor in Strasbourg (ceded to the new German Empire in 1871), Giessen (the home of a famous mid-nineteenth-century teaching laboratory under Justus Liebig) and, from 1888, Würzberg, in north Bavaria. Röntgen's laboratory was full of the kinds of apparatus that exemplify the interplay between physical science and new industry. From the mid-nineteenth century, physicists such as Julius Plücker at Bonn, Johann Wilhelm Hittorf, and Eugen Goldstein, working under the great Herman Helmholtz in Berlin, as well as William Crookes in London, had investigated the phenomena exhibited when currents were passed through glass tubes that had been filled with gas, or evacuated to produce a vacuum, and then subjected to magnetic fields. The components were all the stuff of new industry: electrical currents, high voltages, strong glass, electrical wire, magnets, new gases. Conversely, the control of electrical discharges, along with any related phenomena, was of great interest to the burgeoning industries of electrical power generation. The apparatuses designed by these scientists in their laboratories isolated and displayed often beautiful phenomena: 'cathode rays' that made the glass walls glow or 'canal rays' that travelled in straight lines. Röntgen's laboratory could only have been assembled in a world which interconnected academic physical science and electrical industry.

On the afternoon of Friday, 8 November 1895, Röntgen was experimenting with cathode ray tubes. German physicists understood electromagnetic phenomena in terms of vibrations and movements in ether, an elastic fluid that pervaded all space. Helmholtz had predicted high-frequency oscillations in the ether that might be detectable in cathode-ray discharges.[8] Philipp Lenard, a senior and highly able experimentalist, had sought to identify radiations from high-voltage cathode ray tubes from the late 1880s but had been stopped when, in 1894, Hertz's sudden death impelled him to take over the directorship of the physical laboratory at Bonn. Röntgen later refused nearly all invitations to reflect on his discovery, but he did state tersely that 'There were some aspects of the Hertz–Lenard experiments with cathode rays that needed further investigation.'[9] Röntgen started by reproducing all of Lenard's experimental results. Now, in late 1895, he had the techniques mastered and refined.

In a darkened room, Röntgen set up a sparking interrupter (which would produce a high-voltage spark at the close of a switch), attached it to a cathode ray tube, and wrapped the whole in black cardboard. His thinking was: the voltage will produce a flood of cathode rays, they will be absorbed by the glass, and the glow will be contained behind the cardboard; any high-frequency ethereal emanations might be revealed as a glow in a barium platinocyanide screen. This last instrument had been devised by the British mathematical physicist George Stokes, who had sought explanations of crystal fluorescence in terms of ether theory.[10] Röntgen was planning to place the screen almost touching the cardboard, where Lenard's phenomena had been found. But, before he did so, something caught his eye. He noticed that, as he flicked the interrupter switch, the barium platinocyanide screen was glimmering.[11] (Indeed, so faint was this glimmer that had Röntgen looked directly at the screen he might have missed it; as astronomers know, the indirect vision of the eye is more sensitive.) The glimmer remained as he put matter of increasing thickness between the tube and the screen: cardboard, a thousand-page book, a wooden shelf. Röntgen, by now trembling, ate a short, late dinner.

Afterwards he returned to the laboratory. Through the evening he varied his experiment. Lead and aluminium cast dark shadows – clear evidence of a ray emanating from the tube. With the screen close to the tube, it lit up, shuddering with green light, while as far as 2 metres away a dim glimmer remained. This ray could leap across distances. Yet the next observation shook Röntgen to the core. Moving a small lead disc across the screen he saw a skeleton's hand – his own. Is this a trick, he thought? A prank? Or, if real, a manifestation of the spirit world? Many of the cathode ray physicists – Crookes, in particular – believed in spiritual phenomena and saw clear parallels between the manifestation of effects in the séance and the laboratory. And just as the spiritualists had sought to capture objective evidence of the spirit world through photography, so Röntgen immediately began fixing the skeletal images on photographic plates.

Röntgen now worked feverishly. He must substantiate his discovery if he was to claim priority and the credit of the scientific community. November and December 1895 were spent in the laboratory. He sometimes did not go home to sleep or eat. His wife, Bertha, in fact visited him in the laboratory, and it was the bones of her hand, wearing a wedding ring, that were captured in a fifteen-minute exposure taken three days before Christmas. He persuaded a friend, the president of the local Physical Medical Society in Würzberg, to include a handwritten paper, announcing the discovery of his 'new kinds of rays', as a very late addition to the society's proceedings. He had priority by publication. Now, to spread the word about his 'X-rays', Röntgen bundled up preprints with copies of his photographs and sent them to the leading physicists of Europe, mailing the packages on New Year's Day 1896.[12] Any laboratory equipped with similar cathode-ray instrumentation could soon replicate the X-ray effect. By Sunday, 5 January, a report on X-rays was the headline story of Vienna's leading newspaper. Röntgen was, very quickly, extraordinarily famous.

But what had Röntgen found? He had been looking for high-frequency oscillations in the ether. He had found something that did not fit expectations – hence the 'X'. If anything, it seemed a phenomenon of the séance rather than the physics laboratory. Was it even nothing new, but just an extension of Lenard's cathode rays? (Lenard certainly thought so.) The Royal Society didn't even seek to resolve the issue when it awarded its 1896 Rumford Medal to both men for the discovery of 'Lenard or Röntgen rays'.[13] Much has been made of the serendipity involved in Röntgen's discovery, but perhaps the most serendipitous aspect was the choice of that 'X' – far more newsworthy than if the X-rays had been called 'extended Lenard rays'. Even more impressive than the name were the photographs, an astonishingly powerful statement in themselves – morbid, reproducible, suggesting science's ability to see inside the body, to uncover secrets. Röntgen had the photographs, Lenard did not.

While the X-ray experiments were rapidly replicated, there was no consensus on their nature. Indeed, for nearly two decades the rays were interpreted sometimes as particles, sometimes as waves. And within each category there was plenty of room for interpretation; perhaps, for example, X-rays were impulse waves produced by the collision of cathode rays with glass or metal? Indeed, not until 1912 and 1913, when Max von Laue reported on experiments that seemed to show X-rays diffracting and interfering, did the view that X-rays were extremely short waves predominate.[14]

By then the work of the theoretical interpretation of X-rays was almost an academic sideshow, dwarfed by the growth of X-rays as an industrial and medical technology. It was electrical industry that packaged Röntgen's laboratory practices up so that they could be made to work almost anywhere. Within four months of Röntgen's announcement, the Edison Decorative and Miniature Lamp Department, of Harrison, New Jersey, was placing advertisements for the sale of 'Complete outfits for X-ray work', 'for professionals and amateurs', including Ruhmkorff Coils (to discharge the sparks), an A/C generator, Crookes tubes, and 'Fluoroscopes and Fluorescent Screens'.[15] X-rays were being used to treat cancer within two months of Röntgen's announcement. Yet, despite the efforts of industry and the interest of physicians, early X-ray tubes, glass vessels evacuated to very low pressures and subjected to high voltages, were very fragile and unreliable. The turning point for X-rays as a medical procedure was in 1913, when William David Coolidge, at General Electric's corporate laboratory, designed an improved, robust X-ray tube. Manufactured in quantity, and predictable in its operation, the Coolidge X-ray tube allowed X-ray machines to become a routine part of American hospital medicine after the First World War.[16]

The industrial production of X-ray apparatus and the excitement around the possibilities of the penetrating X-rays as tools for industry combined to encourage one of the most important scientific techniques of the twentieth century: X-ray crystallography. In 1912 the nature of X-rays, in particular whether they were particles or waves, was disputed. In February, Max von Laue, a Munich *Privatdozent* who had been educated at the centres of German physics and mathematics, including Göttingen and Berlin, reflecting on the passage of X-rays through crystal lattices, intuited that the similarity between the size of the crystal cells and the purported wavelength of the X-rays created the conditions for interference phenomena – if, of course, X-rays were waves. There was opposition. As von Laue recalled in his Nobel speech:

> [The] acknowledged masters of our science, to whom I had the opportunity of submitting [a proposal to test the idea], entertained certain doubts . . . A certain amount of diplomacy was necessary before [we] . . . were finally permitted to carry out the experiment according to my plan, using very simple equipment at the outset. Copper sulphate served as the crystal, since large and regular pieces of it can easily be obtained. The irradiation direction was left to chance. Immediately from the outset the photographic plate located behind the crystal betrayed the presence of a considerable number of deflected rays, together with a trace of the primary ray coming directly from the anticathode. These were the lattice spectra which had been anticipated.[17]

Max von Laue then wrote out the mathematical theory that related the crystal structure to the interference patterns. Moving in reverse, the interference patterns could now be used to make deductions about the unknown structure of a crystalline material. X-ray diffraction therefore produced not only evidence, which nearly all physicists regarded as decisive, for interpreting X-rays as waves, but also a tool for investigating the structure of matter at scales of a thousand millionth of a metre. X-ray crystallography was rapidly developed, in Germany, in Sweden and, especially, in Britain, by the father and son William and Lawrence Bragg.

Röntgen's X-rays were only one of several new rays, radiations and particles announced in the closing years of the nineteenth century, and all of them were indications of working world concerns. At the Cavendish Laboratory in Cambridge, the successor to Maxwell and Rayleigh as professor of experimental physics, Joseph John Thomson, investigated the cathode ray phenomena, using – as had Röntgen – the improved vacuum pumps 'that had recently been developed to meet the demands of the electric light bulb industry'.[18] J. J. Thomson used this advantage to intervene in a debate about the nature of electricity, over which there was a considerable range of opinion in the 1890s. The Maxwellians pictured current as the breakdown of 'electrical strain', while on the continent physicists preferred traditional particle models, and the Dutch physicist Hendrik Antoon Lorentz had, by 1892, a well-worked-out theory of current as the movement of tiny discrete electrical charges ('electrons') that he hypothesized could be found in nature; Joseph Larmor in England pictured electrons as 'singularities in a rotational ether'. Altogether, these electron theories seemed to account rather well for electrical phenomena, even making verifiable predictions, such as the splitting of spectral lines under magnetic fields found by Pieter Zeeman, Lorentz's student at Leiden, in 1896.[19]

Thomson used his industrial apparatus to support arguments concerning the nature of cathode rays. First, he set up the cathode tube so that the rays passed though metal slits, which were connected to an electrometer, an instrument that measured charge. As the visible rays were seen to pass through the slits, the electrometer registered increasing negative charge. When deflected away from the slits by a magnetic field, the charge too disappeared. For the second round of experiments, the vacuum pump was all important. Thomson showed that the cathode rays could be bent by an electric field, an effect destroyed if the vacuum of the tube was much less than perfect. Finally, Thomson carefully measured the energy of the cathode rays. Altogether, he argued, between 1897 and 1899, that his results could only be explained if cathode rays were real particles (he called them 'corpuscles', we call them 'electrons') carrying mass and negative charge.[20]

While the discovery of Thomson's electron depended on industrial techniques shared with Röntgen, a different class of strange new radiations were first identified as part of an inquiry into what caused Röntgen's apparatus to work at all. Like many others across Europe, Henri Becquerel had received Röntgen's paper in January 1896. If the X-rays were emitted from the fluorescent spot on the glass of the cathode ray tube, he reasoned, perhaps other fluorescent crystalline materials might emit rays. One such material was potassium uranyl sulphate, which when exposed to the sunlight would fluoresce (the colour depended on the mineral). Becquerel intended to expose the potassium uranyl sulphate to varying amounts of sunlight and then measure the strength of fluorescence. Delayed by cloud, he put away the plates for the day. But, on starting again, Becquerel found that a Lumière photographic plate was already exposed, despite being hidden from sunlight, while wrapped around the uranium salt. Becquerel swiftly announced his discovery of spontaneous and constant radiation that was strong in uranium salts and even stronger in refined pure uranium metal.

Soon there were several centres of research into this 'radioactivity'. In Paris, only a carriage ride away from Becquerel, Marie

Skłodowska-Curie and Pierre Curie had secured a supply of uranium and were working out means to measure the strength of the radiation. They soon found their technique by measuring the electrical conductivity of air as it changed with the passage of radiation.[21] With a means of measurement, the Curies could distinguish between the different signatures of different radioactive sources. An early discovery was that the 'radioactivity' (as Marie coined it) of thorium had different characteristics from the radioactivity of uranium. A young researcher at the Cavendish Laboratory, Ernest Rutherford, having arrived from New Zealand in 1895, labelled the radiation typical of the two elements 'alpha' and 'beta'. (Paul Villard would add 'gamma' in 1900.) Meanwhile, in 1898, in the pitchblende ore of uranium, Marie and Pierre isolated new sources – new elements – polonium and radium.

New elements opened up the interesting possibilities of a chemistry of radioactive materials in addition to the physics. Between 1898 and 1902 the emerging field of radioactivity research proceeded by classifying and characterizing the radiations and by deploying the techniques of analytical chemistry to investigate the elements further.[22] Centres were in Vienna, Wolfenbüttel (in the centre of Germany, 40 miles east of Hannover), Paris and Montreal, where the physicist Rutherford, who arrived in 1898, and the chemist Frederick Soddy, who joined in 1900, were employed at McGill University. Soddy and Rutherford combined their skills to investigate the chemical nature of the gases formed by radioactive processes; they argued in 1902 that radioactivity was a process of disintegration, in which heavier elements could be transmuted into lighter ones. They measured disintegration, inventing the concept of the 'half-life' when they noticed that the time taken for half a quantity of a radioactive element to decay was specific and constant. The language was the stuff of alchemy, but chemists objected to the theory on the grounds that it sounded like the Swedish chemist Svante Arrhenius's controversial ionic theory.[23]

By the early years of the new century, radioactivity had become a distinct and stable research field, and it had spawned an industry. Monograph titles and the foundation of journals are good indicators of new fields. Within a span of five years, several of the pioneers produced monographs, such as Marie Curie's *Radioactivité* (1903), Soddy's *Radioactivity from the Standpoint of Disintegration Theory* (1904) and Rutherford's *Radioactivity* (1905), and journals such as *Le Radium* (1903) and *Ion* (1908) were launched. Historian Jeff Hughes draws the conclusion from the subtitle of *Ion – A Journal of Electronics, Atomistics, Ionology, Radioactivity and Raumchemistry*

– that the 'intellectual place of the field' was 'on the one hand . . . between orthodox physics and chemistry but also, on the other . . . part of a new cluster of analytical practices associated with electrons, ionism, and physical and spatial chemistry'.[24]

In Paris, the Curies had a vision of radioactivity that encompassed not only laboratory analysis of small samples of radioactive material but also building up an industrial-scale manufacture and marketing of radioactive products, of which system the laboratory would be part. A company, the Société Centrale de Produits Chimiques, was delegated this task of production. But it was Marie Curie, in a profound sense, who spun out a new working world, a 'radium economy', in which the discoveries of the laboratory could be translated into commercial propositions and circulated back as laboratory material or out as forms of medicine. Radium, an element unknown before 1898, needed to be produced in macroscopic quantities if it was to be put to therapeutic use in the radium institutes. Tons of pitchblende, purchased from the Joachimsthal mine in Bohemia, were reduced first at the Société Centrale de Produits Chimiques, under André Debierne, then at the factory of Émile Armet de Lisle in the Paris suburb of Nogent-sur-Marne, then back at Marie and Pierre's laboratory at the Paris School of Physics and Chemistry, until, by 1902, a tenth of a gram of radium chloride was isolated.[25] The first gram was not available in France until 1910. By then, Curie's Radium Institute was being planned on the rue d'Ulm; funded by the University of Paris and the Institut Pasteur, it contained two pavilions, one for Marie's laboratory, the other for the biomedical research of Claudius Regaud, whose focus was on the treatment of cancer. Further institutes were built across Europe, including the Radiumhemmet in Stockholm, the Marie Curie Hospital in London (from 1929) and the Memorial Center in New York.[26]

A key service that the laboratories provided to the working world of the radium economy was the articulation and measurement of standards.[27] Much of the early work on radioactivity standards was done in Vienna. Nevertheless, in March 1912, 20 milligrams of radium, held in a glass tube, was named the International Radium Standard, and was deposited at the International Office of Weights and Measures at Sèvres, outside Paris.[28] The unit of radioactivity, that produced by five hundred times (i.e., 1 gram) the standard, was named the 'curie'. Standardization improved the practice of radiotherapy. Even with the minute quantities of radium available in the 1900s, therapies were attempted: placing samples next to skin cancers or into body cavities to reach other cancers. While radium was so scarce,

24

the gas radon was used too. A review in 1906 recommended radium therapy for skin diseases (including hypertrichosis, lupus, eczema, psoriasis and even acne), diseases of the nervous system, and some infectious diseases, such as pulmonary tuberculosis.[29] Metrology, laboratory research and medical production all supported each other in the radium economy: the physical laboratories, for example, would never require enough radium to allow the companies, such as Armet de Lisle's, to survive, so the Curies 'worked hard to expand the clinical usefulness of radium'.[30]

Meanwhile, the radioactivity research centres were multiplying. Otto Hahn, who had travelled to Montreal to learn from Rutherford, was installed in Berlin, where he would be joined by Lise Meitner. William Ramsay, who had worked with Soddy, studied gaseous radioactivity at University College London. The Montreal team split. Soddy came to Glasgow University, while Rutherford in 1907 was tempted by the promise of the resources necessary to build a thriving physics laboratory and research school at Manchester University. This project was boosted when Rutherford received the Nobel Prize for chemistry in 1908. (Nobel prizes were first awarded in 1901, with Röntgen the first recipient for physics.) In Vienna, Stefan Meyer was provided in 1910 with a new Institut für Radiumforschung, which remained one of the centres of production of radium – Rutherford's team at Manchester, for example, immensely benefited from generous access to Viennese radium.[31] Meetings tied the groups together. 'Far from being instances of benign internationalism, as many subsequent commentators have assumed', corrects Hughes, 'the numerous meetings of this sort were essential': the groups had to share material resources, ideas, and standards for the discipline to thrive.[32]

Meyer's Institute for Radium Research was the centre for the identification of another new and puzzling type of radiation. Meyer's assistant was Victor Hess, who had been investigating why the atmosphere seemed to be being made conductive by an ionizing radiation. The expectation was that such radiation would decrease away from its presumed source, radioactive materials in the earth. However, Hess, in the 1910s, carefully measured the levels of ionization at different heights, including risky experiments borne by balloon, and found the source of ionization was from above rather than from below. The source would be named 'cosmic rays' in the mid-1920s, by which time Hess had emigrated to the United States.

As the centres split and developed, styles of research became ingrained. In positivist Paris, the Curie's eschewed theoretical abstraction while, in Manchester, a theoretical reductionism was encouraged.[33] Rutherford and Thomas Royds made use of the Viennese radium to identify the 'alpha' particle with ionized helium. Now in the Manchester laboratory, Hans Geiger and Ernest Marsden, from 1908, turned the alpha particle into a probe: by bombarding a gold foil target with alpha particles, they mapped the directions taken by the ricocheting material. Most sailed through, or were deflected a couple of degrees; but every so often an alpha particle would be deflected through 45 degrees or more.

'It was quite the most incredible event that has ever happened to me in my life', Rutherford recalled. 'It was almost as incredible as if you fired a 15-inch shell at a piece of tissue paper, and it came back to hit you.'[34] His conclusion, articulated in 1910–11, was that the mass in a gold atom was not distributed evenly but concentrated in a centre, a 'nucleus'. This was a different picture of the atom to that of his Cavendish mentor, J. J. Thomson; however, as Hughes notes, the nuclear model did not make a strong immediate impression on the world of physics, and 'there was no firm consensus as to the "best" atomic model' until after the First World War.[35] Nevertheless, one Manchester physicist who did take a keen interest in the nuclear model was a visiting Dane called Niels Bohr. He melded Rutherford's nuclear picture with some odd theoretical ideas, known as 'quantum' theories, to produce a new description. The predictive success of Bohr's quantum nuclear atom would strengthen both quantum and nuclear physics.

Working worlds and the quantum theory

The origins of quantum theory can be thought of as the result of a collision between sciences relevant to the working world of steam engines and empirical measurements extracted from the working world of electrical light and power. Specifically, Max Planck, who articulated quantum theory in 1900, was seeking to preserve a thermodynamic worldview (thermodynamics was the science of efficient working engines) while explaining the blackbody spectrum (the intensity of electromagnetic radiation as emitted by a perfectly absorbing body, a surrogate for a furnace or an electric light filament). Gustav Kirchhoff in 1859 had proposed that the ratio of a body's emission to absorption of heat radiation was constant, while between 1879 and 1884 Josef Stefan and Ludwig Boltzmann had refined this idea into a law, relating the energy emitted by a blackbody to the fourth power of the temperature.

The Stefan–Boltzmann law was directly of interest to the electrical industries and to the Physikalisch-Technische Reichsanstalt (PTR), a national institute of a new type concerned with standards and careful measurement. (The institutional model of the PTR was soon copied across the world, including in Britain, the United States and Japan.) Already the German Society for Gas and Water Specialists had asked the PTR 'to develop better light-measuring devices and a better unit of luminous intensity', prompting a search for 'a scientific basis for the improved photometric devices and to develop a universally acceptable unit of luminous intensity; more generally, to furnish the German scientific and industrial communities with more accurate temperature measurements based on an absolute temperature scale'.[36] Willy Wien, of PTR, found a shortcut when in 1893 and 1894 he derived the result that the product of the wavelength and absolute temperature of a blackbody was a constant: now an accurate measurement for one temperature could be generalized. 'Wien's scientific results were seen as potentially useful to the German illuminating and heating industries', notes historian David Cahan, and the PTR redoubled its work, 'as important for technology as for science', noted the leader of the PTR's optical laboratory.[37] There followed an intense period in which increasingly precise empirical measurement of the blackbody energy as a function of absolute temperature alternated with the proposal of mathematical formulae that described the same curves. Wien had proposed one such formula, which worked well at high frequencies but not at low ones.

Enter Max Planck. Planck, like many of the PTR physicists concerned with blackbody radiation, had studied under Helmholtz, as well as under Kirchhoff, whom he had succeeded at the University of Berlin. Now he had in front of him the freshest, most accurate measurements yet, produced at the PTR by Heinrich Rubens and Ferdinand Kurlbaum. Rubens had brought the results direct to Planck's house. 'That very day', 7 October 1900, 'Planck devised a new formula that accounted for all experimental results'.[38] Only the PTR had the resources and the mission, both supported by appeal to a burgeoning empire of German science-based industry, to produce such painstakingly accurate measurements. When it was confident in the correctness of Planck's radiation law, the PTR used the formulae to establish a new exact, absolute thermometric scale.

Planck, on the other hand, was puzzling over why his formula took the form it did. It was not plucked from nowhere, although it was, in a sense, plucked from the ether. He had, since 1895, sought an electromagnetic foundation for thermodynamics, one which avoided the statistical approach of Boltzmann and from which might be derived the form of the blackbody spectrum.[39] Yet now he was forced to draw on Boltzmann's statistically derived thermodynamic laws to interpret his new radiation law. The formula now contains the term 'hv' to describe the energy of 'resonators' – the 'v' represents frequency, while the 'h' is labelled 'Planck's constant' and is regarded as one of the most profound numbers in physics. (Put crudely, setting h as zero turns all quantum theory into classical physical theory. A finite h introduces quantum effects. The size of h is therefore now regarded as a measure of the 'quantumness' of the universe.) Historians sharply disagree about the significance of this moment. Martin Klein regarded Planck's quantum work of 1900 as the moment when classical electrodynamics was left behind,[40] while Thomas Kuhn noted that Planck did not insist that the term 'hv' should be restricted only to certain quantities; that restriction, the start of a true 'quantum' theory, was introduced by a young German physicist, Albert Einstein.[41]

Einstein and Bohr

Albert Einstein was born on 14 March 1879 in Ulm, a small city in the south of Germany, half-way between Stuttgart and Munich. His family was Jewish but not especially religious.[42] When Albert was one they moved to Munich, where Albert's father and uncle established a factory specializing in plumbing and electrical goods. The company was not a success, losing out to Siemens on a crucial electrification contract. Nevertheless, the family firm provided ample context to encourage a lifelong interest in gadgets and inventions.[43] In 1894, the family moved again, over the Alps to Milan, where Albert joined them later: he did not like school, and if he had remained in Germany he would have been conscripted into compulsory military service. In 1896, young Albert entered the Swiss Federal Polytechnic (Eidgenössische Technische Hochschule, ETH) in Zurich, with the limited ambition to train to be a secondary school maths and physics teacher. At ETH he began a tempestuous romance with a fellow pupil, a Serbian named Mileva Marić. By 1901, Marić was pregnant, but the poor couple could not afford a family, and the child was given up to adoption. Despite the domestic drama, Einstein was showing extraordinary talent in physics, but hopes of an academic career were dashed through a combination of anti-Semitism and conflicts sparked by Einstein's own radical and anti-authoritarian streak.

In 1902, with strings pulled by family friends, Einstein took up a post, 'Technical Expert, Third Class', as a patent examiner at the Federal Patent Office in the Swiss capital, Bern.[44] With job security, Albert and Mileva married. They had two sons, born in 1904 and 1910. Popular histories of Einstein, written in the knowledge that he would become by far the most celebrated, iconic, instantly recognizable scientist in the twentieth century, have contrasted his lowly position at the Bern Patent Office with the incredible heights achieved in 1905, when he published four papers that, it is no exaggeration to say, mark a watershed in the history of the physical sciences. The contrast suggests that true talent – genius, indeed – will overcome any everyday obstacle. In fact, recent revisionist history by Peter Galison has shown that Einstein's seat at the Patent Office was an extraordinarily advantageous perch from which to review the innovations of the working world, while his physical training (and, yes, native talent) gave him the intellectual tools with which to reflect on them.[45] By staying in one place, the patent system channelled a stream of early twentieth-century electric technologies past Einstein's eyes.

Among the inventions seeking patents were plenty that depended on a deep grasp of what novelty meant in the fields of electric light and power and in the related field of the synchronization of electric clocks. For all the revolutionary consequences of Einstein's work, the irony is that he was firmly attached to a mechanical worldview that was far more Newtonian than the theoretical frames of many of his contemporary physicists. Newton had described a physical universe explainable in lawful terms of the movement of particles under forces. In the hands of his eighteenth- and early nineteenth-century disciples, a Newtonian universe had been likened to the operation of an immense machine, and they had sought mechanical laws to explain all physical phenomena – not merely the movement of astronomical bodies, but also phenomena such as heat and light.

But in the nineteenth century two different approaches to accounting for the physical world became increasingly attractive. So, to the Newtonian worldview, we must add, second, an energetic worldview (rooted in thermodynamics) and, third, an electromagnetic framework.[46] Under the energetic frame, the hypotheses that mechanics made – for example, that matter was made up of invisible, indivisible atoms in motion – were set aside in favour of restricting physical law to describing the relationships between macroscopic, measurable thermodynamic quantities, such as temperature, entropy or energy. The physical chemist Wilhelm Ostwald pleaded this case passionately. The electromagnetic framework we have already encountered. Drawing on the work of Faraday, Maxwell and Hertz, the electrical

29

and magnetic phenomena were understood in terms of fields mani-
festing from a continuous massless ether that pervaded the universe.[47]
The movements of ether would somehow explain all physical forces
– a confident start had already been made to explain the origins of
electromagnetic waves – and all physical properties, including charge
and mass.

One of the good reasons to reject the mechanical worldview,
argued contemporary physicists, was its inability to explain the irreversibility paradox. Newtonian laws describe a reversible system:
a machine that will work just as well in forward gear as in reverse.
Picture the solar system and run time backwards, and the movements
of the sun, planets and moons obey Newton's laws without a hint of
delinquency. Yet not all of the physical world was so orderly. An ice
cube left in the sun will melt. No one has ever seen a melted ice cube
refreeze itself. Yet if heat was a Newtonian phenomenon it should
work backwards as well as forwards. Here was the irreversibility
paradox. Ludwig Boltzmann in Germany and Willard Gibbs in the
United States had resolved the paradox by appealing to statistics.
Consider the ice cube as a statistically describable collection of water
molecules.[48] Its transition from ice to water, as it absorbed heat in a
fully mechanical manner, was just the transition from a statistically
unlikely state (all the molecules in the ice cube) to a statistically more
likely state (the molecules moving in many directions as liquid water).
This was how the increase in entropy (a thermodynamic quantity, a
measure of disorder) was understood in mechanist terms.

Between 1902 and 1904, Einstein, apparently at this stage unaware
of Boltzmann's and Gibbs's work, retraced their arguments and
derived similar results.[49] Certainly when Einstein read Boltzmann's
papers he felt a sympathetic affinity at work. By now Einstein, the
patent examiner, would have been thoroughly familiar with the
combined empirical and theoretical triumphs of the PTR and Planck.
Any patent protecting instruments to measure temperature – crucial
to industry – would be judged against the background of the PTR/
Planck conclusions. In 1905, Einstein computed the entropy of dilute
thermal radiation, and unlike Planck he enthusiastically, not reluc-
tantly, embraced Boltzmann's probabilistic approach.[50] The quantity
'hv' appeared in his calculation, as it had for Planck. But Einstein,
approaching from a probabilistic perspective, interpreted the term
as representing discrete portions (the portions one, in a probabilistic analysis, would average over). The equation, he felt, only made
physical sense if radiation is emitted or absorbed in discrete lumps,
as quanta.

'In fact, it seems to me', wrote Einstein in his first great paper of 1905, 'that the observations on "black-body radiation", photoluminescence, the production of cathode rays by ultraviolet light, and other phenomena involving the emission or conversion of light can be better understood on the assumption that the energy of light is distributed discontinuously in space.'[51] In particular, Einstein reasoned that the discrete emission and absorption of light quanta explained the photoelectric effect. This was a phenomenon described by Hertz in 1887 and known to precise measurers of the properties of cathode ray tubes (such as Philipp Lenard), electro-technical inventors such as Nikola Tesla, and manufacturers familiar to the more arcane properties of electric light bulbs and early vacuum tubes: light striking the filament of cathode ray tube would generate a current but only once the frequency of the light had been increased beyond a certain point, no matter what intensity the light source. Lenard had discussed it in detail in the *Annalen der Physik* in 1902. Now, in the same journal, Einstein argued that the transfer of energy from light to electrons in discrete quanta explained the onset of the photoelectric effect (because the energy of the light quantum increased with frequency) and why it was independent of intensity (only single quanta were absorbed by single electrons). Putting in the experimental data, noted Einstein, produced a 'result which agrees, as to order of magnitude, with Mr Lenard's results'.

Einstein explained how quantized light made sense of other physical phenomena, such as the ionization of gases by ultraviolet light and the odd anomaly of the decrease of specific heats at low temperatures. 'No major theorist', Darrigol summarizes, nevertheless, 'approved of Einstein's light quantum'; only some of the experimentalists, such as Johannes Stark in Germany and W. H. Bragg in Britain, approved, while the physical chemist Walther Nernst launched an 'extensive program for studying specific heats at low temperature in the light of Einstein's theory'.[52] Einstein himself returned to the topic in 1906 and 1907, showing in greater detail that Planck's law depended on viewing radiation in quantum terms. Nevertheless, the 'quantum', mathematical device introduced in desperation by Planck, and interpreted in real terms by Einstein, was not yet taken to the heart of physics.

Let us now return to Niels Bohr, visiting Rutherford's laboratory in Manchester. Rutherford had begun to picture the atom not as J. J. Thomson's 'plum-pudding', with negative electric charges smeared through a positive charged mass, nor like Thomson's second model of 1904, in which a sphere of positive charge contained within it

31

electrons held stable in orbit by artful positioning, but as a positively charged nucleus surrounded by electrons, orbiting like a planets around a sun. 'Metaphor', comments philosopher Arthur Miller, 'is a tool by which scientists can pass between possible worlds.'[53] But why, wondered Bohr, did the nuclear atom not just collapse? What stopped the planets falling into the sun? These concerns pushed Bohr into writing a paper, published in 1913, which provided a synthesis that would prove foundational for physics and chemistry in the years after the First World War.[54] The older ways of atom model building had, wrote Bohr,

> undergone essential alterations in recent years owing to the development of the theory of the energy radiation, and the direct affirmation of the new assumptions introduced in this theory, found by experiments on very different phenomena such as specific heats, photoelectric effect, Röntgen &c. The result of the discussion of these questions seems to be a general acknowledgment of the inadequacy of the classical electrodynamics in describing the behaviour of systems of atomic size.

The new ingredient that Bohr added to the theory of the atom was the 'elementary quantum of action' of Planck and Einstein. 'This paper', he wrote, would be 'an attempt to show that the application of the above ideas to Rutherford's atom-model affords a basis for a theory of the constitution of atoms. It will further be shown that from this theory we are led to a theory of the constitution of molecules.'[55] In particular, Bohr picked up on a series of papers by the young King's College mathematician John W. Nicholson, who had noticed that the discrete emission lines found in spectra of the solar corona and certain nebulae made sense in terms of multiples of Planck's 'hv'.[56] To Bohr this was a vital clue. He built a model of the hydrogen atom, understood as a single electron in circular orbit around a positively charged nucleus, in which the electron could move between high and low orbits only by emission (or absorption) of discrete quanta of energy. Here was an explanation of atomic stability. Furthermore, the calculation of the frequencies of light emitted as the electron descended orbits in the hydrogen atom produced the empirically known Balmer series with stunning accuracy. The English astronomer James Jeans remarked that 'the only justification at present put forward' for Bohr's quantum assumptions was 'the very weighty one of success'.[57] And, lest we are distracted by the astronomical data, Bohr reminded his readers that the 'reformation of a hydrogen atom, when the electron has been removed to great distances away from the nucleus', was also 'e.g. by the effect of electrical discharge in a vacuum tube'.

The back and forth between empirical data derived from astronomy and those derived from the artificial surrogates in earthbound physical laboratories, mediated and accounted for by quantum theory, was continued in Bohr's analysis of a helium ion. In the emission series 'first observed by Pickering in the spectrum of the star ζ Puppis, and the set of series recently found by Fowler by experiments with vacuum tubes containing a mixture of hydrogen and helium ... we shall see that, by help of the above theory, we can account naturally for these series of lines if we ascribe them to helium.'[58] In further papers, Bohr, who returned to Copenhagen in 1916, would extend these specific insights into a synthesis: into the notions of 'isotopes' (chemically similar atoms with different masses, proposed by Soddy) and a quantized atom in which the chemical nature of the atom depended on atomic charge and radioactivity was the decay of the nucleus.[59] 'This synthesis', summarizes Hughes, 'would have a significant impact on postwar matter theory, the relations of physics and chemistry, and, through the ascendance of reductionism, the broader development of science itself in the middle of the twentieth century.'[60]

The vertiginous strangeness of quantum mechanics would not be felt with full force until the 1920s. Nevertheless, its articulation and reception, alongside relativity, indicated a jump in status of theory, and of theoretical physicists, in the first decades of the twentieth century. In 1905, theoretical physics was a 'new and small profession', involving a tiny group (there were only sixteen faculty members in Germany, which itself was the centre of theory), and was considered a second-choice career after experimental physics; this status in itself made it relatively more accessible for Jewish scientists in anti-Semitic times.[61] Some have seen an ivory tower remoteness in theoretical physics. 'Like his good friend and colleague, the distinguished Berlin theoretical physicist Max Planck', suggests historian David Cassidy, 'Einstein viewed theoretical physics as a high calling, not unlike the priesthood, a calling to search for truths and absolutes that survive the personal and transitory realities of everyday life.'[62] Martin J. Klein argues that Einstein deliberately slummed at the Bern patent office because it gave him the freedom to think: Einstein 'liked the fact that his job was quite separate from his thoughts about physics so that he could pursue these freely and independently, and he often recommended such an arrangement to others later on.'[63]

In fact, Einstein's special theory of relativity, published in 1905, was not produced in monk-like isolation, or even despite his patent work, but was prompted by close scrutiny of patentable electrotechnics. To see why, it is best to follow the steps taken by Einstein in his special relativity paper, titled 'On the electrodynamics of moving bodies', quite closely.[64] Electrodynamics, the mathematical description of the movement of electrically charged bodies, as exemplified by Maxwell's equations, gave different accounts of what seemed to be the same situation. Imagine you have a coil of wire and a magnet. The wire might be attached to a light bulb so that you can tell if current is flowing. What happens when, first, a moving magnet approaches a coil of metal wire, and, second, a moving coil approaches a magnet at rest? In the first case, if the coil moved and the magnet remained still in the ether, Maxwell's equations indicated that the electrons in the coil experienced a force as the electricity traversed the magnetic field. The bulb lights up. While, in the second case, if the magnet moved and the coil stayed still in the ether, as the magnet approached the coil the magnetic field near the coil grew stronger, and this changing magnetic field (said Maxwell) produced an electric field that drove electricity around the stationary coil. Again the bulb lights up. We have the same result, but we also have two explanations of the same phenomenon.

Surely, said Einstein, one phenomenon should have a single explanation? 'Examples of this sort', he wrote, 'together with the unsuccessful attempts to discover any motion of the earth relatively to the "light medium" [the ether], suggest that the phenomena of electrodynamics as well as mechanics possess no properties corresponding to the idea of absolute rest.' Instead, he argued, start from two 'postulates'. The first said that the laws of nature (in this case, electrodynamics) should look the same in any uniformly moving frame of reference. This 'Principle of Relativity' meant that there was no way of telling which frame might be 'at rest'. There would be no appeal to absolute – Newtonian – space. The second postulate, 'which', teased Einstein, 'is only apparently irreconcilable with the former', was that the speed of light was constant, no matter which frame it was measured from. But surely these postulates were irreconcilable? Suppose in one frame a light was shining. To an observer moving in the direction of the light, wouldn't the light appear slower? And, if a difference in velocity was observable, didn't that mean one could tell the difference between frames at rest and those in motion, contradicting the principle of relativity?

Einstein argued that there was no contradiction so long as the procedure by which time was measured was reconsidered. We must look again at the 'relationships between rigid bodies (systems of co-ordinates), clocks and electromagnetic processes', he insisted. 'Insufficient consideration of this circumstance lies at the root of

34

the difficulties which electrodynamics of moving bodies at present encounters.' Einstein spelled out the procedure by which we could judge whether two events happened at the same time – or, in other words, a procedure for judging 'simultaneity'. In fact, two events that might appear simultaneous in one frame may not be so when viewed from another. Those two events might be clocks tolling the hour. Heard in one frame, the tolls would be simultaneous – noon everywhere – but heard from another the tolls would fall out of step. Time, said Einstein, was relative, not absolute.

Isaac Newton's absolute time had already come in for severe criticism by the late nineteenth century, in works such as Ernst Mach's *The Science of Mechanics* (1883) and Karl Pearson's *Grammar of Science* (1892), both texts devoured by Einstein soon after arriving in Bern in June 1902.[65] In thinking of simultaneity as conventional – as having meaning only when an agreed procedure was in place – Einstein also echoed the earlier work of Henri Poincaré, who had argued in 1898 that we collectively agree to procedures that create common hours and minutes so that we can work together. 'How', asks historian and philosopher Peter Galison, 'did Poincaré and Einstein both come to think that simultaneity had to be defined in terms of a conventional procedure for coordinating clocks by electromagnetic signals?'[66] The answer is that both were at the centres of immense campaigns to coordinate time, for industry and for empire. In the nineteenth century local time had been replaced by uniform time (on a regional or national basis) largely due to the spread of railways. Uniform time was necessary for the safe operation of railways and was transmittable via the technological system that spread alongside the rails: the telegraph. The coordination of time was essential to railways, schools, industries and armies. As electrical supply networks grew in the late nineteenth century, time coordination could be achieved through electromechanical systems: imagine each mechanical clock being kept in check by an electrical signal emitted from a central master clock.

With the intensification of trade, inaccurate maps became increasingly problematic. And good maps required good measurement of longitude, which in turn demanded the accurate determination of time – on a global scale. In 1884, at a meeting in the United States, the Greenwich meridian had been chosen as longitude zero. Standard time would be in reference to this one meridian. The French had retaliated by resuscitating the longstanding revolutionary proposal to rationalize – indeed, decimalize – time. Poincaré was charged with the evaluation of this French plan and served in the French Bureau of Longitude. As the plans developed, Poincaré became 'the administrator of a global network of electrical time', in which the task was to coordinate clocks.[67] Therefore it is no surprise to find him considering simultaneity as an issue of procedure.

Einstein in Bern, reviewing patent after patent in the area of precision electromechanical instrumentation, was also fantastically well placed to observe and understand the spread of time coordination procedures and technologies. Indeed, if Galison is correct, a clue can be found if we just sit at Einstein's desk and glance out of the window. 'Einstein must also have had coordinated clocks in view while he was grappling with his 1905 paper, trying to understand the meaning of distant simultaneity', he says. 'Indeed, across the street from his Bern patent office was the old train station, sporting a spectacular display of clocks coordinated within the station, along the tracks, and on its façade.'[68] 'Every day Einstein stepped out of his house, turned left, and made his way to the patent office', tracks Galison, 'he passed the myriad of electric street clocks recently, and proudly, branched to the central telegraph office'; 'there is much thinking to be done' on such walks, Einstein wrote to a friend.[69] Patents for coordinated clocks awaited him at his office; one that arrived in April 1905, just two months before the special relativity paper would be published, was a patent application for an 'electromagnetically controlled pendulum that would take a signal and bring a distant pendulum clock into accord', a set-up that could have come straight from the paper.[70]

Remembering Einstein

What an extraordinarily different picture of the historical Einstein emerges from this contextualization! Einstein here is not the retiring philosopher, and nor is he the protagonist in dream-like thought experiments – Uncle Albert riding a light-beam, and so on. Einstein, and the special theory of relativity, is examining intimately, and reflecting profoundly, on the working world of clocks coordinated for industry, empire and city life. Just as, likewise, Planck and Einstein and the quantum theory of light was generated within a working world of electric light industries and its demands for precise measurement and standardization.

Indeed, the simplicity of looking at what Einstein was doing to understand what he was thinking is such that the question becomes: why did we not see this before? There are three answers for why an alternative image has survived so long: self-fashioning; an understandable emphasis on the shock of profound, new ideas; and the demands of pedagogy. Later in their careers both Einstein and Poincaré would project an image of themselves as disengaged philosophers of ideas, and later commentators may have taken them at their word. Certainly as Einstein grew older, and his fame spread, he grew into the image of the kindly, otherworldly, abstracted scientist (while still an active proponent of international peace). But this image certainly did not fit the energetic, pushy, prickly youth of 1905. The overturning of absolute space also caught the eye, and historians have spilt much ink tracing continuities or identifying discontinuities with the preceding 'classical' physics. McCormmach, for example, argues that the 1905 papers, diverse as they are, express a project to reform the nineteenth-century mechanical worldview, not a revolutionary attempt to jettison it.[71] On the other hand, the kind of potted histories found in physics textbooks play up the extent Einstein's 1905 paper marked a radical break with everything that had gone before; the sentiment that Einstein's 1905 papers set the 'standard of greatness' is echoed at length by others.[72] Everyone knows the equation '$E = mc^2$'. This statement of the equivalence of mass and energy is present in the 1905 papers, although it is somewhat buried. Only after Hiroshima would the relationship between minuscule amounts of mass and immense quantities of energy be retrospectively identified – at least popularly – as Einstein's achievement.

The ways that pedagogy – specifically here the teaching of special relativity as a topic within university physics courses – has shaped our picture of the development of Einstein's thought has been studied by the historian Gerald Holton. Almost everyone who has been taught the special theory of relativity has first been introduced to the Michelson–Morley experiment. Albert Michelson was born in Prussia, but moved to the United States at an early age. He taught at the United States Naval Academy in the 1870s and was able to travel around scientific Europe on a trip between 1880 and 1883, during which he visited the universities of Berlin and Heidelberg (and therefore saw at first hand how physics prospered in the German system) as well as the Collège de France and the École Polytechnique in Paris. On returning to American academia, Michelson built a series of 'interferometers', the first in 1881, designed to measure the flow of ether produced by the earth's movement through it by the very precise measurement of the velocity of light. (Professor Edward W. Morley collaborated in the experiments.) Michelson was extremely disappointed when, even after the sensitivity of the interferometer was improved in 1887, he still found no difference in the speed of light as the earth ploughed through the ether.

Perhaps, argued the Dutch theoretical physicist Hendrik Antoon Lorentz in 1892, this null result could be explained if any object that moved through the ether was contracted in the direction of travel. The reduced time the light took to travel the shorter contracted difference would explain why no ether had been detected. Now Einstein read Lorentz – indeed, Einstein admired Lorentz for having carved out this new career of theoretical physicist. And the special relativity paper of 1905 reproduces and discusses the same equations as Lorentz's transformations. But there is no discussion of the Michelson–Morley experiment – despite textbooks' insistence that there was a causal, genetic connection between Michelson's experiment, the demolition of the ether, and Einstein's special theory of relativity paper of 1905.

Close examination of the documentary evidence shows that the link was absent or very tenuous at best. Holton uncovers several reasons why the connection was made.[73] The first purpose of science textbooks is the presentation of an inductive, clear, logical method of science, in this case involving the step from crucial experiment to theory change. Pedagogy has trumped historical correctness. Einstein was presented with ample opportunity to deny publicly the influence of Michelson's null result on the genesis of special relativity, but took none up, although he stated plainly his views when asked by correspondents. Holton writes that this specific reticence squared with Einstein's wider reluctance publicly to correct others' views on his work: 'he tolerated [for example] even the most vicious printed attacks on his work and person by Nazi scientists . . . with astonishing humor'.[74]

One particular opportunity for Einstein to speak publicly on the relation of the 1905 paper to Michelson's experiment was the first and last meeting between the two, in January 1931, when Einstein was invited to Pasadena, California. Holton tells the story well: 'The occasion must have been moving. Michelson, twenty-seven years his senior, was much beloved by Einstein from a distance . . . But Michelson was well known to be no friend of relativity, the destroyer of the ether.'[75] Michelson, himself, was sure that his own experiments had prompted special relativity. At the grand dinner, Robert Millikan introduced both great men. He set up the moment: science has, 'as its starting point, well-authenticated, carefully tested experimental facts' – for which read Michelson: 'the man who laid [special relativity's] experimental foundations'. Einstein's speech in response resisted this invitation to state publicly that Michelson's null result gave rise to special relativity, despite the immense pressure to do so.

(Nevertheless, it is often recorded – incorrectly – that he did do so, so great was this pressure.)

To complete this story we must see why Californians wanted, to put it crudely, to hear that without Michelson there would have been no Einstein. As we shall see later in this book, the social context of Californian science, led by Robert Millikan and George Ellery Hale, encouraged the funding of expensive instruments and the organization of science around them. In his autobiography, published in 1950, Millikan recalled that he had moved from the University of Chicago to the Californian Institute of Technology because there 'science and engineering were merged in sane proportions'. He set forth his ideological basis as follows:

> Historically, the thesis can be maintained that more fundamental advances have been made as a by-product of instrumental (i.e. engineering) improvement than in the direct and conscious search for new laws. Witness: (1) relativity and the Michelson–Morley experiment, the Michelson interferometer came first, not the reverse; (2) the spectroscope, a new instrument which created spectroscopy; (3) the three-electrode vacuum tube, the invention of which created a dozen new sciences; (4) the cyclotron, a gadget which . . . spawned nuclear physics [and so on].[76]

So we think of the Michelson experiment when we think of special relativity partly because good textbooks sometimes demand bad history, but also because a science characterized by investment in expensive instrumentation was beginning to flourish – particularly in California – at the same time as histories were fashioned. Millikan was right to say that the special theory of relativity sprang from a working world, but he picked the wrong instrument: it was patentable, coordinated electric clocks that came first, not Michelson's interferometer.

General relativity

After considering how the laws of science must be invariant when moving between so-called inertial frames (frames of measurement moving at uniform linear velocities) and deriving the special theory of relativity, Einstein turned in the years before the First World War to measurement in accelerating and falling frames. The central postulate this time was that there was no means, by local measurement, of determining if one was in a falling frame or in a frame that was under no acceleration. Alternatively, an accelerating frame would re-create in indistinguishable ways the effects of gravity. (Imagine you are in a falling lift. You are floating weightless in the air. In the brief time you have left before the lift smashes into the ground there is no physical experiment you could perform, within the lift, that would tell you that you were not floating in gravity-free space.) This generalization of the postulates of special relativity led to the general theory of relativity. The process was not smooth and involved the testing of many candidate equations – indeed, a close examination of Einstein's notes, especially the so-called Zurich notebook, shows that he stumbled on the 'correct' equations in 1912, three years before he would come back to them and publish the gravitational field equations, in 1915.[77]

The ideas, again, were profound. The theory is, undeniably, beautiful. But many physicists in 1915, when confronted with Einstein's gravitational field equations, would have scratched their heads. The general theory of relativity does not only introduce a new theory of gravity, it also introduces a set of new mathematical tools, the tensor calculus. (Mathematical notation is mere shorthand representing relationships and tools for manipulating those relationships; it is no surprise that new notations accompany new ideas, just as Newton's calculus accompanied his articulation of a new natural philosophy.) So, for example, a physicist reading the proceedings of the Prussian Academy of Sciences in 1915 would stumble across a term such as:[78]

$$G_{im} = -x(T_{im} - \tfrac{1}{2}g_{im}T)$$

The terms with the subscripts were tensors, a generalization of vectors (variables with both direction and magnitude). The mathematical techniques of using tensors to represent geometrics, and then to perform calculus, were developed, as absolute differential calculus, by an Italian school at the University of Padua under Gregorio Ricci-Curbastro. Initially dismissed as 'abstract symbolism that failed to produce concrete geometrical results', the methods were elaborated by Ricci and his student Tullio Levi-Civita into the modern tensor calculus – by careful application to elasticity theory and hydrodynamics.[79] There was a working world context, therefore, even for the seemingly abstract and arcane mathematical tools of general relativity. We will see later how the general theory of relativity was tested in the aftermath of the First World War.

Uses of the new physics

A curious indicator of the rise in status of theoretical physics, and the working world context for the new physics as it was hammered out at the turn of the century, was the Solvay conferences. In the summer of 1910, Walther Nernst convinced Max Planck of the advantages in holding a meeting of physicists to discuss radiation theory.[80] Nernst also persuaded Ernest Solvay, a Belgian chemist who had made a fortune with a soda manufacturing process, to bankroll the project. (Solvay, who had funded an Institute of Physiology in 1895 and an Institute of Sociology in 1909, was persuaded that a further institute for physics, launched on the back of an international meeting, would further Belgian science.) In October 1911, the plush Hôtel Métropole in Brussels welcomed twenty-one physicists to the First Solvay Congress in Physics, including Lorentz, Wien, Marie Curie, Henri Poincaré, Max Planck, Arnold Sommerfeld, Ernest Rutherford, Paul Langevin and Einstein, as well as the organizer, Nernst, and patron, Solvay.

The First Solvay Congress has traditionally been viewed as a moment when the quantum theory broke out from being the specialized concern for a handful to become a central, if contested, component of modern physics. It has also been celebrated as a vector of influence of ideas, with for example Rutherford reporting back to Bohr enough of the proceedings to shape the quantum model of the atom. Both historical judgements are misleading.[81] Indeed, if anything, Solvay 1911 catches the major physicists still making up their minds about quantum theory, Einstein included. Nevertheless, the Solvay conferences were more than an attempt by Nernst to garner confirmation of his Heat Theorem (the so-called Third Law of Thermodynamics – that, as absolute zero is approached, so entropy is minimized), as Kuhn has suggested.[82] Neither was the congress the spectacular intellectual success (as the invited press were encouraged to write up) or the damp squib recorded privately by Einstein: 'Nothing positive was accomplished', he wrote to his friend, Michele Besso. 'I benefited little, since I heard nothing that was not known to me.'[83] Instead, the Solvay Congress is best understood as marking a moment of institutional ambition on the part of physicists, to continue the interplay of experimental and theoretical argument and to tap the patronage of the working world to do so.

The most telling example of this patronage came from the United States, where the younger generation of physicists excitedly re-created the quantum experiments and embraced the quantum theory.

41

These physicists included Owen Richardson and Karl Compton at Princeton, where Einstein's photoelectric effect was reproduced in 1912, Robert Millikan at Chicago, where Planck's constant was independently measured, and Theodore Lyman at Harvard, where Bohr's predicted hydrogen emission lines in the ultraviolet range were recorded.[84] This generation soon took over the institutions of American physics (such as the American Physical Society, with the journal the *Physical Review*, from 1913), but struggled initially to find patrons who would support research into the new physics. 'The income of no single research institution', noted Robert S. Woodward of the Carnegie Institution, 'can come anywhere near meeting the wants of the great army of competent investigators now pressing for financial assistance.'[85] Nor could the American equivalent of Germany's PTR, the National Bureau of Standards, step in to provide federal funds. This comparative reluctance illustrates how important the standards bureaux directors' relatively broad or narrow vision of 'physics' was for shaping research programmes.[86] Finally, the new physics, depending as it did on novel mathematical techniques and counter-intuitive assumptions, lacked popular intelligibility, as historian Daniel Kevles illustrates with a quotation from *The Nation* from 1906:

> Physics has outgrown the old formulas of gravity, magnetism, and pressure; has discarded the molecule and atom for the ion, and may in its recent generalizations be followed only by an expert in the higher, not to say transcendental, mathematics ... In short one may say not that the average cultivated man has given up science, but that science has deserted him.[87]

The answer to the patronage problem was industry. The corporate research laboratories at American Telephone & Telegraph, Du Pont, Standard Oil of Indiana, Westinghouse, General Electric, and so on, demanded trained physicists as part of the general trend to move invention in house, to routinize research and development, and to expand corporate capacity to read each other's technical products and patents. The connections with the new physics were direct. The AT&T laboratory, for example, established in 1911, was headed by Frank B. Jewett, a student of Michelson, and employed Millikan as a consultant. The laboratory's first strategic task was to find ways to turn Lee de Forest's triode, patented in 1907, into a means to amplify and repeat telephone messages across the vast distances of the United States. The concentration of inventive effort at this troublesome point in the system is an excellent example of a general phenomenon: of

American system-builders' inventiveness to facilitate flow – in this case, of voice messages.[88]

Industrial patronage supported not only physicists within the corporations, in industrial laboratories, but also academic physicists. General Electric, for example, provided funds for the X-ray spectra work underway at Harvard University, while AT&T endowed MIT's physical laboratory.[89] In general, we should not be surprised by physicists securing such patronage. Indeed, rather than regarding the new physics as a novelty scrambling to find financial support, it is more accurate to see the new physics as being produced by, and in turn assisting, working worlds.

— 3 —

NEW SCIENCES OF LIFE

But what of the living world? Across space, organisms exhibited patterns of similarity and difference. Horses were similar to zebras. Zebras differed from buffalo or crocodiles or baobab trees. In the eighteenth and nineteenth centuries the spatial diversity of the living world had been classified, measured and mapped. The results, biogeography, were the achievement of scientists such as Alexander von Humboldt, whose travels through South America were inspirational to the host of explorers who followed in his footsteps. Humboldt had not only recorded the spatial variations of the living world, but had mapped out such variation alongside other measurable quantities of the natural world: height, temperature, rainfall, and so on. (And, of course, in order to obtain measurements he had had to take standard instruments and measures. Comparability across space had required the movement of similar values.)

One such Humboldtian follower had been the young Charles Darwin, who, as a companion to the captain of HMS *Beagle*, had witnessed the global variety of the living world as the vessel crossed the Atlantic via Tenerife (pausing for Darwin to climb Mount Teide in emulation of his hero, Humboldt), travelling along the South American coastline and stopping at the isolated Galapagos isles, before returning home to London via the Pacific. Darwin's theory of natural selection, finally published in 1859, offered an explanation of the variety found in the natural world.

Darwin's theory accounted for spatial variation, among, for example, the finches of the Galapagos, by proposing that variations in organisms, passed on between generations, were selected for, if they were to survive, by the environment, living and non-living, in which they dwelled. The competition for resources in different locations produced different living forms.

Heredity and Mendelism

Darwin's theory was less satisfactory as an explanation of temporal variation. He proposed that changes between generations were gradual – smooth, and continuous. But what was the mechanism that carried or caused these changes? In later writings, Darwin had proposed a mechanism for accounting for similarity and difference across the time-gap of generations. He called it pangenesis, a theory of heredity that proposed that every part of the body – and by the late nineteenth century the unit part was taken to be the cell – shed tiny particles, 'gemmules', which collected in the reproductive organs and represented features of the parent. If the nineteenth century was sliced away from the twentieth century and we asked what made the living world of the old century similar and different to the new, Darwin would say it was the gemmules leaping the gap that carried any similarity or difference.

Notice that such a mechanism for heredity made this later Darwin rather like Lamarck: organisms could acquire characteristics that could be passed on. Nevertheless, Lamarckism was vehemently opposed in some quarters. The German zoologist August Weismann, who worked for much of his life at the University of Freiburg, argued, from investigations of hydrozoa (such as the *Hydra* of ponds and aquaria), that there must exist some hereditary substance – he called it 'germ plasm' – that was passed on between generations. Germ plasm was not affected by changes in the body, so acquired characteristics could not be inherited. This 'hard heredity', the absolute independence of germ plasm from the rest of the body, was demonstrated, argued Weismann, by experiments in which the tails of mice were cut for many generations: since the tails of the last-born mice were just as long as the first, there was no evidence for the acquisition of the characteristic of 'short-tailedness'. Germ plasm threading across the nineteenth to the twentieth century explained why twentieth-century organisms were like nineteenth-century organisms. Different combinations of germ plasm in a new generation explained why individuals differed.

Many scientists in the late nineteenth century found Darwin's or Weismann's accounts implausible. Perhaps the greatest difficulty they had was with the role of gradual continuous change between

45

generations. Take, for example, an organism that was ever so slightly better equipped to survive the rigours of the natural world. It may even be able to pass on this characteristic to its offspring. However, to reproduce, it would have to mate with an organism with lesser talents, and any advantage would be reduced. Repeat this process a few generations and it was clear, at least to many, that any advantage would be irretrievably diluted: 'swamped' was a term used. Another dissenting argument came from the towering figure of William Thomson, Lord Kelvin, who had pronounced with all the authority of a physicist and a peer of the realm that the relatively rapid cooling of the earth had not left time enough for evolution by natural selection.

The success of such arguments left Darwin a relatively marginal figure by 1900. While the fact of evolution was largely accepted, many scientists – including the later Darwin – were non-Darwinian in two senses. First, they may be Lamarckian, at least in the sense of accepting some kind of acquired characteristics as an important route by which features could be passed to later generations. Most scientists around the turn of the nineteenth into the twentieth century accepted evolution and preferred Lamarckian mechanisms to Darwinian natural selection.[1] Second, Darwin's theory of the natural selection of continuously changing characteristics was rejected in favour of selection over discontinuously changing characteristics.[2] A 'sport', a member of a new generation that was radically different in size or shape or performance from the parent generation – an observed and familiar phenomenon in artificial breeding – was seen as the source of evolutionary steps. These jumps were also called mutations or 'saltations'.

On the other hand, a new group of scientists and mathematicians, who looked and measured different organisms, clung to the Darwinian picture of continuous variation. If the heights of a hundred men are measured and plotted on a graph, numbers against height, then a peak is observed. Few men are very short, and few men are very tall. More men are of heights in between. Take enough measurements and the graph begins to look quite smooth. Indeed the more measurements taken, the smoother the graph. Every range of height between the shortest and the tallest is eventually taken, and the peak reveals an average. A similar pattern can be found when many natural variables are measured. As an image of the natural world, such graphs powerfully support the idea of continuous gradual change across a population. The scientists and mathematicians who made such measurements, 'biometricians' ('measurers of life') such as Karl

Pearson, as well as botanists such as W. F. R. Weldon in London, backed Darwin and continuous variation.

The potential for a fierce dispute between the biometricians and the saltationists turns out to be critical to how the work of an obscure mid-nineteenth-century monk, Gregor Mendel, was received. One outcome of the debate would be a new science, genetics, which provided the twentieth century with its dominant language for talking about heredity. However, the reworking of Mendel happened in two phases or contexts.[3] The first concerns the so-called rediscovery of Mendel around 1900, and the second is the debate between the biometricians and the saltationists. We will examine both in turn. But, first, let's rewind to say something about who Mendel was and what he did.

Johann Mendel had been born into a family of peasant farmers in 1822.[4] When novices enter a monastery they can change their name, and, on admission to an Augustinian monastery at Brünn in Moravia (now Brno in the Czech Republic), Johann Mendel became Gregor Mendel. Rather unusually, the novice was allowed to study at university, attending in the early 1850s the University of Vienna, where he was lectured by Franz Unger in cytology (the science of cells), evolution and even hybridization experiments, and in physics by Christian Doppler. After returning to the monastery, Mendel grew plants in the gardens and investigated hybrids. In his famous experiment, he noted seven pairs of character differences in *Pisum* (pea) plants – some seeds were yellow and others were green, or some seeds were wrinkled and others were smooth – and he cross-bred plants with opposing characteristics. In results presented to the Brno Natural History Society in 1865, and published in their journal the following year, Mendel reported that, while a character might entirely disappear for one generation, it would reappear in a fixed 3:1 ratio in the second generation. Characters seemed to be carried by discrete 'elements' passed to the next generation.

The later mythologizers of Mendel would hold that the monk's work was ignored by the scientific world, only to be triumphantly vindicated thirty-five years later. Loren Eiseley, in *Darwin's Century* (1958), for example, dwells on the moment of Mendel's presentation: 'Stolidly the audience had listened . . . Not one had ventured a question, not a single heartbeat had quickened . . . Not a solitary soul had understood him.'[5] Had Mendel been neglected, it would have been for good reasons: the paper contained mathematics that would have been forbidding to an audience interested in hybridization, and the paper being presented was by a scientist of little status and was

47

published in an obscure journal. Retrospectively, Mendel's work has been accused (for example, by Ronald Aylmer Fisher) of being too clear-cut, the numbers almost unbelievable.[6] In fact Mendel's work drew him into correspondence with some of the best botanists of the day, men such as Carl Nägeli. So Mendel was not ignored. A rather different factor in the fate of his science was Mendel's own changing research interests: he increasingly left peas for another organism – *Hieracium* (hawkweed), a small dandelion-like plant common across Europe but possessing (we now know) an immensely complex process of reproduction, involving self-pollination and hosts of micro-species. Furthermore, Mendel was elected abbot of the monastery in 1868, which filled his time.

Mendel's work was known in the scientific world, but it was not obvious that it was of great importance. Only later, retrospectively, would its relevance seem clear. As historian Peter Bowler has summarized, the 'traditional image of Mendel is a myth created by the early geneticists to reinforce the belief that the laws of inheritance are obvious to anyone who looks closely enough at the problem.'[7] The scientific world needed a positive reason to return to Mendel.

In 1900, a Dutch botanist, Hugo de Vries, published a paper in the *Comptes Rendus*, France's premier scientific journal – far more prestigious than a provincial natural history newsletter. The Netherlands, of course, was a world centre of horticulture, especially the breeding of flower-bearing plants. Since 1866, De Vries had been growing and comparing a plant called the evening primrose, in particular a species called *Oenothera lamarckiana*. A tall plant with large yellow flowers, *Oenothera*, in fact, was rather like *Hieracium*, in appearing in many different forms, or micro-species. Also, it was the kind of organism that frequently threw up sports, and De Vries was a convinced believer in evolution by discontinuous variation. Indeed, in *Oenothera* he thought he saw mutations in the process of becoming new species in front of his eyes. In 1899, he had written up some of his views in *Intracellular Pangenesis*, which had divided Darwin's pangenesis concept into, first, a concept of material particles responsible for each character of an organism (which he accepted) and, second, the transportation hypothesis, the idea that these particles moved around the body, which he rejected. As De Vries's plant-breeding experiments expanded, he began to observe a 3:1 ratio in the second generation. The 1900 paper set out these results in support of his notion of the segregation of characters. But he did not cite Mendel.

Meanwhile, Carl Correns, a German botanist, received an advanced version of Hugo de Vries's first article, and read it with interest and gathering alarm. He, too, had experimental evidence for a 3:1 ratio, the product of years of painstaking labour. What should he do? Should he concede defeat gracefully, as scientists are supposed to do, acknowledging that De Vries had pipped him to the post, or could he find a compromise that recorded his contribution? Correns chose to publish his results, but, unlike De Vries, he cited the work of the obscure monk, Gregor Mendel. What Correns was saying was this: Correns may be second to De Vries, but De Vries and Correns together were following an older ancestor. As Augustine Brannigan, a historian of scientific discovery, explains, Correns framed

> his announcement so as to indicate that though he had lost priority to de Vries, both had lost out to an earlier researcher, even though the initial intent of that research . . . [was] somewhat less than identical. In other words he neutralises his loss . . . This is accomplished decisively by labelling the discovery 'Mendel's law'. This is perhaps the most important fact in the reification of Mendel as founder of genetics.[8]

The reason why Mendel is famous now has much to with the way that this altercation between Hugo de Vries and Carl Correns was settled. Had Correns chosen a different strategy it is quite likely that 'Mendel's Law' would be known as 'De Vries's Law', or 'Correns's Law', or even 'von Tschermak-Seysenegg's law', after an Austrian botanist who also interpreted Mendel's ideas around the same time as De Vries and Correns but in terms of largely pre-Mendelian concepts of heredity.

Following the 'rediscovery' of Mendel's ideas, Mendel's original paper was republished and probably read by many for the first time. If the initial resurrection of Mendel had been brought about by a priority dispute between botanists in continental Europe, his fame was firmly secured by appeals to his name in a fierce British controversy, between the biometricians and the saltationists.

Recall that the biometricians saw patterns of smooth variation within organisms of the same type. They saw themselves as Darwinian. On the other hand, the saltationists argued that evolution happened by the selection of qualitatively different 'sports', dramatically different-shaped individuals thrown up once in a while. William Bateson, a Cambridge zoologist, thought such discontinuity was typical in populations. He rejected Darwinism, and wrote books such as *Materials for the Study of Variation* (1894), saying so. Around 1900 he too read the now publicized work of Mendel, and was convinced not only of Mendel's evidence with regard to peas, but that Mendel's laws were universally valid, and that in the 3:1 ratio there was evidence for fundamental discontinuous variation. Bateson marshalled all the evidence he could lay his hands on and argued this general case in *Mendel's Principles of Heredity* (1902). Note how far 'Mendel' had now come.

Why genetics? Breeding and eugenics

But Bateson added something else of significance: a new name. Frustrated that the science of heredity, defined along saltationist lines, was attracting little support or patronage, he put clear blue water between his kind of science and that of the biometricians by naming his work 'genetics'. Ironically, then, 'genetics' was named and promoted as a weapon against Darwinism. Several other core terms of genetics were invented during these exchanges. The Danish botanist Wilhelm Johanssen, for example, took 'pangene', a term of De Vries's, and shortened it to 'gene'. (Yet what Johanssen thought the term referred to was an energy state within the organism, probably unobservable, and quite different from what 'genes' might mean later in the century.) Johanssen also insisted that the 'genotype' (another of his new words) was the correct subject for the study of heredity, as opposed to the 'phenotype', or the body. Genetics can be seen emerging as part of explicit attempts to rule out other experts, such as biometricians or embryologists, who did know how to measure and study 'phenotypes', from being able to speak authoritatively on heredity.

But why was it worth fighting over?

There are two, intimately connected answers. First, in the working world of agriculture, the possession of skills, or even merely the reputation, backed with the prestige of science, to be able to breed better plants or animals was lucrative. The possessors of a science of heredity would expect to be encouraged and be able to attract patronage from the working world of animal and plant breeding. Second, the idea that these methods of 'good breeding' could be applied to human reproduction, the project of eugenics, greatly appealed to many groups and individuals in the late nineteenth and early twentieth centuries. Let us consider each of these interests in turn.

The organized application of science for agricultural and horticultural improvement dates at least from the seventeenth century. In the mid-nineteenth century Liebig's research school at Giessen had systematically moved between new chemistry and methods of increased production of foodstuffs. The spirit of the eighteenth-century improving gentleman-farmer had motivated the establishment of the Rothamsted Experimental Station in Hertfordshire in 1843, where John Bennet Lawes's long-term experiments on fertilizers and crop yields began; they have run continuously ever since.[9] A trust fund set up on Lawes's sale of his fertilizer business in 1889 provided funds to keep Rothamsted an independent concern, collaborating but not controlled by government or farmer interests. In the United States, however, agricultural research was an early and highly significant recipient of federal and state funding.[10] In 1862, the Morrill Act had provided federal land, which in turn could be sold to fund the expansion or creation of 'land-grant colleges'. In the mid- to late nineteenth century, the United States was an industrializing nation, which in turn created a demand for increased agricultural productivity. In 1887, the Hatch Act repeated the Morrill Act ploy, granting federal land for sale for states to fund agricultural experiment stations. Many of these stations were attached to land-grant colleges. The result was a large number of institutions, each with a research and teaching staff.

In 1903, members of the American Association of Agricultural Colleges and Experiment Stations established the American Breeders' Association, attracting as members a broad audience of 'commercial breeders, professors at agricultural colleges and experiment stations, and researchers at the US Department of Agriculture', with the latter two providing the 'great majority'.[11] Its leader was Willet B. Hays, of Minnesota experimental station and, from 1904, assistant secretary of agriculture in Theodore Roosevelt's administration. Hays had a strong preference for broad cooperative alliances, including ones between researchers, plant breeders and animal breeders; the 'producers of new values through breeding', Hays summarized in a 1905 speech to the association, 'are here brought together as an appreciative constituency of their servants, the scientists'.[12]

The breeders had a primary and overwhelming interest in improving stock, to which they brought two established methods, the selection and breeding of best individuals and hybridization to breed in desired characteristics. This combination, for example, was behind the formidable successes in plant breeding at the Vilmorin company in France.[13] However, unlike the French, the American breeders rapidly embraced Mendel's theory. Between 1903 and 1909 the proportion of papers presented at meetings of the American Breeders' Association mentioning Mendel increased from one-ninth to one-fifth, and 'virtually all such citations corroborated the theory'; the 'breeders particularly appreciated the predictive value of the Mendelian pattern of inheritance of various traits', which chimed with their practical interests.[14] The practical and professional interest in encouraging desirable characters encouraged the breeders to embrace other new methodologies, for instance the biometry of Pearson, which promised predictability and control. In contrast, the breeders were far less keen on the emphasis on unpredictable sports in Hugo de Vries's mutation theory. The professional breeders were among the first, for example when compared to elite academic scientists, to distinguish sharply between the two theories; the working world of the breeders therefore played a substantive role in the shaping of genetic science.

The second answer to the question as to why the science of heredity was worth fighting for in 1900 can be uncovered by following another saltationist, Darwin's cousin Francis Galton. Something of a child prodigy, very much a sport, Galton had read mathematics at Trinity College, Cambridge, had a nervous breakdown and travelled in Africa before reading his relative's great work, *On the Origin of Species*, in 1859. Galton was possessed by the project to quantify Darwin's theory, to place the theory of evolution by natural selection on a firm mathematical basis.[15] He began a life's work collecting data on human variation, developing methods of measurement and mathematics to make sense of it. Galton published some of his quantitative arguments, such as the claim that the children of eminent fathers tended to inherit superior qualities, in *Hereditary Genius* in 1869. 'Galton's achievement in founding statistical biometry should not be minimised', writes historian Theodore Porter. 'He did so by merging evolutionary theory with "statistics" in the nineteenth-century sense of the term – that is, social numbers.'[16] In his *English Men of Science: Their Nature and Nurture* of 1874, Galton also contributed, via a subtitle, an oppositional framework for debates on heredity that lives on still.

Galton reflected in these works on the pairing habits of the English upper classes, a topic that encouraged attention to notions of good breeding. He was convinced that society must ensure the more active reproduction of its fitter members if it was not to be swamped by the 'unfit'. He proposed, for example, that members of families of high rank should be encouraged to marry early. In 1883, he coined a term 'eugenics' – or 'good in birth' – to capture this programme: the improvement of the human race along the lines of the plant or animal breeding. Discouragement or prevention of human reproduction judged undesirable is called 'negative' eugenics, while programmes such as the Galton's, to encourage the higher ranks to breed early and more often, to multiply reproduction judged desirable, is termed 'positive' eugenics.[17] Overall, Galton wrote in 1904, 'the aim of Eugenics' was to 'cause the useful classes in the community to contribute more than their proportion to the next generation'.[18]

Eugenics was decisively shaped by the developing sciences of heredity in the late nineteenth and early twentieth centuries. Given that Weismann had undermined Lamarckian evolution, and had insisted that there was a hard separation between the germ plasma (the vehicle of heredity) and the somatic cells, then how could a race be improved? Any improvement to the cells of the body during the life of the organism would not be passed on to the next generation. Therefore, the only source of improvement was the germ plasm – the genes, in the new jargon.

And many thought there was an urgent need for improvement. The sprawling nineteenth-century metropolises such as London, Paris or New York, with their mobile and sometimes violent populations, were theatres of fear. Social mobility threatened those on top. Fears of decline or degeneration were often nationalistically inflected. Eugenic campaigns, therefore, were often national campaigns to preserve the national race. Overall, it was a powerful strategic move to claim expertise in good breeding, which is why it was worth fighting for the right to speak authoritatively on heredity. Of course, someone – of some social group – would have to choose who were and were not good breeders. This fact that the eugenic official – 'a compound', noted biologist J. B. S. Haldane, 'of the policeman, the priest, and the procurer' (or pimp) – would be socially self-interested was eugenics' political advantage and flaw.[19]

Eugenic movements were active in Britain, the United States, Germany, Scandinavia and elsewhere. Members were mostly white and middle class, with large numbers of professionals, scientists (particularly, later, geneticists) in addition to non-scientists.[20] Eugenics movements provided 'a legitimation of the social position of the professional middle class, and an argument for its enhancement'.[21] Their aims were to prevent social 'degeneration'. Problems of industrial society – such as crime, slums, prostitution – were attributed to biological causes, first poor 'blood' and later poor 'genes'. In all eugenic programmes there was a special focus on those identified as 'feeble-minded', as well as certain medical conditions (such as epilepsy and diabetes) and perceived social burdens (alcoholism, prostitution, criminality, poverty). In particular, eugenicists distinguished and described underclasses for special attention and action. 'Defective delinquency' theory, for example, equated 'high-grade feeble-mindedness' with criminality.

In Britain, Galton encouraged the foundation of a Eugenics

Records Office at University College London in 1904, followed by a Eugenics Laboratory in 1907, directed by Karl Pearson. In 1911, Galton's bequest of £45,000 created a chair of eugenics, held by Pearson, followed by R. A. Fisher. A Eugenics Education Society – later shortened to the Eugenics Society – was formed in 1907, and journals were produced: *The Eugenics Review* and *Biometrika* (from 1901). The political achievements of the British eugenics movement included family allowance policies, a positive eugenic measure that rewarded the right kinds of families, and the Mental Deficiency Act of 1913, which 'allowed for eugenically motivated segregation' and remained in force until 1959.[22]

In the United States, the most active centre of eugenic research was the Eugenics Record Office, at Cold Spring Harbor, Long Island, New York. Its director, Charles B. Davenport, was a keen Mendelian.[23] Davenport was also, with Willet Hays, one of the leaders of the American Breeders' Association, a strong indication that ideas about 'good' human, animal and plant breeding were considered of a piece in the early twentieth century. The association set up a Committee on Eugenics in 1906 and pushed eugenic programmes in its *American Breeders' Magazine*; Hays in 1912 addressed the National Farmers' Congress on the subject of 'The farm, the home of the race', outlining a vision in which corn-fed 'genetically efficient people' would be encouraged to breed, all justified by appeal to Mendel.[24] Meanwhile, Davenport's *Heredity in Relation to Eugenics* of 1911 attributed to single genes many mental and behavioural characteristics, including insanity, epilepsy, alcoholism, 'pauperism', criminality, 'shiftlessness' and even 'thalassophilia', or a love of the sea, found in families of seafarers. According to research conducted at the Eugenics Research Office by Henry H. Goddard, feeblemindedness was a 'condition of mind or brain' which was 'transmitted as regularly and surely as color of hair or eyes'.[25] The office also sponsored in-depth studies of families, tracing the appearance of undesirable traits across whole genealogies.[26]

The political impact of eugenic research in the United States was extensive and lasting. Starting with Indiana in 1907, sterilization laws, negative eugenic methods that brutally prevented undesirable breeding, were introduced in twenty-eight states by the 1920s; 30,000 people had been 'sterilized on eugenic grounds' by 1939, nearly half of them in California.[27] Such laws were challenged but declared constitutional in a Supreme Court decision (*Buck* vs. *Bell*, 1927), in which Justice Oliver Wendell Holmes declared that 'three generations of imbeciles is enough'. The United States led the way with eugenic laws. Many American states were already practising sterilization before they were joined by European states: the Swiss canton of Waadt (1928), Denmark (1929), Germany (1933), Norway (1934), Sweden and Finland (both 1935), among others.[28]

Until the 1930s, eugenics programmes and policies in Germany were broadly comparable with those in Britain, the United States and Scandinavia. Indeed, German eugenicists looked to states such as California, which sterilized more people than any other, for inspiration and example. The active centres included the Kaiser Wilhelm Institute for Research in Psychiatry, established 1918 in Munich, and the Kaiser Wilhelm Institute for Anthropology, Human Heredity and Eugenics, founded in Berlin in 1927. German sterilization laws were in place before the Nazis took power in 1933.

The technocratic appeal of eugenics – the opportunity to claim the rights to steer social policy through an appeal to technical expertise – and the credit, status and resources that might follow partly explain the intense interest in sciences of heredity in the late nineteenth and early twentieth century. It was not a prize that could be shared, and the competition to secure the right to speak authoritatively was the context in which key terms – not least 'genes' and 'genetics' – were used. Nor was it only a say in human heredity that was at stake: authority over plant and animal breeding would also pay dividends. Nevertheless, despite the pressures to compete and challenge, in the 1910s the controversy between biometrical and Mendelian sciences of heredity was significantly, if not completely, resolved. The triumph of techniques of two kinds – statistical and biological – was the heart of the resolution.

Galton's protégé was Karl Pearson, a fierce supporter of Darwinian gradualism, who developed the mathematical techniques to understand correlation between variables, techniques that became standard statistical methods. His χ-squared test, for example, was proposed in 1900 as a replacement for the established normal curve. Sociologist Donald MacKenzie considers Pearson's statistical methods to have been shaped by his eugenic interests, while historian Eileen Magnello denies any influence between the two central projects of Pearson's life.[29] Nevertheless, writes historian Ted Porter, it is in Pearson's projects that the ambition and reach of biometrics can be clearly witnessed:

His *Grammar of Science* (1892) was, after all, a philosophical tract designed to extend the reach of science into every domain. He idealized statistics, a field long nourished by similar ambitions, as the true form for all scientific knowledge. Statistics was descriptive and positivistic, not causal and metaphysical. Pearson's philosophy, which was positiv-istic in the sense of Ernst Mach, construed statistics as a fine model for all scientific knowledge. Statistical methodology implies the omnipres-ence of chance, or at least uncertainty; Pearson held that uncertainty in the social and political realm is of the same character as in physics, though it is often greater in magnitude. In short, Pearson hoped to unify knowledge under a positivistic philosophy and statistical form. The metric of biometry was to embrace sociology and politics as well as the dimensions of pea-pods and shrimp-skulls.[30]

It was precisely because 'genes' could not pass the positivist criterion that Pearson thought them metaphysical, and to be rejected. The debate between Mendelians and biometricians was therefore simul-taneously about the proper method of science – in particular what could be measured and how such measures should be compared – and about what scientific objects existed.

Pearson, the first Galton Professor of Eugenics at University College London, was an ideologue, highly unlikely to concede ground to his enemies, and while he held power the controversy was not going to end. However, his successor in the UCL chair, Ronald Aylmer Fisher, was more flexible. In 1918, Fisher published a paper, 'The correlation between relatives on the supposition of Mendelian inheritance', which gave a sharp mathematical sense to the concept of variance. Now defined as the square of the standard deviation, dif-ferent variables could be identified and measured in convenient and clear ways.[31] Deploying variance, Fisher argued that the gradually changing data found by the biometricians could be produced by the action of combining large numbers of discontinuous variables such as Mendelian genes. In other words, Mendelian genetics could produce Darwinian results.

This mathematical argument was not enough, in itself, to resolve the controversy. Just because Mendelian genes could produce smooth changes in the characteristics of large populations did not mean that Mendelian genes existed, except as a useful mathematical fiction. Instead, the emergence of a scientific consensus of the material reality of genes required a solution that was part material technique, part social innovation.

Making and circulating *Drosophila* genetics

Recall that there was considerable disagreement in what scientists took 'genes' to be, even if they admitted their existence. Johanssen had considered genes to be energy states, yet De Vries thought them material, and in the hands of the mathematicians they were a manipulable variable with unknown reference. The identification of genes with a material substance – the 'chromosomes' that microscopists had found in cell nuclei – is usually considered the achievement of Thomas Hunt Morgan, although, as we shall see, it was the creation of an extraordinarily rich and productive material culture that enabled him to sustain the claim. In fact the identification of chromosomes as the hereditary substance was another aspect of the emergence of genetics where the working world of plant and animal breeders at least prepared the ground.

The idea that pairs of chromosomes, observable in cytological studies, might be associated with pairs of Mendelian factors occurred to many scientists after 1900. Evidence was put forward in 1902 by two scientists in particular; the venerable German embryologist Theodor Boveri, working with sea urchin eggs at Naples, argued that normal development required full sets of chromosomes, while in Columbia, a young postgraduate, Walter S. Sutton, drew similar conclusions based on studies of an agricultural pest, the grasshopper, in Kansas. Nevertheless, the theory remained controversial. However, papers on breeding, genes and chromosomes appeared frequently in the *Proceedings of the American Breeders' Association*. The United States Department of Agriculture scientist W. J. Spillman argued for a connection in 1907; the Illinois swine breeders Q. I. Simpson and J. P. Simpson enthusiastically attributed Mendelian factors to chromosomes; while W. Lawrence Balls, of the Khedivial Agricultural Society, argued in 1909 from cotton research for the same attribution.[32] The 'readiness to entertain' such theories was encouraged by the breeders' practical concerns and illustrate how the working world was ready to receive, and in turn maintain, Morgan's chromosomal theory.

Morgan had benefited from training at the American university that most closely followed the German model, Johns Hopkins University. He had won his PhD in 1890. In 1904 Morgan was appointed professor of experimental zoology at Columbia University. At Columbia, he chose an organism that seemed ideal for biometric studies and, most relevantly in a university following the German model, for teaching by research.[33] The fruit fly, *Drosophila melanogaster*, was common, was easily sustained – a population would happily breed in a glass jar with some bananas as food – and multiplied rapidly. Large populations of flies provided plenty of material for measurements. Indeed, *Drosophila*'s cheapness made the fly suitable student project material.

It was, however, an unlikely organism on which to stake a professional career as a geneticist: it had none of the status or literature of experimental mammal organisms, nor was it domesticated, and nor did it seem to show many visible characters or mutations.

However, the choice was fortunate. The cells of the salivary glands of *Drosophila* have four chromosomes of immense physical size, some of the largest among all organisms.[34] They were morphologically distinct, and their size made them easier to dissect out and examine. Furthermore, as the biometrical studies were scaled up, the populations under study became so large that mutations were more likely to appear. From 1908, Morgan increasingly used *Drosophila* as a model organism in Mendelian research. (Indeed, he had started the investigation thoroughly sceptical of Mendelism.) His methods encouraged the building of a research school. Students could be trained and fellow researchers could work together, collaborating around the jars of *Drosophila*. The identification of new phenotypes, and the breeding of separate populations that displayed interesting traits, was doable but productive science – 'normal science', in Kuhn's terms. Furthermore, Morgan's method could travel: small jars of breeding fruit flies were exchanged with like-minded colleagues around the United States, and eventually across oceans.[35]

Many eyes looking down microscopes are better than one. In 1910, Morgan's school noticed rare phenotypic variations and mutations, including male flies that had white eyes as opposed to the typical red. This seemed odd: why should a trait such as the colour of eyes be confined to one sex? Furthermore, some sex-linked traits were not inherited together, while others were. This prompted Morgan to speculate that genes that were closer together were more likely to be inherited together – in doing so he drew on the studies of the skilful Belgian cytologist Frans Alfons Janssens, who had observed through the microscope chromosomes approaching, breaking and recombining. From this insight, Alfred Henry Sturtevant, who had been trained in Morgan's fly room and shared the work of identifying and isolating the white eye mutants, drew up 'chromosome maps', which showed which genes were close and which were distant. The results were published in 1913. Notice what was happening here: the articulation and spread of material techniques that gave Morgan and his school the ability to isolate and quantify traits in populations were accompanied by the identification of the genes with a specific location in a material substance – the chromosome – and, indeed, a measurable position within it.

In 1915, Morgan, with students Sturtevant, Calvin Blackman

Bridges and Hermann Muller, published the methods and results in *The Mechanism of Mendelian Heredity*. This work of synthesis defined the components of the classical theory of genes, including the claims that: 'chromosomes are the bearers of hereditary material', '"genes" (not "unit characters") are the fundamental units of heredity', 'genes are arrayed linearly on chromosomes', 'the number of linkage groups of genes . . .equals the number of chromosomes', 'each distinct gene, though it may have many alleles, remains unchanged except by mutation', 'environmental factors (e.g. temperature and nutrition) can influence the effects of some genes', 'some genes can modify the effects of other genes, sometimes quite specifically', 'genes themselves are not altered when their effects are changed by modifier genes', 'genes must cooperate in large numbers to yield observable traits', and 'many mutations have large effects, but many more have small effects'; altogether, 'even though the pathways from genes to characters are wholly unknown, Mendel's principles, interpreted via gene theory, provide "a scientific explanation of heredity [that] fufills all the requirements of any causal explanation"'.[36]

At a growing pace throughout the twentieth century, to be a certain kind of geneticist was to be part of the network that sustained and exchanged bottles of fruit flies and interbred, dissected, observed and correlated the features of their model organism. Similar people, in the sense of trained students, passed through Morgan's fly room and research school, and carried similar techniques and organisms with them. With the network, and similar flies, people and techniques, in place, a statement or measurement about the chromosome was more likely to be comparable, repeatable and stable – or, where necessary, indeed, disputed in agreed, contained and productive ways. The multiplication and export of similar entities were the secret to Morgan's success. In application to animal and plant breeding, too, new genetic knowledge would contribute to other circulations of similar things.

Furthermore, the heredity that genetics was mastering was increasingly an aspect of an artificial world. The mutants were chosen and favoured in reproduction. The fruit fly became a standard in order that mutations might be identified and isolated, and so that measurements could be made and compared in different places and at different times. The fruit fly was not really *Drosophila melanogaster* any more, at least in the sense that the population of the fly room diverged from the population in the 'wild', but might be called, perhaps, *Drosophila laboratoriensis*, a creature of a particular artificial context. Just as in physics, the remarkable new natural sciences of the turn of the nineteenth into the twentieth century were the result of the study of

an artificial world. And the United States was where such study flourished. Morgan, at Columbia and from 1928 at the California Institute
of Technology, was a world leader in the science of genetics, while
Drosophila remained the leading model organism for genetics into
the twenty-first century. William Bateson, the Englishman who had
coined the word 'genetics' in 1905, paid homage to the shift in power
at the gathering of scientists at the American Association for the
Advancement of Science in December 1921: 'I come at this Christmas
Season to lay my respectful homage before the stars that have arisen
in the West.'[37]

Another consequence of the sheer success of Morgan's *Drosophila*
work was a narrowing of Mendelism as a scientific programme.
'Since the development and evolutionary consequences of Mendelism
were not readily tested', note historians Richard Burian and Doris
Zallen, 'those issues were gradually dismissed as speculative and set
aside.'[38] The roles of genetic change in evolution would not return
as research questions until the period of the so-called Evolutionary
Synthesis beginning in the 1930s. The separate, difficult, if profound,
questions about the relationship between development and genetic
make-up would return as highly active research topics towards the
end of the century.

Working worlds of the life sciences

We have seen that agricultural experimental stations and agricultural
colleges provided institutional homes for the new genetics. The first
American genetics departments – at Berkeley, Cornell and Wisconsin
– were located in the agricultural faculties.[39] University-trained
plant physiologists, employed at the United States Department of
Agriculture's Bureau of Plant Industry, began the development of
hybrid corn deploying Mendelian theory.[40] The Agricultural College
in Berlin housed genetics in Germany – although it was unique in
Germany as a place where genetics researchers established strong
links with breeders.[41] Many early Mendelians found employment
at agricultural institutions – such as Hermann Nilsson-Ehle at the
Swedish plant-breeding station at Svalöf, Erich von Tschermak-
Seysenegg at the Agricultural College in Vienna, William Bateson
at the John Innes Horticultural Institution in Britain, and Raymond
Pearl at the Maine Agricultural Station in the United States.[42] This
working world context for genetics would be continued during the
Evolutionary Synthesis era, when, for example, R. A. Fisher worked

at Rothamsted and Sewall Wright was employed at the Madison agricultural station.

But genetics was not the only life science that was encouraged within this working world. 'Biochemistry', notes historian of biology Jonathan Harwood, 'also took root in agricultural soil.'[43] Eduard Buchner had been appointed professor of general chemistry at the Agricultural College in Berlin in 1898, after passing through lectureships at Munich, Kiel and Tübingen. His research showed that the liquid obtained from ground and pressed yeast cells would ferment sugars into alcohol. Buchner would be awarded the Nobel Prize for chemistry in 1907 for the discovery of this cell-free fermentation, and he is remembered as a father of a new discipline, biochemistry. Biochemistry would flourish in the twentieth century, although less so in the United States, where it was located in medical schools in which it was encouraged – and constrained – to develop analytical methods useful in the clinic and in medical science; at Cambridge, under F. Gowland Hopkins, and in Berlin, under Otto Warburg, biochemistry was located away from medical institutions and targeted basic biological problems with success.[44]

As well as at the Agricultural College, fermentation research was carried out in Berlin at the Institute for Fermentation (*Gärungsgewerbe*), where Max Delbrück, uncle of the famous mid-twentieth-century physicist turned biologist, was also exploring yeast. Delbrück saw the organism in engineering terms: at a conference in 1884 he had begun a paper with the simple statement 'yeast is a machine'.[45] Outside Berlin, two other cities where fermentation research was being conducted in close contact with industry were Copenhagen and Chicago. Denmark, indeed, would carve out a reputation as home to the most scientifically shaped agriculture in the world. The Copenhagen Polytechnic School employed Sigurd Orla Jensen, who had spent several years as director of the Central Institute of Cheese, perfecting the production of holes, as professor of the physiology of fermentation and agricultural chemistry. At Copenhagen, the Carlsberg Brewery established an institute under E. Hansen in 1876, while, at Chicago, John Ewald Siebel built up a brewing school, the Zymotechnical College.[46] This zymotechnology would eventually be renamed 'biotechnology'.[47] Indeed, Orla Jensen's chair was renamed professor of biotechnical chemistry in 1913.[48] The working world context here was the manufacture of beer and cheese, and the science was the first 'biotechnology'.

Finally, interconnections between the working world of agriculture and the German universities had provided the context for the development of a new botany – laboratory-based plant physiology. Two major research schools had grown up under Julius Sachs at the University of Würzberg and Anton de Bary at the university at Strasbourg. The intersection of practical agricultural research and academic botany lay in interest in the 'nutritional requirements of crop plants [which had] provided a strong impetus to the development of plant physiology'.[49] Sachs, for example, had taught in Germany's agricultural colleges before setting up his Würzberg laboratory. De Bary's students, the Russian Sergei Winogradsky and the Dutchman Martinus Willem Beijerinck, who had also taught at an agricultural school, explored the wild occurrences of micro-organisms, identifying new types of microbes in soils and plants, respectively. In 1898, Beijerinck named the tobacco mosaic disease-causing substance which passed through the fine filters that removed known micro-organisms a 'virus'. Winogradsky's techniques, on the other hand, illustrate how scientists related to the working world of agriculture: by isolating microbial components within the soil, or working with simplified surrogates of the natural world (the common pedagogical tool for illustrating bacterial ecology, the Winogradsky column, for example), he was able to refine and describe processes of interest to academic microbiology (chemoautotrophy, for example) while generating findings, for instance the identification of iron and sulphur bacteria, of direct interest to, and taken up by, the working world of agriculture.[50] The new life sciences, with new laboratories, entities ('virus', 'gene') and theories, were intimately connected to working worlds.

— 4 —

NEW SCIENCES OF THE SELF

It would be tempting, but wrong, in writing a history of the sciences of the twentieth century, to emphasize the novelty of four clusters of events and their relevance for particular branches of sciences: the announcement of the discovery of new radiations, Planck's quantum theory and the new physics, the rediscovery of Mendel's laws and new sciences of life, and new sciences of the psyche and the publication of Sigmund Freud's *The Interpretation of Dreams* (1900). The new physics of the turn of the twentieth century rode continuities as much it marked discontinuities with nineteenth-century physics. We also saw that any novelty was recognized only slowly, and was accommodated within previous frameworks of understanding. Likewise, the meaning of Mendel's laws after 1900 was made in a context of continuing nineteenth-century debates about, for example, evolution as a pattern of smooth or jumpy change, or in a continuing intense concern for identifying and improving 'good breeding'. Furthermore, both physical and life sciences took place, typically, in thoroughly nineteenth-century spaces: the field of the explorer, the industrial workshop or laboratory, the German model research university, the institute, or the metrological laboratory. In the human sciences, the pattern, again, is one of radical change, but also of change thoroughly instigated and constituted within social and intellectual spaces inherited from the nineteenth century.

In this chapter I will examine several new sciences in the light of this continuing inheritance. First, I will review the insights gained and lost by considering Sigmund Freud as a nineteenth-century 'biologist of the mind'.[1] Many species of psychology have to be understood as responses to working world demands for classification, control, and perhaps treatment of the inhabitants of asylums and hospitals.

Psychoanalysis, as a talking cure, was both a biologistic contribution to this working world and a cultural, interpretative branch of the humanities. Nevertheless, other sciences, such as Alfred Binet's methods of measurement, contributed more directly to these working worlds.

Human sciences such as anthropology, child study and certain trends in psychology shared an interest in development, in change over time, and asked questions about whether the drivers of change were internal, external or both. They shared the intellectual questions asked by embryology, one of the most dynamic of sciences in the mid- to late nineteenth century.[2] If the embryo recapitulated evolutionary sequences in its development, was it therefore a natural model for how human societies and human natures moved forward, or, as was realized with a Victorian chill, perhaps moved backwards?

Physiology, another leading laboratory-based science of the nineteenth century, also promised extensive insights into the nature of the higher animals, including humans. I will trace the emergence of a physiologically inspired science of behaviour through the work of Charles Sherrington, Ivan Petrovich Pavlov, and the distinctively American science of John Watson's behaviourism. Seemingly trivially, behaviourism presented one of the popular images of twentieth-century science: the white rat finding its way through a maze, followed attentively by white-coated scientists. In fact, physiology and behaviourism, in its various flavours, offered many important stances to twentieth-century science, from the temporary disavowal of 'mind' as subject for objective scientific investigation, to the promotion of the reductive but usefully flexible concept of the 'reflex' as a model of action, to the imagination of social engineering.

Finally, the extraordinarily productive nineteenth-century collaboration between academic organic chemistry and the chemical industries of dyestuffs and pharmaceuticals, along with the development of serotherapy – treatment of a disease using an immune serum or antitoxin – as a commercial business, provided the context in which new biological sciences of the self would flourish in the new century, the foremost example being immunology, the science of the body's defences. Already the subject of fierce controversy, between the rival microbiological camps of Pasteur and Koch, Paul Ehrlich offered, early in the new century, a new theory of immunology centred on a specific theory about the formation of antibodies. While Ehrlich's so-called side-chain theory was itself controversial, it also crystallized immunology as a distinct discipline.

Psychoanalysis

Sigmund Freud in the late 1870s was an unexceptional, if mildly promising, product of the German model of science. Between 1876 and 1882 he could be found working in the Viennese laboratories of Ernst Brücke, a prestigious professor, remembered, along with Émil du Bois-Reymond, Hermann von Helmholtz and Carl Ludwig, as figures that from the 1840s had revolutionized physiology by reforming it in the image of physics. Freud wrote and published scientific papers on the neuroanatomy of primitive fish, the gonads of the eel, the nerve cells of the crayfish, and methods of nerve cell microscopic preparation.[3] He was therefore well indoctrinated in the practices and values of laboratory physiology.

However, Freud's career path in Brücke's institute was blocked – the few permanent places he could aspire to were already filled by recent appointments. Brücke therefore advised him to look elsewhere. Freud chose to practise medicine and gained three years experience at the Vienna General Hospital, where he soon flourished as a would-be neuropathologist, fostered by the professor of psychiatry, Theodor Meynert. Nevertheless, Brücke provided one final boost to Freud's career when his enthusiastic intervention swung in the young man's favour an application to travel to Paris to spend time studying under Jean-Martin Charcot at the Salpêtrière. Freud would be travelling a well-trodden path. Whereas eighty years previously the huge Parisian hospital had been a site of pilgrimage for students of medical analysis, now Salpêtrière was a 'Mecca of neurology', where visitors attended Charcot's Tuesday and Friday lectures to acquire the skills of neurological diagnosis.[4] In the 1870s Charcot had boldly struck out in new directions, first drawing on the controversial techniques of hypnotism and second, most importantly, recognizing – and thereby granting a measure of respectability to – the condition of hysteria. Hypnotism seemed to be a way of artificially inducing hysteria, and therefore a clue to its understanding. (Notice immediately here – again – the importance at the turn of the century of artificial states standing as experimentally tractable surrogates for a natural world that cannot be scientifically understood directly; the hypnotized body stood to the complexities of the human body as the light bulb/cathode ray tube stood in for the complexities of the physical universe.)

Freud picked up Charcot's methods and categories – hypnotism and hysteria – and paraded them on his return to Vienna, lecturing in 1886 on a rare case of male hysteria. Two people of crucial importance to Freud entered to shape his life. The first was Josef Breuer,

65

a fellow physiologist under Brücke, now a doctor with a private practice serving some of Vienna's intellectual elite. One of Breuer's clientele was the remarkable Bertha Pappenheim, better known by her case name, Anna O. She had displayed a series of strange and debilitating symptoms, including paralysis, occasional deafness, tics and phobias. During the day the symptoms would overwhelm her, while in the evening Anna O. managed to put herself into a state of autohypnosis. Breuer diagnosed hysteria, and noted that repeating the words spoken by Anna O. during the daytime back to her in the evening seemed to relieve symptoms. The astonishing Anna O., collaborating with Breuer on this 'talking cure', found ways to recall and recount, in order, all the times a symptom had appeared. Talking through them in reverse order made the hysterical symptoms disappear – at least for a time. Breuer discussed the case over with his neurological friend Sigmund Freud. Freud began to find and describe similar cases, and similar cures.

Diagnoses of hysteria are very rare today. Yet in Vienna in the 1880s hysteria was rampant. 'Reading the case histories recorded by Breuer and Freud', notes historian Frank Sulloway of the pair's *Studies on Hysteria* of 1895, 'is therefore to re-enter a lost world of incredible and often bizarre behaviour which is preserved for us only through the printed word.'[5] In the transition from nineteenth- to twentieth-century frames, many models and institutions were carried forward, but some things – such as hysteria – were lost (or perhaps unmade). Yet moulded in the hysteria literature was a theory and set of practices – psychoanalysis – that, institutionalized and translated – would be carried through much of the twentieth century with considerable force.

In addition to particular case histories, Freud's contribution to *Studies on Hysteria* was a proposed mechanism for the process whereby hysterics suffered from their reminiscences. Excess 'excitation' – mental energy, conceived as a quantifiable entity that could flow, and that was kept in the nervous system at a ideally constant level – would in normal individuals be disposed of, for example through sex. If this discharge was thwarted, it would leave through alternative, inappropriate channels, creating the symptoms of hysteria. By uncovering the original block to the discharge – a traumatic moment in the patient's past, recoverable by talking backwards through the occurrence of the symptoms – the block could be removed and, in the cathartic moment, a cure made.

Many commentators have noted the debt this model of excitation has to economic and electrical models.[6] Indeed an electrical model for psychology was widespread in the 1880s; William James, for example, wrote:

> The whole drift of recent brain-inquiry sets towards the notion that the brain always acts as a whole, and that no part of it can be discharging without altering the tensions of all the other parts. The best symbol for it seems to be an electric conductor, the amount of whose charge at any point is a function of the total charge elsewhere.[7]

I agree with those who recognize Freud's debt, but suggest a deeper contextualization, one that has to distinguish between two historical phases of electrical and economic imagery. The notion that nervous excitation might flow and be discharged draws straight from a view of electricity as an imponderable fluid, articulated a century before Freud. However, the principle of constancy – that the nervous system aims to restore excitation to a constant level, and the associated notion that the nervous system, a network of managed channels, might have a 'total load' – is the language of electrical power systems. Through the efforts of Edison, and others, the techniques of management of the flow of electrical power through networks were being invented from the 1880s.[8] This inventive work included the articulation of load management as a critical tool.[9] Freud's mechanism, his primary theoretical contribution to *Studies on Hysteria*, is reminiscent of such techniques. But Freud in Vienna is not like Einstein in Zurich – sitting at a desk where the working world was presented to him in direct, unavoidable physical forms, as patents or electrically synchronized railway clocks. Nevertheless, electrical light and power was an unmissable public spectacle in the great urban cities of the late nineteenth century. Vienna, indeed, hosted in 1883, while Freud was working with Brücke on the nervous systems of animals, one of the first great electrical exhibitions, after Paris's own of 1881 and Munich's in 1882.[10] In the twentieth-century theory of psychoanalysis, then, we can trace another flicker, albeit a dim refracted flicker, of the working world of late nineteenth-century electrical empires.

Tensions between the two authors of *Studies on Hysteria* increased to the point where, by 1895, Freud and Breuer were no longer on speaking terms. On the surface, the cause of the rift was Freud's obstinate defence of a radical view, that the aetiology – the study of causes of disease – of the neuroses, including hysteria, revealed the fundamental role of childhood sexuality. Sulloway argues that to understand Freud's 'growing fanaticism' requires recognizing his commitment to nineteenth-century developmental biological models, in particular Lamarckism, as a means of providing a causal account of the acquisition of neuroses.[11] (We saw in the last chapter that this commitment was unexceptional among scientists around 1900.) It also requires recognizing what alternative explanations of neuroses were available. Most neurologists, such as the great Charcot, viewed neurotics as cases of hereditary degeneration. Freud, in opposition, argued that neuroses must be acquired, nearly always during childhood, in a manner that could be passed on to later generations under the Lamarckian theory of acquired characteristics passing from parent to child, but more likely during the individual's early life. Freud, writes Sulloway, 'psychoanalytically resolved the issue of "bad heredity" into the one recurrent cause of neurosis – repressed and unsatisfied sexuality – which he deemed to be civilization's congenital burden, increasing from generation to generation by Lamarckian inheritance'.[12] Gloomy for civilization, but a conclusion that offered a brighter outlook for patients: hereditary degeneration could not be undone, whereas Freud's theory, and techniques such as the talking cure, opened the door to a solution.

In 1895 Freud filled notebooks in an attempt to write out a full detailed account of how nervous energy could be channelled and directed, blocked and thwarted, through three systems – of perception, memory and consciousness – grounding the theory within two kinds of biological frameworks, a mechanical physiology and evolutionary biology. These notebooks were not published until after Freud's death, when they were given the title *Project for a Scientific Psychology* by his English editor, James Strachey. The problem, Freud found, was that he could not account for the repression of childhood sexual experiences. He then switched from his so-called seduction theory, in which neuroses were caused by the repression of passive sexual trauma, to a recognition of active and spontaneous early childhood sexuality, the existence of which he accounted for by appealing to an evolutionary story.

This theory is at the core of Freudian psychoanalysis and, despite his acolytes' later claims that their leader had invented a new, independent psychological discipline, is freighted with nineteenth-century evolutionary developmentalism. The key figure here is Ernst Haeckel, the German champion of Darwin, who argued that embryological sequences retraced evolutionary timelines. This argument, powerfully presented visually by series of growing embryos passing through gill or tail stages, for example, was summed up by the slogan 'ontogeny recapitulates phylogeny'. Informed and inspired by the ideas of his closest friend during the years around 1900, Wilhelm Fliess, Freud argued that, as our bodies play back the record of past evolutionary stages, so our minds (usually) sweep the traces into the subconscious; in particular, whereas civilized adults will have repressed the erotic zones, patterns and cycles that drove and excited our ancestors, they are at play in our childhoods.

Freud was by no means unusual to interpret psychological development in evolutionary terms. Darwin himself had written on the subject, in *The Expression of Emotions in Man and Animals* (1872), and published 'A biographical sketch of an infant' in the new journal *Mind* in 1877. One of Darwin's inner circle, George Romanes, wrote in *Mental Evolution in Man* (1888) that the 'emotional life of animals is so strikingly similar to the emotional life of man – and especially of young children – that I think that the similarity ought fairly to be taken as direct evidence of a genetic continuity between them.' Romanes drew charts of how the emotional and intellectual lifetime of a human being recapitulated major footholds on an evolutionary tree, while, in a book read by Freud, the American evolutionist and psychologist James Baldwin concluded that the 'embryology of society is open to study in the nursery'.[13] Evolution, and in particular developmental recapitulationism, swept through anthropology and child psychology, the nascent field of sexology exemplified by Freud's Viennese colleague Richard Kraft-Ebbing and his Berlin rival Albert Moll, and decisively shaped psychoanalysis.[14] Freud's theory, he wrote in 1919, opened up 'a phylogenetic childhood – a picture of the development of the human race, of which the individual's development is in fact an abbreviated recapitulation influenced by the chance circumstances of life', and would therefore 'claim a high place among the sciences which are concerned with the reconstruction of the earliest and most obscure periods of the human race'.[15]

Freud's later published work, including *Totem and Taboo*, published just before the First World War, and *Beyond the Pleasure Principle*, which introduced the death instinct in 1920, argued from the basis of humans as evolved beings. Yet psychoanalysis, as presented in Freud's two most read texts, *The Interpretation of Dreams* (1900) and *The Psychopathology of Everyday Life* (1901), did not trumpet their debt to evolutionary thought, which went far beyond a shared conviction that the past contains the keys to understanding the present. The former argued that 'a dream is a (disguised) fulfilment of a (suppressed or repressed) wish'. Wishes, from the unconscious, where they are repressed from childhood, are usually censored by the 'ego', wrote Freud; but during sleep the ego relaxes, and the wishes emerge. Nevertheless, the wishes are too troubling to be comprehended directly, so they undergo 'dream-work', which combines, reshapes and generally disguises them. Nevertheless, dreams provided a 'royal road to the unconscious'. Likewise, Freud argued in *The Psychopathology of Everyday Life*, by paying the closest of attention to the making of small mistakes ('parapraxes') – lapses in memory, slips of the tongue, doodled blots – the repressed wishes could be uncovered.

The Viennese context shaped psychoanalysis. Vienna was the capital of the Austro-Hungarian Empire, a deeply conservative society, a patriarchy with the emperor, Franz Josef, at its head. The constrained life of women, victims of double standards when it came to matters of sex, prompted, at the very least, the wave of hysteria. Freud, like the sexologists, was a whistle-blower; his theories deliberately provoked as well as analysed Viennese society. His clientele also stood apart, a small leisured intellectual elite, with few material worries and perhaps, as Wassermann argued, greater time to develop anxieties over sex.[16] (Anna O.'s analysis alone had taken many, many hours of her and Breuer's time.) Opposition in turn encouraged the psychoanalysts to consider themselves separate, a clique with a sacred message to protect and pass on. Within the group, the psychoanalysts were unstinting in their microscopic observation of each other, making psychoanalysis a peculiar science of a leisured, urban, unworking world. Freud's translator, A. A. Brill, recalled:

> We observed and studied and noted whatever was done or said about us with unfailing patience and untiring interest and zeal. We made no scruples, for instance, of asking a man at table why he did not use his spoon in the proper way, or why he did such and such a thing in such and such a manner. It was impossible for one to show any degree of hesitation or make some abrupt pause in speaking without being at once called to account ... We had to explain why we whistled or hummed some particular tune or why we made some slip in talking or some mistake in writing. But we were glad to do this for no other reason than to learn to face the truth.[17]

Without this defensive yet inventive clique, creating analysis by observing themselves, psychoanalysis, had it been created at all, would not have developed so distinctive a theory, methodology and image, nor been so jealously protective of them (while splitting, as all sects do, to protect a particular version of message). Freud, having shrugged off co-workers Breuer and Fliess, gathered acolytes in the new century, starting with a regular Wednesday meeting in 1902. Two years later, a group of psychologists at the Burghölzi clinic in Switzerland, including Carl Jung, made contact and were drawn into the circle. At the same time the Vienna clique expanded, drawing in Ernest Jones from England, Sándor Ferenczi from Hungary and the American Brill, each crucial to propagating Freud's theories to new audiences.

It was the acolytes, the Freudians, who turned psychoanalysis from neurology, a evolutionary biology-inflected theory of neurosis, into something independent, a whole new self-sustaining branch of psychology, as distinct from earlier disciplines as the Freudians saw themselves as distinct from wider society. The followers believed this distinction, while Freud, it appears, encouraged the illusion. Freud himself had his own reasons to play along and to play down his intellectual debt to the father disciplines of evolutionary biology and physics-inspired physiology, becoming what Sulloway calls a 'crypto-biologist'.[18] When repackaged by the Freudians, psychoanalysis was more easily transported away from Vienna; in the United States, in the years after the First World War, psychoanalysis would flourish as a seemingly autonomous form of psychological themes.

Measuring mind

The Salpêtrière, 'more like a small town than a hospital', had the resources to support a diverse array of psychological approaches.[19] Working there in the 1880s, at the same time as Charcot, was Alfred Binet. In 1889 a new psychological laboratory was founded at the Sorbonne, and Binet moved to it in 1891. This was a period of discipline definition and boundary marking. Between 1880 and 1903, for example, American psychology gained a professional society (the American Psychological Association, founded 1892), specialist journals (such as the *American Journal of Psychology*, from 1887, and the *Psychological Review*, from 1894), and nearly forty psychological laboratories, many of them becoming factories of doctorates.[20] In France, a younger generation of psychologists distinguished their researches from philosophy-oriented approaches, hypnotism and psychical studies by emphasizing experimental methods. They were also making a political point. 'French conservative philosophers believed in the transcendental self, a reality that could not be the subject of scientific psychology', writes historian Roger Smith, and when Pierre Janet, chair of experimental and comparative psychology at the Collège de France, 'suggested that cases of multiple personality proved that there could be more than one self, then, so his point went, psychology achieved status as a scientific subject'.[21]

71

In the early nineteenth century, 'intelligence' meant a faculty, quickness of understanding or a communication, more often a relation rather than a thing, and very rarely something to be measured. The mid-century phrenologists had offered one means of measuring faculties, including intelligence. But, from the late nineteenth century, measurement of intelligence came together with statistical techniques to offer a powerful hybrid, one associated with the names of Alfred Binet, Théodore Simon, Charles Spearman and Cyril Burt. Binet was wealthy enough to work unpaid at the Sorbonne laboratory of physiological psychology, although he established himself swiftly and securely enough to become its director on the retirement of its founder, Henry Beaunis, in 1894.[22] Through the 1890s, Binet, with his co-workers, developed techniques to compare and measure the mental characteristics of individuals.

There was a clear working world context for this work. From the mid-nineteenth century, the French state had supported universal primary education, a policy to encourage a common national identity. The administrators of primary education wanted a means to distinguish the slower from the faster students, in order that the two classes could be separated and the streams treated differently.[23] In 1904 Binet was invited to advise a government commission on the matter. Binet, with Théodore Simon, offered a test that measured children's achievements in thirty tasks that they claimed generated a 'mental age' for each child, and offered a standard to distinguish the 'normal' from the 'subnormal' students the administrators were so keen to identify. They published the test in *L'année psychologique* in 1905, with revisions in 1908 and 1911, the year of Binet's death.

Indeed the working world interest was broader than education administration. As early as 1895, Binet, with his colleague Victor Henri, had written that their work had 'importance ... for the pedagogue, the doctor, the anthropologist and even the judge'; 'Management of children, the mentally disordered, differences of race and criminals gave psychology both its subject matter and its social significance', comments Smith, all of which 'reverberated well into the twentieth century'.[24] And this was territory contested between disciplines, old and new. Medicine, in particular, pressed a strong claim to expertise. This created a professional interest among would-be psychologists for a technique that clearly offered utility to potential clients, that responded to widespread fears about degeneration, and, by possession and training, that distinguished psychologists from rival professionals.

Binet avoided defining 'intelligence' as a singular entity, even while

72

providing administrators with the tests to distinguish between normal and subnormal schoolchildren. The German psychologist William Stern took the Binet–Simon test and, in work published in 1912, created an 'Intelligence Quotient' by dividing mental age by actual age. This seemed to be a measure, although Stern denied it, of a discernible entity, 'intelligence', shared by all of us and measurable on a single scale. The British psychologist Charles Spearman had argued in the 1900s, independently of Binet, that a 'general intelligence' could be objectively measured, and indeed that a 'general' or 'g factor' explained why performance across different mental activities was correlated – and he claimed such correlation could be demonstrated with statistical methods.[25] Spearman, professor of psychology at University College London, in the 1920s offered a detailed two-factor – general and specific – theory of psychology. British schools used 'general intelligence' as a means of streaming schoolchildren.[26] Cyril Burt, Spearman's successor at UCL from 1932, had, earlier in his career, developed tests for the London County Council in order to map subnormality across the capital.[27]

Similarly, in the United States, Alfred Binet's tests were welcomed by professionals charged with administrative tasks. Henry Herbert Goddard, eugenicist and future author of *The Kallikak Family: A Study in the Heredity of Feeble-Mindedness* (1912), as research psychologist at the New Jersey Training School for Feebleminded Boys and Girls, found that the Binet–Simon test amply met his institution's 'needs' for classification.[28] During the twentieth century, in a manner that plots the growth in influence of psychology as a discipline, psychological categories and measurements would increasingly be applied to wider society. From the 1910s, the Stanford psychologist Lewis Terman promoted modified Binet–Simon methods as a test of 'normal' as well as 'subnormal' children. This was an important step in the general psychologization of society.

Psychology as a synthetic science

While providing tests to clients was one way that psychology could establish some professional status, another was to cleave very closely to the established discipline of physiology, which was secure because it contributed to medical training or, in the case of Germany, already held prestigious professors' chairs in its name. The study of the physiology of reflex actions from the mid-nineteenth to the mid-twentieth century illustrates this strategy. In the 1820s, observers such as Charles Bell in London and François Magendie in Paris had distinguished between posterior nerves that channelled sensory information inward in the nervous system from anterior nerves that channelled motor responses outwards. This pattern had been named the 'reflex'. Through the nineteenth century, the specificities of the nervous system were traced, with controversies raging over the extent that locatable regions of the brain were associated with particular movements. Two central techniques were vivisection – removing, for example, parts of the brains of dogs – and records of impairments after brain injuries. The research also charted the complexity of nerves (Santiago Ramon y Cajal and Camillo Golgi would argue about the details of the map). In order to make sense of the complexity, Carl Wernicke in Berlin pictured the brain as being like a telegraph exchange office, with incoming and outgoing wires and different locations of the brain responsible for motor projections and others for processing or storing sensory data.[29]

A second synthesis was offered around the turn of the century by Charles Scott Sherrington. A shining undergraduate student at Cambridge in the late 1870s, Sherrington learned as he travelled in the 1880s from Michael Foster's physiological laboratory to Goltz's laboratory in German Strasburg, where he vivisected the cortexes of dogs, visited Venice to study cholera victims, and journeyed on to Berlin to see Rudolf Virchow and Robert Koch. His early work on reflex actions, at St Thomas's Hospital, London, addressed the classic crossed-leg knee-jerk.[30] But it was his later vivisections of dogs, carried out at Liverpool University, that provoked Sherrington to offer a synthesis that viewed the nervous system as nested levels of reflex arcs – with higher levels modifying the actions of lower. The synthetic nature was indicated by the title of Sherrington's published Silliman Lectures, given at Yale, *The Integrative Action of the Nervous System* (1906).

While offering the new century its first thorough account of the brain, Sherrington kept reflections about mind, aside from 'some speculations on sensory perception', 'more especially for poetry and for philosophical reflection in retirement'.[31] A late example of the latter reflection can be found in the Gifford Lectures, given in Edinburgh in 1937–8, and published in 1940 as *Man on his Nature*. Sherrington asks:

> But what of Mind? Minds knows itself and knows the world: chemistry and physics, explaining so much, cannot undertake to explain Mind itself. It can intensify knowledge of Nature but it cannot be shown that

Mind has hitherto directed the operations of Nature. In that sense Mind and Nature are different.[32]

Reflex research was taken up by Ivan Petrovich Pavlov in St Petersburg. The son of a priest, in 1870 Pavlov switched study from seminary to medical school at St Petersburg. In the early 1890s he had the good fortune that a connection to Russia's 'most eminent physician', Sergei Botkin, brought him to the attention of Prince A. P. Ol'denburgskii, a 'member of the extended tsarist family and heir to a tradition of medically oriented philanthropy' who was founding the Imperial Institute of Experimental Medicine; Pavlov became head of the institute's division of physiology and 'thereby master of the country's largest and best-equipped physiological laboratory'.[33] Coincidentally, the tsarist bureaucracy had decreed that physicians should seek laboratory training, sweetening the deal with a number of benefits. Pavlov suddenly had, if he chose to direct it, an army of laboratory workers. He did, directing it to study the nervous physiology of digestion.

Between 1891 and 1904, around a hundred people worked in Pavlov's laboratory, most of them the temporary physician-students, and of these all but one was male; these 'praktikanty [as they were called] conducted thousands of experiments . . . painstakingly collecting, recording, measuring, and analyzing the dogs' secretory reactions to various excitants during experimental trials that often continued for eight or ten hours at a time'.[34] Pavlov also had a management team, and assistants who attended to the dogs. He assigned research topics. Typical experiments included implanting a fistula – an artificial channel – so that the features of the digestive process could be continued to be measured when the dog had recovered from surgery. Gastric secretion was achieved by teasing a dog with food. This brought matters of psyche into the equation: the dog's thoughts of food and their physiological consequence.

Pavlov's laboratory, as historian Daniel Todes tells us, was like a factory. Whereas previous physiological laboratories were more like workshops, with each stage of the production process under one person's control, at the Institute for Experimental Medicine the doing was separated from the directing. The result was a highly disciplined, very productive laboratory, churning out papers (all passing under the editorial eye of Pavlov) and responding quickly to challenges. And the picture of the digestive system generated – that it was 'a purposeful, precise, and efficient factory' – mirrored the greater whole.[35] The factory also produced Pavlov's reputation, and in 1904, after wrestling with the difficulties of squaring its ideal of the scientist as lone hero with the scale of Pavlov's enterprise, the Nobel committee awarded its medicine or physiology prize in 1904 to the Russian.

By then Pavlov was directing his workforce to the study of the differences between unconditioned reflex action, the salivation of a dog at the smell or sight of food, for example, and a conditioned reflex, such as salivation after a sound replaces direct stimulus. Western scientists were exposed to a limited version of Pavlov's science in the 1900s, after his Nobel Prize. But from the 1900s to the 1930s Pavlov built up from his conditioned dog experiments an immensely ambitious programme, investigating learning, replacing mentalist statements about 'ideas' with physicalist statements about procedures with conditioned animals.[36] In total, here was new unified science. Only when his monograph *Conditioned Reflexes: An Investigation of the Physiological Activity of the Cerebral Cortex* (1926) was published, did Western scientists begin to grasp the scale of Pavlov's science. 'While anglophone neurophysiologists followed Sherrington and studied the minutiae of inter-neuronal activity, Pavlov speculated about the . . . whole structure of the higher brain', summarizes Roger Smith. 'The division of labour in Western science, with few exceptions, fostered highly specialized research and created barriers between the disciplines of neurophysiology and psychology. By contrast, Pavlov and his school, at least in principle, attempted to achieve a unified science of human nature.'[37]

Pavlov's conditioned reflex set the stage for behaviourist psychology in the early twentieth century, 'approaches that, ironically enough', writes historian Anne Harrington, 'would largely eliminate considerations of brain and biology from the experimental picture altogether'.[38] Behaviourism flourished in the United States, for good contextual reasons. In the late nineteenth century the United States was a rapidly changing country, as the expansion of industry and the migration of people spurred the growth of cities. Psychologists, members of a growing profession, not only offered the tests helpful to administrators of schools and asylums, as we saw above, but also offered themselves as experts in facilitating individuals' adjustment to the challenges of modern urban living. Behaviourism, as a 'science of what people do', was picked out among branches of psychology because it exemplified this practicality.[39]

The manifesto for behaviourism was written by John Broadus Watson, a South Carolinian, who in 1903 had written a dissertation at Chicago, in John Dewey's philosophy department, on 'animal education', or, rather, how the white rat learned despite induced brain damage. Watson was later appointed to a chair of psychology and director of the psychological laboratory at Johns Hopkins University. Watson's manifesto, 'Psychology as the behaviourist views it', was published in the *Psychological Review* in 1913. He set out his argument in no uncertain terms:

> Psychology as the behaviorist views it is a purely objective experimental branch of natural science. Its theoretical goal is the prediction and control of behavior. Introspection forms no essential part of its methods, nor is the scientific value of its data dependent upon the readiness with which they lend themselves to interpretation in terms of consciousness. The behaviorist, in his efforts to get a unitary scheme of animal response, recognizes no dividing line between man and brute. The behavior of man, with all of its refinement and complexity, forms only a part of the behaviorist's total scheme of investigation.[40]

In order to be a 'purely objective experimental branch of science', the behaviourist should record only stimuli and behavioural responses, thereby distinguishing the science from the practice in psychology, as derived from philosophy, of drawing evidence from the results of introspection of mental states. 'Psychology', Watson summarized, 'needs introspection as little as do the sciences of chemistry and physics.' Naturally, such a science would exclude any reference to mind.[41] As Watson put it:

> The time seems to have come when psychology must discard all reference to consciousness; when it need no longer delude itself into thinking that it is making mental states the object of observation. We have become so enmeshed in speculative questions concerning the elements of mind, the nature of conscious . . . that I, as an experimental student, feel that something is wrong with our premises and the types of problems which develop from them. There is no longer any guarantee that we all mean the same thing when we use the terms now current in psychology.[42]

With introspection discarded, animal experiments had as much validity as human experiments – no 'dividing line between man and brute'; indeed, they were potentially more useful because the ethical restrictions were laxer. The way was clear for the behaviour of the white rat to stand in for the human psyche. And, in turn, when psychology was freed from constantly having to justify its experimental selections in terms of what they might tell about human consciousness, space was opened to work out, for the first time, a 'scheme for the prediction and control of response in general'.[43] With prediction and control would come psychology as a tool for adjusting individuals in modern society.

With his student Rosalie Rayner, Watson demonstrated that the behaviour of a toddler, known as Little Albert, at the Phipps Clinic at Johns Hopkins University was alterable by conditioning. Just as Pavlov's dogs would salivate at a conditioned sound, so Little Albert learned to fear furry creatures – starting, perhaps inevitably, with a rat – by associating their appearance with a crashing noise; a suitable environment made humans pliable, even determined human nature:

> Give me a dozen healthy infants, well-formed, and my own specified world to bring them up in and I'll guarantee to take anyone at random and train them up to become any type of specialist I might select – doctor, lawyer, artist, merchant-chief, and, yes, even beggar-man and thief, regardless of his talents, penchants, tendencies, abilities, vocations and race of his ancestors.[44]

Watson's affair with Rayner led to his academic downfall in 1920. His wife, Mary, was the sister of Harold Ickes, a leading figure in Republican and Progressive party politics, and the divorce case was splashed over the front pages of the Baltimore newspapers. Watson moved on to advertising, and if anything increased his standing as a public intellectual with the publication of *Behaviorism* in 1924. Rayner and Watson also co-wrote a child-care guide, *Psychological Care of Infant and Child*, published in 1928. I do not recommend it to prospective parents.

In fact psychologists, when they adopted the self-description 'behaviorists' in the 1920s, tended to take from Watson the insistence on admitting observable data, with concepts defined by operations, 'rather than acceptance of Watson's extreme environmentalism', in the psychological sense shown by his child learning boast above, 'or his denial of the existence of mind'.[45] Edward C. Tolman at Berkeley tweaked Watson's overly atomistic categories of stimulus and response and brought back purpose as a valid category; Clark L. Hull at Yale sought to build behaviourist psychology into a unified theoretical edifice, as if he was a latter-day Newton of psychology; while from the late 1930s B. F. Skinner returned to some of the stark anti-theoretical, anti-mind positions of Watson in developing operant psychology.[46] Running through all this work – but especially Tolman's – were rats in mazes. 'I believe that everything important in psychology', Tolman announced in his presidential address to the American Psychological Association, 'except perhaps such matters as the building up of the super-ego, that is everything save such matters as involve society and words', he added parenthetically, 'can be investigated in essence through the continued experimental and theoretical analysis of the determiners of rat behavior at a choice-point in a maze.'[47] The rat joined the fruit fly as a model organism of the twentieth century.

Nevertheless, some of the behaviourists did not just study rats in the laboratory. Watson, for example, argued in his 1913 manifesto for the complementarity of evolutionary and behaviourist explanations: evolution theory would explain functionally why an inherited adaptation survived in an environment, while behaviourist study would provide an account of how habits were learned. Watson cites his experience with a 'certain species of birds' (in fact, terns). Only when Watson visited the Dry Tortugas, islands in the Florida Keys, did he find 'the animals doing certain things: some of the acts seemed to work peculiarly well in such an environment, while others seemed to be unsuited to their type of life'. As he reported it:

> I first studied the responses of the group as a whole and later those
> of individuals. In order to understand more thoroughly the relation
> between what was habit and what was hereditary in these responses, I
> took the young birds and reared them. In this way I was able to study
> the order of appearance of hereditary adjustments and their complex-
> ity, and later the beginnings of habit formation. My efforts in deter-
> mining the stimuli which called forth such adjustments were crude
> indeed. Consequently my attempts to control behavior and to produce
> responses at will did not meet with much success. Their food and water,
> sex and other social relations, light and temperature conditions were all
> beyond control in a field study.[48]

In order to achieve 'control', 'work of this kind must be supplemented by carefully controlled laboratory experiments.' The field, thought Watson, while a necessary source of phenomena, must be made secondary to the laboratory if psychology was to become a useful tool of individual adjustment. And, lest we forget that lesson, Watson immediately followed the tern story with the statement: 'Had I been called upon to examine the natives of some of the Australian tribes, I should have gone about my task in the same way.'[49]

Indeed, Watson was probably right when he argued that the branches of psychology flourishing in the early twentieth century were those that did not merely offer tools, but fed voraciously off the working world:

> What gives me hope that the behaviorist's position is a defensible one is
> the fact that those branches of psychology which have already partially

withdrawn from the parent, experimental psychology, and which are consequently less dependent upon introspection are today in a most flourishing condition. Experimental pedagogy, the psychology of drugs, the psychology of advertising, legal psychology, the psychology of tests, and psychopathology are all vigorous growths. These are sometimes wrongly called 'practical' or 'applied' psychology. Surely there was never a worse misnomer. In the future there may grow up vocational bureaus which really apply psychology. At present these fields are truly scientific and are in search of broad generalizations which will lead to the control of human behaviour.[50]

Watson's colleague, Karl Lashley, also visited the Tortugas, but his work concentrated on the white rat. Lashley wanted to trace the physiology of rat learning. However, in the 1920s, he was frustrated in his attempts to identify specific sites within the rat cortex where the memory of a learned behaviour might be localized.[51] Lashley rejected localization, returning to a Sherrington-style 'mass action' theory in which the brain as a whole was integrative. The evidence recalled Watson's thesis methodologies, asking what a rat could still learn after parts of its brain had been selectively destroyed. Lashley 'inspired a large number of students with this type of experimental approach to psychological questions'.[52]

Immunology and conflict

While psychology developed, at least through its strains of behaviourism, as a twentieth-century science that rejected 'mind' when accounting for the individual in the environment, the parallel emergence of a science of immunology would articulate a different language for how the self coped with a challenging, even aggressive, environment. The decades before and after 1900 featured a scientific controversy waged between the two great microbiological research centres, a Parisian camp around Louis Pasteur and the institute opened in his name in 1888 and a German camp around Robert Koch. The controversy revolved around theories of immunity, of how the body fought off unwelcome invaders. Notice, however, how the metaphors of immunology are the images of military campaigns – fighting, invasion, repelling attacks; the language was formed in the aftermath of the Franco-Prussian war and carries its inflections. If anything, such metaphors were deepened and extended through the twentieth century. Emily Martin, in a rather wonderful anthropological study of the metaphor in 1980s popular treatments of immunology and also, crucially, of the consequences and alternatives, quotes some very telling examples:

> The immune system evolved because we live in a sea of microbes. Like man, these organisms are programmed to perpetuate themselves. The human body provides an ideal habitat for many of them and they try and break in; because the presence of these organisms is often harmful, the body's immune system will attempt to bar their entry or, failing that, to seek out and destroy them . . . When immune defenders encounter cells or organisms carrying molecules that say 'foreign', the immune troops move quickly to eliminate the intruders.[53]

Our own body (the self) was cast as a nation-state at war against a foreign non-self, a 'total war', where the wounds themselves were 'battlefields' and the protection was the body's immune system, written as an army. Martin quotes from another popular account:

> The organization of the human immune system is reminiscent of military defence, with regard to both weapon technology and strategy. Our internal army has at its disposal swift, highly mobile regiments, shock troops, snipers, and tanks. We have soldier cells which, on contact with the enemy, at once start producing homing missiles whose accuracy is overwhelming. Our defence system also boasts ammunition which pierces and bursts bacteria, reconnaissance squads, an intelligence service and a defence staff unit which determines the location and strength of troops to be deployed.[54]

Only some of this martial language would have been meaningful to the protagonists of the Franco-Prussian war – highly mobile regiments, yes, but not homing missiles or tanks. The metaphors of immunology changed alongside the techniques of warfare in the twentieth century.

The two theories at stake in the controversy between Paris and Berlin were named 'humoral immunity' and 'cellular immunity'; 'humoralists believed that immunity to infectious diseases was the result of the bactericidal action of substances called antitoxins, or antibodies, found in sera' and, while their opponents emphasized the action of cells, phagocytes, in the view of the humoralists, 'merely scavenged the corpses of microbes killed by the chemical substances of the humours'.[55] In Berlin the supporting evidence came from observations of the remarkable effects of diphtheria antitoxin. In 1888 Émile Roux and Alexandre Yersin, at Pasteur's lab, had uncovered a toxin, a chemical substance that brought on the effects of diphtheria, while in response, following up in Berlin two years later, Emil Behring and Shibasaburo Kitasoto discovered the production of antitoxins. By injecting the antitoxin, a ten-year-old, raging feverish with diphtheria, had been saved. Antisera production became a major pharmaceutical enterprise soon after. In terms of theory, antitoxins seemed to support dramatically the notion that immune responses were the production of chemical substances.

Nevertheless, a Russian working at the Institut Pasteur, Ilya Ilyich Mechnikov, disagreed. While Mechnikov was born near Kharkov, in the Ukraine, his early research life was spent in Germany, at Giessen and Göttingen, and at the field stations at Heligoland and Naples. When he returned to Russia, at Odessa and St Petersburg, he married twice. His first wife, Ludmilla Feodorovitch, died of tuberculosis, and his second, Olga, of typhoid. (Each time the distraught Mechnikov, also suffering ill health, attempted suicide.) After leaving Russia, he set up in 1882 a private laboratory at Messina, in Sicily, where he continued his marine biological studies through the microscope, now with an interest in how organisms responded to invasive 'foreign bodies'. 'Certain of the lower animals, transparent enough to be observed alive, clearly show in their midst a host of small cells with moving extensions', he reported in his 1908 Nobel speech. 'In these animals the smallest lesion brings an accumulation of these elements at the point of damage. In small transparent larvae, it can easily be shown that the moving cells, reunited at the damage point, do often close over foreign bodies.'[56] Mechnikov focused his microscope on the 'floating larvae of starfish ... called *Bipinnaria*. Large enough for several operations, they are transparent and can be observed alive under the microscope.' He recalled what happened next:

> Sharp splinters were introduced into the bodies of these *Bipinnaria* and the next day I could see a mass of moving cells surrounding the foreign bodies to form a thick cushion layer. The analogy between this phenomenon and what happens when a man has a splinter that causes inflammation and suppuration is extraordinary. The only thing [different] is that in the larva of the starfish, the accumulation of mobile cells round the foreign body is done without any help from the blood vessels or the nervous system, for the simple reason that these animals do not have either the one or the other. It is thus thanks to a sort of spontaneous action that the cells group round the splinter.[57]

'Disease', Mechnikov said, summarizing the cellular immunity theory, 'would be a fight between the morbid agent, the microbe from outside, and the mobile cells of the organism itself. Cure would come from the victory of the cells and immunity would be the sign of their acting sufficiently to prevent the microbial onslaught.'[58]

Despite this observation of phagocytes, cells that engulf or 'eat' invaders, the evidence published in the 1890s swung the debate towards the humoralists. In Koch's laboratory in 1895, Richard Pfeiffer observed how cholera microbes were degraded when placed in cell-free anti-sera, while Jules Bordet, a Belgian working at the Institut Pasteur, found in 1898 that cell-free sera split red blood cells, a process called haemolysis, in a way strikingly similar to the action of anti-sera. Bordet developed methods by which microbes could be identified by the specific actions of the sera they produced. This work was published early in the new century, and was immediately seized upon by the Koch Institute for Infectious Diseases in Berlin. In 1905 Fritz Schaudinn and Eric Hoffmann at the Berlin Charité hospital identified a spiral microbe, which they named *Spirochaeta pallida*, as the causative agent of syphilis. Three Koch researchers, August von Wassermann, Carl Bruck and Albert Neisser, soon identified a serological test for syphilis. This 1906 diagnostic procedure, named the 'Wassermann test', would become the central tool in massive public health campaigns against venereal disease in Europe and the United States before the mid-twentieth century.[59]

More than a test was needed, however. The excitement that met the announcement of the Wasserman test was redoubled, therefore, when Paul Ehrlich, working with Sahachiro Hata, announced in 1909 that a chemical 'magic bullet' had been found, a substance that not only identified syphilis but treated it too. Ehrlich, who had passed through Koch's laboratory in the 1880s and 1890s, now had his own institute in Frankfurt am Main. An immense series of syphilitic rabbit experiments, where the creatures were tested against different candidate substances, culminated, in experiment 606, in success with arsphenamine. 'With the discovery of Salvarsan', as the drug form of arsphenamine was named, writes historian Allan Brandt, 'Ehrlich initiated the modern age of chemotherapeutics.'[60] In fact, since Salvarsan required intravenous injection, a procedure with which American general practitioners were unfamiliar in the 1910s, and because Salvarsan generated some toxic side-effects, only when a second substance, the 914th product tested by Ehrlich, called Neosalvarsan, became available did a real alternative to mercury – the maddening traditional treatment for syphilis – emerge. 'The instrumental ideals of efficiency, prevention, and cure soon began to influence professional thinking concerning the sexually transmitted diseases', summarizes Brandt, and 'venereal disease came be perceived as a scientific problem with a scientific solution.'

The Ehrlich story – and especially the dream of the 'magic bullet' against disease – provided one of the popular images of science that was particularly powerful in the first half of the twentieth century. 'The magic bullet' may be, as Paul Ehrlich described it, antibodies 'which find their targets by themselves', but this selfless narrative had plenty of room for their discoverers as heroes in the human struggle against disease, and was also very congenial to the growing pharmaceutical companies which manufactured, marketed and depended commercially on chemical medicine. The Koch, Pasteur, Mechnikov and Ehrlich stories featured in one of the first history of science bestsellers, Paul de Krief's *The Microbe Hunters* (1926), while the life of the discoverer of Salvarsan was made into an Oscar-nominated film, *Dr Ehrlich's Magic Bullet* of 1940 (tagline: 'Two women stood by him while all the world jeered . . . a world he was trying to save!').

Paul Ehrlich's search through chemical candidates was conducted in the light of a theory of immunity – and practice of serology – that he was co-developing. Practically, his emphasis was on quantification and standardization of serum antitoxins. 'The words of the gifted natural philosopher Clerk Maxwell, who said that if he were required to symbolise the learning of our time he would choose a metre measure, a clock, and a kilogramme weight, are equally apposite in reference to progress in the field of inquiry in which we are at present interested', Ehrlich told the Royal Society in his Croonian lecture of 1899, linking practice to the announcement of his 'side-chain theory'. 'And so at the very beginning of my theoretical work on immunity I made it my first task to introduce measures and figures into investigations regarding the relations between toxine and antitoxine.'[61] He defined the 'toxic unit' as 'that quantity of toxic bouillon which exactly sufficed to kill, in the course of four days, a guinea-pig of 250 grammes weight'.[62] This metrological work – vital for the export of similar serological procedures from the laboratory to working world settings – was matched by the development of his theory of antibody formation. Ehrlich proposed that cells had on their surfaces 'receptors' which ordinarily would react specifically, normally to particles of 'food-stuffs', and on occasion to antigens. But these receptors, or 'side chains', could also break off and enter the serum, where they would act as 'antibodies'.[63]

Ehrlich's 'side-chain' theory 'provided the humoralists with a comprehensive framework they had previously lacked'.[64] While it was repeatedly attacked in the 1900s, side-chain theory contributed to immunology emerging as a distinct discipline, shaping a set of central problems – it gave an answer to the questions of the nature and specificity of the antibody–antigen reaction and offered a mechanism of antibody formation, while simultaneously creating doable research problems suitable for postgraduate work, such as figuring out the specific physical structure of particular antibody molecules. Specialist journals, beginning in Germany, provided an outlet for the disciplinary products of immunology.

One set of criticisms in particular, from Karl Landsteiner, based first in Vienna and then at the Rockefeller Institute in New York, served to sharpen up the side-chain theory. Landsteiner in 1918 argued that the shape of the electrostatic charge of the antigen mattered, rather than its precise chemical nature, in provoking the production of antibodies.[65] Ehrlich had proposed that each of his side chains, being a different molecule, would be specific. In Landsteiner's theory, only the electrostatic outline was specific. This theory explained, for example, why antibodies would be made in response to substances that would not have been encountered naturally by an organism. Template theories of the specificity of antibody–antigen reactions were developed further in later decades.

Immunology and psychoanalysis were radically different sciences of the self, although both spoke, in albeit very different ways, about how the self was shaped defensively against a hostile world. Binet and Watson's psychologies offered something directly practical. In the next chapter we will see how a global conflict opened up opportunities to apply some of these techniques extensively.

Part II

Sciences in a World of Conflict

— 5 —

SCIENCE AND THE FIRST WORLD WAR

Like other volunteers at the outbreak of the First World War, many scientists wanted to fight. Some, such as Henry Gwyn Jeffreys Moseley, were recruited direct to the front line, employed as soldiers.[1] Born in 1887, Moseley came from an influential Manchester landowning family, several of whom had been scientists: one relative taught natural philosophy at King's College, London, while his father, a zoologist, had travelled on the *Challenger* expedition and become Linacre professor of human and comparative anatomy in Oxford.[2] Harry Moseley had studied physics at that university, achieving a modest second-class degree in 1910, before joining Rutherford in Manchester as a demonstrator of experiments. His research, first in collaboration with another scientific scion, Charles Galton Darwin, turned to the new field of X-ray science. This was expensive research, slightly aside from Rutherford's main lines of inquiry. The great physicist needed persuasion. His trust was soon vindicated when Moseley, building on the work of William and Lawrence Bragg at Leeds University, used the father and son's design of X-ray spectrometer to record five sharp X-ray emission lines from platinum.

Following his return to Oxford in 1913, Moseley's research programme was clear: to measure the spectra of emission lines when the platinum target was replaced by other elements. Nevertheless, there were many practical obstacles: he had to beg scarce laboratory space as well as equipment – a Gaede pump from Balliol and Trinity's joint laboratory, a spectrometer from the Clarendon; he was reliant on expert glass blowers to produce the bespoke X-ray tubes he required; and the rare earth elements had to be secured through the networks of scientific cooperation that supported European physics. After begging equipment and several months' toil in the laboratory, Moseley had measured the K and L emission line frequencies of X-rays produced by elements from the light (aluminium) to the heavy (gold). Plotting the square root of the frequency against an integer allocated to each element produced, on one page, readable at a glance, a remarkable graph: the elements arranged on straight lines, with a few odd gaps.

Moseley's interpretation was as follows. The integer, which increased from the lightest (aluminium, 18) to the heaviest (gold, 92), was the atomic number, or number of positive charges in the atom. When the number was construed as the number of electrons normally filling the orbits of an atom, the relationship between frequency and atomic number seemed to make sense according to Bohr's theory of atomic structure, which in turn garnered support. (The Oxford physicist Frederick Lindemann disputed this contribution.) Moseley's results also helped resolve the 'riddle of wrong positions' in the periodic table, for example providing new justification for the relative placing of iodine (53) and tellurium (52),[3] as well as swapping the positions of cobalt (27) and nickel (28). Finally, the gaps, at integers 43, 61 and 75, suggested the existence of as yet undiscovered elements with these atomic numbers (later found and named technetium, promethium and rhenium). Applying Moseley's results in reverse provided a means of discovering the elemental make-up of an unknown sample. X-ray research, so dependent on the working world of industry, would return the favour by providing a major technique of analysis in the form of this non-destructive X-ray investigation of materials.

When war was declared in 1914, Harry Moseley was in Australia, where a meeting of the peripatetic British Association for the Advancement of Science was under way. He rushed back to England, was granted a commission in the Royal Engineers and later was made a signalling officer of the 38th Brigade of the First Army. He left in June 1915 for the Dardenelles, where an ill-prepared invasion of Turkey via the beaches of Gallipoli was being planned. On 10 August 1915, only a few days after landing at Suvla Bay, while telephoning an order to his division, Moseley was shot in the head by a Turkish sniper. He died instantly.

Just as the death of the war poets Rupert Brooke, Isaac Rosenberg and Wilfred Owen came to stand for the untimely slaughter of young artistic talent, so the death of Moseley was held by contemporaries to be a wasteful sacrifice of the brightest and best in science. 'To use such a man as a subaltern', bewailed Rutherford, was 'economically equivalent to using the *Lusitania* to carry a pound of butter from Ramsgate to Margate.'[4] The American Robert Millikan called Moseley's death 'one of the most hideous and most irreparable crimes in history'.[5] George Sarton, the Belgian historian of science who had fled the war, emigrating to the United States, wrote that Moseley's death represented the tragic fate of 'genius':

> It must necessarily occur that men of genius die before having been able to justify themselves and to give out the treasures that were in them, and these are perhaps of all events the most tragic. Just think of the loss which mankind is thus suffering, for in the last analysis everything great and really worthwhile is due to the genius of individuals – and think also of the pity of having been called to the human stage to play an exceptional part and being swept off before having begun. The tragedy thus is not to die young but to die before having done what one was prepared to do.[6]

'Moseley's death', Sarton consoled, was 'tragic enough', but 'our grief is assuaged by the thought that his fame was already established . . . that his memory will be green forever. He is one of the immortals of science . . . He died in beauty', his death a 'consecration' of individual genius.[7] The glassware of his experiments, once temporary assemblages, became treasured relics of a secular saint. (Some of the apparatus can be seen in the Museum of the History of Science in Oxford.)

But science's role in the First World War was not, primarily, as a sacrificial victim, nor was it best characterized as the contributions of individual genius. Instead, along with all other factors in the economy deemed to be strategically important, science was collectively mobilized in an increasingly organized fashion. Rutherford's lament reveals far more of contemporary attitudes than Sarton's conventional elegy: the waste of Moseley was a waste of resources. Just as great civilian steamers should be requisitioned as supply ships and used in the most efficient manner, so scientists should be organized and put to work. Moseley, who had been offered scientific work at home, chose to fight abroad, responding to a patriotic fervour he shared with many young men in 1914. The complaint was not that bodies should be sent to the front to fight and, if necessary, die, but rather, by putting his brain in the line of fire of a Turkish bullet, it was not being deployed efficiently or effectively enough.

Organizing science for war

Historians agree that the First World War accelerated a trend towards increased organization in the modern world. As hopes for a quick conflict faded, so states found that they had to improvise a constant state of war preparation. Armies in the field had to be fed, equipped, trained, healed and buried.[8] This situation favoured large-scale industrial manufacturers, with which states learned to liaise, plan and coordinate. It also encouraged state intervention in matters of health, food supply and childcare. Furthermore, all sides were afflicted by shortages of key strategic materials. We will shortly see perhaps the most famous case: how the Royal Navy blockade of the Chilean nitrate mines prompted the German mass production of artificial ammonia and nitrates through the Haber–Bosch process. Likewise, Britain, France and the United States were cut off from high-quality optical glass (used for gun sights), synthetic dyes (used to dye soldiers' uniforms) and a host of other products of Germany's peerless chemical industries. German dominance of pharmaceuticals, another aspect of the strength in chemicals, created alarming shortages on the outbreak of war; the painkiller acetanilide, for example, rose in price from 20 cents a pound to nearly $3, while the fever-reducing medicine antipyrine went from $2 to $60 a pound.[9] These shortages in turn encouraged strengthened industry–academic linkages in the United States.

The overall process, argues historian William McNeill, was to encourage large-scale industries, often in the form of state-orchestrated national cartels, to manage innovation, with two very important consequences: the mass production of nearly everything and the institutionalization of planned invention. In general, the First World War intensified the organizational revolution – a gathering concern for scale, organization and efficiency that was discernible in the mid- to late nineteenth century, as described by Robert Wiebe, Alfred Chandler and Louis Galambos.[10] However, this increase in organization and management of innovation was not imposed on science from above, but was also called for from below, as the following case studies of scientists in Britain, the United States and Germany will show.

Since science does not exist independently of the working world, scientists have always had to justify their activities to political powers and social institutions to garner 'good will, patronage and cooperation'; from the 1870s, argues historian Frank Turner of the British 'public scientists' who led such justification, the rhetoric shifted 'from the values of peace, cosmopolitanism, self-improvement, material comfort, social mobility, and intellectual progress towards values of collectivism, nationalism, military preparedness, patriotism, political elitism, and social imperialism'.[11] There were several causal factors behind this shift, including perceived failures of government support for science education and trepidation at the success of the anti-vivisectionist movement, but, perhaps most importantly, the demonstrable advance of Germany as an industrial nation and imperial threat. Specifically, the army and navy, the largest and most generous state sponsors of science in Britain,[12] were targeted by organizations such as the British Science Guild, a body set up in 1905, the purpose of which was

> to stimulate not so much the acquisition of scientific knowledge, as the appreciation of its value, and the advantage of employing the methods of scientific inquiry, the study of cause and effect, in affairs of every kind ... [for example the problems of] the statesman, the official, the merchant, the manufacturer, the soldier and the schoolmaster ... [The] value of a scientific education lies in the cultivation which it gives of the power to grasp and apply the principles of investigation employed in the laboratory to the problems which modern life presents in peace or war.[13]

A British Science Guild prize question of 1910, set against the background of a dispute over rearmament, specifically the order of Dreadnoughts for the navy, asked for essays on 'The best way of carrying on the struggle for existence and securing the survival of the fittest in national affairs' – for example, 'whether a system of party government is sufficient to secure all the best interests of the State in those directions in which brain power and a specific knowledge are needed ...'.[14] Norman Lockyer, editor of *Nature*, prime mover of the British Science Guild, who had lectured the British Association on 'The influence of brain power in history', urged the mobilization of science. In October 1914, *Nature* editorialized that 'This war, in contradistinction to all previous wars, is a war in which pure and applied science plays a conspicuous part', and throughout the war public scientists charged the government with inadequate attention to science and scientists. No wonder Moseley was made into a symbol of waste.

But how did practice compare to rhetoric? In Britain, as the hopes for a quick victory faded, confirmed by the disaster at Gallipoli, the public scientists stepped up their pressure, clamouring for greater direction (and offering, of course, themselves as guides). In June 1915, H. G. Wells wrote to *The Times* in protest against the 'very small part we are giving the scientific man and the small respect we are showing scientific method in the conduct of war'; Sir Philip Magnus, an expert on education, chimed in: 'Our scientific men are in no way inferior to those of Germany, but they work independently and are not in close co-operation'; while electrical engineer and the inventor of thermionic rectifier, John Ambrose Fleming, echoed the sentiment: 'There is no want of ability, but there is an entire absence of external directing power.'[15] The government responded to these complaints in July 1915, when two advisory boards of civilian scientists were set up, one for the War Office and one for the navy.

These boards provided a route by which civilian scientists could contribute to the organized war effort. Their establishment was a defensive sop to public pressure, as the first lord of the Admiralty, Arthur J. Balfour, of scientific stock himself, privately acknowledged ('Although I attach no very great value to public sentiment as such . . . the establishment of a Board . . . [would] do much to satisfy public demand').[16] In fact the armed services, and especially the navy, had long-standing intramural traditions of technical expertise and innovation (witness Admiral Jacky Fisher's Dreadnoughts). To a great extent the civilian public scientists were not required.

Nevertheless, a civilian science body such as the Royal Navy's Board of Invention and Research is interesting to us for three reasons. First, the creation of the board, and the arguments that supported it, are evidence for the First World War as intensifying the trends towards greater organization of research, towards the management of strategic resources such as science and scientists. Likewise the frequent complaints that the various research branches of government were duplicating military work – 'several or all of them, . . . working on the same problems, with incomplete knowledge, or none at all, by each of them of what the others are attempting or have accomplished' – should be taken as evidence for the trend towards increased organization and management rather than merely complaints about inefficiency.[17] Second, the board was significant for what it achieved, in particular innovating techniques of anti-submarine defence. Finally, the involvement of civilian scientists via the board is interesting when compared to similar trends and tensions in other countries.

The board, chaired by the great enthusiast for naval invention Admiral 'Jacky' Fisher, was composed of some of the most eminent scientists and engineers in civilian life: physicists such as J. J. Thomson, Ernest Rutherford, William H. Bragg and Lord Rayleigh, chemists such as George Beilby (inventor of a cyanide production process, critical to gold extraction), William Crookes and Percy Frankland, and engineers such as Charles Parsons (inventor of the steam turbine), electrical engineer William Duddell and metallurgist Henry Carpenter. But the navy's problem was not lack of ideas, but too many. Not only were the in-house technical experts pushing ideas but many suggestions were flooding in from outside. The board therefore was useful to the navy not just as a means of tapping eminent civilian scientific brains, but also for screening out the public wheat from the chaff. Between July 1915 and December 1916 the board considered almost 20,700 inventive ideas; by mid-1917, the figure had reached 40,000.[18]

The board considered inventions relating to many branches of warfare, including aircraft, balloons, the construction of ships, torpedoes, the storage of oil and the saving of lives. But by far the greatest number of ideas were responses to the greatest threat of the war: the submarine. Contrary to international law and the Hague convention, submarine warfare was practised by both sides (indeed, Fisher had been instrumental in building up the British experimental fleet), but in 1915, when Germany declared merchant shipping as legitimate targets, and especially following the sinking of the ocean liner *Lusitania* in May 1915, the submarine was established as a terrifyingly effective weapon of war. The unprecedented loss of shipping would justify some unusual scientific responses, among them the conditioning of sea lions, in a manner akin to Pavlov's dogs, to locate U-boats.[19] The successful techniques were, if anything, just as outlandish. In-house naval research, at HMS *Vernon*, developed torpedoes, while the Board of Inventions and Research began developing underwater acoustic – hydrophone – methods to listen out for the chug of the U-boats' engines. In 1916 this work, directed by the physicist William H. Bragg, pushed the limits of detection out to 4 miles and began to improve methods of determining the direction of the submarine. Staff and resources grew. The civilians made a small but significant contribution to the greater research and development being conducted by the Admiralty in general, and the Admiralty's Anti-Submarine Division in particular.

Furthermore, the organization of research was transnational. From March 1915, Paul Langevin in France developed the idea of using sound detection to locate submarines by drawing upon an effect he had studied in Paris under Pierre Curie: the capability of quartz crystals to produce tiny currents of electricity when put under pressure.[20] Since sound is a pressure wave, this piezoelectric effect could be the basis of a sound detector. Quick results were achieved in France. In 1917, work by two Rutherford students, the Canadian Robert W. Boyle and the British mainstay of acoustic detection, Albert Wood, took the French ideas and experimented with detection by measuring ultrasound echoes. Cooperation with the French turned this experiment into ASDIC, an early form of sonar. ('ASDIC' does not stand for Allied Submarine Detection Investigation Committee, a story invented in the 1930s, but the moniker contains a truth.)[21]

The board, a body independent of the Royal Navy, inevitably met with distrust and opposition and was eventually all but dissolved in 1917.[22] The record of the Munitions Inventions Department, set up in 1915 to play a similar role for non-naval military invention, reveals the same tensions.[23] Nevertheless, these tensions are best seen as the usual organizational friction rather than resistance to science. The channels of scientific advice were retained, even as the Admiralty reverted to the intramural invention it preferred.

The story of the relationship between the civilian scientists and the British military had direct parallels in the United States. Furthermore, as historian Daniel Kevles has shown, the Great War became a moment when an older, highly successful tradition of innovation was challenged by a newer, science-based one; the challenge is best seen in the contrast between the Naval Consulting Board and the National Research Council.[24] Before the First World War, American military laboratories concentrated on 'simple testing of materials and devices' and the 'cut-and-try improvement of guns, cannons, engines and gadgetry', while civilian inventors and industrial firms were looked to as the source of new weapons.[25] Elmer A. Sperry, for example, had provided the gyrostabilizer technologies that the US Navy had introduced in 1912; during the First World War the links between the Sperry Gyroscope Company and the US Navy intensified to such an extent that the historian Thomas P. Hughes spies a well-formed 'military-innovation complex' in operation.[26]

Following the sinking of the *Lusitania*, the secretary of the US Navy, Josephus Daniels, oversaw the appointment of a Naval Consulting Board. His plan was to channel the enthusiasm and ideas of independent inventors – Elmer Sperry, Leo H. Baekeland (the inventor of bakelite plastics), Frank J. Sprague (electric transport) and, particularly, the great Thomas Alva Edison, installed as the board's leader. Despite the presence of Willis R. Whitney, of General Electric, home of the influential corporate laboratory, the members of the Naval Consulting Board saw military invention as the preserve of engineers, and the board itself as a clique of the nation's 'very greatest civilian experts in machines'.[27] Physicists, and other professional scientists, were to be excluded, 'because', explained Edison's chief engineer of the great man's motives 'it was his desire to have this Board composed of *practical* men who are accustomed to *doing* things, and not *talking* about it'.[28]

The physicists' response was led by George Ellery Hale. Hale was a past master at persuasion – we will see the results of his success in converting the industrial wealth of philanthropists into funding for American science in a later chapter. In 1916, Hale was not only one of the world's leading astrophysicists and director of the Mount Wilson Observatory in California (see chapter 8) but also a reforming president of the National Academy of Sciences. The Academy was an odd, anomalous, almost moribund body: 'a private organisation with a federal charter, created in the middle of the Civil War to provide expert advice to the government'; Hale sought to make it more relevant, cooperating with other bodies, dispensing grants to young researchers, but his attempt to persuade the Carnegie Corporation to pump the grant scheme failed, and he was on the look out for other patrons.[29]

A week after Woodrow Wilson issued a final ultimatum to Germany, in April 1916, Hale was at the White House pitching an idea. His proposal was for a National Research Council – a body of scientists drawn from academia, industry and government that would encourage pure and applied research for 'the national security and welfare'. It would be run under the National Academy of Sciences, breathing life into the old institution. While the pacifist psychologist James McKeen Cattell described it as 'militaristic', and the US Navy ignored it in favour of its Naval Consulting Board, Hale's creation of the National Research Council on 9 June 1916 drew warm applause.[30]

The differences in constitution and values between the Naval Consulting Board and the National Research Council are best illustrated in the bureaucratic turf war fought over anti-submarine defence. U-boats were devastating Atlantic traffic, and formed the greatest threat to American supply to support intervention in Europe. While merchant shipping losses dropped by a third following the adoption of Admiral William S. Sims's convoy proposals, U-boats still sank hundreds of thousands of tons. The Naval Consulting Board, in the absence of its own workshop (a proposal for a laboratory was held up by a dispute over 'purpose, control and location'), authorized the Submarine Signal Company to establish an official experimental station at Nahant, Massachusetts, to be shared with General Electric; again academic physicists were excluded, this time on the grounds that they would 'complicate the patent situation'.[31]

The National Research Council responded by enthusiastically receiving an allied British–French scientific mission that passed on devices for submarine detection. These instruments were demonstrably better than developments at Nahant, and Hale and Robert

Millikan successfully persuaded the US Navy to set up a second laboratory, at New London, Connecticut, that would be staffed by physicists and build on the French–British inventions. Research and development at both Nahant and New London led to successful anti-submarine methods. Nevertheless, it was New London 'which steadily acquired more equipment and more physicists [and] was by the spring of 1918 virtually absorbing Nahant', as in turn the National Research Council was 'overshadowing' the Naval Consulting Board.[32] The stock of physics was rising, that of the independent-inventor engineers falling.

Physicists demonstrated their usefulness to military patrons in other fields too. For the American Expeditionary Force, Augustus Trowbridge and Theodore Lyman, for example, tested varieties of sound ranging equipment (for locating the position of enemy artillery), as well as flash ranging techniques.[33] Another consequence of the mobilization of civilian science was that military research projects began to filter into the academy: in the United States forty campuses hosted highly secretive military research 'under the constraint of tight security regulations for the first time in American history'.[34] Overall, the Edisonian model (independent invention, backed by workshop laboratories) failed where the Hale/Millikan model (incremental physicist-led invention, planned in a manner following the pattern of the corporate laboratories) succeeded. 'For thoughtful military observers, concludes Kevles, 'the meaning of it all . . . was clear: The advance of defense technology required the organized efforts of scientists and engineers' drawing on 'fundamental physical truths and engineering data'.

In Germany too the First World War promoted the militarization of academic research, important contributions by civilian science to the development of new weapons and defences of war, and an intensification of the organization of industrial production, of which planned research was part. Range-finding of artillery by sound detection employed physicists such as Max Born as well as psychologists such as Kurt Koffka and Max Wertheimer, whom we will meet in the next chapter as instigators of the new Gestalt theory, from 1915.[35] The most notorious example of academic-military-industrial science, however, was the systematic waging of chemical warfare. Chemical warfare was no aberration. Instead, as tracing the life and career of Fritz Haber will show, chemical warfare was the expression of organized science and managed innovation.

'No nation can withdraw from economic competition, the pursuit of technology and advancement of industry, without putting its very existence at risk', Carl Engler, the rector of the Technical University in Karlsruhe, had argued in a speech in 1899, continuing: the 'struggle for existence – the fate of the nation – is decided not just on bloody battlefields, but also in the field of industrial production and economic expansion.'[36] Such social Darwinian rhetoric – war as an inevitable component of the struggle for existence between nations – was not uncommon in the 1910s.[37] Engler went on to echo the warning cry and call to arms made by the English scientist William Crookes, who had told the British Association for the Advancement of Science in 1898 that, unless chemistry could turn the nitrogen of the air into fertilizing nitrates, the world would starve and Western civilization would end; in Engler's audience was the chemist Fritz Haber, whose life and achievements illustrate the linkages between the working worlds of science, industry and the military.[38]

Haber was born in 1868 to a Jewish family in the Prussian city of Breslau. Set on an academic career, opposed by his father, who knew the extent of anti-Semitism in German universities, he nevertheless studied chemistry in Berlin, a patriotic choice, since the growth of chemical industries was driving Germany's industrial economy at breakneck speed.[39] Specializing in physical chemistry, Haber was rebuffed when he applied to join the field's leader, Wilhelm Ostwald, in Leipzig. Instead, he began a series of industrial placements and low-status academic work. He also renounced his Jewishness, accepting the arguments of the historian Theodor Mommsen that patriotic duty demanded it.[40]

In 1894, Haber was employed as an assistant at the chemistry institute of the Technical University of Karlsruhe, which had close working links both with the state government of Baden and with the great chemicals firm BASF, 40 miles north on the Rhine. He married another Breslau chemist, Clara Immerwahr, and cracked on with research. Around 1903, the Margulies brothers of Vienna, owners of chemical works, contacted Haber and offered him consultancy money to investigate an intriguing indication of ammonia production in their factories. Haber attempted to replicate the conditions, heating nitrogen and hydrogen to 1,000 degrees Celsius, while introducing iron filings to act as a catalyst. Barely any ammonia was produced, and he informed the Margulies brothers of the disappointing result. Here the intensely competitive and hierarchical nature of German academia is relevant. The ambitious Haber already felt himself in competition with his superiors, Wilhelm Ostwald and Walther Nernst. In 1907, Nernst publicly trashed Haber's account of the ammonia equilibrium point.

Haber redoubled his efforts in the laboratory. He marshalled all his resources – setting his best chemical engineer, an Englishman named Robert de Rossignol, to work, acquiring the best in gas compressors to drive the pressure as high as could be achieved, and drawing on his industrial allies, with BASF funding a research programme to the hilt. Finally, in what seems at first glance to be an extraordinary moment of contingency, but in fact illustrates just how important the working world of electro-technical and chemical industry was to science, in 1908 a Berlin gas and electric light company, Auergesellschaft, had passed to Haber rare earth elements to test as materials for light-bulb filaments.[41] Haber tried one of these rare earths, osmium, as a catalyst. The combination worked. 'Come on down! There's ammonia', shouted Haber. 'I can still see it', his assistant later recalled. 'There was about a cubic centimetre of ammonia. Then Engler joined us. It was fantastic.'[42]

Haber swiftly contacted BASF. Buy the world's osmium, he advised. But BASF were not convinced that this laboratory demonstration, with its immense pressures – two hundred atmospheres – and rare elements, could be scaled up. Here the BASF chemical engineer Carl Bosch's judgement of what was possible was decisive. The scaling up of the Haber–Bosch process, a product of the German working world of the 1900s, was arguably the most influential contribution of science to the twentieth century.[43] Cheap ammonia can be used to make cheap nitrates. The so-called Green Revolution in the second half of the twentieth century, in which new high-yielding crop varieties combined with the widespread application of pesticides and fertilizers, including nitrates and ammonium sulphate, would by some counts feed 2 billion people in China and India who would otherwise starve.[44] This was the substance behind Max von Laue's elegy for Haber: that he had 'won bread from air'. The Green Revolution will be discussed in chapter 14.

Nevertheless, the mass production of ammonia was of more immediate consequence for sustaining the fighting power of Germany in the First World War. Nitrate, a powerful oxidizer, was a component of gunpowder, while nitric acid was essential to the production of modern explosives. While Chilean mines dominated the world's pre-war supply of nitrates, the supply lines linking Chile to Germany were cut early in the conflict. If the war had been short and swift, like the Franco-Prussian War of 1870–1, then the German command would have had nothing to worry about: existing stockpiles would have sufficed. But as the mobile war stalled, and as massive bombardment became the main battering ram of advance, so supplies of ammunition were rapidly running out for lack of nitrates. Walther

Rathenau, Germany's foremost organizer of technological systems, mobilized industry, including BASF, to the patriotic cause. Bosch, chivvied on by Haber, transformed BASF's ammonia factories into nitrate factories. By May 1915, BASF was producing 150 tons of nitrate per day and sustaining Germany's industrialized warfare.

The strong ties between German industry, government and academia had been tightened in the years prior to 1914. Specifically, in 1910, the theologian Adolf von Harnack, one of the most eminent scholars in Germany, had pitched an idea to Kaiser Wilhelm: the imperial nations were locked in competition and the quality of scholarship, which in turn could only be secured by generous funding, would determine the outcome.[45] In the United States, the immense wealth of the second industrial revolution – oil, steel, and so on – was being recycled by philanthropists such as Rockefeller and Carnegie into research institutes against which Germany's academic institutes could not compete. 'This cannot, this dare not remain the case', Harnack concluded, 'if German science and with it the fatherland – its inward strength and its outward image – are to avoid grave damage.'[46] Swayed, but unwilling to commit scarce imperial funds, Wilhelm summoned Germany's industrialists to Berlin on 14 May 1910. For the Fatherland, the captains of German industry would fund the Kaiser-Wilhelm-Gesellschaft zur Förderung der Wissenschaften (Kaiser Wilhelm Society for the Advancement of Science), which in turn would manage the channelling of German industrial wealth into new, large research Kaiser Wilhelm Institutes. They became power-houses of twentieth-century German science.

Institutes founded before the First World War, mostly in the south-western suburb of Berlin of Dahlem, included the Kaiser Wilhelm Institute (KWI) for Biology (1912), the KWI for Biochemistry (1912), the KWI for Brain Research (1914, not in Dahlem but the other side of the city, in Buch), the KWI for Chemistry (1911), a KWI for Coal Research (1912, in Mülheim in the Ruhr) and the KWI für Arbeitsphysiologie (1912, work or occupational physiology). Further Kaiser Wilhelm institutes were opened during the war, including a KWI for Experimental Therapy (1915), a KWI for German History (1917, Berlin), a KWI for Iron Research (1917, Aachen, near the border with Belgium) and a KWI for Physics, which was founded, with Albert Einstein installed as director, in 1917.

Nevertheless, the institute of direct interest to us now was the KWI for Physical Chemistry and Electrochemistry, founded in 1911 and built in Dahlem, Berlin. The funds were supplied by Leopold Koppel, financier and controller of assets that included the Auergesellschaft gas and electric light manufacturer, which in turn had forged consulting links with Fritz Haber. It was Koppel's support that swung Haber the directorship of the prestigious KWI for Physical Chemistry and Electrochemistry away from his more established rival, Walther Nernst.[47] Haber, with family, moved from Karlsruhe to Berlin in July 1911. He found the city at a peak of imperial anxiety and patriotic excitement: Germany and France were squabbling over Morocco, and the appearance of a gunship, the *Panther*, off the port of Agadir seemed, for a moment, to be the opening shot of a new European conflict.

In fact, by the time war did break out, Haber, from his base at the KWI, had become the central node of a network of contacts that spread across imperial Berlin. As we have seen, he was able to help Walter Rathenau and Carl Bosch turn the BASF ammonia factories into producers of nitric acid for explosives. This network would enable Fritz Haber, an academic chemist with a successful consultancy, to become a military organizer of chemical warfare, a powerful scientist-soldier, a man of high status in the imperial hierarchy.

Germany was not the first country to use chemical weapons in the First World War. Britain, France and Germany were each researching tear gas weapons in 1914: the French tried tear gas in August 1914, the Germans used a chemical that induced sneezing in October 1914, near Neuve-Chapelle, and the British prepared to launch a few so-called stink bombs.[48] None of these attacks was effective – or even, at the time, recognized by the enemy. The shift in attitudes to chemical warfare occurred in late 1914 and early 1915 as the war stalled. In December 1914, desperate for a new tactic, the chief of staff, Erich von Falkenhayn, consulted Emil Fischer, professor of chemistry at the University of Berlin, about the possibility of chemicals that might put soldiers 'permanently out of action'; Fischer seems to have been reluctant, but Fritz Haber was keen.[49] He suggested chlorine, a gas that was heavier than air and would drift and fill trenches, a gas that attacked the lungs, asphyxiating a victim.

In January 1915, Fritz Haber was authorized to start a substantial programme to develop chlorine as a chemical weapon. His scientific group included three future Nobel laureates: James Franck and Gustav Hertz (who together won the physics prize in 1925) and Otto Hahn, the future discoverer of nuclear fission.[50] Haber also collected and trained hundreds of specialized gas troops, who would learn how to handle the chlorine containers and how to deploy the gas on the battlefield. In March 1915 the development and training was over, and the troops were sent to the Ypres salient, a bulge in the lines of trenches of the Western front. The German commander at Ypres was initially reluctant to use this new, unfamiliar and unpredictable weapon, which he knew would raise charges of atrocity. However, an accidental release, caused by an enemy bombardment, which killed three German soldiers, convinced him of chlorine's potency. On 22 April 1915, the gas valves were loosened, and clouds of chlorine, like yellow smoke, drifted over Canadian, French and Algerian troops. Soldiers, completely unprotected, suffocated or ran.

Both sides were surprised when the first use of chlorine punched a 4-mile gap in Allied defences. If the German army had been prepared, then it could have poured through and made advances of hundreds of miles, perhaps turning the war. But General Erich von Falkenhayn had seen the Ypres salient as 'merely a diversionary move; it would help "to cloak the transportation of the troops to Galicia"', assisting the eastern front.[51] In fact, 1 mile was gained. In London, newspaper editors at first played down the attacks, but as soon as the physiologist John Scott Haldane submitted an official report, published on 29 April, confirming the lethality of the gas, the editorials and letters pages filled with accusations of atrocity; *The Times* called it 'an atrocious method of warfare', the use of which would 'fill all races with a new horror of the German name'.[52]

In fact the other nations responded in kind as soon as research, development and production permitted. The British deployed chorine at Loos in September of the same year. Chorine dispersed quickly and had other disadvantages. The chemists therefore quickly developed new compounds. Phosgene, isolated by the French chemist Victor Grignard, was highly toxic, colourless and smelled only faintly (of hay).[53] Defensive technologies, such as gas masks, partly countered chorine and phosgene. Germany responded in 1917 with mustard gas, a liquid, which lay on the ground for weeks, blistered the skin on contact and, if it got into the lungs, caused deadly bleeding. 124,500 tons of poison gas was used in battle in the First World War, roughly in equal quantities by Allied and Axis powers, mostly chorine and phosgene. If anything, chemical weapons favoured the Allies, at least on the western front, where the prevailing wind was at their backs.

Each side built a considerable alliance between chemical producers, scientists and the military in order to develop and produce these new weapons. The British, French and Americans each expanded their chemical warfare branches, mobilizing chemistry. Remarkably, in the United States, more than 10 per cent of the country's chemists eventually would aid the work of the army's Chemical Warfare Service.[54] A contemporary American journal found only one chemist who had refused to assist the gas project.[55] Haber's Kaiser Wilhem Institute, under military command from 1916, was at the centre of an empire of 1,500 staff, including 150 scientists.[56] The gas produced was used both for killing soldiers and for killing vermin, the lice and insects that infested the front. As the historian Edmund Russell argues, this connection between chemical warfare and insecticides continued through the twentieth century.[57] Haber enjoyed the military life. More specifically, he had no regrets about developing chemical weapons: it was his patriotic duty as a German, and he felt, in a sentiment that he shared with other commentators, such as J. B. S. Haldane in Britain, that the weapons, when their psychological effects were compared to the alternative of being ripped apart by an explosive shell, were, if anything, cleaner and more modern forms of killing. His wife, Clara, disagreed, and shot herself with Haber's military pistol in May 1915.

Aircraft and tanks were two other technologies that drew heavily on organized scientific expertise and promised breakthroughs in the stalled, trench warfare. Aircraft, which could fly over the lines, started the war barely different from the Wright brothers' designs and ended the war as a sophisticated and powerful technology. Despite exaggerated claims for impact on battle, both before and after the Great War, aircraft were actually used for niche functions: reconnaissance, artillery sighting, aerial photography, and combating the enemy's planes. Aeroplanes often crashed, were unreliable and, like gas, could be countered with defensive techniques, such as improved listening. Nevertheless, flying aces, such as the American Eddie Rickenbacker and Manfred von Richthofen, better known as the Red Baron, became celebrated, not least because their individuality and bravery stood out in contrast to the reality of industrialized warfare and the unknown soldier.

Nevertheless, the increased stability, reliability and speed of aircraft were the outcome of rounds of intensive improvement and invention. In Britain an Air Inventions Committee was set up, complementing the Royal Aircraft Establishment, Farnborough, one of the most important sites of British science in the twentieth century. Sciences such as metallurgy were shaped by this military inquiry – the investigation of cracks, for example, prompted broader speculations about the mechanical properties of solids in the post-war years; in particular, research at Farnborough during the First World War on the propagation of cracks as a mode of plastic deformation of metals 'in turn became an immediate predecessor to modern dislocation theory'.[58] X-rays were one application of the new physics of direct relevance: they were used to probe the structure and frailties of materials.

As the length of the war encouraged the institutionalization of continuous invention, so strategists disputed the best use of aircraft, tanks, trucks with internal combustion engines, and other novelties. In 1918, a junior British staff officer, J. F. C. Fuller, urged replacing the war of attrition with a new strategy, a direct attack on the head-quarters and supply lines of the German command with tanks, sup-ported by aircraft. This scheme, the basis for 'Plan 1919', relied on the metaphor of the army as body: rather than bleed the enemy to death slowly with a million cuts, Plan 1919 aimed at the 'brain'. Fuller later called his plan 'brain warfare', and Plan 1919 a 'shot through the brain'. Plan 1919 was neuropathology as military strategy. McNeill, locating a different emphasis, argues that the 'remarkable feature of the "Plan 1919" was that its feasibility depended on a weapon', the fast, manoeuvrable tank, 'that did not exist when the plan was drawn up'.[59] Here was the epitome of 'deliberate, planned invention'.

Mobilizing human science

Health was also managed. Preventative medicine on the front kept many infectious diseases at bay; indeed, 'inoculation and other sys-tematic precautions against infectious diseases, which in all earlier wars killed far more soldiers than enemy action', argues McNeill, 'made the long stalemate of the trenches possible.'[60] Nevertheless, the influenza epidemic of 1918–19 killed more people throughout the world than fell on the fronts of the Great War. The mobilization of medical and scientific expertise to support the management of war society created opportunities for ambitious disciplines. Psychology in the United States provides a clear example.

We saw in an earlier chapter how psychologists, such as Alfred Binet, Théodore Simon and Lewis M. Terman, had developed and promoted tests of intelligence to various clients as a means of secur-ing status and resources for a young discipline. Nevertheless, outside professional circles, before the First World War, 'mental tests usually met with skepticism, if not outright hostility'; but the entrance of the United States into the war gave a chance to the psychologists 'to prove themselves', and their techniques, to be respectable.[61] One such opportunist was Robert M. Yerkes, president of the American Psychological Association, who in 1917 appealed to his colleagues that 'Our knowledge and our methods are of importance to the military service of our country, and it is our duty to cooperate to the fullest extent and immediately toward the increased efficiency of our Army and Navy'; in particular, psychologists should not restrict themselves merely to the identification of mental incompetents, but rather offer a comprehensive service, a 'classification of men in order that they be properly placed in the military service'.[62]

Yerkes chaired the Psychology Committee of Hale's National Research Council and won the support of another member, surgeon general of the army William C. Gorgas, who had witnessed the potential of science deployed in a military context as part of Walter Reed's campaign against yellow fever in Cuba. In the summer of 1917, Yerkes, with Terman, trialled a test, 'examination A', and analysed the results at Columbia University; he was soon urging its deployment as the basis of separating officer-class recruits from others. Yerkes soon had rivals. Walter Dill Scott and Walter V. Bingham, at the Carnegie Institute of Technology, who had developed tests of potential salesmen, viewed Yerkes as a dilettante, and offered a businesslike scheme of even greater generality. However, Yerkes and Scott found themselves collaborating for the greater good: 'we shall do much more for our science, as well as for national defense', wrote Yerkes, by doing so.[63] They soon had an 'Alpha' test, for literate recruits, and a 'Beta' test, for illiterates. Yerkes persuaded the army to open a school of military psychology, at Fort Oglethorpe, Georgia, to train examiners; by May 1918 the examiners were handling 200,000 intelligence tests a month.[64]

There was no doubt that the army was under strain: the US Army had ballooned from 6,000 officers and 200,000 soldiers in March 1917 to 200,000 officers and 3.5 million soldiers in November 1918; the secretary of war had conceded that 'Some system of selection of talents which is not affected by immaterial principles or virtues, no matter how splendid, something more scientific than the haphazard choice of men, something more systematic than preference or first impression, is necessary to be devised.'[65] Kevles stresses the opposition of the military to the psychological tests; to 'the critics', he summarizes, 'the intelligence testers were interfering with serious business, saddling the army with questionable personnel practices, and, above all, undermining traditional military prerogatives'.[66] The test results conflicted with the judgement of experienced officers. Furthermore, the military suspected the scientists of opportunistically gathering data, useful to their science, but wasting the valuable time of an army marching to war. John Carson, in a more recent historical study, detects a more nuanced situation, a 'story of negotiation and transformation', in which both sides, faced with an unprecedented moment of mass mobilization, made concessions.[67]

Despite the language of objectivity and the systematic application of science, it is clear that the tests, as they were devised, carried out and analysed, were freighted with assumptions. Kevles contrasts Yerkes's faith that he was measuring 'native intelligence' with the implicit knowledge that lay behind such Alpha questions as 'The Knight engine is used in the – Packard – Stearns – Lozier – Pierce Arrow'.[68] Other Alpha questions included:

4. Why is beef better food than cabbage?
   Because
   □ it tastes better
   □ it is more nourishing
   □ it is harder to obtain
6. If someone does you a favor, what should you do?
   □ try to forget it
   □ steal from him . . .
   □ return the favour
10. Glass insulators are used to fasten telegraph wires because
    □ the glass keeps the pole from being burned
    □ the glass keeps the current from escaping
    □ the glass is cheap and attractive
16. Why should we have Congressmen?
    □ the people must be ruled
    □ it insures truly representative government
    □ the people are too many to meet and make their laws.[69]

Franz Samelson, a third historian of this episode, agrees that one of the 'root problems was the belief of the psychologists that they were scientifically measuring essentially "native ability rather than the results of school training", a belief for which they had no real grounds except their awareness that this was what they had set out to do, and, of course, their prior assumptions about the nature of intelligence.'[70] According to contemporary testimony, the examiners of the Beta tests for illiterates, 'for the sake of making results from the various camps comparable', had been 'ordered to follow a certain detailed and specific series of ballet antics, which had not only the merit of being perfectly incomprehensible and unrelated to mental testing, but also lent a highly confusing and distracting mystical atmosphere to the whole performance.'[71] Bored and mystified recruits 'dozed off en masse'. The measured intelligence of illiterates duly fell.

Yerkes's programme was abolished in 1919. But psychological testing remained as a feature of personnel management in the

army. Furthermore, Yerkes, and other psychologists, had assidu-
ously trumpeted their discipline's patriotic contribution to the war,
and celebrated its apparent successes. The Rockefeller Foundation
stumped up $25,000 to develop the tests for use in schools; the result-
ing National Intelligence Test was given to 7 million children in the
1920s.[72] Carl Brigham made the army Alpha test the basis of the
Scholarly Aptitude Tests (SATs), a hoop through which university
students in the United States, and increasingly elsewhere, would be
made to jump in the twentieth century. Business, too, was interested.
Psychology, declared Terman in his 1923 presidential address to
the American Psychological Association, had been turned from 'the
"science of trivialities" into "the science of human engineering". The
psychologist of the [earlier] era was, to the average layman, just a
harmless crank ... no psychologist of today can complain that his
science is not taken seriously enough.'[73] What had made the difference was psychology's well-publicized mobilization in the working
world context of the war.

Finally, intelligence testing in the First World War was encouraged
by, and contributed to, broader trends. First, intelligence testing, as
devised by the psychologists, was one more example of the methods
to judge similarity in a world, forged in the processes of industrialization and urbanization, marked by the export of similar things. An
early critic of tests, Justice John W. Goff of the New York Supreme
Court, had ruled that 'Standardizing the mind is as futile as standardizing electricity'.[74] But Goff was wrong on both counts. Just as
the spread of electrical networks rested on the agreements of negotiated electrical standards, so method of management in mass society
could be seen to depend on negotiated standards of equivalent minds.
Twentieth-century institutions – educational, military, workplace –
would all be shaped by such arguments.

Second, feelings of difference were just as important as judgements
of similarity; indeed, they were two sides of the same coin. In the
United States, the progressive emphasis on efficiency went hand in
hand with eugenic visions of a well-ordered society. Officers who
imbibed this culture were more ready to accept the psychologists'
claims. 'At the very least', concludes Carson, 'Progressivism's effects
on the military were to predispose officers, when confronted with
problems not amenable to traditional army solutions, to look toward
techniques that could be deemed "scientific" as the answer to their
needs.'[75] The results of intelligence testing fed directly into the eugenics debate. Well-publicized results included the average mental age
of the American soldier (thirteen, when twelve was considered the

108

'upper limit of feeble-mindedness') and the relative performance of black and white soldiers. Both would be challenged in the 1920s and shown to rest on mistakes and hidden assumptions, but not before being accepted as scientific 'fact'.[76]

Psychology, then, rose in esteem as it contributed to the management of putting an army into the field. Other medical sciences were mobilized to keep soldiers at the front. Psychiatrists, for example, were called upon to make expert judgements in suspected cases of malingering. They were crucial to shaping the category of 'shell shock' in the First World War. The term was coined by Charles S. Myers, a psychologist who had built up experimental psycho-physics at Cambridge and travelled on the 1890s Torres Strait expedition, and later in the war would devise tests to help select the operators of acoustic submarine detectors.[77] Symptoms were extremely diverse, including uncontrollable shaking and the reliving of experiences, and some were sympathetic, such as stomach cramps inflicting soldiers who had been knifed in the gut and snipers who lost their sight. Nor was shell shock restricted to those who had experienced shelling by artillery, or even frontline experience. Interestingly, suffering varied according to relative agency: a man in an observation balloon, fixed, was much more likely to be a victim of shell shock than a pilot, who was mobile and self-directed. An intense debate raged between those who saw shell shock as a physical condition, a degradation of the nerves, and those who saw it as the result of damaged minds. Such a debate had clear and obvious relevance for old and new disciplines (psychology, physiology, psychoanalysis), as well as for military authorities, who wanted the men back in the field.

In Britain, one particular effect of the huge numbers of soldiers diagnosed as suffering shell shock was to break down pre-war attitudes, in particular the firm line drawn between the insane and the sane. If 'healthy young males' could so suddenly exhibit symptoms of neurasthenia, then fixed categories were called into question; 'shell shock' was the label for this 'no-man's-land'.[78] The trend was from physiological to psychological understanding. Physicians deploying psychoanalytic techniques, such as W. H. R. Rivers (another Torres Strait veteran) and William Brown at the Craiglockhart War Hospital in Scotland, treated shell shock with modified talking cures, replacing rest or electric shock therapies. Rivers, in *Science* in 1919, noted how the extraordinary conditions of war had both 'shown the importance of purely mental factors in the production of neurosis' and also led to a wider recognition of 'the importance of mental experience which is not directly accessible to consciousness'.[79] The psychoanalytic argument ran that war enabled many unconscious desires and violent wishes to surface, which the repressing mind struggled to contain; indeed, 'all sorts of previously forbidden and hidden impulses, cruel and sadistic, murderous and so on', were being officially sanctioned in the soldier.[80] The result of this struggle was mental disintegration and shell shock. Collectively, the war was exposing civilization's repressed unconscious horrors. The First World War marked the entry of psychoanalysis into the working worlds of British medicine, where it was contested, while at the same time prompting an anguished national debate over morality, cowardice and war.

The ratchet effect of war

Wars have a general tendency to sharpen debates between old and new ways of doing things. It is important to see science enrolled in both sides of the debate. It is entirely arguable that none of the horrific novelties of the First World War – the gas, the tanks, improved aircraft, the submarines and their defences – were as important as the contributions made by chemists and biologists to maintain traditional supplies. In a war where men and materials were brought to railheads but then were moved forward by horse, the feeding and supply of horsepower was critical. Max Delbrück, the elder, organized the industrial production of yeast as a means of producing well over half of Germany's animal feed during the war. The process used substrates nutrified by ammonia produced using Haber's methods. Indeed, it is in similar schemes that the historian spies early articulations of 'biotechnology'; specifically in Austria-Hungary the economist Karl Ereky copied and expanded Danish methods for huge pig farms, and reflected on the whole in a book titled *Biotechnologie*.[81] Another contribution to the production of bulk materials was made by Chaim Weizmann, who, drawing on methods developed at the Pasteur Institute, found a biotechnological way of turning starch into acetone and butanol, chemicals essential to the production of smokeless powder.[82] 'There is no truth in the story', writes historian Robert Bud, however, 'that the British Balfour Declaration, offering Palestine as a national home for the Jews, was made out of gratitude to Weizmann.'[83]

But perhaps there is truth in a generalization, of which the Balfour Declaration of 1917 is just one case, that the fluid, destabilizing, crisis conditions of war allow the articulation and introduction of policies that would be almost inconceivable in peacetime. Furthermore there exists a recognized 'ratchet effect': measures introduced in wartime, often labelled temporary and justified by appeals to national security, have a habit of remaining in place after the conflict has ended.[84] Wars therefore also have the tendency to expand the state. When science and scientists were mobilized during the First World War, the result was not just a redirection of research programmes but also an increase in the interconnections between state and scientific institutions. (Indeed, institutions like the National Research Council in the United States were fashioned, by public scientists such Hale and Millikan, in a manner that preserved an apparent autonomy of science while nevertheless harmonizing science towards military agendas.) But did the ratchet effect apply to science policies? Were changes in science that were prompted by the conflict retained after the Armistice?

Generally, the civilian scientists mobilized during the First World War returned to civilian life afterwards. One interesting effect of this trend concerns women scientists. Less than 1 per cent of academic scientists in post in 1900 were women. However, in parallel with other sectors, a combination of the pressing demands for labour during a total war and the patchy social fluidity typical of wartime, plus the creation of new roles in a drawn-out, transformative conflict, formed the conditions encouraging scientific jobs for women. Thus it was that 'pressing manpower needs' created jobs for women in chemistry and engineering in countries such as Canada, Australia, England and Germany, while new applications, such as military X-ray medicine, were invented and staffed by women, including Marie Curie and Lise Meitner.[85]

At the war's end many scientists did indeed revert to pre-war occupations: Marie Curie to her radioactivity projects, William H. Bragg to University College London to continue X-ray crystallography, George Ellery Hale to California. Yet the return of civilian scientists to civilian posts should not be at all surprising, and should not be taken as evidence that the First World War had little lasting effect on science. Many civilian scientists retained linkages (such as Bragg consulting for the Admiralty) or, more importantly, fought to retain institutional gains. The fate of Hale's National Research Council is instructive. The council's success in coordinating research had increased government and industry's respect for academic science, physics in particular.[86] The Carnegie Foundation, previously reluctant, now raised the possibility of an endowment, placing the council on a secure, permanent financial base. Furthermore, scientists, such as Hale, who were deeply suspicious of government support for science, fearing government control of research agendas, enthusiastically embraced both the notion of philanthropic funding and the institutional format in which the National Research Council was kept private and independent yet retained its unique status and constitutional responsibilities. Hale manoeuvred to persuade Woodrow Wilson to make the peacetime council permanent in this form. His wish, with a modification or two, was granted. 'We now have precisely the connection with the government that we need', crowed Hale.[87]

In Britain, war encouraged the making and retaining of an institutional landscape with some broad similarities to that in the United States. In 1913, to disburse the one penny per person collected under the introduction of national health insurance in 1911, a Medical Research Committee (MRC) had been set up. The committee decided to fund long-term medical research rather than target 'short-term specific problems', such as tuberculosis.[88] Before and during the war the government was subject to a constant tirade from public scientists who argued that their expertise was not being tapped sufficiently deeply or efficiently. These scientists formed a Committee for the Neglect of Science in May 1916 to press their case and to offer their services to government. However, the sting had been drawn from this would-be British NRC by the formation of a body closer to the centre of the state apparatus. First an advisory body, the Privy Council on Scientific and Industrial Research, was set up, in July 1915. This body mutated into the Department of Scientific and Industrial Research (DSIR), established on 1 December 1916. It was a complex, unusual body, half advisory council, half tiny government office. However, the DSIR could channel government funds to civil science while retaining, through its advisory council, a measure of autonomy.

But these civil bodies, created during the war, were dwarfed by institutions that channelled military expertise. Many military institutional innovations were retained, particularly in the technological fields opened up and expanded over the years of conflict. In Britain the Admiralty and War Office retained the expanded experimental establishments, such as the Admiralty Experimental Station at Parkeston Quay, Harwich, and the Royal Aircraft Establishment at Farnborough. In the United States, the National Advisory Committee on Aeronautics (NACA), which had been established in 1915 after the Smithsonian Institution, led by its secretary Charles D. Walcott, persuaded Congress, was retained. (NACA, much later, would transform into NASA.) While the budgets of many of these bodies were slashed in the difficult economic years of the aftermath, the more important point is that they were kept.

In general, historian William McNeill is right to see the industrialized, drawn-out conflict of the First World War as promoting a shift towards organization and, if not permanent mobilization, then towards further coordination and interconnection. Even if, as Alex Roland notes, following Carol Gruber, disharmony in government–military or science–industry relations 'arrested the enterprise [of cooperation and increased patronage] well short of its potential', we must also take such cries of dismay as evidence of the encouragement of unusual cooperation.[89] In the United States, and to some extent in Britain and France, the military had gained access to new techniques and expanded facilities; scientists had gained kudos from contributing to the war effort and prestige without sacrificing control. We will examine the situation in Russia later.

Science and nationalism

However, Germany, imploding in 1918 and castigated as the aggressor, presented a profoundly different picture. The consequences for science in the new Weimar Republic will be explored in the next chapter. One of the conditions under which Weimar scientists worked was isolation caused by the breaking of international scientific links after 1918. In October 1914, in the first nationalistic spasms of the war, a manifesto carrying ninety-three signatures had appeared in German newspapers. It read, in part, as follows:

> To Civilization!
> As representatives of German sciences and arts, we hereby protest to the civilized world against the lies and slander with which our enemies are endeavouring to stain the honour of Germany in her hard struggle for existence – in a struggle that has been forced on her.
> . . .
> It is not true that Germany is guilty of having caused this war. Neither the people, the Government, nor the Kaiser wanted war . . .

And after denying the war atrocities in Belgium, including the destruction of the library at Louvain, these ninety-three 'heralds of truth' concluded:

> It is not true that our warfare pays no respects to international laws. It knows no undisciplined cruelty. But in the east, the earth is saturated with the blood of women and children unmercifully butchered by the wild Russian troops, and in the west, dumdum bullets mutilate the breasts of our soldiers . . .

113

It is not true that the combat against our so-called militarism is not a combat against our civilization, as our enemies hypocritically pretend it is. Were it not for German militarism, German civilization would long since have been extirpated ... The German army and German people are one ...

We cannot wrest the poisonous weapon – the lie – out of the hands of our enemies. All we can do is proclaim to all the world that our enemies are giving false witness against us ....

Have faith in us! Believe that we shall carry on this war to the end as a civilized nation, to whom the legacy of a Goethe, a Beethoven, and a Kant, is just as sacred as its own hearths and homes.

Alongside the names of prominent scholars and artists, signatories included some of the most eminent names in German science, among them Fritz Haber, Max Planck, Wilhelm Wundt, Adolf von Baeyer, Paul Ehrlich, Emil von Behring, Emil Fischer, Ernst Haeckel, Philipp Lenard, Walter Nernst, Wilhem Ostwald and Wilhelm Röntgen. A counter-manifesto – a 'Manifesto for Europe' organized by the physiologist Georg Friedrich Nicolai – which drew the support of precisely four signatories, albeit including Albert Einstein, went unpublicized.

The scientific institutions of the Allied countries retaliated. Fellows of the Royal Society of London demanded that all Germans and Austrians be struck from the list of foreign members, while the French Académie des Sciences expelled those who had signed the manifesto.[90] George Ellery Hale, in communication with the French, vehemently agreed with the ostracization of German scientists, but also spotted an opportunity to reconstruct the institutions of international science in a manner against the interests of Germany. In early 1918, Hale proposed that each of Germany's enemies should set up a body along the lines of his own National Research Council; these bodies would then federate as an Inter-Allied Research Council.[91] The result in 1919, after much politicking, was the creation of an International Research Council, a body composed of Allied and neutral countries, specifically designed to exclude the Axis powers. Japanese scientists were in an unusually delicate position since their institutional links were often to Germany, while in the war Japan had fought for the Allied cause.

True internationalists, such as the Dutch astronomer Jacobus C. Kapteyn, deplored this division of science, 'for the first time and for an indefinite period, into hostile political camps'.[92] Internationalism, structured by technology and held to be the hallmark of science, had also been the theme of Nicolai and Einstein's 'Manifesto for Europe':

114

Never before has any war so completely disrupted cultural co-operation. It has done so at the very time when progress in technology and communications clearly suggest that we recognize the need for international relations which will necessarily move in the direction of universal, worldwide civilization . . . Technology has shrunk the world . . . Travel is so widespread, international supply and demand are so interwoven, that Europe – one could almost say the whole world – is even now a single unit . . . The struggle raging today can scarcely yield a 'victor'; all nations that participate in it will, in all likelihood, pay an exceedingly high price. Hence it appears not only wise but imperative for men of education in all countries to exert their influence for the kind of peace treaty that will not carry the seeds of future wars, whatever the outcome of the present conflict may be.[93]

One 'man of education' who set out to exert influence was the English astronomer Arthur S. Eddington, who organized an expedition to Brazil in 1919 to observe the light-bending effect on starlight, visible at the moment of solar eclipse, predicted by Einstein's general theory of relativity. 'The war had just ended, and the complacency of the Victorian and Edwardian times had been shattered. The people felt that all their values and all their ideals had lost their bearings', recalled Ernest Rutherford. 'Now, suddenly, they learnt that an astronomical prediction by a German scientist had been confirmed by expeditions . . . by British astronomers . . . [It] struck a responsive chord.'[94] Yet the expedition was no foregone conclusion: it faced intense anti-German opposition and only succeeded because of the driving belief of Eddington, a Quaker, in the pacifism exemplified, he thought, by international science. Furthermore the 'collective memory of this test of Einstein's theory as a straightforward and harmonious cooperation between scientists from nations embroiled in political conflict was not solidified until many years later', notes historian of the expedition Matthew Stanley, and 'it was only through Eddington's deliberate presentation of the expedition as a milestone in international scientific relations that it came to have that valence.'[95] Nevertheless, the symbolism was perfect, and the results further heightened Einstein's towering reputation: 'A prophet who can give signs in the heavens', wrote J. B. S. Haldane, 'is always believed.'

In a controversial attempt to reassert the spirit of scientific internationalism, the Swedish Academy of Sciences pointedly awarded the first two post-war Nobel prizes to Germans. Furthermore, both – Max Planck (Physics, 1918, for energy quanta) and Fritz Haber (Chemistry, 1918, for the synthesis of ammonia) – had signed the 'Manifesto of Ninety-Three'. Haber, in particular, was reviled as the father of gas warfare. In contrast, German and Austrian scientists were banned from many international gatherings, such as the League of Nations' Commission internationale de coopération intellectuelle, which met in 1922, or the Solvay conferences, which restarted in the same year.[96] Likewise, the fifth international congress of mathematicians took place in Strasbourg, now of course a French city, and excluded German mathematicians.[97] However, in some fields, such as radioactivity research, professional and personal bonds were so strong that a truly international community survived the war; Rutherford, for example, organized the purchase of radium to support Meyer's Vienna institute.[98]

Science, wrote the American philosopher John Dewey in 1916, 'has not only rendered the enginery of war more deadly, but has also increased the powers of resistance and endurance when war comes'; yet, Dewey argued, the response to the catastrophe of war should be not to reject but to extend science:

> The indispensable preliminary condition of progress has been supplied by the conversion of scientific discoveries into inventions which turn physical energy, the energy of sun, coal and iron to account ... The problem which now confronts us ... is the same in kind, differing in subject matter. It is a problem of discovering the needs and capacities of collective human nature as we find it ... and of inventing the social machinery which will set available powers operating for the satisfaction of those needs.[99]

Dewey recommended that pupils be educated in the process of scientific inquiry, 'the only method of thinking that has proved fruitful in any subject', as he had emphasized in *How We Think* (1910). The method, once learned, applied to everything; teaching could start with 'varnishes or cleansers, or bleachers, or a gasoline engine'; 'without initiation into the scientific spirit one is not in possession of the best tools which humanity has so far devised for effectively directed reflection'.[100] Dewey's philosophy shaped interwar science education in the United States.

In general, however, the Great War was seen as a catastrophe for Western civilization, and some, citing gas and other technologies, blamed science for contributing to the horrific character of modern industrialized warfare. Herbert Kaufman, writing in the *Boston Sunday Herald*, to just pick one example, wrote:

> For half a century we have liberally endowed, supported and encouraged the scientists. Community funds paid for the institutions in which they were educated and underwrote their experiments.

116

And all the while, we believed that these endeavors were promotions in
the interest of civilization . . .

To-day we stand horror-stricken before the evidence of inhumanities
only made possible through scientific advancement.

Chemistry, you stand indicted and shamed before the Bar of History!
. . . You have prostituted your genius to fell and ogrish devices . . . You
have turned killer and run with the wolf-pack.

But we will reckon with you in the end.

This particular diatribe was shrugged off in the August 1916 letter
pages of *Science*: 'To blame chemistry for the horrors of war', one
commentator noted, was 'a little like blaming astronomy for noctur-
nal crime';[101] yet the wider point was conceded: 'science has increased
the amount of suffering war inflicts'. That science could be so implicated in the catastrophe of civilization of the First World War contributed to a general sense of crisis that marked the post-war years, no
more so than in Germany.

— 6 —

CRISIS: QUANTUM THEORIES AND OTHER WEIMAR SCIENCES

After four years of attritional warfare on two fronts, Germany imploded in the late summer of 1918. The army and navy mutinied and communist agitators fought the authorities for control of Berlin. In November, Kaiser Wilhelm II abdicated and went into exile in the Netherlands, and a new German republic was born. The Weimar Republic, which lasted from 1919 until the Nazi Party's *coup d'état* in 1933, was never stable or secure. An immense fracture ran through German politics, with a weak centre constantly pressed by the extreme right and the extreme left. Hyperinflation in the early 1920s, a cause and consequence of this instability, further fuelled a sense of crisis that pervaded Weimar culture. Political crisis and economic crisis fed a sense of moral crisis and intellectual crisis, which in turn encouraged feelings of a 'Krisis der Wissenschaft', a 'crisis in learning'.[1] In early 1918, most Germans had confidently expected victory; by December their society had collapsed. The perception of crisis was deep in proportion to the shock of the fall.

Responses to crisis

German science can fairly said to have been leading the world in the decades before the First World War. Even so, it would not have been surprising if the cataclysm of war, civil conflict, and the near destruction of the economy in rapid succession had destroyed the glories of German science. In fact, German science entered a period of remarkable creativity. The culture of crisis, when it did not disrupt utterly the organization of science, provided the intellectual products of the Weimar period with a distinctive context.

118

There are several possible responses to a perceived crisis if it cannot be ignored. First, existing institutions might be defended, protected and expanded if they provide an oasis of order. Technological systems and giant industrial combines provide examples of such institutions. Second, a search might be made for new, sounder foundations. In mathematics a century of anxieties over foundations culminated in profound interventions by Weimar mathematicians. Likewise, in experimental psychology, resuscitated research programmes premised on new basic concepts promised the means of healing a fractured discipline, and perhaps a fractured society. In general, Weimar culture was marked by what the historian Peter Gay calls a 'hunger for wholeness', a longing to reject division, reduction and analysis and to seek instead holism and synthesis.[2] Third, and finally, scientists might embrace crisis and develop a science that reflected the divisions of the wider culture. The historian Paul Forman has argued that the cultural crisis strongly imprinted itself on physical theory, constituting central features of the quantum theory as articulated by physicists during the Weimar period. Other historians have disagreed. His claim – the 'Forman thesis' – has been the subject of lively debate, and is so central to historians' interpretation of Weimar science, and indeed to the social history of scientific knowledge more generally, that, before reviewing the arguments on both sides of the Forman debate, we will start by surveying the extraordinary developments of quantum theory and then turn to examine mathematics, psychology and chemical engineering.

Quantum physics

Let us start by recalling what we have learned of the emergence of quantum physics.[3] In 1900, Max Planck had introduced mathematical terms, interpretable as quantized resonators, into his equations in order to make sense of the empirical measurements of spectra being made in Berlin at the Physikalisch-Technische Reichsanstalt. Albert Einstein in 1905 had not only interpreted these resonators in quantized terms but had applied the idea of a discrete packet or 'quantum' of light to account for anomalous physical phenomena, including the frequency-dependence of the photoelectric effect and the decrease of specific heats at low temperatures.[4] Einstein's intervention had divided physicists: no major theorist approved of the light quantum, yet several experimentalists did, such as Johannes Stark in Germany and W. H. Bragg in Britain. The idea that the physical

119

world possessed some fundamental, perhaps irreducible, lumpiness or discontinuity was up in the air in the 1900s. Even as they gathered at the Solvay conference of 1911, the jury of elite physicists was still out, at least in private.

Niels Bohr's use of the quantum theory of light to account for the stability of his quantum model of the atom in 1913 generated interest and further research. On the eve of leaving to fight, Harry Moseley had generated X-ray results that seemed to lend powerful support to Bohr's model. Einstein, too, during the war, in Berlin, drew on Bohr's theory to discuss the processes producing thermodynamic equilibria between matter and electromagnetic radiation.[5] Nevertheless, the war had been a traumatic distraction for some of Europe's physicists, while also an opportunity for patriotic application of their skills for others.

In 1916 Niels Bohr had taken up a permanent post at the University of Copenhagen and was soon busy on two fronts: organizing the establishment of his Institute for Theoretical Physics, which was successfully opened in 1921, and deepening his synthesis of quantum theory. His theoretical analysis of the quantized hydrogen atom, from which he had so stunningly derived the formulae for hydrogen emissions, was extended. First, the director of the theoretical physics institute at Munich, Arnold Sommerfeld, introduced a second quantum number to explain why the spectrum of hydrogen atoms exhibited the fine structural details, a number interpreted by Bohr as relating to the once-in-40,000-turns precession of the electron orbit.[6] Second, closely following this result, Karl Schwarzschild, writing from the eastern front, and Paul Epstein, a Russian by birth who had been suffered by the German state to remain in Munich, offered an account of why the hydrogen spectrum was subtly transformed by an electric field, a phenomenon accurately measured by Johannes Stark. Third, in 1917 Sommerfeld and his ex-student Peter Debye published an account of Zeeman splitting (the multiplication of spectral lines in magnetic fields) that used Bohr's theory.

But what was the relationship between classical and quantum pictures? Bohr, emboldened by these results and 'by pursuing further the characteristic connection between the quantum theory and classical electrodynamics', now set out his 'correspondence principle'. In his Nobel Lecture of 1922, he stated that, according to the principle,

> The occurrence of transitions between the stationary states accompanied by emission of radiation is traced back to the harmonic components into which the motion of the atom may be resolved and which,

120

according to classical theory, determine the properties of the radiation to which the motion of the particles arise.[7]

This seemingly overly specific 'principle' in effect spoke of the ability to move from descriptions made in the mode of classical electrodynamics to descriptions in the mode of quantum theory. Bohr, with his Dutch student Hans Kramers, worked feverishly to show how larger atoms could be understood in terms of the addition of more electrons into orbits whose interpretation could now be guided by the correspondence principle. Electrons filled successive shells: two in the inner shell, four each in two second-level shells, and so on. In effect they were working out a language that translated between the classical and the quantum atom, a language that still contained many of the idioms of the old talk of orbits and perturbations. Nevertheless, the table of elements derived from Bohr and Kramers's calculations mightily impressed the world of physics. The regularities of Mendeleev's table could now be partly explained, through quantum mechanics. Bohr won his Nobel Prize in 1922. 'This picture of atomic structure', he informed his Swedish audience, having shown them the table, 'contains many features that were brought forward by previous investigations', from the idea of J. J. Thomson, Irving Langmuir and G. N. Lewis to look for clues in the arrangement of electrons, to Harry Moseley's classification of elements by atomic number. Bohr and Kramer were offering a synthesis of the old and the new.

But at the heart of this success was a mystery. What, precisely, was a quantum? Were quanta, such as the quantum of light, the photon, real entities or useful mathematical fictions? How could a lump, a particle, a photon, create the phenomena of waves, such as interference? For all the empirical success of Bohr's Old Quantum Theory there existed troubling questions unanswered. Furthermore, the answers, when they came, were so strange that physics has yet to come fully to terms with them.

Quantum theory had its own geography. Anyone wanting to understand Bohr and Kramers's complex methods of orbits, perturbations and harmonies had to travel to Copenhagen. It seemed that only at the master's side could the 'correct' application of the correspondence rules be made. However, there was an alternative. In Munich, Arnold Sommerfeld worked with his extraordinarily talented students, including Werner Heisenberg and Wolfgang Pauli. Together they built up an account of *Atombau* – atom-building – that relied less on talk of orbits and more on abstract quantum rules. They

121

were wading further into the ocean of mathematical formalism and away from the reassuring shore of familiar physics. But Sommerfeld's accounts were 'clear and easily exportable', certainly when compared to Bohr's, and these virtues helped them move to other places of theoretical physics.[8]

The period 1922 to 1925 was marked by movement and gathering crisis. Pauli moved from Munich to Göttingen, swapping Sommerfeld for Max Born. Heisenberg, too, travelled to Göttingen, where he heard Bohr give a series of guest lectures in 1922. Born, assisted by the talents of Pauli and Heisenberg, took on Bohr and Kramers's orbital mathematics of atoms and, replacing intuitive fudges with derivations from first principles, made the theory as robust as possible. But, crucially, the theory now made predictions – regarding, for example, the helium atom – that 'flatly contradicted Bohr and Kramers's description and thereby threatened Bohr's explanation of chemical periods'.[9] The old quantum physics seemed to be running into the ground.

Pauli made a fundamental break. The picture of atom-building made sense if electrons in the same state (possessing the same 'quantum number') could not coexist, for example in the same atom. There was no mechanical reason to believe this 'exclusion principle' other than it seemed to operate. As Pauli wrote: 'We cannot give a closer foundation for this rule; yet it seems to present itself in a very natural way.'[10] Pauli urged that talk of orbits be dropped. If electrons moved, and obeyed laws of movement, then they were unvisualizable in the normal fashion.

Meanwhile, Bohr, in an attempt to discard once and for all the light quantum, wrote out, with Kramers and the American John Slater, a new account of the interaction of atoms with electromagnetic fields in which the strict conservation of energy was given up in return for excluding talk of photons. This B–K–S theory was published in the *Philosophical Magazine* in 1924. The fact that physicists were willing to sacrifice one of the cherished principles of physics in order to preserve a classical image – light as electromagnetic waves – suggested a science in crisis. Also, B–K–S soon ran into troubles of its own, with empirical evidence, such as measurements of the conservation of energy as light was scattered, and theoretical reservations undermining the edifice. Bohr himself began to favour the strategy of ditching the familiar, visualizable guides from classical physics in favour of an exploration of quantum rules, numbers and abstract analogies.

Those skilled at abstract mathematics were at an immediate advantage. And mathematics, no matter how abstract, can sometimes be a local affair. In this case the university at Göttingen, home to David Hilbert's school of mathematics and the mathematical physics of Max Born, became the birthplace of a new quantum theory. Between 1923 and 1925, Werner Heisenberg had accepted his doctorate in Munich, moved on to Göttingen, where he completed the final stage of German postgraduate training, a *Habilitation* under Born, and travelled to Copenhagen for a stint with Bohr.[11] Once back in Göttingen, the theoretical physicist more than fulfilled Bohr's call for a quantum theory divorced from mechanical representations, writing a paper in which he replaced familiar notions of motion and mechanics with something far more abstract. It was ready in July 1925. When Heisenberg revealed it to Born and his younger student Pascual Jordan, the latter two noticed that Heisenberg's mathematics could be represented in an elegant way by drawing on the pure mathematical theory of matrices. Born and Jordan swiftly set about writing up Heisenberg's insights in this new, even more abstract notation. Even as they were finishing the new theory, a young mathematician in Cambridge, Paul Dirac, who had read Heisenberg's first paper closely, also wrote out the new quantum rules in terms of mathematical algebras.

The new quantum theory would result from the merger of two quite distinct, seemingly contradictory streams of work. The first, as we have seen, was the abstract matrix theory, built by Heisenberg, Born, Jordan and Dirac, which had been a leap away from familiar, visualizable mechanics. The second, known as wave mechanics, which often bears Erwin Schrödinger's name, was a reflection on continuing experimental work on the statistical thermodynamics of gases and careful measurements of radiation. A good place to start to trace the emergence of wave mechanics is in Paris, where since the end of the Great War the aristocrat Maurice de Broglie had built up a lavish laboratory for X-ray spectroscopy. Maurice's younger brother Louis – or, to give him his full name, Prince Louis-Victor de Broglie – had followed him into physics. In 1922, the pair were arguing about X-rays, about how sometimes they appeared as particles and sometimes as waves. Louis, with the independence that aristocratic leisure can bring, pushed the argument as far as it could seemingly go. Having written out equations describing material 'corpuscles', he found a wave-like component and, rather than dismiss these as theoretical figments, held out that, if matter was accurately represented, mathematically, as waves, then matter could indeed be waves. But surely 'matter waves' were madness?

In Berlin in 1924, Einstein, reflecting on a paper he had encouraged

123

to be written by the Indian physicist Satyendra Nath Bose, spotted more terms that described matter as wave-like. (This paper marked the first description of what today's physicists call the Bose–Einstein statistics.) Einstein made the connection to de Broglie's startling proposal, and endorsed and publicized it. Other physicists took note. One was Erwin Schrödinger. Born of Austrian and English parents, Schrödinger had also, so far in his career, travelled the German-speaking academic lands – although not perhaps at the heights of Heisenberg: from Vienna, where he had worked on radioactivity and the physiology of colour vision, to Jena and Breslau, both briefly, and then on to Zurich, where he took up a chair in theoretical physics in 1921. In late 1925, Schrödinger got hold of Louis de Broglie's studies, and wrote to Einstein to say he found them 'extraordinarily exciting'. By early 1926, Schrödinger had written out his 'wave equation'.

Schrödinger's wave equation represented matter, for instance an electron, by a wave-like term and showed how it would change over time, for example when placed in an electric field that represented the positive charge of a nucleus. Indeed, Schrödinger's first paper of 1926 solved the wave equation for a hydrogen atom (the nucleus represented by a potential well – a steep-sided electric field) and derived accurate predictions of the hydrogen emission spectrum. It was a marvellously general but clean equation. The response from many of Germany's physicists was ecstatic. Planck wrote, on receiving a reprint, that he had read it 'like an eager child hearing the solution to a riddle that had plagued him for a long time', and later as 'epoch-making work'; Einstein, who received the papers from Planck, wrote to Schrödinger that 'the idea of your work springs from true genius!', and ten days later: 'I am convinced that you have made a decisive advance in the formulation of the quantum condition, just as I am convinced that the Heisenberg–Born method is misleading.' Paul Ehrenfest, in Leiden, wrote in May 1926, that he and his colleagues were awestruck: 'I am simply fascinated by the . . . theory and the wonderful new viewpoints that it brings. Every day for the past two weeks our little group has been standing for hours at a time in front of the blackboard in order to train itself in all the splendid ramifications.'[12]

Two immediate questions arose. First what was the relationship between Schrödinger's theory and the matrix mathematics of what Einstein labelled the 'Heisenberg–Born method'? Was wave mechanics a replacement for quantum mechanics? Many physicists perhaps hoped so: the matrix formulation, although powerful, was beyond the grasp of many of them, while Schrödinger's wave mechanics looked familiar. Matter waves might jolt common sense, but they were visualizable, in a manner, and wave equations were the common currency of optics and ordinary mechanics. Discontinuous quanta, however, described by arcane matrix methods, daunted them. While Schrödinger privately regarded the two formulations as interchangeable, to many physicists it seemed that only one representation must be right. Nevertheless, Dirac, Jordan, Heisenberg, Hermann Weyl and a young Viennese mathematical prodigy, John von Neumann, demonstrated that the two representations were indeed interchangeable: anything expressible in one mathematical language could be translated into the other. This solution satisfied many, especially since they could choose to work in the mathematical form that suited them. However, a second set of questions remained. What, exactly, were the waves of the wave equation? Waves of what? If the result of calculations were probabilities, should they be understood as facts about lots of events, or as facts about single events? And what were quanta exactly? If they were real, did it mean that nature was fundamentally discontinuous?

Explaining the quantum break

The change in worldview implicated by the quantum theories was so abrupt and so strange that some historians have argued that only peculiar and extreme historical circumstances could have created the conditions in which it was imaginable. In particular, the American historian Paul Forman set out this case in a landmark paper, 'Weimar culture, causality and quantum theory, 1918–1927: adaptation by German physicists to a hostile intellectual environment', which was published in 1971. Forman focuses on one peculiar and distinctive aspect of the new quantum theory, its rejection of strict causality, such as that found in the Newtonian world of masses subjected to forces. Forman's thesis is that

> In the years after the end of the First World War but before the development of an acausal quantum mechanics, under the influence of 'currents of thought', large numbers of German physicists, for reasons only incidentally related to developments in their own discipline, distanced themselves from, or explicitly repudiated, causality in physics.[13]

Forman's argument has three stages. First he characterizes what he calls the 'intellectual milieu': the surrounding society and culture within which physicists and mathematicians had to work. The features of this milieu had been traced by historians in works such as

Fritz Ringer's *The Decline of the German Mandarins*. In particular, it had been argued that

> the dominant intellectual tendency in the Weimar academic world was a neo-romantic, existentialist 'philosophy of life', revelling in crises and characterized by antagonism towards analytical rationality generally and toward the exact sciences and the technical applications particularly. Implicitly or explicitly, the scientist was the whipping boy of the incessant exhortations to spiritual renewal, while the concept – or the mere word – 'causality' symbolized all that was odious in the scientific enterprise.[14]

The cultural atmosphere of Weimar Germany is effectively sampled by reading Oswald Spengler's *Decline of the West*, a work published at the end of the First World War and devoured by millions of readers. Spengler envisaged the rise and fall of Western civilization as being akin to the turning of great seasons – and it was already winter. The modern sciences – narrow disciplines, which dismembered and analysed their subjects – were a symptom of wintery decline. The 'principle of causality', he wrote in his bestseller, was 'late, unusual' and, in effect, 'soul-destroying'. Spengler's hostility spread far, and physicists and mathematicians would have heard such sentiments being expressed in every café, beer hall or public debate. Forman's best evidence that the hostile milieu was felt by scientists comes from the records of contemporary scientists themselves. So, for example, the venerable chemist Wilhelm Ostwald reports in the mid-1920s that 'In Germany today we suffer again from a rampant mysticism, which, as at [the time of Napoleon's victory] turns against science and reason as its most dangerous enemies'; while Arnold Sommerfeld, one of the architects of quantum theory, reflecting on a Germany where 'more people get their living from astrology than are active in astronomy', drew the conclusion that the 'belief in a rational world order was shaken by the way the war ended and the peace dictated; consequently one seeks salvation in an irrational world order'.[15]

In the second stage of his argument, Forman traces the responses of scientists to the culture of crisis. He argues that there was 'a strong tendency among German physicists and mathematicians to reshape their own ideology toward congruence with the values and mood of that environment'.[16] Faced with hostility, then, German physicists and mathematicians presented an image of science in public debates that fitted the prevailing mood. A key feature of this mood, of course, was crisis talk, rhetoric the scientists now began to share, not least, notes Forman, because it provided 'an entrée, a ploy to achieve instant "relevance", to establish rapport between the scientist and

126

his auditors. By applying the word "crisis" to his own discipline the scientist has not only made contact with his audience, but ipso facto shown that his field – and himself – is "with it", sharing the spirit of the times.'[17] As evidence, Forman cites the public speeches and writings of physicists Wilhelm Wien, Richard von Mises and Sommerfeld, as well as the titles of texts, all composed between 1921 and 1922 – that is to say, on the eve of the new quantum theories, such as *On the Present Crisis in Mechanics* (von Mises), *The Present Crisis in German Physics* (Johannes Stark), *Concerning the Crisis of the Causality Concept* (Joseph Petzoldt) and *On the Present Crisis in Theoretical Physics* (Albert Einstein).

Finally, Forman asks whether this 'tendency toward accommodation' extended 'into the substantive doctrinal content of science itself' This third stage, the demonstration of social context helping to constitute the content of science, if true, is the most profound. There is no doubt that a breach with traditional conceptions of causality was part of the quantum break. In the words of Werner Heisenberg: 'the resolution of the paradoxes of atomic physics can be accomplished only by further renunciation of old and cherished ideas. Most important of these is the idea that natural phenomena obey exact laws – the principle of causality.'[18]

Forman sketches a route by which acausality was embraced. First, he finds figures, such as the statistical physicist Franz Exner and the mathematician Hermann Weyl, announcing that a rejection of – or at least a 'relaxation' of the demand for – causality was necessary, with no reference to quantum physics. Weyl's new conviction was that 'physics in its present state is simply no longer capable of supporting the belief in a closed causality of material nature resting upon rigorously exact laws'.[19] The next station on Forman's route are heavyweight physicists – Robert von Mises, Walter Schottky and Walther Nernst renouncing, to a greater or lesser degree, causality in their work. Their words echo Spengler's (for example, there is a repeated depiction of causality as '*starr*', 'rigid'). Erwin Schrödinger followed in the autumn of 1921. In the discussion of the new ideas brought by Slater to Germany and used in the Bohr–Kramers–Slater paper of 1924, Slater 'became persuaded that the simplicity of mechanism obtained by rejecting a corpuscular theory more than made up for the loss involved in discarding conservation of energy and rational causation, and the paper . . . was written'.[20] In total, Forman writes, there is evidence for a 'wave of conversions to acausality', a 'capitulation' that was felt by the participants as 'a quasi-religious experience, of a rebirth, of contrition for past sins – in a word . . . a conversion'.[21]

Some physicists held out, including two of the most powerful figures in German physics, the elderly Max Planck and the younger Albert Einstein, who were drawn closer together in opposition to the new wave.[22] Nevertheless, by the period of the synthesis of quantum theory of 1925 and 1926, including the presentation and recon-ciliation of Schrödinger's and Heisenberg's alternative descriptions of quantum theory, German physicists could ditch causality, confi-dent that they were moving with sentiments of wider culture. They had been 'impelled to alter their ideology and even their content of their science to recover a favourable public image'.[23] To summarize, Forman's threefold argument is that a post-war Weimar culture of crisis encouraged an embrace of new philosophies which rejected traditional scientific values; faced with this hostility, the scientists first adopted the new values in public talk and then introduced them into their scientific thought. The result was a context that enabled a break with traditional scientific values of law-like, causal, mechanical accounts of the physical world. Weimar culture acted as midwife to the birth of quantum mechanics.

Forman's sociological thesis has provoked reactions ever since publication. Scholars P. Kraft and P. Kroes wondered whether the scientists were just displaying 'mock-conformity' by playing along to fashionable, if hostile, trends.[24] John Hendry regards the Forman thesis as a 'case not proven'.[25] The weaknesses Hendry identifies make quite a list, although not all are telling. For example, the pos-sibility that scientists had all the while held anti-utilitarian values, and the shift in cultural wind marked a release not an accommoda-tion, does not affect Forman's overall argument, for the milieu still encouraged a rejection of traditional scientific values in public and in work. Likewise, Hendry is right when he points out a problem of cultural specificity: Forman's thesis relies on an appeal to a Weimar culture, but how can this affect non-German physicists?[26] The 'first serious attempt to reject causality', note Kraft and Kroes, 'did not stem from scientists who lived in the German cultural sphere, but from outside . . . the BKS paper was written by a Dane, a Dutchman and an American.'[27] Nevertheless, an answer, I suspect, can be built in two ways: either by conceding that the German lands were the sun around which orbited all theoretical physics in this period, and so the culture of the German lands was especially relevant, even in the post-war years of apparent international isolation, even for visitors from Denmark, the Netherlands and the United States; or, by constructing a specific argument for other cultures, a parallel, Forman-style argu-ment could be built up.

An apparently better objection of Hendry's is that there were 'strong internal reasons for the rejection of causality' – that is to say, the intellectual issues raised by the physics alone were sufficient to prompt the rejection of such core principles as energy conservation and causality.[28] In turn, the sociologists' answer would be that, in situations where physical theory is changing rapidly, or where there is no guide to answer definitively whether a set of empirical evidence is sound or not, there is no rule – nor are there 'internal reasons' – against which core principles can be judged. At such moments social factors can predominate. But Weimar culture was part, not all, of the mix.[29]

Of course the reason Forman's thesis is controversial is because it suggests that social or cultural factors shaped one of the glittering jewels of twentieth-century physical science. It is not only a worked case study in the social history of scientific knowledge, but it also tackles one of the prima facie hardest cases. If the core of modern physical theory can be shown to be socially shaped, then surely all science can be? This is not the place to debate the sociology of scientific knowledge in general. However, it is significant that Forman's thesis was not even original with Forman: as he was happy to concede, speculations about the relationship between milieu and knowledge were raised by his 1920s protagonists.[30] Weimar Germany was the setting for the articulation of powerful theories of the sociology of knowledge, of various political stripes, ranging from the Heidelberg sociologist Karl Mannheim's *Ideology and Utopia* to Spengler's *Decline of the West*. In both of these texts, knowledge – even mathematical knowledge in the case of Spengler – was held to be relative to society or culture. Sociology of knowledge flourished in interwar Germany and in the West of the 1960s and 1970s; both cultures were marked by 'crisis talk'.

Uncertainty and complementarity

But the best attack on Forman by Hendry is the argument that Forman picked a minor aspect of the great quantum revolution. Putting to one side, for the moment, Heisenberg's assessment that acausality was at the heart of the quantum theories, there are other candidates. Whether or not physical theory should draw on visualizable processes was one item for debate, especially after Bohr's solar system model collapsed around 1925.[31] On one hand Heisenberg argues that mathematics, not visualizable images, should be the guide, while on the other Schrödinger felt 'discouraged, not to say repelled, by the methods of the transcendental algebra, which appeared very difficult to me and by lack of visualisabilty'; Heisenberg thought such sentiments 'trash'.[32] 'To most of those concerned with the fundamental problems raised by the quantum and relativity theories', Hendry writes, the requirement to accept causality was 'clearly secondary to the criterion of a consistent description of phenomena in ordinary space and time, and to that of an objective (observer-free) theory, and this was true in particular for those physicists most closely involved with the development of a new quantum mechanics, for Pauli, Heisenberg and Bohr.'[33]

We have already seen how Wolfgang Pauli had proposed his exclusion principle in 1925 as a means of making sense of the arrangement of electrons in Bohr's atoms. Heisenberg, too, was working closely with Bohr when in 1926 he began to present qualitative and mathematical arguments for why a pair of observable entities, position and momentum, could not both be measured to arbitrary precision. If position was known precisely then momentum would be uncertain, and vice versa. The degree to which this effect appeared correlated to the size of Planck's constant, h. If h was large, then these paired quantum uncertainties were large; if h was small, then the uncertainties were small; while, in an imaginary universe where h was zero, there would be no quantum uncertainties at all. Heisenberg's argument was called 'the uncertainty principle'. The notion that physical theories where h is negligible should look like classical physical theory was behind Bohr's 'correspondence principle'.

Position and momentum were not the only observables that were paired in this way. So, for example, were energy and time. Furthermore, rather than resolve the issue of whether light, say, was fundamentally a particle or a wave, the two modes could be seen as complementary perspectives on the same entity. Bohr made the existence of such pairs the primary clue to a new philosophical framework for making sense – or at least some sense – of quantum theory: 'complementarity', presented first in Como in 1927. Together, Heisenberg's and Bohr's philosophies make up the so-called Copenhagen Interpretation. (Heisenberg coined the label in 1955.) While later commentators disagree about precisely what should be included in the Copenhagen Interpretation, core elements are the correspondence principle and complementarity.[34] Another element, the strong role of the observer, was promoted by Heisenberg, in the 1950s, but not by Bohr.

In 1935, Albert Einstein, now at Princeton (having not returned to Germany after the rise to power of the Nazi Party), along with two Institute of Advanced Study colleagues, the Russian-born Boris

Podolsky and Nathan Rosen, published a thought experiment ('EPR') that cast doubt on whether quantum mechanics could be considered a complete theory.[35] They started with some assumptions about how a physical theory should relate to physical reality. In particular, the authors wrote that, in any 'complete' theory, 'every element of the physical reality must have a counterpart in the physical theory'. With this definition they considered the measurement of two of the quantities – such as position and momentum – that are subject to quantum uncertainty. Their conclusion was that the 'description of reality' given by quantum mechanics was not complete, while leaving 'open the question of whether or not such a [complete] description existed'. It was in correspondence with Einstein exploring the difficulties of Bohr's Copenhagen Interpretation in 1935 that Schrödinger described the surely paradoxical situation of a cat that was neither alive nor dead (or both) until observed. Bohr's response to the EPR debate was to sharpen his language: when he talked of uncertainty it was not because the particles had inherent definite quantities that were somehow obscured (that is to say, they remained classical bodies at heart) but that such quantities were fundamentally indeterminate (quantum bodies are not classical bodies).

Historians have sought contextual arguments for why the Copenhagen Interpretation of quantum mechanics won out over alternatives, such as Einstein's erstwhile assistant David Bohm's hidden variables interpretation, in which what seems to be indeterminacy is in fact determinacy at a deeper, holistic level. James Cushing argues that the Copenhagen Interpretation had the immense but contingent advantage of being articulated first.[36] David Bohm's interpretation did not see light of day until 1952, by which time the Copenhagen Interpretation was entrenched. Mara Beller argues that Bohr's considerable authority carried the Copenhagen Interpretation to acceptance when in fact it was riddled with inconsistencies and obscure thought.[37] In this reading only the prestigious, celebrated Bohr could have got away with it. If indeed the Copenhagen Interpretation was not a unified position at all until the 1950s, then these arguments need retooling.

Crises and foundations: mathematics, philosophy and psychology

Physics was not the only subject to experience a period of crisis in which old foundations fell apart and new ones were built. In both mathematics and, to a lesser extent, psychology questions about foundations, which had begun to accumulate in the late nineteenth century, prompted profound re-examinations of the subjects that were most forcefully debated in Europe, perhaps especially in Weimar Germany.

In mathematics, the ancient Greek Euclid's *Elements* had, for centuries, offered a model of deductive reasoning. By starting from five postulates – statements about the properties of lines and points whose truth seemed self-evident – more complex geometrical propositions were proved. Because the certainty of the postulates was (apparently) unshakeable, then the same confidence could be held in the truth of later theorems. In *Elements*, the culmination of chains of foolproof deductive reasoning was Archimedes' theorem: that the square of the hypotenuse of a right-angled triangle is equal to the sum of the squares of the other two sides. Such was the magic of deduction that this mathematical statement must be a universal truth – despite the fact that it could not be grasped at a glance – so long as the postulates are true.

Most of the postulates look simple and are very hard to doubt. But the fifth is different: it is longer, less immediately obvious. Here it is, the so-called parallel postulate: 'That, if a straight line falling on two straight lines makes the interior angles on the same side less than two right angles, the two straight lines, if produced indefinitely, meet on that side on which are the angles less than the two right angles.' In early modern Europe, many attempts were made to prove this postulate from the first four, with no success. Then, in the first years of the nineteenth century, two obscure mathematicians, János Bolyai and Nikolai Ivanovich Lobachevsky, and one mathematical titan, Carl Friedrich Gauss, asked what mathematics would follow if a different fifth postulate was chosen. The mathematics, it turned out, was indeed different but, crucially, not self-contradictory.

The repercussions of this move were ultimately shattering. Geometry, the science of Euclid, had been the model of certainty. Not only was truth guaranteed in a manner that the inductive sciences could never approach, but it was also the branch of mathematics that most clearly told truths about the physical world. Engineering and architecture depended on true deductions about triangles and rectangles, just as optics, and therefore trust in telescopes and gunsights, depended on true knowledge of lines and points. Yet, by the mid-nineteenth century, mathematicians had mapped out many geometries. Foundations were beginning to shake.

In the last decades of the nineteenth century, especially in Germany, the axioms of geometry came under sustained investigation. Moritz Pasch demonstrated that extra axioms were necessary in 1882, while in Turin Giuseppe Peano set out a new axiomatization in 1888.[38] In August 1900, at the International Congress of Philosophy, Peano, flagged by his students, gave a triumphant demonstration that mathematics might be described in the compressed language of logic. In the audience was Bertrand Russell. The young English philosopher identified in Peano's programme a solution to unstable foundations: the various branches of mathematics might be rooted in the certainties of logical relations and notation ('logicism'). He also soon found a major paradox, relating to how to make sense of sets that contain themselves, that lay in wait of followers of this programme ('Russell's paradox' of 1901, which had been identified two years earlier by Ernst Zermelo).[39] Nevertheless, Russell himself, with Alfred North Whitehead, set out to write the immense *Principia Mathematica* (1910–13) to demonstrate that logic could ground mathematics. Published in the years before the First World War, its complexity and compromises made sure that Russell fell short of his Newtonian-scale ambitions. Henri Poincaré dismissed the whole approach as 'barren'.

Another sceptic was the Dutch mathematician Luitzen Egbertus Jan Brouwer. L. E. J. Brouwer had turned from remarkably creative ideas about set-theoretic topology to dig further and further at the foundations of mathematics. He promoted an alternative to logicism. Called by Brouwer 'intuitionism', and by later commentators 'constructivism', his programme was an ascetic rejection of all but a handful of starting points. He rejected the axiom of choice, which spoke of the ability to choose elements of infinite sets. He also rejected, in 1907, the logical Principle of the Excluded Middle (the principle that mathematical statements are either true or false).

In Göttingen in 1899 David Hilbert had presented the first version of his *Grundlagen der Geometrie*. This foundation of geometry rested on twenty-one axioms, but it was Hilbert's approach to axioms, not their number, that was important. Hilbert was a 'formalist' whose attitude to axioms was to treat them like rules to a game: you can choose different rules and have a different game, but what matters is that the game works rather than to worry whether there is one best game (or one true geometry, or one sound foundation, as in logicism). In a sense, he wrote, it did not matter what mathematics was about, just what form of game was derived; points, lines and planes could just as well be called 'tables, knives and beer-mugs' – the game would remain the same.

Hilbert may have been most interested in the form of mathematics but he was also very confident that mathematics, so understood, could meet any challenge offered. Indeed he set out a series of unsolved problems, first at the Paris congress of mathematicians (they followed the philosophers) in 1900, adding to them in later years, which would be just such a challenge. His confidence shone through in a famous pronouncement:

> However unapproachable ... problems may seem to us and however helpless we stand before them, we have, nevertheless, the firm conviction that their solution must follow by a finite number of purely logical processes. Is this axiom of the solvability of every problem a peculiarity characteristic of mathematical thought alone, or is it possibly a general law inherent in the nature of the mind, that all questions which it asks must be answerable? ...The conviction of the solvability of every mathematical problem is a powerful incentive to the worker. We hear within us the perpetual call: There is the problem. Seek its solution. You can find it by pure reason, for in mathematics there is no ignorabimus [we will not know].[40]

Such optimism began to waver in the Weimar years. In 1921, Hermann Weyl, in his essay 'On the foundational crisis in mathematics', drew a comparison between the instabilities of current mathematics and politics. In both there was a 'threatening dissolution of the regime'; 'Brouwer', Weyl wrote, 'is the revolution!' 'No Brouwer is not ... the revolution', Hilbert wryly commented a year later, 'but only the repetition of a Putsch attempt with old means.'[41] In 1928 an immense squabble erupted between Hilbert and Brouwer. The Göttingen mathematician had supported Brouwer's career, helping him secure a position at the University of Amsterdam and appointing him to the editorial board of the *Mathematische Annalen* in 1914. Now, as if to show that mathematics could also undergo a Weimar crisis, the two fell out, with others, such as Einstein, brought into the argument.[42]

But this political and institutional crisis was dwarfed by an intellectual surprise three years later. One problem that Hilbert had set was about the completeness of mathematics: given a suitably broad set of axioms, was every reasonable mathematical statement capable of being proved either true or false? In 1931 the Viennese mathematician Kurt Gödel published a paper, 'On formally undecidable propositions of *Principia Mathematica* and related systems'. Gödel's result was as devastating for Russell's logicist programme as it was astounding for formalists. It showed that any mathematical system sophisticated enough to contain arithmetic – a reasonable enough bare minimum for an axiomatic system – would still leave out math-

134

ematical statements that were perfectly sensible but impossible to prove. (Strictly speaking, Gödel showed that either this was the case or that mathematics was inconsistent; the latter is a greater horror.) Mathematics, in the form Hilbert hoped it took, was therefore incomplete. Interesting mathematics could never be reduced to deductions from axioms.

The English logician Alan Turing and the American Alonzo Church soon closed a major loophole in this otherwise devastating blow to hopes of finding one best foundation. In 1935–6 the pair independently showed that there was no way of deciding whether rogue propositions were true or false. To build his proof Turing introduced a remarkable imaginary device – a 'universal machine' capable of following the mechanical methods of calculation – which would later be materialized in the stored-program computer.[43]

Nevertheless, one response to dodgy foundations is carefully to rebuild from scratch. In mathematics, philosophy of science and, arguably, psychology this response can be identified in the projects of three groups: the 'Bourbaki' mathematicians, the philosophers of the *Wiener Kreis* (the 'Vienna Circle') and the Gestalt psychologists. The Bourbaki mathematicians were young French scholars, including Henri Cartan, Claude Chevalley, Jean Delsarte, Jean Dieudonné, René de Possel and André Weil, who met at the École Normale Supérieur and the Café Capoulade in Paris.[44] From 1935 they began an immense project to produce a published account of mathematics in the form of a logical construction that moved from the most general to the more specific. Thus, for example, set theory came before the real numbers, because the former was more general than the latter. They chose to remain anonymous, crediting their publications to a fictional 'Nicolas Bourbaki'. The first text appeared in 1939, and the project would continue for decades.

The Vienna Circle was a cluster of intellectuals, including the sociologist Otto Neurath and philosopher Rudolf Carnap, led in the early years by the aristocrat physicist Moritz Schlick, who met in the Austrian capital in the 1920s. Others, such as Kurt Gödel, were on its fringes. A congregation of outsiders in their early years, when it went by the name the Ernst Mach Society, the Vienna Circle, as proponents of logical positivism, would become the most influential school of philosophy of science of the mid-twentieth century. Its members embraced Ludwig Wittgenstein's project of the *Tractatus Logico-Philosophicus* of 1921, which they read twice, collectively, sentence by sentence, in the mid-1920s, nodding in approval at the claims – for example, 'That the meaning of a sentence is through

135

its method of verification'.[45] The Vienna Circle's own project was a reconstruction of the 'scientific world-conception' in the face of attacks by a mystical, nationalist, conservative and Nazi right, some of whom claimed to ground their own philosophies in modern science – a stand disputed by others. The project was well expressed in a manifesto of 1929:

> First [the scientific world-conception] is empiricist and positivist: there is knowledge only from experience, which rests on what is immediately given. This sets the limits for the content of legitimate science. Second, the scientific world-conception is marked by an application of a certain method, namely logical analysis.[46]

A new stable structure (an *Aufbau*) would be built on the firm foundation of verified scientific statements about experience. Things would not fall apart, because the centre would hold. Historian Peter Galison notes how constructional and architectural imagery pervades Carnap's own statement of the Vienna Circle project: a 'long, planned construction [*Aufbau*] of knowledge upon knowledge', a 'careful stone-by-stone erection of a sturdy edifice upon which future generations can build'.[47] A core project was the writing of an *International Encyclopedia of Unified Science*. Furthermore, this rationalist, secular project, this Unified Science, which built upwards from the firm foundations of experience, countered the rival entities of the right – the *Volk*, the state, God.

The Vienna Circle shared its outlook with the architects of the Bauhaus. 'Both enterprises sought to instantiate a modernism', writes Galison, by 'building up from simple elements to all higher forms that would, by virtue of the systematic constructional program itself, guarantee the exclusion of the decorative, mystical, or metaphysical ... [By] basing it on simple, accessible units, they hoped to banish incorporation of nationalist or historical features.'[48] Both would be dispersed under Nazi persecution but flourish anew in the land of exported similarities, the United States.

The Vienna Circle intersected with the paths of psychologists as well as architects. Karl Bühler, psychologist of the 'aha!' moments of sudden insight, encouraged his students and assistants to attend the Circle's meetings in the Red Vienna years, while American psychologists embraced thier philosophy of science, at least in a domesticated, behaviourist form, in the expatriate period.[49] 'What American psychology took from European philosophy', writes historian Roger Smith, 'was its authority rather than its substance', as well as an invaluable rule for demarcating themselves: 'for both groups ...

anybody who did not present knowledge in the form of empirically verifiable statements was simply not a scientist, and this made it possible to draw boundaries around the science of psychology.'[50]

Precisely these questions, concerning the boundaries and foundations of psychology, were at the centre of a highly charged debate in German academia in the first third of the twentieth century, out of which was forged a new approach to studies of the mind, Gestalt psychology. Unlike physics and physiology, with their grand imperial institutes, experimental psychology had received threadbare support in the late nineteenth century. Furthermore, psychology was classed as part of philosophy, so every psychologist who wanted to use laboratory methods faced entrenched scepticism from philosophers. A fierce fight between philosophers and natural scientists was sparked whenever professorial chairs fell empty. While Wilhelm Wundt at Leipzig from the 1870s and Carl Stumpf in Berlin from the 1890s built local empires in experimental psychology, philosophers continued to object to the intrusion of science. Edmund Husserl, for example, in an essay of 1911, denounced the 'experimental fanatics' for their 'cult of facts'.[51]

Therefore, when the new generation of experimental psychologists looked around in the 1910s and wondered how to build a career, they saw a discipline that had to demonstrate that it was 'filled with philosophical interests'; in contrast to the United States, where the behaviourists, as we saw in chapter 4, could promise practical benefits (even social control), German psychologists had to promise engagement with moral and metaphysical discourse.[52] One area that promised philosophical relevance was the fundamental nature of consciousness: could it be broken down into elements and be atomistically analysed? Or was it rather a stream, a process, something that worked as a whole that could not be broken down? The process view was urged by psychologists and philosophers such as William James, Henri Bergson, Wilhelm Dilthey and Husserl. But while the whole – the 'Gestalt' – was undoubtedly of philosophical interest and attracted some experimental attention, the Gestalt problem seemed beyond the capacity of experimental psychology to handle.[53]

Historian Mitchell Ash has shown how the challenge was met by three psychologists of the new generation, all three students of Carl Stumpf: Max Wertheimer, Kurt Koffka and Wolfgang Köhler. The oldest of the trio, Wertheimer, had already made a name for himself in the 1900s as an inventor of a method, based squarely in associationist psychology, of identifying criminals by their reactions to words. But by 1910 he was thinking and writing about melody.

137

In particular – and here is an excellent example of the importance of the export of similar things – working at the Phonogramm-Archiv in Berlin, he listened to the music of the Vedda tribe of Ceylon – recorded as wholes, transported as wholes, listened to as wholes – and came to the conclusion that the 'melody is not given by specific, individually determined intervals and rhythms, but is a Gestalt, the individual parts of which are freely variable within characteristic limits'.[54] Telephony and phonography also formed the backdrop to Köhler's research in psycho-acoustics.

Notice again how so much of the new science of the twentieth century was prompted, at least initially, by commentary on the working world of the late nineteenth: quantum mechanics and the light bulb, special relativity on electrical clocks, Gestalt psychology on phonograph recordings. In some accounts, too, Gestalt psychology as a commentary on railway systems could be added. Wertheimer, travelling on holiday between Vienna and Frankfurt, was struck by the appearance of movement generated by alternating railway signals. At Frankfurt, armed with a toy stroboscope that created the same effect, Wertheimer contacted Köhler, who was working in the city, and, joined by Koffka, a classic Gestalt experiment began. Ash insists that the 'conceptual developments came first', but the broader point is that the Gestalt psychology, even while designed to be able to address higher philosophical debate, was rooted in a working world context.

The fundamental claim of Gestalt psychology was that structured wholes were the primary units of mental life: wholes were grasped as wholes, not assembled from bitty sensations. In the case of apparent movement of alternately flashing lights, one grasped movement as a whole. Wertheimer built experimental apparatus, including a tachistoscope that teased out how observers distinguished (or, crucially, could not distinguish) between real and apparent motion. At times observers saw motion without any moving object (an illusion labelled the 'phi' phenomenon). Wertheimer wrote up the experiments in 1912 and 1913. Here was a promise of a new foundation: if perceptual wholes, accessible to laboratory investigation, were prior to sensation, then both physiology and philosophy might be revised under a new light.

Before the First World War, experimental Gestalt science was the preserve of a handful of young peripheral scholars. Nevertheless, they developed an expansive and ambitious vision. Koffka argued that action, in addition to perception, featured Gestalts. During the war, Köhler was productively marooned on Tenerife, where he had taken up a post as second director of the anthropoid research station of the Prussian Academy of Sciences.[55] Chimpanzees, he concluded after close and sustained observation, grasped actions – such as stacking boxes to reach food – as complete solutions, as wholes. At the war's end, both chimps and Köhler moved to Germany. The apes succumbed to the cold and wet of Berlin zoo. Köhler, shrugging off feelings of being 'chimpanzoid' himself after so much close contact, turned to write on Gestalts in the physical world: physical systems, too, must be analysed in terms of Gestalts rather than as isolated events.[56]

Gestalt psychology, primed by the necessity of addressing philosophy, offered concepts and arguments of great appeal in crisis-torn Weimar Germany. In turn, the Weimar context, in which a rattled middle and upper middle class saw around them fragmentation and in turn felt a 'hunger for wholeness', encouraged the flourishing of Gestalt science. When Max Wertheimer gave a lecture 'On Gestalt theory' at the Kant Society in Berlin in 1924, he defended science not as the dry dead practice of analysis but as a vivid enterprise capable of grasping wholes in flow. 'Who would dare, scientifically, to attempt to grasp the rushing stream?', he asked his audience. 'And yet physics does it all the time.'[57] And, now, so could psychology. Wertheimer not only described what Gestalts were in psychology, pedagogy and music, but offered his approach as a response to the 'problem of our times'. 'It is more natural', Ash interprets Wertheimer as concluding, 'for people to work together for a common goal than to be in opposition to one another' as individuals.[58] Wertheimer was a progressive, a friend of Einstein. Nevertheless, the Gestalt 'whole' could be, and was, mobilized in support of political projects of the right and the left, as well as of the unstable middle way.

Organizers and organization

The 'hunger for wholeness' shaped other sciences too. The great pathologist Ludwig Aschoff, at the University of Freiburg, for example, urged a holistic view of disease that rejected the atomization implicated in tissue and cellular pathology.[59] Conversely, the life sciences were interpreted as having political implications. Theodor von Uexküll, for example, promoted a combined theory of biological organisms, as they interacted with an environment, with an analogous theory of the state. His *Biology of the State: Anatomy, Physiology, Pathology of States* of 1920 took the family as the natural cellular unit, while different occupations were functional and akin to organs.

He concluded that 'the only form of organisation demonstrated by every [healthy] state is necessarily the monarchy'.[60] In contrast, Weimar Germany's new parliamentary system was likened to 'the tapeworm that has killed a noble war horse, [and] then tries to play the horse itself'.[61] Parasites could be external (English parliamentary democracy) or internal – and here it is clear that Uexküll's readers, picking up on standard euphemisms, would have read a reference to the Jews of Germany.[62]

German genetics was decisively shaped by the social and cultural context, right down to how the cell was interpreted. Like Gestalt psychologists, geneticists could only find secure institutional positions in the traditional universities if they addressed broad topics and cultivated the higher values. These generalists (or 'comprehensives', as historian Jonathan Harwood labels them), such as Richard Goldschmidt and Richard Kühn, were quite different to the narrow American fly geneticists of the Morgan school.[63] However, another way of getting ahead in German genetics was to be a 'pragmatic' and to pursue a utilitarian science, in which genetics as a specialized field would aid agriculture, industry and social engineering. Erwin Bauer, for example, took this path from the Berlin Agricultural College to the Kaiser Wilhelm Institute for Breeding Research at Münchberg. The comprehensives considered themselves above politics, but shared with other mandarins the sense of crisis in the face of modernization. The pragmatists were quite politically active, on left and right. As Harwood shows, the scientific interpretations supported the political values held by the scientists, and vice versa. The understanding of the relationship between cytoplasm and nucleus, for example, correlated to the view held about the proper relationship of academia to the political order. One side saw power-sharing while another saw nuclear control.

Another powerful message from biology came from embryological research. Across the campus from Aschoff in Freiburg worked Hans Spemann, whose school developed microsurgery techniques capable of manipulation and rearrangement of amphibian embryos, using fine glass instruments.[64] One student, Hilde Pröscholdt, transplanted the 'dorsal lip', part of the early, gastrula stage of the newt embryo, on to another host embryo. These embryos responded as if being governed or organized by the dorsal lip material, developing for example a second body axis, including a second central nervous system. Spemann claimed, to the astonishment of the scientific – and wider – worlds, to have found a natural 'organizer'. This 'organizer' was, notes historian Nick Hopwood, 'a metaphor for the restoration of social order'.

Metaphors have no necessary fixed application to the working world. Indeed, the most substantial instance of organization in Weimar Germany concerned chemistry and industry rather than biology and the state. We have already discussed the importance of the German dyestuffs and pharmaceutical firms of the nineteenth century. In the twentieth century these companies continued to grow, to specialize to some extent, but also to compete, nationally and internationally. On Christmas Day 1925, three already powerful chemical firms – BASF, Bayer and Hoechst – along with three smaller enterprises – AGFA, Chemische Fabrik Griesheim-Elektron and Chemische Fabrik vorm Weiler Ter Meer – merged to form Interessen-Gemeinschaft Farbenindustrie AG. IG Farben encompassed 100,000 employees. British and French companies had to copy the German titan in order to compete. Imperial Chemical Industries (ICI) was founded in 1926 out of British Dyestuffs, Brunner Mond, Nobel Industries and the United Alkali Company, while Rhône-Poulenc was formed two years later. Neither matched IG Farben as a leader in chemical research and industrial production. We will meet several examples of IG Farben's innovative products – the first antibiotic sulfonamide, the plastic polymer polyurethane, synthetic petrol – in forthcoming chapters.

141

— 7 —

SCIENCE AND IMPERIAL ORDER

The fundamental problem in the working world of imperial administration is construction and maintenance of order. An emperor truly has no clothes if someone can say that there is no order in his empire. Imperial order has to be known and maintained, and depends, as historian James Scott has demonstrated, on building simplified representations of complex and messy worlds.[1] Well before the twentieth century, powerful techniques for classifying and counting imperial subjects and objects had been developed. French and Dutch administrators conducted censuses. Administrators of the East India Company and, from 1858, the British Raj repeatedly wrestled with the problem of finding indicators of Indian populations, culminating in the supposedly complete and uniform census of 1881. Furthermore, techniques for counting subjects were transferred back and forth between those developed for estimating human populations and those used for building simplified representations of the natural world. The census of population can be seen as a natural history of human populations, answering questions such as what types?, where?, and how many? Censuses also reduce an immensely complex and geographically spread out phenomenon – the shifting population – to something that can be held in one place, perhaps a shelf in a colonial office. This reduction, which it was in many senses, is crucial to the power of censuses. No longer did the imperial eye have to be everywhere a population might be – an impossible task, of course. Now, the imperial power need only stand in one place with the books and tables open to survey and to command. In practice, of course, an administrator could sit in the emperor's seat and inherit the same powers of knowledge and direction.

Natural histories are means by which stocks can be taken. The working world of an empire will put value on some items while classing others as worthless or even destructive of imperial order. Order, as described, for example, through natural history has to be revealed if it is to serve its imperial purpose. So, for instance, natural history museums displayed the useful, decorative, threatening or destructive organisms of empire, arrayed systematically. Likewise, botanical gardens set out the imperial plant world for the eye to see; the botanical bed was directly analogous to the census table.

Order, having been identified, has to be maintained. Maintenance is where the double geography of systematized imperial knowledge becomes especially powerful. The first geography is that of the whole empire. The second geography is that of the representations of empire held within museum, botanical garden, census office or laboratory. The messy and immense complexity of the first geography was reduced, simplified and reordered to make the second geography. The reduction, simplification and reordering was expensive and time-consuming; few have the resources to replicate the effort. The second geography is graspable, more manipulable, measurable and accessible only by imperial permission.

Order has to be maintained in the second geography – museum specimens preserved and catalogued, botanical beds weeded and labelled, census books protected from termites, and laboratories equipped and staffed. At the very least a well-maintained second geography acted as a reference for classifying entities in the first geography. Laboratories, however, could take an even more active role. Threats to imperial order, such as tropical diseases, could not merely be isolated and classified but manipulated and defeated.

Tropical medicine

To illustrate this process, preventative medicine in military and colonial settings is a good topic to start with. The complex life cycles of protozoan parasites as causes of human disease were described in colonial and military medical laboratories, rather than in those of academic biology.[2] Laboratory methods, for example, were essential to the identification of the aetiology of malaria. Alphonse Laveran was a professor of military diseases and epidemics, posted in 1878 to Bône in Algeria, where he began to research malaria. In Algeria, and subsequently in Italy, Laveran claimed to find parasites in the blood of malarial patients. This was a remarkable and controversial finding, and was criticized and debated before being eventually accepted in

143

the 1880s. The full aetiology – the causal account of how a disease is acquired – of malaria was completed by another military medic. Ronald Ross, who was born in India into an army family and worked in the Madras branch of the Indian Medical Service from the 1880s, had been shown Laveran's parasites by physician Patrick Manson. Ross shared with Manson the notion that mosquitoes played a role in the spread of malaria. In various field stations in India, some in malarial areas, others clear of the disease, Ross, with Indian assistants, was able to dissect mosquitoes and identify and isolate unusual cells. By 1898 he was convinced that the parasite was transmitted via the bites of mosquitoes. He communicated this fact to Manson. The outcome was knowledge of the aetiology of malaria; but, to get there, the complex malarial world had had to be stripped down to manipulable, isolatable components.

After 1900, the parasite-vector model was found to apply to other diseases too.[3] In 1902–3, David Bruce, another Scot, identified trypanosomes as parasites and the tsetse fly as the vector in sleeping sickness; the disease was a major factor shaping the geography of agriculture in southern Africa, since cattle farming was excluded where sleeping sickness raged. Likewise, Walter Reed, an American army doctor, confirmed the theory of the Cuban Carlos Finlay that mosquitoes were the vector of yellow fever. The immediate implication was that yellow fever epidemics could be controlled by programmes to eliminate the mosquito. In the construction of the Panama Canal, between 1905 and 1914, General William Gorgas drew on military engineering and knowledge of the parasite-vector models to assault malaria and yellow fever in the canal zone. These two diseases had devastated workers on the previous French project in Panama.

William Boog Leishman, yet another Scot and military medic, identified parasites in samples of spleen drawn from victims of the disease 'kala azar'. Again processes of simplification, isolation and identification were crucial. Leishman had not only arranged for just one extract of this disease to be studied – the spleen samples – but had also used a method of staining of his own devising. In the process, starting in the early 1900s, entities became renamed: kala azar became 'Leishmaniasis', the new causative agent 'Leishmania', and the stain 'Leishman's stain'. The vector was found to be sand flies. Some life cycles of parasites were very complex. The Brazilian doctor Manuel Augusto Pirajá da Silva, for example, in the 1900s, traced the life cycle of the *Schistosoma* trematode as it passed through freshwater snails to humans to water again. In humans the trematode causes the disease bilharzia.

144

These successes supported the growth of new specialties. Within medicine, tropical medicine condensed around the laboratory study of malaria, sleeping sickness, yellow fever and other parasite-vector diseases. The Liverpool School of Tropical Medicine and the London School of Tropical Medicine were both established in 1899. A Hamburg Institute for Naval and Tropical diseases appeared in 1901. Within biology, entomology, parasitology, protozoology and helminthology (the study of worms, including flatworms such as the trematodes) attracted direct imperial interest and institutional support. Entomology was transformed from being an amateur to a professional science, not least by the impact of lavish imperial investment.[4]

Ecology and empire

Another specialty that grew to be a major new discipline in the twentieth century, partly forged in a working world of empire, was ecology. As a study of how nature husbands resources, ecology has always had great appeal as a scientific language with which to naturalize claims for systems of management of human as well as non-human resources. Long histories of ecology have demonstrated this connection.[5] Nineteenth-century studies had investigated the patterns of communities of organisms, coining several terms: von Humboldt's plant geographies of South America, August Heinrich Rudolf Grisebach's 'phytogeographical formations', Karl August Moebius's 'biocoenosis' and Stephen Alfred Forbes's 'microcosm', the 'little world within itself' of a lake or pond.[6] Forbes, in 1887, framed ecology as economy and vice versa: 'Just as certainly as the thrifty business man who lives within his income will finally dispossess the shiftless competitor who can never repay his debts ... the well-adjusted aquatic animal will in time crowd out his poorly-adjusted competitors for food and the various goods of life.'[7]

In the last years of the nineteenth century Henry Chandler Cowles was finishing his doctorate, writing his thesis on the 'ecological relations of the vegetation on the sand dunes of Lake Michigan'. The idea of focusing on sand dunes – on which the plant-life was temporary, changing, variable – to gain insights into the dynamics of plant communities had come from the great Danish botanist Eugenius Warming, but Cowles emphasized a particular process, that of 'succession'. Walking inland from the shore, Cowles noted the transition from sand grains, to low flowering plants, to substantial woody plants, to cottonwoods and pines, to woodland of oaks and hickories, and finally to beeches and maples. He argued that a natural succession of plant forms in time could be traced by moving through physical space.[8] Tracing succession in different landscapes provided a suitable and productive graduate activity, and Cowles's research school flourished, leading to the foundation of the Ecological Society of America.

The amiable, collegiate Cowles had a rival in the arrogant, priggish and generally disliked Minnesota ecologist Frederic E. Clements.[9] Both saw plant communities as marching in time through stages of succession, and both argued that any location had an identifiable 'climax'. Nature exhibited order. 'Nature's course', environmental historian Donald Worster writes, summarizing Clements's views, was 'not an aimless wandering to and fro but a steady flow toward stability that can be exactly plotted by the scientist'.[10] Mixed in with this progressivism was an organicist view of ecological entities: like an organism, plant communities developed, possessed structure and reproduced. 'As an organism', wrote Clements in his influential book *Plant Succession* (1916), 'the formation arises, grows, matures and dies.'[11] Succession was botany written as embryology. Also, just like an organism, a plant community would have physiological processes; indeed, Clements appealed to the contemporary physiology of stimulus and response to explain how plant communities responded to environments and to each other.[12]

In the 1910s ecology was a science restricted largely to the study of plant communities. Yet by the 1930s it had been extended to be a science of systems that included animals, even humans. Furthermore, a core term, the 'ecosystem', which was defined in opposition to the organicism of Clements, had been coined, and was gathering supporters. This historical shift, argues historian Peder Anker, was framed by the administrative and political working world of the British Empire.[13] The central protagonists of this shift were the British botanist Arthur George Tansley, the Oxford ecologist Charles Elton, and the ecologists that clustered around the South African leader Jan Smuts.

In Britain in the 1910s ecology revolved around two poles, in London and Scotland. At Scottish universities, the science and politics was largely conservative. (Although an exception to this generalization must be made in the case of the unclassifiable Patrick Geddes.) At University College London the followers of Edwin Ray Lankester fostered a research school sensitive to the interplay and parallels of ecology and economy in studies of plant geographies, within a pro-

146

gressive, slightly countercultural climate, that supported female as well as male students. Tansley, a Cambridge ally of the UCL crowd, founded the progressive *New Phytologist* in 1912, which quickly became a mouthpiece for Mendelism. Tansley also served on the Central Committee for the Survey of British Vegetation, the creation of which in 1911 has been described as the 'formative event for modern plant ecology in Britain'.[14] Perhaps most importantly, this committee combined the performance of a professional function for the British state (a classification of vegetation types) of a kind that could be extended to be a professional function for the British Empire. Furthermore, such as service created a channel of resources that was used to found institutions. The British Ecological Society was established in 1913.

The political minefields of British ecology scuppered Tansley's desired appointment as the Sherardian Professor of Botany in Oxford. Devastated, Tansley turned inwards. He read Freud, visiting the psychoanalyst in Vienna in 1922. He resigned his Cambridge lectureship and practised as an analyst himself. It was a brief but profound interlude for the botanist. Crucially, Tansley imbibed a systemic theory of human psychology, picturing the human mind as a mixture of instincts, stimuli, responses and Freudian forces in equilibrium. He imported this picture back into ecology. In 1924, returning to botany as chair of the British Empire Vegetation Committee at the Imperial Botanical Conference of that year, Tansley oversaw the planning of a complete botanical survey of the empire. Britain, argued the plant ecologists, was like a manager of a general store who was unaware of the 'stock at his disposal'; furthermore, professional ecologists could guide the hand of colonial administrators so that changes in the environment were made in the interest of the British crown.[15] Tansley, with T. F. Chipp, wrote in detail about how the survey must be done in *Aims and Methods in the Study of Vegetation*, published for the British Empire Vegetation Committee and the Crown Agents for the Colonies in 1926.

But at the conference the British ecologists were confronted, much to their surprise, by a rival proposal from South Africa. There were good reasons why the South African project was at odds with the British one: rather than being involved in a stock-taking exercise framed by an existing empire in which administration and efficient exploitation were the working world aims, South African ecologists were inventing accounts of the natural world around them at the same time as 'South Africa' was being forged as a new whole. The South African project was stamped with the ideas and personality of its patron, the political leader and philosopher Jans Christian Smuts. Smuts's extraordinary life began on Empire Day – Queen Victoria's birthday – 1870. By 1910, he had not only passed through Cambridge, where he wrote a brilliant thesis on the poet Walt Whitman, but had fought as a general in the Boer wars, and, with Louis Botha, united South Africa as a new political entity. Smuts articulated a philosophy, holism, which supported his projects of unification. His holism saw the physical and living world as an evolving whole, with some (white political philosophers) at a higher stage than others (blacks, women). In the 1910s, Smuts's philosophy had been circulated privately. But in 1923 he announced holism to the world in a sermon from the mount – in this case a speech on Table Mountain, outside Cape Town. Its sentiments, that all matter was alive, evolving, with parts serving a harmonious whole, were further publicized in his book *Holism and Evolution* (1926):

> In the cells there is implicit an ideal of harmonious co-operation, of unselfish mutual service, of loyalty and duty of each to all, such as in our later more highly evolved human associations we can only aspire to and strive for. When there was achieved the marvellous and mysterious stable constellation of electrical units in the atom, a miracle was wrought which saved the world of matter from utter chaos and change.[16]

Smuts the celebrated politician made for an attractive patron, and ecology returned the favour by offering means of describing natural worlds, including human societies, as wholes. John William Bews had carried Scottish ecology to South Africa when he had joined Natal University College in 1909. Bews brought Clementsian succession stories to Smuts. A second botanist, John Phillips, moved from South Africa to Edinburgh, returning to study forest ecologies. Bews and Phillips's *Botanical Survey of South Africa*, a 'grand list of all the species of the young nation', including a study of ecological inter-relations, was a picture of a unified South Africa, much influenced by Smuts.[17]

In 1934 and 1935, Phillips published a three-part paper, on 'Succession', 'Development and the climax', and 'The complex organism', an analysis of ecological concepts, in the *Journal of Ecology*. He appealed extensively to Smuts's philosophy. Tansley, now installed in his dream job as Sherardian professor, was appalled. He was not entirely comfortable in Oxford. His critics, enemies even, were idealists, among them Alexander Smith, the philosopher of science R. G. Collingwood, C. S. Lewis, J. R. R. Tolkein, John Frederick

Wolfenden and Thomas Dewar Weldon. These idealists had invited and welcomed Smuts when he had visited Oxford in 1929, preaching holism and justifying the Rhodes policies of segregation, the separation of black and white settlement, by appeal to ecology.[18] Phillips had done some of the ecological legwork for these segregationist arguments. In an earlier paper, he had contrasted a 'holistic road', of man knowing his place in a biotic community, with a 'mechanist road', in which mechanistic science sought control of nature for material wealth, weapons and environmental destruction . . .

> uncontrolled and intensified firing of vegetation of all kinds; reckless use of the axe in savannah, scrub, bush and forest; primitive and more advanced but wasteful agricultural practice; indiscriminate, heavy stocking of grazing and browsing lands; . . . destruction of plants and decimation or extinction of animals; introduction of alien plants and animals, such sometimes creating biological and habitat problems of the greatest economic importance; wrongly designed public works – such as roads, railways, drains, irrigation schemes – responsible for unnecessarily accelerated run-off of water and consequent soil erosion. In a word, [an environment] disturbed by man, to his ultimate detriment.[19]

In this account black South Africans courted destruction if they were allowed to adopt a European lifestyle, while, if segregated biotic communities were preserved by wise holistic, ecologically informed government, all races would prosper. (One also can't help reading Phillips and remembering the context of Smuts's invitation to Oxford to have called to mind Tolkein's description of the destruction of forest by Saruman's army in *Lord of the Rings*.) Tansley, a realist opposed to the idealists, a materialist man of the left opposed to holist conservatism, was therefore primed for a fight that had local significance as well as an imperial context. 'Phillips' articles remind one irresistibly of the exposition of a creed – of a closed system of religious or philosophical dogma', wrote Tansley in the journal *Ecology* in 1935. 'Clements appears as the major prophet and Phillips as the chief apostle'; in total it was a sermon on Smuts.[20] So, to oppose a holistic, organismic concept of the 'biotic community', the ecological foundational unit of his enemies, Tansley offered an alternative: the 'ecosystem':

> I cannot accept the concept of the biotic community. This refusal is however far from meaning that I do not realise that various 'biomes', the whole webs of life adjusted to particular complexes of environmental factors, are real 'wholes', often highly integrated wholes, which are the living nuclei of systems in the sense of the physicist. Only I do not

think they are properly described as 'organisms' . . . I prefer to regard them, together with the whole of the effective physical factors involved simply as systems.[21]

Ecosystems included the living organisms (the biome) as well as 'inorganic factors'; human activities could radically change where the equilibrium of an ecosystem lay. Tansley's essay was on the limits of words, and he thought that to describe his ecosystem as an 'organism' was to stretch the word 'organism' too far. Furthermore, just as a chemist would admit that the properties of water could in principle be deduced from the properties of hydrogen and oxygen, and just as an inventor would not claim that a new machine had properties beyond the component materials, so an ecosystem was just the sum of its scientifically analysable parts, and no more. 'Ecosystem' was invented as a carefully circumscribed term: useful for stock-taking, inviting for future research, but shorn of 'illusory' commitments to nature as super-organism.

Humans were drawn into ecological models in the 1920s and 1930s. 'Regarded as an exceptionally powerful biotic factor which increasingly upsets the equilibrium of preexisting ecosystems and eventually destroys them, at the same time forming new ones of very different nature', wrote Tansley, 'human activity finds its proper place in ecology.'[22] A pivotal figure in the articulation of animal and human ecology was the young member of a revitalized zoology at Oxford, Charles Elton. Colleagues included the biologists Julian Huxley and J. B. S. Haldane and the economic historian Alexander Morris Carr-Saunders. Huxley and Carr-Saunders shared a keen interest in eugenics, with Carr-Saunders, for example, arguing in *The Population Problem: A Study in Human Evolution* (1922) that over-population was the result of over-breeding of a human race with low mental capacities.

Elton's career was founded in the unlikely setting of the polar islands of Svalbard, including Spitzbergen and Bear Island.[23] After the First World War the national status of Spitzbergen had been resolved in an extraordinary way. The 1920 treaty, signed by several countries including the United Kingdom, Norway and, slightly later, the Soviet Union and Germany, granted Norway full sovereignty on condition that all signatory nations and peoples had a right to explore, exploit and settle. This settlement opened a window to a minor imperial competition in which undertaking science was encouraged to demonstrate an active claim to future polar riches. Organized by George Binney, and led by Huxley and Carr-Saunders, the first Oxford Spitzbergen expedition left in the summer of 1921. Elton and Victor Samuel Summerhayes were two students on the expedition.

The communities of animals and plants in their physical settings were simpler in the far north than in the temperate or tropical south. Indeed, almost the complete assemblage of organisms could be identified and described. Nitrogen taken in by bacteria passed to plants, which in turn fed worms and lemmings, geese and flies. Small insects and spiders were devoured by ptarmigan, snow buntings and purple sandpipers. A parallel track took the nitrates into algae and then to small aquatic creatures, which in turn fed kittiwakes, guillemots, auks, divers, ducks and petrels. Finally, a handful of carnivores – arctic foxes and polar bears on land – consumed nitrogen in the refined form of other animals. All the while, dung and decay returned nitrogen to the environment. Elton and Summerhayes's diagram of this complete process, drawn while they were students and published in 1923, presented to the world the nitrogen cycle, a fundamental ecological process.

Elton returned on the second Oxford Spitzbergen expedition in 1923, this time as chief scientist, as well as on the third in 1924.[24] Out of this study came his account of lemming behaviour that has lived on in a curious manner as popular knowledge. The notion that lemmings have a suicidal urge to throw themselves off cliffs can be traced to Elton's publication in which he wrote that the small creatures marched 'with great speed and determination into the sea'.[25] The jump of the lemmings was later animated by Disney and in that cinematic form witnessed by millions. However, it is a myth. Lemmings do experience peaks in populations, and can suddenly seem to be ubiquitous when the snow covering their tunnels melts. Lemmings can appear to be running everywhere in these peaks, but they do not make for cliffs.

In fact, Elton's lemming population studies and the rest of Oxford's comprehensive ecological stock-taking make sense when related to the interests of two patrons: the administrative departments of the British Empire and the private companies exploiting the far north for profit. First, for the British colonial agencies the Oxford scientists offered techniques that could generate comprehensive charts of the natural resources of an area. These included not only the cycles drawn by Elton but also the use of airplanes as tools of ecological research which literally provided a 'grand overview' of land to be administered.[26] The second patron, private businesses confronted with the uncertainties and dangers of working in the north, were interested in any new methods of climatic prediction. Elton's plots of lemming populations interested British Petroleum, for example. Likewise, the Hudson's Bay Company, which had a direct interest in good analyses of the ecological stock of arctic regions, because it exported all sorts of natural materials, such as furs, supported Elton financially from 1926 to 1931.

This combination of supportive interests and arctic research enabled Elton to synthesize a new kind of ecology. Animal ecology was always a harder proposition than plant ecology: animals move around and can be difficult to find. Plant ecology already had its exemplary programme of research: the investigation of succession. There was no equivalent to succession in animal ecology. But Elton had produced for the Hudson's Bay Company diagrams that represented populations in a pyramid of numbers: thousands of fish might support tens of seals which might support a single polar bear. Food chains, cycles and pyramids were a mouth-watering prospect for an ambitious ecologist. Now, informed by economic historian Carr-Saunders's account of economic cycles and conflict in industrial England, Elton projected the Spitzbergen vision of pyramids onto the entire animal world. His *Animal Ecology* of 1927, published as part of the 'Science for All' movement, made the study of food chains central to the new specialty. Elton also introduced the concept of the ecological niche, drawing a direct analogy from specialized functions in human societies. 'When an ecologist says, "There goes a badger"', Elton wrote, 'he should include in his thoughts some definite idea of the animal's place in the community to which it belongs, just as if he had said, "There goes the vicar".'[27] Not only did human society provide inspiration for animal ecologies, but ecologies began to include humans as part of the mix. Anker describes one of Elton's diagrams, showing sealers, polar bears and seals, as 'probably the first published ecological illustration that includes human beings'.[28]

Elton subsequently pushed not only for an animal ecology but for a human ecology that included the rest of the natural world as well as humans in the cycles it described. In the 1920s, a small group of sociologists were using the term 'human ecology' to describe a science of interactions between peoples; they would henceforth be corrected. Nor was Elton alone. Soviet ecologists in the 1920s shared similar views. In the 1930s and 1940s, ecologists reflected on human activities, drawing methodological and sometimes moral conclusions. Tansley, working in a British landscape continuously reshaped by human work, argued that human activity be considered an ordinary ecological process among many. Paul Sears, surveying in *Deserts on the March* (1935) the destruction of the prairies that resulted in the Dust Bowl of Oklahoma and Texas, argued that human interference, destroying nature's checks and balances, was the cause. Frederic Clements agreed, maintaining that a new 'ecological synthesis' was needed to guide land use. Sears, who backed the idea of appointing permanent expert land advisers, referred to the British Empire as the model in its deployment of expertise.[29] Finally, Aldo Leopold, in a beautifully composed essay in his *Sand County Almanac* of 1949, offered ecological ways of thought for public consumption.

With the focus on food chains and animal populations in place, other scientists – and mathematicians – developed powerful techniques of analysis. Just as Elton's Spitzbergen had offered a simplified exemplar of population structure and dynamics, so mathematical models could, in simplified form, reproduce features of observed changes in population. The American biologist, and eugenicist, Raymond Pearl rediscovered the logistic curve in 1920. In a logistic curve, the population can rise approximately exponentially before levelling off at a ceiling. It was a new mathematical take on Malthus. A compatriot of Pearl's, the physicist Alfred James Lotka, wrote out a differential equation that linked the rates of growth of two populations, of predators and a prey. When solved, the resulting graph – showing peaks and dips – looked like the kind of relationship between, say, populations of lemmings and snowy owls. Lotka had an unusually varied background, including spells as a chemist and in insurance, which perhaps encouraged his striking out on a new path.[30] Nevertheless, the Italian Vito Volterra, concerned with fishing in the Adriatic, developed Lotka's mathematical models along similar lines.[31] The working world context for this science was therefore the administration of populations as stocks.

Colonial management, disease and vitamins

Similarly, research programmes in tropical medicine were often driven by the demands of management of specific colonial problems.[32] The very high mortality of mine workers recruited from Central Africa and employed in the gold mines of the Transvaal was partly due to pneumonia. In a brutal cost–benefit calculation, the cost of developing an anti-pneumonia vaccine was calculated to be less than increasing minimum living space – £40,000 for a bacteriological laboratory compared with £250,000 for new accommodation for the miners.[33] The working conditions of the mines also caused many other lung problems – such as tuberculosis and silicosis – that generated research programmes that paralleled those in the global North.[34] Veterinary science followed this pattern too. The administration of land required travel by horse. But horses in southern Africa suffered from horsesickness. The need to maintain the working world of the horse-borne official therefore prompted research into immunization against horsesickness, for example at the Onderstepoort Veterinary Institute near Pretoria from the 1900s. However, the immunization succeeded only in the 1930s, when careful methods of standardization were exported.[35] The export of similarity went before success.

Research prompted by the colonial working world could break received theories. The colonial administrators of the Dutch East Indies were well aware of the debilitating disease of beriberi, and of how the swellings in the legs could weaken a workforce. In the late nineteenth century, with germ theory of disease riding high, the discovery of a bacterial cause was expected.[36] However, Christiaan Eijkman, taking fowl polyneuritis as an animal model, argued from the 1890s that beriberi was a nutritional disorder, caused by too much milled white rice in the diet. By the time of the meeting of the Far Eastern Association of Tropical Medicine, held in Manila in 1910, most colonial authorities shared the nutritional view. Research programmes supported by this imperial working world identified a specific chemical, thiamine (Vitamin B_1), and, in the 1930s, a chemical synthesis was provided. 'Vitamins' were a new and promising idea. And with burgeoning research programmes came credit; Eijkman shared the 1929 Nobel Prize for medicine with a fellow 'vitamin' man, Frederick Gowland Hopkins.

Colonial management was also one of the major sources of conservationist thinking, encouraging decisions to set aside tracts of land to conserve natural resources, which in turn depended on a vision of nature as a stock that could be depleted and must be accounted for. Historian Richard Grove argues that such policies had originated in island administration in the eighteenth century.[37] The colonial necessity to preserve a supply of wood and water had protected upland forests in Mauritius or Tobago. In the nineteenth century these policies had been transferred to larger administrations. Natal's forest code of the late nineteenth century, for example, was imported from Mauritius.[38] Transvaal set up game preserves in the 1890s. In the twentieth century, zoologists such as Julian Huxley defended the setting aside of huge preserves for African megafauna. The ecologist John Phillips was another important advocate for conservation in the 1930s.

The laboratories and field stations of colonial science were major sites of employment for scientists being trained in the first two-thirds of the twentieth century. According to Jon Harwood, one-quarter of science graduates went into colonial service.[39] But there was a shift in style as tensions grew in the imperial projects, pointing towards post-colonial worlds.

Models of science and empire

Historians of science have repeatedly tried to model how science fitted into the geopolitical world of colony and empire. As a first-order description it is clear that the geopolitical map was highly uneven: there existed centres, in particular metropoles – Paris, London, Tokyo, Berlin, Amsterdam – that were privileged with respect to peripheral regions. While an unqualified language of centre and periphery is problematic, it is the language of the first influential model of science in empires: that of George Basalla, a young historian working in Austin, Texas, who proposed a three-stage model of the development of colonial science in the journal *Science* in 1967.[40] Basalla's model is one of 'diffusion': it assumes that science is made in centres and moves outwards without change in form. Basalla's first phase is 'characterized by the [Western] European who visits the land, surveys and collects its flora and fauna, studies its physical features, and then takes the results back to Europe.' This phase describes European activities in, for example, South America from the sixteenth century, North America from the seventeenth century and the Pacific Ocean from the eighteenth century. It also captures Germany's scientific relations with its newly acquired colonies in the late nineteenth century.

Basalla's second phase is that of 'colonial science', a situation in which there are a larger number of scientists working on the periphery, but these scientists remain dependent on the centre. A colonial scientist would be trained in the centre and, while working on the periphery, would seek to publish in the journals of the centre, join the institutions of the centre, and have work validated by peers at the centre. The phase is a misnomer, in fact, because Basalla states explicitly that there does not need to be a colony for there to be colonial science: the crucial relation is that of dependency. So, in Basalla's model, the United States remained in a state of colonial science well after political independence. Likewise, independent Meiji Japan featured 'colonial science': between 1868 and 1912, 'over 600 students were sent abroad for special training in the scientific and technological centers of America and Europe'.[41]

In Basalla's third phase, colonial scientists become dissatisfied with their dependent relationship with centres and seek independence. 'Ideally', he wrote, a third phase scientist will

> (i) receive most of his training at home; (ii) gain some respect for his calling, or perhaps earn his living as a scientist, in his own country; (iii) find intellectual stimulation within his own expanding scientific community; (iv) be able to communicate easily his ideas to his fellow scientists at home and abroad; (v) have a better opportunity to open new fields of scientific endeavor; and (vi) look forward to the reward of national honors – bestowed by native scientific organizations or the government – when he has done superior work.[42]

In Basalla's view, 'the non-European nations, after a long period of preparation, have only recently approached the supremacy of Western Europe in science.' In particular, the United States was demonstrably independent in the fields of genetics and optical astronomy by the interwar period, with physics close behind. More grudgingly, Basalla granted the Soviet Union parity, too. China has sent 1.2 million students and scholars abroad since the 1980s – contributing to a significant brain drain since only a quarter have returned – and would be classified by Basalla as in this phase.[43]

Basalla's model was shaped by its times. To see this, let's note some other features. It is a three-stage model of universal application: while 'making a preliminary survey of the literature', recalled Basalla, 'I discovered a repeated pattern of events that I generalized into a model which describes how Western science was introduced into, and established in, Eastern Europe, North and South America, India, Australia, China, Japan, and Africa.' It's hard to see where the model would not apply. Such 'one size fits all' models of international social change were a feature of the first two decades after 1945. Walt Rostow, the American economist, for example, proposed a single model of development that every industrializing country would pass through. Just as Rostow's theory became prescription in the hands of policy-makers in the post-war field of 'development' – your country must do this, then that, then that, to achieve economic take-off – so Basalla's model invited prescriptions in science policy: a 'program for scientific development familiar to students of bilateral and international aid policies in the 1970s'.[44]

But it is also significant that Basalla's model was articulable at all. Basalla attributed the absence of any comparative study of the development of science in different national settings to the 'widespread belief that science is strictly an international endeavor'.[45] Or, in other words, if science was the same everywhere, then there could be no non-trivial sociology of science. Basalla held to a different sociology of science that admitted that a local setting, which, while it could 'not decisively mold the conceptual growth of science, could at least affect the number and types of individuals who are free to participate in the internal development of science'.

Studies in science and imperialism post-Basalla have started with different notions of empire and different sociologies of science. 'Empire' has become a complex term in modern social and cultural historiography, and historians of science have reflected this trend. MacLeod, in an article originally published in 1982, reeled off a list of complaints against Basalla's model:

1 It generalizes all societies, regardless of cultural context, into a single scheme.
2 The scheme is linear and homogeneous, assuming that there is a single Western scientific ideology which is disseminated uniformly. It does not take into account 'south–south', or intercolonial movement, or movement between the colonies of one country and other European countries.
3 It alludes to, but does not explain, the political and economic dynamics within a 'colony' that make for change and that occupy a shaded area between phases one and two or between two and three.
4 It fails to account for the relations among technological, social, and economic developments on political forces that may arrest development in other spheres.
5 It does not take into account the cultural dependence that lingers long after formal colonial political ties have been thinned or cut.
6 It does not take into account the wider economic interdependencies that have, since empire, contributed to the plight of the third world, for which science alone scarcely offers consolation.[46]

Other historians have added further evidence to support these complaints. Inkster, for example, in support of complaint 5, notes that, even when Australian scientists had reached Basalla's phase 3 (certainly so by 1916), cultural dependency on the metropole had not diminished.[47]

But it is the sociology of science held by Basalla's critics that lies at the root of their argument with him. What 'connection, if any, can be found between patterns of economic and political evolution and patterns of intellectual development in the various sciences?', asks MacLeod. Contra Basalla, these historians do wish to know how imperial settings could 'mold the conceptual growth of science'. Indeed, allegiance to sociologies of science frames the literature on science and empire. The historian of science Lewis Pyenson undertook a substantial research programme in the 1980s and 1990s examining the exact sciences (astronomy, mathematics, physics) in areas of German, Dutch and French imperial influence.[48] Pyenson argued that the exact sciences are not moulded by context, and he claimed his imperial history demonstrated this fact. Paolo Palladino and Michael Worboys disagreed. Starting from a contextual history of science, they argued that a different conclusion could be drawn from the apparent stability of the exact sciences in different imperial contexts. Rather than being a consequence of the independence of the exact sciences from a social context,

> this relative immunity of physicists and astronomers to political pressure ... was the consequence of institutional positions established through social negotiation ... Colonial physicists' and astronomers' relative immunity to the mundane politics of the Age of Empire may simply have been a reflection of the social and political position of metropolitan physics and astronomy; that immunity does not stand as evidence against the social history or sociology of science.[49]

Likewise, historian Roy MacLeod's five-phase account of science and empire stresses how each supports, shapes and transforms the others. The first three phases move from dictation from the centre (for example, London in the case of Australian science) through degrees of colonial autonomy until a dynamic relationship emerges. In the fourth phase, imperial unity was promoted even further. In the case of the British Empire this phase was intensified by the shock of the Boer wars. British critics of how this conflict against white Afrikaners was being fought offered a technocratic solution: measures to promote national efficiency. Science would be the politically neutral, efficient guide to imperial administration: scientific method would 'unite empire in unity of truth, of tradition, and of leadership, from Curzon's India and Lorne's Canada to Smuts's South Africa'.[50] Not only would science be a model for administration, but the discovery of techniques would materially transform the empire, not least through tropical medicine. As Joseph Chamberlain put it in 1898:

The man who shall successfully grapple with this foe of humanity and find the cure for malaria, for the fevers desolating our colonies and dependencies in many tropical countries, and shall make the tropics liveable for white men ... will do more for the world, more for the British Empire, than the man who adds a new province to the wide dominions of the Queen.[51]

MacLeod's fifth phase, from the First World War to after the Second World War, featured an emphasis on imperial and scientific unity, typical of high imperialism, giving way to a more organic coordination and cooperation between scientists and the scientific institutions of empire.

I have summarized two models of how science fits into a geopolitical world of empires. Basalla's three-stage model was diffusionist: science started in centres (in Western Europe) and was transplanted into peripheral colonies, which eventually grew to become centres in their own right. The context – a geopolitical map of centres of political power and peripheries of colonial activity – only directs the movement of science and does not shape the content of scientific knowledge. On the other hand, Basalla's critics have stressed that science could be a model for empire, just as empire could not only direct but also shape the content of science produced. Other historians, such as Deepak Kumar and David Arnold, who have written on science and India, or Stuart McCook, who proposes the category of 'creole science' in his study of the science in the Spanish Caribbean, provide further frameworks for reassessing and reformulating centre and periphery in colonial and imperial science.[52] While in Basalla's model centres and peripheries are unproblematic, centres are metropoles where science originates and peripheries are colonies where science can end up, in MacLeod's the terminology of centre/metropole and periphery/colony is becoming problematic:

a dynamic conception of imperial science gives a fresh outlook to the study of imperial history. There is no static or linear extrapolation of ideas; there are multiple autochthonous developments that have reverberating effects ... [The] idea of a fixed metropolis, radiating light from a single point, is inadequate. There is instead a moving metropolis – a function of empire, selecting, cultivating intellectual and economic frontiers. In retrospect it was the peculiar genius of the British Empire to assimilate ideas from the periphery, to stimulate loyalty within the imperial community without sacrificing its leadership or its following.[53]

Recent scholars, noting the complexity and extent of such reciprocity, have redrawn the picture of 'centre' and 'periphery' altogether.

Centres change peripheries and peripheries change centres. The point of this historiographical reflection, and the histories of tropical medicine and ecology that precede it, is that the working worlds of both generate science.

— 8 —

EXPANDING UNIVERSES: PRIVATE WEALTH AND AMERICAN SCIENCE

The United States in the nineteenth century was a land of immense distances, of energetic exploitation, of a mobile and growing population, of new towns, and of towns becoming cities. The inventors and entrepreneurs that built the systems technologies of the second industrial revolution – chemicals and oil, electric light and power, telephony, and so on – could be found in old Europe as well as the New World. But the United States promised bigger markets for whomever could integrate markets and technological systems, and so the opportunities for making wealth, and the spur to innovation, were particularly strong in the United States.

This chapter examines the consequences of this wealth for twentieth-century science. There are three parts. First, the colossal nineteenth-century fortunes, concentrated in the hands of a few individuals, became, for a number of reasons – not all of them philanthropic – fountains of patronage in the twentieth century, dispensing money for scientific research. The choices made by Andrew Carnegie or by the trustees of the Rockefeller Foundation distinctively shaped twentieth-century science, both in their direct impact on American science and in the policy responses provoked across the Atlantic. Second, private business continued to be a direct patron of science in its own interest, which in turn supported scientific theory-building of an apparently more pure kind. My core case study, chosen because of the centrality of oil to understanding the twentieth century, will be petroleum science. My example of theory-building supported, in a secondary fashion, by this working world will be continental drift. Finally, the working world of industry itself became the subject of scientific inquiry in a new and deepened fashion in the twentieth century. So my third perspective on private wealth and American science will be the scientific study of industry. The spread of scientific values that seems to be entailed by the above trends and topics was not without opposition and contention. I will briefly review the famous Scopes trial, in which the teaching of evolution was challenged, interpreted as a contest of new and old values in the United States.

Wealth and philanthropy

Industrial wealth already supported individual careers in American science. William Henry Osborn had built his fortune managing the Illinois Central Railroad. His son, Henry Fairfield Osborn, led the American Museum of Natural History's department of vertebrate paleontology in its famous collection of dinosaurs from expeditions to Wyoming and elsewhere. In this way, carboniferous wealth was translated into fossil bones, prestigious institutions and sparkling careers.[1] But the Osborn family riches pale in comparison with those of the tycoons Andrew Carnegie and John D. Rockefeller.

Carnegie's wealth came from steel. Scottish born, Andrew Carnegie emigrated to the United States as a young boy, later training (like Edison) in telegraphy before being promoted to a position of railroad manager. Any spare income was invested – very wisely it turned out, for Carnegie was rich and independent by the age of thirty. However, his vast fortune was made by investing his wealth in the importation of the Bessemer process of steelmaking in the 1870s. His company was innovative, not only importing the best techniques from Britain but also improving on them. The Carnegie Steel Company, consolidated in 1889, exploited economics of scale made possible by the large American market as well as vertical integration and efficiencies, achieved through purchasing iron ore deposits. By 1890, United States steel output surpassed Britain's. Economic power would later be translated into scientific power.

In 1901 Carnegie's company was sold to J. P. Morgan's United States Steel Corporation for $250 million. Andrew Carnegie had already spelled out his feelings on the causes, opportunities and duties of the very rich in an essay of 1889, titled 'The gospel of wealth'. He wrote that a man who accumulates great wealth had a duty to spend it during his lifetime on projects to benefit and improve mankind. It was a philosophy he saw as rooted in the protestant work ethic: that money was the reward for virtuous work. As men of great accumulated virtue, it followed, the rich were peculiarly best suited to decide where their philanthropy should be directed. Carnegie distributed

162

$350 million in his lifetime, some of which went to the British Empire (every town in Scotland, it seems sometimes, has a Carnegie library), but the greater part, $288 million to be exact, was spent in the United States. Beneficiaries of Carnegie largesse included whole new institutions – such as the Carnegie Institution of Washington (1902) and the Carnegie Corporation of New York (1911) – and, as we shall see below, the projects of favoured and persuasive individuals. The endowment of the Carnegie Institution of Washington alone, some $10 million, was more than the total sums dedicated to research at all American universities at the time.

Rockefeller's wealth, on the other hand, came from oil. 'Pious, single-minded, persistent, thorough, attentive to detail, and with both a gift and fascination for numbers, especially numbers that involved money', Rockefeller's character would shape not only the oil business but also, through its example, modern industry and the modern corporation.[2] In 1865, Rockefeller had bought out his partner in a company that dealt with many commodities – wheat, salt, pork – but which also had an oil refinery in Cleveland, Ohio. Slowly but surely Rockefeller expanded oil operations, always keeping a beady eye on costs. As he did so, Rockefeller negotiated ever better deals on transportation. Rival refiners who could not be bought out were bullied. Nor, argued Rockefeller, should he be thought of as doing anything wrong, or at least anything unnatural. If nature was red in tooth and claw, then so was business. The fittest, the best, survived. 'The American beauty rose can be produced in the splendour and fragrance which bring cheer to its beholder only by sacrificing the early buds which grow up around it', announced Rockefeller in a speech to a Sunday school class, adding, in case the social Darwinist point was missed, 'And so it is in economic life. It is merely the working out of a law of nature.'[3]

The genius of Standard Oil was to control, almost completely, the refining and transportation of oil, leaving its extraction from the ground to secondary dependent producers. The distinctive achievement of American industrial capitalism, as Thomas Hughes reminds us, was the innovative management of flow.[4] And business was booming: the discovery of Pennsylvanian and Ohio oil contingently coincided not only with demands for a 'new light', to replace smoky tallow and whale oil, but also, as electric light threatened to succeed it, a new unexpected demand for petroleum as fuel for internal combustion engines.

The dominance of Standard Oil, the largest corporation in proportion to the overall economy ever, attracted intense criticism. The

163

anti-trust Sherman Act of 1890 targeted the monopolies. When his enemies achieved the break-up of Standard Oil in 1911, Rockefeller's personal wealth jumped to $900 million. By then Rockefeller had begun dispensing philanthropy. In the nineteenth century most of the funds had gone to the Baptist Church, including, he thought, a substantial sum to found a Baptist university.[5] This 1889 donation created the University of Chicago, which, founded when the idea of the German research university was highest in vogue, became an important twentieth-century scientific institution in its own right. In 1901, the Rockefeller Institute for Medical Research, an American copy of the French and German research institute model, was founded in New York. The following year Rockefeller provided the funds, first $50 million and later increased to $180 million, for a General Education Board, which in turn dispensed funds for schools in the American south (including schools for black children) and for agricultural modernization. However, the biggest influence on science was through the Rockefeller Foundation, established in 1913 to channel the wealth garnered through the break-up of Standard Oil. Altogether Rockefeller gave away $550 million.[6]

The motivations for philanthropy were diverse, from benevolence or the pursuit of interests akin to a particularly well-financed hobby to the following of self-interest in the forms of tax evasion, the assuagement of guilt, or the fashioning of a kindly if ultimately post-humous reputation. Many of these reasons dictated that the benefici-aries of philanthropy should be publicly visible. Indeed, the historian of science Michael Dennis describes the competitive philanthropy of Carnegie and Rockefeller as a form of conspicuous consumption – or, in other words, philanthropy was a means of spending wealth to fashion a public reputation.[7] Astronomy and biomedical science were the big winners in this competition.

Turning wealth into telescopes

George Ellery Hale, historians of twentieth-century astronomy concur, was 'perhaps the first leading American scientist to under-stand the importance of the new philanthropic foundations as patrons for pure research';[8] he was 'the pivotal figure in the shifts of the ways that US astronomers ran observatories and planned, built and operated big instruments'.[9] A graduate of MIT, Hale had joined the new University of Chicago in 1892. The new universities in the mid-west and west provided opportunities to build new sciences and to challenge, ultimately, the Ivy League establishment. Hale persuaded the local streetcar and railroad magnate Charles T. Yerkes to fund a new observatory dedicated not to the traditional astronomical programmes of star surveys but to the new professional and interdisciplinary area of astrophysics: the application of physical instrumentation, especially photography and spectroscopy, to astronomical subjects. The Yerkes Observatory housed a refracting telescope with a 40-inch aperture, the largest built. Like ecology, astrophysics flourished in the new universities. Hale was a leading light of the new specialty, devising a new instrument between 1890 and 1894, the spectroheliograph, for viewing the sun through narrow spectral lines, and founding the *Astrophysical Journal* in 1895.

In the new century Hale moved west to California. He brought his superb skills in astrophysics and in the tapping of patrons for financial support for science. The very largest telescopes, visually impressive and the surest path to astronomical leadership, the very image of lofty aspirations, appealed to the philanthropists. James Lick, who made his fortune from San Francisco real estate, had been persuaded to fund a telescope on Mount Hamilton in California in the 1870s. Hale, likewise convinced that the mountains of the western United States possessed the clear air necessary for good astronomical observation, lobbied the Carnegie Institution of Washington, and Andrew Carnegie directly, to support his plan for an observatory at Mount Wilson, near Los Angeles. In 1904 the Carnegie Foundation donated the funds to build an observatory. An extraordinarily large telescope – the 60-inch reflector – was designed by George Ritchey as an immense camera: its first purpose was not for naked-eye observation but for photographic recording of the cosmos. The completion of the 60 inch in 1908 marked the opening of a new astrophysical era in astronomy. Pushing on, Hale lobbied, successfully, for an even bigger instrument. The 100-inch reflector, the largest in the world, was completed in 1917. While the mirror was paid for by a Californian businessman, John D. Hooker, nine-tenths of the costs were covered by Carnegie. In his 1911 letter to the trustees of the Carnegie Institution of Washington, the old man had made his wishes clear:

> So great has been the success of the Institution that I have decided to increase its resources by adding Ten Millions of Five Per Cent Bonds, value Eleven and a Half Millions, which will ultimately give you Five Hundred Thousand Dollars a year increased revenue. ... I hope the work at Mount Wilson will be vigorously pushed, because I am so anxious to hear the expected results from it. I should like to be satisfied before I depart, that we are going to repay to the old land some part of

the debt we owe them by revealing more clearly than ever to them the new heavens.

The traditional leaders of science in Europe would indeed be shown new heavens, for the possession of the largest and most sophisticated telescopes in the world, made possible by the recycling of American industrial wealth, would enable the United States to overtake the 'old land'. From the completion of the 100-inch Mount Wilson telescope, all major astronomical debates in the twentieth century would be settled using data generated first and foremost by American instruments. Astronomy was the first science to manifest American leadership. Others would follow.

Resolving island universes

The resolution of the debates around the astounding astronomical discoveries of the 1910s and 1920s will illustrate the point. In the nineteenth century the Milky Way had been considered to be, to all intents and purposes, the universe. There were no 'island universes'. 'No competent thinker', wrote Agnes Clarke in 1890, summarizing the consensus view of nebula astronomy, 'with the whole of the available evidence before him, can now, it is safe to say, maintain any single nebula to be a star system of co-ordinate rank with the Milky Way.' Yet, in less than a generation, astronomers would accept that their picture of the visible universe must be redrawn in three ways: the size of the Milky Way galaxy was larger than they thought by a factor of ten, the existence of external galaxies was accepted and, most surprising of all, the motions of the galaxies revealed that the universe itself must be expanding.[10]

The primary focus of the new astrophysical astronomy was to understand the course of stellar evolution. This in turn suggested the close scrutiny of spiral nebulae, which seemed to be stars in formation. James Keeler at the Lick Observatory, also in California, used the 35-inch refractor there to show that spiral nebulae could be found in their thousands. German research, by Julius Scheiner, showed that the spectrum of the Andromeda nebula resembled that of the sun, while observations by Max Wolf at Heidelburg supported the implication that spirals were star systems. Wolf's investigations, notes the historian Robert W. Smith, were the 'last important observational contributions to the island universe debate by an astronomer working outside the United States'.[11] Measurements of the radial velocities of nebulae, made possible by the astrophysical techniques of measuring the blue or red shifting of spectral lines, also contributed to a revival of the island universe theory in the 1910s. Vesto M. Slipher at the Lowell Observatory at Flagstaff, Arizona, measured the blue shift of the Andromeda nebula and concluded that it was moving towards us at a speed of 300 kilometres per second. By the outbreak of the First World War, very fast recessions – of the order of 1,000 kilometres per second – had been deduced from red shifts in the light of nebulae. These speeds were much greater than the radial velocities of stars. Perhaps, asked Slipher, the spiral nebulae were fleeing the Milky Way?

Meanwhile, astrophotography at Mount Wilson was being refined to deliver stunning images of the spirals. In 1910, Ritchey's photographs of the spiral M33 resolved it into stars. Also at Mount Wilson, Harlow Shapley, who arrived in 1914, used the 60-inch reflector to measure carefully the colour and magnitude of globular clusters and argued by several different methods that they must be distant – far outside the Milky Way. One of the methods had been given by Henrietta Swan Leavitt, who, while working as a 'computer' at the Harvard College Observatory, had noted that the variable stars in the Large Magellanic Cloud obeyed a relationship between luminosity and period of variation. Once the distance to the nearest examples had been measured by the Dane Ejnar Hertzsprung, the distances to spiral nebulae and globular clusters could be calculated using these Cepheid variable stars as standard 'candles'.[12]

Shapley now argued that the universe should be seen as a multitude of distant globular clusters encircling a giant central Milky Way galaxy. But his picture was opposed in a 'Great Debate' of 26 April 1920 by Heber D. Curtis of the Lick. Drawing on observations of novae and deductions about the size of the Milky Way in relation to the distance of spirals, Curtis argued that the spirals must be island universes. (However, he did so while rejecting the reliability of the Cepheid method.) Into this American controversy stepped a young astronomer at Mount Wilson. Edwin Hubble made systematic use of Henrietta Leavitt's Cepheid method, as well as the power of the 100-inch reflector, to announce in 1925 that the spirals were in fact distant objects. The Andromeda nebula – or rather Andromeda galaxy, as it now appeared to be – was calculated to be a million light years away.

It is fair to say, however, that throughout the 1910s the nature of the spirals was controversial. Equally sound evidence pointed away from the island universe theory. For example, Adriaan van Maanen, also at Mount Wilson, used a blink stereocomparator, which flicked between photographic images of the spirals taken on different dates to discern stellar movement. M101, he announced in 1915, rotated every 85,000 years. Such an object could not be too far away, certainly within the Milky Way. Desk-top astronomers in Europe, such as James Jeans, unable to compete directly with new evidence, nevertheless published commentaries on such data, as Jeans did in his *Problems of Cosmogony and Stellar Dynamics* of 1919, using van Maanen's data to support an interpretation of spirals as proto-globular clusters.[13] Arthur Stanley Eddington in Cambridge, England, meanwhile argued that the stellar movements seen by van Maanen were not real. In this state of irresolution, the data did not determine belief. Not only did data not determine belief in island universes, but the attachment to island universe theory preceded evidence from the Cepheid period-luminosity law.[14] Instead, whether or not an astronomer accepted island universes was as much a matter of 'scientific taste' of its advocates as a decision driven by empirical data.[15]

Nevertheless, those without the means of generating data were relatively powerless. In a period exacerbated by the disconnections of global war, knowing the standards – explicit and tacit – for good research required attendance in a select few American locations. 'For anyone to participate fully in the debate over island universes in the late 1910s, they had to know what was being thought and discovered in the United States, in particular Mount Wilson, Lowell and Lick', concludes Robert Smith – anywhere else and an astronomer 'was likely to find himself isolated, his work ignored because it was outdated and performed with inferior instruments, his perception of theories fogged because he was ignorant of the value judgments by astronomers of other astronomers and their results, unaware too of shifts of the criteria of theory and method selection of the leading researchers in spiral nebulae.'[16]

Hubble and the expanding universe

Hubble's data generated at Mount Wilson was also relevant to an esoteric debate about the overall shape and dynamics of the universe in its largest and most fundamental terms, the subject of the borderline science of cosmology. Einstein's general relativity field equations of 1915 had described mathematically the relationship between, on one side, the distribution of energy and matter and, on the other, the shape of space and time. Adding a constant term (the Lambda

168

term – Λ), as was entirely legitimate in the derivation of the equations, Einstein found one solution for the equations that represented a simplified model universe, one which was static and closed.[17] (An observer in such a universe would not see the galaxies moving away from them, but could, if he or she travelled far enough, return to his or her origin.) The Dutch theorist Willem de Sitter found a second solution in 1917: assuming the universe was empty to make the equations tractable, and throwing away the Lambda term, he found a universe that was open but forever expanding. The Abbé Georges Lemaître, a Belgian who had trained for the priesthood, proposed a third set of solutions in 1927: an expanding universe with a compressed superdense state at the beginning of time. In 1931 Lemaître would term this state the 'primeval atom'. This was cosmology that looked, and read, like a divine creation. Similar mathematical results were published by the Russian Alexander Friedmann. However, at the time, little notice was paid to these papers. 'It is remarkable that neither Friedmann's nor Lemaître's works made any impact at all', comments the historian of cosmology Helge Kragh. 'The reasons for the neglect are not entirely clear, but ingrained belief in the static nature of the universe was undoubtedly an important sociopsychological factor.'[18]

In the 1920s Hubble set about measuring the red shifts of galaxies, now taken to be distant island universes akin to our own Milky Way. He was assisted by Milton Humason, a gifted observer and technician. (Humason had joined Mount Wilson as a hotel bellhop and mule driver, married the chief electrician's daughter, and progressed from janitor, to night assistant, to Hubble's right-hand man.) In a paper of 1929 Hubble and Humason's red-shift results were set out: the further away a nebula was, the faster it receded from us. Double the distance, double the speed. The universe, it seemed to others (Hubble was initially reticent), was expanding. This result, on the face of it more astounding than the island universe theory, was accepted with little controversy. Authority of observation depended on the possession of the world's largest telescope. In a trickle-down effect some of the authority was passed back to the theoreticians: Lemaître's paper was retrospectively declared to be 'brilliant'. The historian Kragh, however, argues for a reversal in this chain of credit: the theoretical speculations and the Hubble data together produced a paradigm shift in cosmology, yet 'Among the three main candidates, Lemaître was the only one who clearly argued that the universe is expanding and drew on both theoretical and observational arguments.'[19]

Elites: philanthropic funding and interdisciplinary teamwork

The 100-inch telescope was not only instrumental in providing data on the basis of which the shape and structure of the universe was reinterpreted, it also contributed to a trend towards new styles of organization of scientific work. Its funding typified an effective means of generating support for institutions that was applied by Hale and others elsewhere: the channelling of both philanthropic and local business interests into the support for pure and applied science. The lowly Throop College, for example, had been turned into the dynamic California Institute of Technology (Caltech) in 1920 by using Rockefeller Foundation and Carnegie money plus matching contributions from a local business elite. In turn Caltech provided the engineers essential to Californian projects of electrical power transmission (Los Angeles, for example, was one of the very first cities to be lit by electric light), as well as in oil fields, shipbuilding, dams and aqueducts.[20] Funds were not sought from the federal government. Hale shared many Americans' suspicions of central government. (Recall that one of the reasons Hale had proposed and designed the National Research Council in the First World War had been to keep government's hands off science.)[21] Not everyone, however, was happy with this cosy and productive local arrangement; 'You believe that aristocracy and patronage are favourable to science', the psychologist James McKeen Cattell wrote, criticizing Hale. 'I believe that they must be discarded for the cruder and more rigorous ways of pervasive democracy.'[22] In a period when American science was marked by the 'advent of the science-based university and the professionalization of the scientific community generally', Hale, in his building up of Caltech and a prestigious cadre of professional astronomers at Mount Wilson, tapped philanthropic funds to promote both.[23] There was a shift in the balance of scientific power not only from Europe to America but also, within the United States, from east to west coast.

Large expensive instruments required different skills to build and maintain; the observatories that protected them also housed an increasingly fine division of labour. Following a trend that can be traced to the calculating house of George Biddell Airy's Royal Greenwich Observatory in the nineteenth century, the Lick Observatory in California and the Harvard College Observatory under William Henry Pickering had become data-processing factories. The new institutions also supported a distinctive style of research: interdisciplinary teamwork. Astrophysics, for example,

combined the techniques of the physics laboratory with the organization and aims of the astronomical observatory. In the life sciences,
too, different combinations of disciplines led to a retargeting of
research programmes in the first half of the twentieth century. This
trend was decisively encouraged by the philanthropic foundations,
especially the Rockefeller Foundation. The most important example
of this trend, examined in detail in a later chapter, was the emergence
in the 1930s of a 'molecular biology': research founded on the conviction that the processes of living organisms were most clearly revealed
in their analysis to the scale of molecules.

As Lily Kay has shown us in her history of the relations between
patronage and the 'molecular vision of life', the Rockefeller Foundation
punched far above its weight.[24] Between 1929 and 1959 it spent $25
million on molecular biology, which sounds impressive but was a
mere 2 per cent of the federal budget for scientific research and development in this period. Its strength lay, first, in concentrating on the
areas that the federal spending left alone – so not defence, nor much
of agriculture; second, its 'effectiveness lay in creating and promoting
institutional mechanisms of interdisciplinary cooperation through
extensive systems of grants and fellowships, and in systematically
fostering a project-oriented, technology-based biology'.[25] Finally, this
web of interlocking finance was promoted by trustees, administrators
and science leaders who shared a social position and worldview. This
consensus amounted, argues Kay, to hegemony.

The attitudes of the science leaders were marked first by their suspicion of federal control, which implied an unwillingness to accept
central government funding, but also by a commitment confirmed
during the mobilization of research of the First World War that successful science required a marriage of science and industry and was
most often interdisciplinary in nature. 'On the one hand, through
the large scale mobilization of science and the diverse cooperative
war projects it spawned – wireless communication, submarine detection devices, chemical warfare, pharmaceuticals, blood banks, and
mental testing – science permanently altered the nature of warfare
and grew indispensable to it', summarizes Lily Kay. 'On the other
hand, these projects shaped the organization of scientific knowledge
by placing a premium on interdisciplinary cooperation and on a
liaison with industry and business.'[26] The leaders included the trio
Hale, physicist Robert Millikan and chemist Arthur Amos Noyes at
Caltech, geneticist Thomas Morgan (who joined the trio), physicist
Max Mason, psychologists Lewis Terman and Robert Yerkes, and
the brothers Simon and Abraham Flexner. 'Ideologically ill-disposed

171

to government control', Kay writes, 'the leaders of American science lobbied successfully for a substantial increase in financial support of science by the private sector, notably the Carnegie Corporation and the Rockefeller Foundation.'[27]

In the nineteenth century, the working world of railroads and other industrial concerns had prompted the formation of the modern corporation – hierarchies of impersonal management and a fine division of labour.[28] The science leaders and the trustees of foundations agreed that this corporate model for industry was the correct one for thriving modern science. 'Just as the multiunit business structure depended on the team player and the coordinating manager', writes Kay, in support of an argument that the style of research encouraged by the Rockefeller Foundation went beyond mere collaboration, 'so the new science relied on management and group projects directed toward interdisciplinary cooperation.'[29] The move towards group-work – teamwork – was one of the outstanding trends of twentieth-century science, and the unusual leverage that the philanthropic foundations had in the first half of the century of American science forms part of its explanation.

The leaders inherited not only a model for corporate organization but also the prejudices and worldview of a social elite. In particular, the United States, which had rapidly industrialized in the nineteenth century, was a society marked by inequalities in wealth and power and by continued external and internal migration. The winners – wealthy, protestant, white – credited their own success to hard work, justly rewarded. They viewed the others – migrants from South and East Europe, blacks, the poor – as sources of disturbance. In the 1890s and 1900s a 'social control' theory argued that conflict in such a society was inevitable and must be mitigated by the modification of individual behaviour to ensure conformity. Social control created roles for scientific experts, peaking in the 1920s, especially in the human sciences of sociology, psychology, some political science and some life sciences that ruled with authority on the processes of socialization; social control offered a technocratic promise.[30]

The trustees of the Rockefeller Foundation, and therefore the people whose choices would guide the funding of research, embraced the social control agenda. There were two important consequences. First, the human sciences that promised to contribute to social control received support. The president of the University of Chicago, Harry Pratt Judson, one of the most influential trustees, wrote that his fellow decision-makers were faced with a choice between two broad policies: uncontroversial ones that satisfied human wants (such as improved medicine or education) and 'challenging ones, which conflicted with human desires', for which 'the real hope of ultimate security lies in reinforcing the police power of the state by training of the moral nature so painstaking and so widespread as to restrict these unsocial wants and substitute for them a reasonable self control'.[31] It was this kind of sentiment that secured funds for the eugenic work of the Rockefeller Bureau of Social Hygiene from 1913, as well as the Carnegie Institution of Washington's support of Charles Davenport's eugenics office at Cold Spring Harbor, on Long Island, from 1902.

The second consequence was the adoption of a policy of 'making the peaks higher'. The Rockefeller Foundation sought out and further funded the places where it judged excellent research to be taking place. Democracy or redistribution of wealth was certainly not on the cards. 'Democracy is the apotheosis of the commonplace', wrote trustee and later president of the Rockefeller Foundation, Raymond B. Fosdick; it was the 'glorification of the divine average'.[32] Several universities benefited from being found to be 'peaks': Chicago, Stanford, Columbia, Harvard and Wisconsin, to name a few. But the biggest winner was Caltech.

Unlike the situation at Chicago, the faculties of life sciences at Caltech did not have to respond to the demands of a big medical school. The emphasis on teamwork and the lavish patronage, from a combination of philanthropic and local business sources, encouraged the growth of influential groups. Thomas Hunt Morgan was persuaded to move his *Drosophila* genetics from Columbia in 1928, and became head of biology. Theodosius Dobzhansky soon joined, as did Linus Pauling, George Beadle, Boris Ephrussi, Edward Tatum and Max Delbrück. Pauling's group of chemists became a major research school. Other teams formed around research on model organisms, such as the mould *Neurospora* (Beadle and others) and bacteriophages (Delbrück). Furthermore, the leaders of science deliberately instilled a corporate structure to Caltech. Indeed, the historian John Servos has called the whole, in a study of Noyes's chemistry, a 'knowledge corporation'.[33]

The philanthropic foundations strongly shaped the sciences in the first half of the twentieth century, through leadership – designating 'peaks' and making them higher – and through the consequences of following the social interests of the trustees. The Rockefeller Foundation channelled money into laboratory specialties – genetics, embryology, general physiology, reproductive biology, biochemistry, biophysics – letting them expand. Buffered from the demands of medical or agricultural faculties, some institutes – the Rockefeller

Institute for Medical Research was the prime example, but the Pasteur Institute in Paris also followed the pattern – launched into novel research programmes, which would receive showers of scientific credit, not least Nobel Prizes.[34] Other specialties – evolution, systematics, ecology – 'received far less support', and depended on rare and specific patronage or survived (and often grew) by integration into the working worlds of agriculture or medicine. 'Thus', concludes the historian of biology Jonathan Harwood, 'the pattern of patronage – be it the supply of funding or the demand for expertise – can explain why some academic fields have flourished and others languished at any given time.'[35] Other countries looked on enviously at the support for research offered by the American philanthropic foundations. We have already seen how the German policy of establishing the Kaiser Wilhelm Institutes had been partly a response to the competition offered by Carnegie and Rockefeller funding.

Sciences *for* the working world of industry: petroleum

The philanthropic foundations spent money accumulated during the remarkable expansion of American business of the second industrial revolution. Science, however, not only benefited from the redistribution of this wealth, it had also contributed, profoundly, to its creation. The history of oil provides the best illustration. Geologists had the skills to identify likely oil deposits – a new role that became professionalized in the new specialty of petroleum geology. 'By the last decades of the nineteenth century', argues historian Paul Lucier, 'geologists had formulated a theoretical and practical science of petroleum, one of the chief intellectual contributions of nineteenth-century Americans.'[36] These contributions included a theory of the structure of oil reservoirs as caps in anticlines replacing a simplistic picture of subsurface pools. Chemists found new ways to turn the flood of crude oil into ever more valuable refined commodities. By the second half of the twentieth century these efforts would produce a staggering array of products, from plastics to fuels. Finally, the spawning from geology of a new discipline of geophysics was sustained by the burgeoning working world of oil.

Historians of science remember 1859 as the year of Darwin's *Origin of Species*. Yet there is a good case for arguing that the discovery of oil in Pennsylvania had a comparable influence on the modern world. In the 1850s, Yale chemistry graduate Benjamin Silliman, Jr., and his partners, a lawyer, George Bissell, and a banker, James

Townsend, employed 'Colonel' Edwin J. Drake to apply a salt-boring technique – drilling – to prospect for oil.[37] Their strike, in August 1859, fed the demand for the 'new light': kerosene as an alternative to animal fats. As we have seen, by the 1870s, Rockefeller's Standard Oil controlled the new industry, with an iron grasp on refining and distribution. As Pennsylvanian production fell, Standard Oil looked elsewhere for sources, such as the Lima field on the border of Ohio and Indiana. However, Lima oil smelled of rotten eggs – a difficult commercial prospect for household lighting! Chemistry provided an answer. In 1888 and 1889 the chemist Herman Frasch, a German employed by Standard, found that refining Lima oil over copper oxide removed the stink.[38] Kerosene from Lima oil could now be sold at considerable profit.

Rockefeller's rivals also tapped scientific expertise. Of the sons of Immanuel Nobel, Alfred, the inventor of dynamite, is synonymous with science through the Nobel prizes awarded in his name. Alfred's brothers, Ludwig and Robert Nobel, however, took a different course: they built an immense oil empire in tsarist Russia. St Petersburg, gloomy in winter, had a special need for the 'new light' of kerosene, and in Baku, in the far south of the Russian Empire, oil seeped from the ground. 'Ludwig Nobel was also a great industrial leader, capable of conceiving a plan on the scale of Rockefeller', writes Daniel Yergin. 'He set about analyzing every phase of the oil business; he learned everything he could about the American oil experience; he harnessed science, innovation, and business planning to efficiency and profitability; and he gave the entire venture his personal leadership and attention.'[39] The Nobel Brothers Petroleum Company turned Baku oil from hand-dug pits to the 'most scientifically advanced [refinery] in the world, the first oil company to create permanent staff positions for petroleum geologists'.[40] When the Rothschilds built a railway out of Baku in the 1880s, accompanied by Marcus Samuel's Shell line of transportation ships from Baku to the east from the 1890s, Russian oil flooded onto a world market.

When Ludwig Nobel died in 1888, some European newspapers reported the event, erroneously, as the death of his famous brother. Reading his own obituaries, which vilified the 'dynamite king' as a merchant of death, Alfred Nobel was traumatized.[41] He rewrote his will, setting aside much of his wealth for the encouragement of peace and science. The Nobel prizes have been awarded – with suspensions during world conflicts – annually since 1901.

Science made the oil industry nimble. Just as the products of Edison's workshops and General Electric's corporate laboratories –

the systems of electrical light and power – threatened to extinguish the 'new light' of oil, a new demand for petroleum products came from the automobile. Californian start-up companies – such as Union Oil – made systematic use of petroleum geologists, employing graduates of the San Francisco universities of Stanford and Berkeley.[42] They grew rapidly in the 1890s. From the turn of the century they were joined by new giants: Sumatran wells from 1899 found by Hugo Loudon, crucially assisted by a team of professional geologists, later managed by Royal Dutch/Shell, and Texan oil, starting with the Spindletop gusher of January 1901. After 1911, Standard Oil was broken up into different regional companies. This new pattern of competition, and new demands for gasoline (petrol for cars) rather than kerosene, encouraged further application of science.

The scientific work at Standard Oil of Indiana, which led to thermal cracking, has a strong case for being the most lucrative consequence of science in the twentieth century. William Burton held a doctorate in chemistry from Johns Hopkins University, the model research university of Baltimore. He had joined Standard in 1889, working on the sulphur dioxide stink of Lima oil. By 1909 Burton had assembled a research team, populated by fellow Johns Hopkins graduates, to investigate the refining of crude oil.[43] Crude oil is a mixture of different hydrocarbons, of which gasoline makes up no more than one-fifth of the total. As car culture grew in the twentieth century, this low proportion of gasoline became more and more significant. Burton's team tried different methods of 'cracking' – of turning heavier hydrocarbons into lighter fractions. The line of research that seemed most promising – 'thermal cracking' – using high temperatures (650°C) and high pressures – was also the most dangerous. (There is an interesting parallel here with the Haber–Bosch process being developed at the same time, and discussed in chapter 5. Both were highly significant achievements of high-pressure chemical engineering.) 'Practical refinery men were frightened', writes historian of oil Daniel Yergin. 'As the experiments progressed, the scientists had to clamber around the burning-hot still, caulking leaks, at considerable personal risk, because the regular boiler men refused to do the job.'[44] Yet success – more than doubling the production of gasoline from a barrel of crude – translated into riches. Thermal cracking stills were in operation by 1913. The main beneficiary was Standard Oil of Indiana, which licensed the process to other companies.

Despite the flood of new sources of oil (Texas, Sumatra, Borneo, Persia) and the remarkable improvements in productivity (thermal cracking), the increased demand for petroleum products resulted in a spike of anxiety over the imminent depletion of global oil resources in the late 1910s and early 1920s. The spike prompted some countries, such as Britain, to investigate the conversion of coal and research into what are now called biofuels – even Jerusalem artichokes were considered a possible source.[45] This anxiety also encouraged petroleum geologists to shift from surface geology to the use of techniques drawn from physics to see deep under the ground. 'Many of the geophysical innovations', notes Yergin, 'were adapted from technology that had been drafted into use during World War I.'[46]

For example, the torsion balance, an immensely delicate instrument that had been developed in the 1900s by the Hungarian physicist Baron Loránd von Eötvös to measure the degree of equivalence of inertial and gravitational mass, was used by German engineers assessing the devastated, but strategically essential, oil fields of Romania during the war.[47] Post-war surveys of variation of the gravitational field provided clues about subsurface geology, and hence the possibility of new oil fields. This was high-quality science. The best pendulum gravimeter available before 1930, according to the assessment of the historian of geology Mott Greene, was 'invented by scientists working for the Gulf Oil Company'.[48]

Location by the triangulation of sound sources had been used by both sides during the war to locate enemy artillery. Now, combining these techniques with nineteenth-century seismography, the science of the measurements of earthquakes, sound waves were used to probe under the earth. German techniques, developed for the East European oil fields, involved setting charges of dynamite, tracing the reflected and refracted sound as it passed through rocks, and drawing conclusions about underground salt domes – potential locations for oil. 'Thus', concludes Yergin, 'a whole new world was opened up to exploration irrespective of surface signs.'[49] Aircraft, another technology that had grown up during the Great War, contributed by allowing aerial surveillance. When, after the 1930s, lightweight, supremely sensitive gravimeters and magnetometers, based on designs by the Texan physicist Lucien J. B. LaCoste, could also be flown in aircraft (and taken down mines), the range and economics of prospecting were transformed. LaCoste's company, founded with his academic colleague Arnold Romberg, dominated the post-Second World War world of gravimetric instrumentation.[50]

From the nineteenth into the twentieth centuries, geologists moved from being typically independent consultants or employees of state surveys to being employed experts. The oil industry became the largest employer of geologists in the twentieth century.[51] The oil companies, in turn, have been rewarded with new fields, higher productivity and further wealth. (Although there was one major exception: Arabian oil was discovered in the 1930s without input – indeed against the expert consensus – of petroleum geologists.) Mining companies, too, following the model provided by Anaconda Copper of Butte, Montana, established geological departments. Geophysical techniques – magnetometers, gravimeters, seismographs, micropaleontology – were central. New specialties – economic paleontology, microlithology, exploration geophysics, sedimentology – formed to induct students in the techniques, and new associations, such as the Society of Economic Geologists (1920) and the American Association of Petroleum Geologists (1918), were formed to represent professional interests.[52] (American invertebrate paleontology, made strong through support by the petroleum business, would later become a major voice in evolutionary sciences.)[53] But this academic and professional branching was kept in order by the common demands set by the working world of geological industry. 'The strategy and structure of twentieth-century geological industries', summarizes Paul Lucier, 'have, in large degree, determined the nature of the earth sciences that have served them.'[54]

Continental drift

It was in this context that a nineteenth-century debate about the nature of the earth's interior was resurrected in the first half of the twentieth century. Was the earth largely solid, as nineteenth-century physicists, such as Lord Kelvin, held, or was it fluid, as some geologists argued? In the 1910s, Alfred Wegener, a German academic, proposed a grand theory to account for the distribution of fossils and the forms taken by land masses.[55] The old theory was that the distribution of organisms could be accounted for by their passing across land bridges, which had since fallen into the sea. Instead, suggested Wegener, perhaps the continents themselves had moved? He speculated that a single continent, 'Pangaea', existed in the earth's deep past. Pangaea subsequently broke up and the parts moved slowly to their present positions. The suggestive contours of the Americas and Africa, fitting together like a planet-sized jigsaw, were one piece of evidence. Wegener's theory of continental drift was published in German in 1915 and translated into English in the 1920s as *The Origin of Continents and Oceans*. Others had suggested continental drift before, but none with the vigour or range of supportive argument of Wegener.

The theory was vigorously rubbished by most earth scientists, especially geophysicists.[56] Geophysics and geochemistry had flourished in the United States in the first half of the twentieth century for three reasons. (Related fields, such as meteorology and oceanography, also prospered.) First, these sciences found greater favour from philanthropists than traditional geology. The Carnegie Institution of Washington, for example, housed a geophysical laboratory from 1907. Under Warren Weaver and Max Mason, the Rockefeller Foundation 'emphasised the intellectual superiority of controlled laboratory work'; the pair, 'both trained in the physical sciences, consciously excluded geology from major gifts . . . declaring it insufficiently "fundamental"'.[57] The Rockefeller Foundation, however, did fund geophysics. Behind this support was, note historians of geophysics Naomi Oreskes and Ron Doel, a high cultural value accorded to laboratory values – 'exactitude, precision and control' compared to the values associated with good field science – 'authenticity, accuracy, and completeness'.[58] They add, thirdly, that the sheer scale of the United States, in which geologically interesting sites are remote and widely scattered, might have contributed to this American prominence: since it was difficult to inspect sites directly, as a geologist might, laboratory approaches were an appealing alternative.

Sometimes the history of continental drift is summarized as follows: Wegener's theory was all very well, but there was no physical mechanism known in his day that could produce the effect of continental drift; in the 1950s, new geophysical investigations revealed not only new evidence for the theory – renamed plate tectonics – but also provided realistic mechanisms. Wegener, however, while admitting the problem of the mechanism, suggested several possible ones: 'flight from the poles, tidal forces, meridional forces, processional forces, polar wandering, or some combination of them'.[59] Indeed, many mechanisms were proposed. In the 1920s, Arthur Holmes, a British geologist who had worked in the oil business in Burma before returning to academia, for example, suggested large-scale convection. All these potential mechanisms were discussed in the interwar decades. Nevertheless, all of Wegener's evidence was disputed, while geophysicists such as Harold Jefferys, in *The Earth* (1924), and the American James Macelwane argued that the rigid earth ruled out most mechanisms.[60]

The fate of Wegener's theory would not be settled until after the Second World War, when the relevance of geophysical evidence was reassessed, and is therefore properly discussed in a later chapter. Nevertheless, there is an interesting and sharp divide in the reception of Wegener's theory in the 1920s between scepticism among academic geologists and a commercially inspired interest piqued among petroleum geologists. If the theory of continental drift was borne out, then it would provide practical clues to the location of oil deposits. Oil in West Africa, for example, might suggest the worth of examining the coast of Brazil. Willem A. J. M. Waterschoot van der Gracht, Dutch geologist and vice-president of the Marland Oil Company, was interested enough to organize the 1926 meeting of the American Association of Petroleum Geologists in New York.[61] It was the 'most famous symposium ever held on continental drift', evidence that Wegener's theory was taken seriously.[62]

Sciences *of* the working world of industry: Taylorism and other human sciences

We have seen in the foundations how industrial wealth translated into support for American science, and we have seen how scientists could be employed as experts by industry. However, industry itself could become the subject of scientific inquiry. Historian Anson Rabinbach has shown that there were two different sciences of the human body in industry before the First World War.[63] A European science of work posed as disinterested research. French and German scientists made laboratory measurements of the human body, which they theorized as being like a motor. They offered their conclusions for public consumption, assuming that their implications for shorter working weeks, accident reduction, efficient training and production would be used by the state in framing legislation. But the European science of work made far less impact than a second science of the workplace that moved much closer to the working world of industry: Taylorism.

While definitely a cut above his blue-collar station, Frederick Winslow Taylor had begun work as a foreman in a Philadelphia steel company in the late 1870s.[64] Rising up the management hierarchy, he set out to end what he saw as 'soldiering': the deliberate slow labouring of skilled workers. In contrast, the workers saw it as their right to choose to use their skill at a pace that suited them.[65] Taylor set out to find the 'one best way' of conducting an industrial task, including the best speed and organization of labour, by making the tasks the subject of measurement and scientific scrutiny. His book, *Principles of Scientific Management*, as well as discussions of his ideas at a series

180

of public inquests into railway pricing, both of 1911, publicized the name for his science of industry. Others had sought ways of making management a system, but Taylor's 'scientific management', or simply 'Taylorism', gave it labels and a set of classic exemplars. The key measuring technology was the stopwatch. 'Timing was not a new practice', notes historian Thomas Hughes,

> but Taylor did not simply time the way men worked: he broke down complex sequences of motions into what he believed to be the elementary ones and then timed these as performed by workers whom he considered efficient in their movements. Having done this analysis, he synthesized the efficiently executed component motions into a new set of complex sequences that he insisted must become the norm ... The result was a detailed set of instructions for the worker and a determination of time required for the work to be efficiently performed. This determined the piecework rate; bonuses were to be paid for faster work, penalties for slower. He thus denied the individual worker the freedom to use his body and his tools as he chose.[66]

Taylorism was a controversial intervention into industrial relations. It generated enthusiasm and hatred in equal parts during its years of introduction – provoking, for example, a famous strike at the Watertown Arsenal in 1911.[67] Historians have debated and contested its influence ever since. However, the importance of Taylorism to a history of science in the twentieth century is threefold. First, it exemplifies how the study of the working world created the spaces, techniques and data for new science. The extraction of skill achieved by the measurement of workers' movements, described by Hughes above, was also a hard-won transfer of knowledge: from the embodied skills of the worker to the explicit knowledge of the planning department. In process, argues Harry Braverman, work was degraded, deskilled.[68] In turn, the scientific managers of the planning department were enskilled, empowered. A distinctive science of work was called into existence: compiling measurements of work, studying the simplified representations of work as abstracted in the planning department, reorganizing work on the basis of this analysis and synthesis. But it was also a familiar pattern for modern science: science of the working world.

Furthermore, scientific management was promoted as a means of escaping from class conflict. By insisting that attention be devoted to efficiency rather than the division of the surplus, Taylor and his followers argued that the wealth of increased production would benefit both workers and owners. In language that anticipates late twentieth-century theories of the knowledge society, industrial strife would wane as knowledge was gathered and wielded by neutral experts. 'What we need', urged Taylor's disciple, Henry Gantt, are 'not more laws, but more facts, and the whole question [of industrial relations] will solve itself'; Gantt again: 'The era of force must give way to the era of knowledge.'[69]

After the First World War, Taylorism and scientific management more broadly were exported and reinterpreted away from their American heartlands. 'Whereas in America the commitment to technological efficiency and productivity pervaded almost the entire culture, in Europe it appeared more selectively', writes historian Charles Maier, adding: 'It is noteworthy that the ideological breakdown between the enthusiasts and the indifferent or hostile did not follow any simple left-to-right alignment . . . [scientific management] appealed to the newer, more syncretic, and sometimes more extreme currents of European politics.'[70] These currents included Italian fascists, German proto-Nazis, liberals, industrial planners such as Walther Rathenau, and leaders of the young Soviet Union.

Scientific management was part of a broader early twentieth-century interest in human engineering, which shaped and encouraged several associated sciences. 'Human engineering' became common currency in the 1910s.[71] Its twin supports were Taylorism and engineering-styled biology, most associated with the German-born biologist Jacques Loeb, who worked at the research universities of Chicago and California and the Rockefeller Institute for Medical Research.[72] Robert Yerkes pursued a human engineering agenda when, with support from the Rockefeller Foundation, he investigated the psychobiology of 'personality' in primates as a step towards the management of personality in human workplaces; likewise, John B. Watson's behaviourism, as 'a new psychology whose "theoretical goal [was] . . . the prediction and control of behaviour"', was a science of the working world of management and social control.[73]

Even where the 'crude human engineering' of Taylor was criticized, the response was to develop supposedly more sophisticated forms of human science, such as the physiological studies of work and fatigue in Europe or the psychotechnics of Hugo Münsterberg at Harvard University.[74] Münsterberg set out his ideas in *Psychology and Industrial Efficiency* in 1913. These sciences had new dedicated sites of research, such as the Kaiser Wilhelm Institute for the Physiology of Work in Berlin and the Fatigue Laboratory at Harvard.[75] Industrial psychology burgeoned as a handmaid of personnel management. In the United States, James McKeen Cattell led the establishment in 1921 of a Psychological Corporation that channelled academic psychology and psychologists into the working worlds of business and government. University programmes boomed. Japan built an Institute for Science of Labour; in Britain a National Institute for Industrial Psychology was set up in the same year as the Psychological Corporation; the Soviet Union had one a year earlier; while in Austria a special train, staffed with psychotechnicians, trundled the tracks testing personnel.[76]

One of the most celebrated – and contested – scientific studies of the working world was the so-called Hawthorne experiments, which ran at the plant of Western Electric in Hawthorne, Chicago, between 1924 and 1933.[77] The workplace was treated as a laboratory, and workers' performance and 'attitudes' measured by psychologists as the working environment, such as levels of lighting, were varied. Its enthusiastic interpreters, such as Elton Mayo at Harvard, argued that the experiments demonstrated the practical and extraordinary contribution industrial psychology made to management. In particular they showed that expert attention to human relations increased productivity. Mayo wrote so in books such as *Human Problems of an Industrial Civilization* of 1933.

The new sciences provided opportunities for women as well as men. Lillian Gilbreth, a doctor of psychology, offered, with her husband, Frank, a form of scientific management that relied on the new tool of the motion-picture camera to conduct time-and-motion studies.[78] As the 'workplaces of rapidly industrializing Europe and America became a new area of interaction between workers, business, the state and experts', so other research areas were encouraged in the last decade of the nineteenth century.[79] Alice Hamilton, who had trained in medical science in Germany and the United States, was a pioneer of the science of 'industrial hygiene'.[80] In 1908 she was appointed overseer of a survey of occupational diseases in the state of Illinois. The survey was a crucial tool of her later work for the federal government in the 1910s and, from 1919, as the first female professor at Harvard University. The surveys established the toxic effects of many substances being used or produced in the science-based industries of the late nineteenth and early twentieth centuries: aniline dyes, radium (in wristwatch dials, among other uses), benzene, the chemicals of the new storage batteries, and the gases – carbon disulphide and hydrogen sulphide – that were by-products of the manufacture of rayon. Her most famous campaign was against lead, in the form of white lead (used for paints) and as tetraethyl lead (used as a petroleum additive). Subsequent changes in practice saved many lives.

The values of American society

While the vogue for efficiency may have 'pervaded the entire culture' in the United States, and was stronger compared to Europe, there were still tensions that could make experts uncomfortable.[81] Sinclair Lewis's novel *Arrowsmith* of 1925 followed the titular hero as he trained in medical science, passing in his journey through all the working sites of American biological and medical research: the medical school, the private medical practice and the medical research institute (closely modelled on the Rockefeller). All have their distractions, which pull our hero away from the dedicated research that brings rewards to humanity. Loeb appears in the guise of Arrowsmith's tutor, Max Gottlieb, distracted by the demands of teaching. In *Arrowsmith*, Sinclair Lewis portrays the American working world as an obstacle to productive science. The opposite was in fact the case.

Furthermore, there was a world of American society that was set against the metropolitan and organized values of science. This was the message of the Scopes trial of 1925. The state of Tennessee had a law, known as the Butler Act, that prohibited the teaching of 'any theory' that went against the biblical story of creation. A teacher, John Scopes, in full knowledge that he was breaching the law, taught Darwin's theory of evolution by natural selection. His trial was prosecuted by the 'Great Commoner', the politician William Jennings Bryan. Scopes was defended by Clarence Darrow, 'who believed in Science like Bryan believed in Christ, [and who] dubbed his opponent "the idol of all morondom"'.[82] (Note use of the eugenic – that is to say, expert – category of 'moron' in this jibe.) Scopes was found guilty, which surprised no one.

However, this was less a trial of science versus religion than a battleground carefully chosen. On one side were 'ordinary Americans, angry that their most cherished beliefs were being undermined with their own tax dollars', who 'resented an educational establishment that made beasts of men, taught human inequality, put eugenics in school textbooks, and portrayed life as a godless bloody struggle',[83] while on the other was an organized campaign determined to confront anti-expert opposition. Scopes was bankrolled by the National Civil Liberties Union. More significantly for us, the Scopes trial marks a milestone in organized science journalism: the new Science Service reported it, photographer-journalists pictured it, and a journalist, H. L. Mencken, gave it a notorious framing label: the 'monkey trial'.[84]

The Scopes trial, rather than being a sign of Darwinian science in

184

danger, in fact came at the beginning of a period in which Darwinian explanations rolled back with ever greater force. To understand this global transition we need to visit the land of ideological polar opposites to the United States, the Soviet Union.

— 9 —

REVOLUTIONS AND MATERIALISM

Science came to Russia in the eighteenth century, when Peter the Great encouraged the import of Western expertise. Since then the Russian state has had a tempestuous relationship with scientific and managerial experts, and Russian history is marked by a pattern of alternating enthusiasm and suspicion. Periods within which experts advanced include the enlightenment regimes of Peter and Catherine, the nineteenth-century tsars Alexander I and Alexander II, the post-war, pre-Stalin 1920s and the glasnost 1980s. Among periods of reaction are the reign of the nineteenth-century tsars Nicholas I and Nicholas II and the years of Stalinist terror. Nevertheless, the story of Russian science is far more complex and more interesting than this first description suggests. One second-order generalization that can be made, for example, is that Russian science has flourished during political oppression where the flow of resources has been strong, but has withered in times of relative freedom when the money has dried up.[1]

Science in Russia is one of the topics that make history of twen-tieth-century science so special. While starting from similar initial conditions, with the import of science from Western Europe in the eighteenth century, the subsequent enormous changes in political and social life make Russia an ideal laboratory for scrutinizing the rela-tionships between science and the working world of the state. Indeed, this very way of asking historical questions has some of its roots in Russia. History of science was being proposed as a discipline as early as the 1890s, by the great systematizer Vladimir Vernadsky. It received state support in the 1920s and was purged in the 1930s, but not before articulating a way of seeing science as intimately connected to a broader external world of work, money, power and technology.

186

After the purges, Soviet history of science was narrow and apolitical, and understandably timid. The history of twentieth-century Soviet and Russian science that I draw upon here is the Cold War work of American historians, especially the eminent Loren Graham, and that of the post-Cold War generation of historians who have accessed closed archives and told new stories.[2]

Science, expertise and the Soviet state

Scientists and engineers, in overwhelming numbers, supported the first revolution of 1917, which brought in Western-style liberal government in February, and were not at all keen on the Bolsheviks who seized power in October.[3] The new Soviet government inherited a scientific establishment that had strengths in some areas – mathematics, soil science, astronomy, physiology, embryology and structural chemistry – and weaknesses in others – including crucial ones such as industrial research and a poor university science structure.[4] The most important, and most conservative, institution was the Academy of Sciences, which still retained the centralized leadership in scientific matters which in other countries had been ceded to a diverse range of bodies. The academy was filled with people who had built their status and influence in the tsarist years, and might have been expected to have been an immediate target for revolutionary levelling. It is, however, as Graham notes, a seeming paradox that the academy and the Soviet state soon found ways to work together.

We now know that, between the revolution and the end of the civil war, the period of highest revolutionary fervour, the future of the academy was up for grabs. Three proposals circulated.[5] The radical peasant revolutionaries of the Northern Communes argued that the academy should be completely reoriented in a move to 'win science for the proletariat': directed science would be harnessed closely to the needs of the revolutionary state, an application that would replace an alleged 'fetishization of pure science', while teaching rather than research would be prioritized. Second, the Commissariat of Education proposed a Soviet solution: scientists would be organized into regional and disciplinary units that would be the base of a pyramid, with a new 'Association of Science', under firm Commissariat control, at the top. Autonomy, under this proposal, would end. Finally, academicians, in particular the permanent secretary of the academy, S. F. Oldenburg, and the geologist A. E. Fersman, were not willing to wait passively to be reformed out of existence. They offered their alternative plan: the academy would be kept and would work for the state, channelling funds for example, but would not cede control to the planners. (Some) autonomy would be rescued. An alternative communist academy would also be established.

In fact none of the three proposals was adopted. There were several reasons why not. First, during the desperate years of war communism, the fate of the academy was just not a priority; then, as the Soviet state began to rebuild from 1921 in the years of the New Economic Plan, the state was too dependent on expertise to chance reform. 'Pressed by the military situation', summarizes Graham, 'the Soviet leaders turned to experts who were often not politically compatible with the new order but who were nonetheless extremely competent.'[6] In autumn 1922, these tensions shaped a 'great debate' in the pages of *Pravda*. At stake was whether Soviet Russia should be run by the proletariat as a worker's state, or whether it should be managed by experts as a different kind of socialism. The proletarian case was put by Vladimir Pletnev, who argued that the generation of scientists, engineers and managers inherited by the Soviet Union must be cast aside for a new generation of proletarian specialists, who would forge a new science and engineering. With full support of her husband (Lenin), Nadezhda Konstantinovna Krupskaya dismissed Pletnev's case: science was the accumulation of facts that must be kept. In her support, Iakov Iakovlev ridiculed Pletnev as irresponsible, insisting that the young Soviet state must learn from, not dictate to, the experts: 'The very existence of Soviet Power is a question of studying at the feet of the professor, the engineer, and the public school teacher.'[7]

Indeed, many specialists did overcome their suspicions to work closely with the Soviet state. Historian James Andrews argues that science popularizers did well by working with the grain of state campaigns.[8] Graham has given us a detailed case study of the tension between expert and state in his account of the extraordinary life of the engineer Peter Akimovich Palchinsky.[9] Born in Kazan, Ukraine, to 'a socially prominent but impecunious noble family', Palchinsky received an excellent training as a mining engineer in St Petersburg in the 1890s. A 1901 study of coal production in the Don basin brought home how necessary it was to analyse engineering projects in their full social dimensions. When, appalled by mine owners' ignorance of miners' living conditions, Palchinsky gathered and published statistics on the matter, he was summarily removed from the investigation and sent to Siberia.

In internal exile, Palchinsky was radicalized, attracted to socialism and Kropotkin's mutual aid anarchism. He eventually escaped imperial Russia to Europe, and for five years, from 1908 to 1913, he pursued a successful career as an industrial consultant. Palchinsky's expertise widened and deepened. He published a four-volume study of the ports of Europe. After returning to Russia in 1913, he set up an Institute of the Surface and Depths of the Earth. He was, by now, a comfortable member of the progressive, democratic wing of Russian socialism, and a strong supporter of the Provisional government established by the February revolution of 1917. Indeed, Palchinsky played a role in the October revolution: he was one of three officials who was charged with the organization of the defence of the Winter Palace. He ordered the remaining guards not to shoot. After spending much of the next year or so in jail, Palchinsky was released and chose to volunteer to work for the Soviet planning agencies. He consulted on the Dnieper River dam, and later rose to be a permanent member of Gosplan, the state planning commission. The Soviet Union needed engineers. Palchinsky, like many other experts, chose to contribute.

The debate on the tapping of tsarist expertise raised the question of whether different political conditions demanded different kinds of knowledge. Interestingly, both sides missed the point. The proletarian movement looked forward to a future proletarian science, different in kind but not yet arrived. Scientists, engineers and doctors were harassed for being identified with the old – 'specialist-baiting' was widespread.[10] 'The assumption in each case', writes Loren Graham, 'was that the old intellectuals were imparting, along with some necessary knowledge, a large amount of dangerous ideological baggage. And not until the old intellectuals had been removed from positions of influence would this danger pass.'[11] Their critics held that science was science, and could – indeed must – be separated from its political cocoon. However, the experiment of communism was already creating a context within which new forms of science were being articulated. Arguably, the distinct Soviet science had already arrived.

New Soviet science?

Graham again is my main source for the history of this new science. His argument has two stages: first, that the Russian revolution was distinct in having a philosophy of science, dialectical materialism, at its heart and, second, that dialectical materialism was, for some, at least in the 1920s, a useful tool to think with, an 'innovative option, not . . . a scholarly dogma'.[12] Dialectical materialism held that 'all

189

nature could be explained in terms of matter and energy', nothing else. Nor was there anything else that needed explanation. What made the philosophy more than just materialism was its opposition to reductionism.[13] Rather than reductively explain society in terms of biology, biology in terms of chemistry, and chemistry in terms of physics, each level had its own laws.

Dialectical materialism took its arguments from the Marxist theory of Engels (his *Anti-Dühring* of the 1870s) and Lenin (the 1908 book *Materialism and Empirio-Criticism*). Yet there was still plenty of wriggle room. Two opposed groups in the Soviet Union, the Deborinites and the Mechanists, clashed over their interpretation of Einstein's relativity theory and quantum mechanics.[14] Both claimed to be Marxist and to flow from dialectical materialism. Relativity was flawed, if not fatally flawed, by Einstein's acknowledged debt to the Austrian philosopher Ernst Mach, singled out for criticism by Lenin.[15] Relativity also looked suspect to the Mechanists because, at least in the hands of Einstein's English interpreters, Arthur Eddington and James Jeans, it offered exceptions to materialism.[16] Quantum mechanics, accused the Mechanists, rejected causality, and therefore could not provide the fully deterministic accounts demanded by materialism. Furthermore, the decisive role that quantum mechanics seemed to allocate to the observer smacked of idealism. In the mid-1920s intellectual and social prestige was at stake in these debates, but within a few years the clash would have lethal repercussions.

Meanwhile, however, science was being shaped by the early Soviet system. Some, such as Vygotsky's psychology or Oparin's biology, were pulled into being; others, such as Vernadsky's biogeochemistry or Vavilov's plant genetics, were pushed in interesting and often conflicting ways by Soviet demands. Overall, the result was a temporary 'symbiosis between Russian science and the Bolshevik state'.[17] The science produced would be significant on a world stage, either in the short term – such as Oparin's influence on the Englishman J. B. S. Haldane or the role that Dobzhansky's transplanted Russian genetics would have in the evolutionary synthesis – or in the long term: Vygotsky's Western influence, for example, peaked in the 1980s and 1990s, while Vernadsky's work was also rediscovered after the end of the Cold War.

Vladimir Ivanovich Vernadsky trained as a geologist, specializing in mineralogy, in the decades before the revolution, and was able to travel to the West – to Italy, Germany and Switzerland – and therefore be shaped within the wider scientific community of the late nineteenth century.[18] In the last tsarist years Vernadsky served the state on the Commission for the Study of Natural Resources, locating and surveying essential raw materials and advising on matters including chemical warfare.[19] In Switzerland, Vernadsky had encountered the geologist Edward Suess, who had proposed the term 'biosphere'. But there was also a source closer to home for Vernadsky's development of this idea of the interrelationship of living and non-living earth systems. His teacher, the great soil scientist V. V. Dokuchaev, had argued in 1898 that 'soil science, taken in a Russian sense, must be placed in the center of a theory of the interrelationships between living beings and inanimate matter, between man and the rest of the world, i.e. its organic and mineral part.'[20] From this Russian soil, Vernadsky developed a vast account of the how the earth had passed from a 'geosphere', dominated by inanimate geological processes, through a 'biosphere', where living and non-living forces shaped each other, through to a 'noosphere', where human thought and action must be added into the picture of interactions between the living and non-living. Furthermore, there was a direct working world inspiration for these arguments: it was, notes historian of global warming Spencer Weart, from 'his work mobilizing industrial production for the First World War' that Vernadsky 'recognized that the volume of materials produced by human industry was approaching geological proportions', reaching the conclusion, profound for our understanding of the earth's history, present and future, that the oxygen, nitrogen and carbon dioxide in the atmosphere were 'put there largely by living creatures'.[21] Vernadsky wrote these arguments up in the early 1920s, *The Biosphere* appearing in 1926.

Vernadsky's writings would be revived, long after his death, as if he had been a proto-New Age speculator on the interconnections of all things, just as ecology would be too. In fact, as discussed in chapter 7, the first decades of the twentieth century witnessed a flourishing of conservationist argument justified by appeal to a scientific ecology.[22] In the 1890s and 1900s, the catastrophic collapse in animal populations on the steppe – such as the sable and the tarpan (a wild horse) – had prompted the establishment of many conservation societies. By 1908, Gregorii Aleksandrovich Kozhevnikov, director of the Zoological Museum and professor of invertebrate zoology at Moscow University, was arguing that 'only a scientific study of nature could provide a firm basis' for conservation. Etalons, pristine reserves, must be set up as standards against which environmental degradation could be measured. The new provisional government of 1917 expressed itself in 'complete sympathy' for these proposals for a pivotal scientific ecology. And, even though peasant demands for land, accompanied by their devastation of princely estates, including virgin steppe, pulled at Soviet interests, scientific ecology flourished under Lenin. The key figure was Vladimir Vladimirovich Stanchinskii, whose students produced studies of the biocoenosis of the South Ukrainian steppe. Describing in detail the energetics of the food chains of the steppe, Stanchinskii was able to contribute to the incipient ecosystem debate.

Psychology was another intellectually creative science of the early Soviet years. Lev Semenovich Vygotsky was born in 1896, graduated in the year of revolution, 1917, and spent his working life in Moscow. He died relatively young, aged thirty-eight, of tuberculosis, which he had caught from his brother. His work on how children develop language argued against the views of Jean Piaget. This Swiss psychologist, working at the Jean-Jacques Rousseau Institute in Geneva from the early 1920s, had proposed a theory of child development as being a series of fixed stages, marking out how the child moves from egocentrism, through magical thinking, concrete thinking and eventually (after age eleven) abstract thinking. The results, published in French as *Le Langage et la pensée chez l'enfant* (1923) and *Le Jugement moral chez l'enfant* (1932), were widely circulated and translated. Piaget saw them as being part of an even grander project, 'genetic epistemology', an attempt to 'make the content of a science, evolutionary and developmental psychology, the same as the conditions of possibility for knowledge'.[23]

Vygotsky also pursued child psychology within a wider framework: his research and writing aimed to create a psychology that was in line with Marxist thought, especially dialectical materialism.[24] One specific quarrel with Piaget was about what happens when a child begins to link words to things. Whereas Piaget posited that a child developed thought first in a private 'autistic' phase, followed by egocentric thought, and only then became socialized, Vygotsky reversed the influence: 'The primary function of speech . . . is communication, social contact. The earliest speech of a child is therefore essentially social'; social speech was then internalized as thought: 'Inner speech is something brought in from the outside along with socialization.'[25] Furthermore, as Vygotsky wrote:

> The nature of the development itself changes, from biological to socio-historical. Verbal thought is not an innate form of behaviour but is determined by a historical-cultural process and has specific properties and laws that cannot be found in the natural forms of thought and speech. Once we acknowledge the historical character of verbal thought, we must consider it subject to all the premises of historical

192

materialism, which are valid for an historical phenomenon in human society.[26]

So, while a child's pre-linguistic thought might have biological roots, as soon as it intersected with language, a social and historical entity, the two – thought and language – developed dialectically. Marxist philosophy of science – dialectical materialism – was here a positive prompt for new science, a new social psychology. It was the basis of a very productive research programme, on areas as broad as pedagogy, 'defectology', informed criticism of IQ tests, and animal and 'primitive' psychologies. However, when Vygotsky's most influential work, the 1934 book *Thought and Language*, was first translated into English in 1962, his publishers systematically stripped out references to Marxism,[27] probably in order to make the text more palatable for a Cold War American readership.

Origins of life

There is still considerable nervousness in discussions of the significance of Marxism as a context for the scientific study of the material origins of life which was reinvigorated in the 1920s by the Soviet biochemist Aleksandr Ivanovich Oparin and the Englishman J. B. S. Haldane. Both were communists. Michael Ruse, not usually a historian to mince words, protests too much:

> In fact, in the 1920s, neither Oparin nor Haldane was yet a Marxist, so the strongest possible connections simply are not there. And when in the 1930s, Oparin did start to put things in Marxist terms . . . much that he said could be translated at once into nontheoretical language.[28]

Ruse seems very relieved to have 'disposed of this ideological red herring', noting that the supposed connection is 'a great favourite of the evangelical Christian opponents to any scientific approach to the origins of life'.[29] Yet this concedes too much. It damages good history to avert our eyes – merely for sake of drawing the sting of fundamentalist argument – from the fact that the two scientists who proposed that a material account of the appearance of life was possible were also politically of the left. 'Communism in Russia in the twenties', wrote Graham, 'provided the kind of atmosphere in which the posing of a materialistic answer to the question What is life? seemed natural.'[30] Haldane, a 'lukewarm' Labour Party supporter at the time of his origins of life work (1929), was already reading Marxist literature in Cambridge, even though he did not describe himself as a convert until 1937–8.[31] Haldane argued that life could be built from within a 'dilute soup' of sugars and simple organic molecules. Oparin argued that the movement from a non-living world to one with life must have taken place in stages, each with its own scientific laws. The similarity between this model and dialectical materialism is no accident, insists Graham. 'Dialectical materialism heavily influenced the very structure of his analysis and the organizational schemes of his books', although it became more evident through the newer editions.[32] (Oparin lived a long and productive life: important versions of his *Origins of Life* appeared in 1924, 1938 and 1966.)

There was a playful materialism in Haldane's science long before he committed himself to Marxism. Early in the spring term 1923, Haldane read a paper to the Heretics, a university society. Subtitled 'Science and the future', 'Daedalus' was a work of 'serious prophecy', but also a provocative review of where inventions come from and of how science challenges morals, ethics and politics. It was published as a slim volume within the year.[33] Haldane revelled in the disgust he anticipated his predictions would inspire, reminding his readers that, objectively, milk and beer would once have provoked the same response. He claimed that a seismic shift had occurred: from physics and chemistry to biology as the 'centre of scientific interest'. Nevertheless, from the old physics and chemistry would come new inventions: lighting so cheap 'there will be no more night in our cities', near instant communication between any two persons on the planet, 'large-scale production of perfume' that might re-educate our sense of smell, and a pollution-free power supply system of windmills and hydrogen storage. Decentralization of industry would follow. Industrial strife would be solved because capitalists would pragmatically hand over control to workers, since continuity of production was worth more than mere profits. We would be fed, before 2143, by a combination of sugars from cellulose and artificial protein made from coal or atmospheric nitrogen. Urbanization would be complete.

Yet it was, wrote Haldane, the 'biological inventions' that mattered more and would create the greatest shocks. An essay within the essay, written, archly, in the style of a 'rather stupid undergraduate member . . . to his supervisor during his first term 150 years hence', projected the development of experimental biology producing the 'first ectogenetic child' in 1951. The fictional scientists Dupont and Schwarz 'obtained a fresh ovary from a woman who was the victim of an aeroplane accident, and kept it living in their medium for five years'. Such ectogenesis, the undergraduate told his tutor, was 'now universal', despite opposition from the Catholic Church. Growing designer babies would have obvious eugenic benefits: 'from the increased output of first-class music to the decreased convictions for theft'. Haldane's pose was that traditional eugenics was not wrong; rather it was unimaginative and inefficient, a poor means to achieve worthy ends. Again profound social effects were imagined to spring from such inventions: with 'reproduction . . . completely separated from sexual love mankind will be free in an altogether new sense', while new medical science would 'control our passions', create 'new vices', manage cancer, and 'make death a physiological event like sleep'.

Haldane's prophecies had one source in the excitement around experimental biology and biomedicine. Techniques were being developed that seemed to fulfil Jacques Loeb's early twentieth-century vision of the engineering of life. One example of the manipulability of life came from animal genetics research targeting cancer. Tumours were transplanted between mice, while mice in turn were inbred to produce strains of identical mice which could be used as regular experimental systems.[34] One outcome was the conviction that cancer had a genetic basis. Another was the export of similar and manageable forms of cancer from laboratory to laboratory – as tumour cell-lines and inbred mice strains. Another series of exciting experiments that seemed to speak of controllable, laboratory-based reproduction was the embryology of Hans Spemann, discussed earlier, in which chemical 'organizers' were found to guide development.

Daedalus prompted two immediate essays in response: *Icarus* (1924) by Bertrand Russell and *Another Future for Poetry* (1926) by Robert Graves.[35] *Daedalus* also echoed through the fiction of his friends, the brothers Julian and Aldous Huxley.[36] Haldane was the model for the protagonist in Aldous Huxley's *Antic Hay* (1923). Haldane's synthetic food, industrialized childbirth and technocratic eugenic engineering all appeared in the same author's *Brave New World* of 1932. In this eugenic future, sex is separated from reproduction, babies are grown in bottles, and the link with science, engineering and politics of the 1920s is repeatedly emphasized by the names of the characters – Bernard Marx (after Claude and Karl), Helmholtz Watson (after Hermann and John B.), Lenina Crowne, and Mustapha Mond (Alfred Mond was a British chemical magnate and recalls Haldane's fictional Dupont). Eugenic castes are labelled like student essays – alpha, beta, gamma, delta, epsilon, plus or minus. In *Brave New World* the consequences of biological invention were normalized – the reader and the misfit protagonists might find them disgusting, but society has adapted and, in its own way, is happy. It is a utopia. As Haldane wrote, biological inventions tend 'to begin as a perversion and end as a ritual supported by unquestioned beliefs and prejudices'.[37]

J. B. S. Haldane is a fascinating figure to follow into the 1930s and beyond because his political and scientific turns illustrate two utterly different ways of continuing a revolution. In the 1930s he would become a committed Marxist, even as Russian communism had mutated into blood-soaked Stalinism. And he would become a prominent figure defending the wide application of Darwinism and Mendelism in the life sciences at the same time as, in Russia, Stalin's patronage was encouraging Lysenko and the purging of Lysenko's rivals. At stake, therefore, were the Darwinian, Mendelian and Russian revolutions.

At the beginning of the twentieth century only a handful of scientists could be described as unadulterated Darwinians.[38] While evolution was widely accepted as fact, natural selection had many, more widely held rivals. Furthermore, Mendelian genetics seemed incompatible with Darwin's description of continuous variation. Nor did other specialists, such as paleontologists, see smooth change among the evidence. Yet in a historical process beginning in the mid-1930s and reaching clarity, if not complete closure, in Princeton at a conference in 1947, Darwinian theory was successfully integrated with Mendelian genetics in a way that accounted, to many scientists' satisfaction, for vast realms of life. This process was labelled the modern evolutionary synthesis.[39] It was an international achievement, albeit in distinctively different regional ways, and one in which the export – perhaps expulsion – of Soviet ideas was catalytic.

The purges

Before we turn to the story of this 'evolutionary synthesis', we need to trace the convulsive and brutal events in the Soviet Union that were fateful for Russian genetics, as they were too for other sciences and expert specialisms. Genetics was one of the hotspots of Soviet science, although its success was achieved perhaps despite, as much as because of, Soviet support. While Graham has provided evidence for how Marxist philosophy of science framed interesting new science, other historians stress the continued autonomy of science, despite the upheavals of the revolution. The patrons of the burgeoning array of research institutes, writes Krementsov, 'rarely interfered in the direction, content, or duration of research, the choice of personnel or equipment, or the structure of institutions; these were largely defined by the scientists themselves ... [The] state's influence on scientific work itself was minimal.'[40] The occasionally generous finance and the minimal interference in research (but not science teaching, where intervention was 'stern and aggressive' to prepare a future generation of truly communist science) encouraged lead researchers to stay and prosper. Such leaders included Nikolai Kolt'sov, Sergei Chetverikov, and plant geneticist Nikolai Ivanovich Vavilov.

Vavilov (and his brother, the physicist Sergey) was born into a 'wealthy merchant family, was urbane and well educated, spoke many languages, and dressed in the tie and starched collar of the old Russian professoriate'.[41] He graduated from Moscow Agricultural Institute and travelled Europe, studying with William Bateson in 1914. As this trajectory suggests, Vavilov's science combined practical agricultural research with theoretical sophistication. In the 1920s, his Institute of Applied Biology became the centre for collection and comparison of the genetic diversity of crop plants extracted from many sites across the world. In *The Centres of Origin of Cultivated Plants* (1926), Vavilov argued that the spots of greatest genetic diversity marked the places of origin of crops. This conclusion had consequences for both understanding human past and plant breeding in the future. The 'promise of practical applications', notes historian Eugene Cittadino of what I call the working world, 'often led to fundamental insights regarding the nature of inheritance and the process of evolution.'[42]

In the early 1920s, Vavilov encountered a youthful practical agronomist, one Trofim Denisovich Lysenko, and encouraged the young man in his career. Working first in Ukraine and Azerbaijan, Vavilov's would-be protégé struggled to grow peas and wheat through the harsh winters. Sometimes he succeeded, and when he did *Pravda* recorded and praised his work. The careers of both men would be transformed – one surging ahead, the other ended – by the rise of Joseph Stalin.

Stalin's grip on power strengthened in the late 1920s until he was indisputably in charge. When the dissident biologist Zhores Medvedev wrote his history of Soviet science, he cast the period between 1922 and 1928 as the 'golden years'.[43] The dark decades of Stalinism provided the historiographical contrast. The Soviet Union was changing as four interrelated policies were pursued. First, the food supply would be modernized by the new policy of collectivization: old estates were broken up and replaced with collective farms, filled with displaced peasants and armed, in theory, with new machinery. Peasants hated the policy, choosing to burn farms and slaughter farm animals rather than hand over their patches of land. For collectivization to work, however, engineers and agricultural scientists would need to be mobilized. It therefore opened up the possibility of patronage to someone who could convince the party patrons that they were agricultural experts. Second, industry was to be driven forward by a series of Five Year Plans. The first plan was to run from 1928 to 1932. Decision-making was centralized. The rapid expansion of heavy industry was prioritized, but success depended on how expertise in production was deployed. Gigantic wasteful projects were pursued, such as the city of Magnitogorsk, built from scratch and producing more steel than Britain, but with no nearby coal, transport or labour, or the White Sea Canal, dug almost by hand by prison labour in double-quick time, but which was too shallow for ships.[44]

Centralization of authority also impacted on the organization of science. Despite managing to maintain some degree of prickly autonomy from the state, the academy, the premier scientific organization in the Soviet Union as it was in Russia, was attacked and overrun by the Communist Party. Some scientists, such as Vernadsky, resisted this move. Others who protested became targets. For, third, the pace of changes and the numerous tensions – between peasant and state, between classes, between party power and expert autonomy – exploded into wave upon wave of denunciations and purges. Scientists and engineers, as we shall see, suffered. Finally, the state philosophy of science – dialectical materialism – became less a flexible prompt for new ideas and much more a 'calcified and dogmatic system', a suffocating language, 'that hobbled science'.[45]

In 1928, at the Eighth Congress of the Union of Communist Youth, Stalin spoke:

> A fortress stands before us. This fortress is called science, with its numerous fields of knowledge. We must seize this fortress at any cost. Young people must seize this fortress, if they want to be builders of a new life, if they want truly to replace the old guard . . . A mass attack of the revolutionary youth on science is what we need now, comrades.[46]

Nearly all specialist and scientific disciplines suffered as the Great Break of the late 1920s became the Great Terror of the 1930s. A scientist, discreetly denounced by hidden accusers (often colleagues equally terrified), would be disappeared into a system of show trials followed by internal exile or execution. (Both resulted in death.) Whole scientific cities were established in the gulag archipelago, as Aleksandr Solzhenitsyn testifies in his novels such as *The First Circle* (1968). The Great Break also gave strength to a communist analogue of the Aryan physics movement: Arkady Timiriazev, anti-Semite and professor of physics in Moscow, intrigued against fellow physicists, especially Jewish theoreticians such as Leonid Mandelshtam, Iavov Frenkel and Lev Landau, because of their mathematical approach to physics and their opposition to the ether.[47]

The ripples felt in the West of the political and intellectual storm in the Soviet Union could have unexpectedly large effects, even within the discipline of history of science. In 1931 London hosted the Second International Congress of the History of Science, a joint production of the French Comité international des sciences historiques, the Newcomen Society for the Study of the History of Engineering and Technology, and the History of Science Society. Excitement gathered around the delegation of Soviet historians, which included the powerful Nikolai Bukharin and the historian of physics and Deborinite intellectual Boris Hessen. History of science in the West at this time tended towards an 'internalist' history of ideas, great men and little social context. Hessen's paper, which directly related Isaac Newton's science to the emergence of bourgeois capitalism in England, electrified the audience. Ever since, Hessen's paper has been held up as a model of the so-called externalist mode of analysing past science. At the end of the Cold War, however, a more complex story was told to Loren Graham.[48] Back in Russia, Hessen had been defending modern physics, especially the work of Einstein, from accusations that it was incompatible with Marxist materialism. Hessen had been denounced by cultural revolutionaries in 1930, on the eve of the London conference. He was directly told where the orthodox line lay: Stalin had written that 'technology in the current stage decides everything'. He had been sent to London, Graham found out from his handler, who defected to the West, as a 'test of his political orthodoxy'.[49] Hessen's paper was a high-wire act: by showing that Newton's ideas related to practice (the technology of the age) he demonstrated his loyalty, and he did so while preserving, just, his intellectual integrity. Newton stood in for Einstein. The audience in London heard a paper about the seventeenth century, and had their eyes opened to what history of science could do. The audience in Moscow heard a paper about the present. Hessen passed. He lived for seven more years before he was purged.

About a third of the 10,000 engineers in the Soviet Union would be imprisoned, or worse. Peter Palchinsky was arrested in 1928. He had, in the early and mid-1920s, promoted a broad vision of engineering as a contextual study of industry and had also expressed technocratic sentiments, drafting a letter, for example, that argued that 'science and technology are more important factors in shaping society than communism itself. This century . . . is one not of international communism, but of international technology. We need to recognize not a Komintern but a "Tekhintern".'[50] Such sentiments would have been seen as a direct challenge to the party. It also aligned Palchinsky with technocrats – labelled by Stalin as an Industrial Party – such as Stalin's then rival Nikolai Bukharin. Stalin's view can be deduced from his comment to the visiting H. G. Wells in 1934: 'The engineer, the organizer of production, does not work as he would like to, but as he is ordered . . . It must not be thought that the technical intelligentsia can play an independent role.'[51] Palchinsky was arrested in April 1928, charged with being the leader of an anti-communist conspiracy (the so-called Industrial Party) and convicted of treason. He was executed by firing squad in 1929.

Vygotsky had an even more direct encounter with Stalin. The Soviet leader 'fancied himself an intellectual', notes Graham, 'and commented without hesitation on many scientific thoughts', including, in this case Vygotsky's ideas regarding pre-linguistic thought.[52] Stalin also might have been angered by his own son's failure to pass psychological tests based on Vygotsky's work. Paedology was purged. Vygotsky, however, died of tuberculosis before he could be arrested. In other sciences, the trials and purges were waged by lesser figures. Kozhevnikov and Stanchinskii's policy of preservation and ecological study was attacked by 'young Turks' in 1932. Isai Israilovich Prezent, director of the biological cabinet of the Leningrad branch of the Communist Academy, denounced Stanchinskii's work on ecological energetics as 'bourgeois formalism' and offered instead the General Plan for the Reconstruction of Fauna in the USSR.[53] Stanchinskii was chased out of his job at Khar'kov University, and his publications were smashed (literally smashed – the casts of the printing blocks of his book were seized and broken up).

Prezent's partner in this ideological cleansing was the young agronomist who had been encouraged by Vavilov: Trofim Denisovich Lysenko. As collectivization tore up the countryside and millions succumbed to starvation, Lysenko offered two things: technical fixes and scapegoats.[54] His patrons, on whose support and protection Lysenko depended, were gratefully generous. Lysenko promised several ways by which crops such as wheat could be made to grow earlier and stronger in the cold conditions of the Soviet Union. He called these techniques 'vernalization'. Whether or not they had any effect is hard to tell, because Lysenko gave no decent statistical assessment. His claim in 1937, for example, to have turned a variety of winter wheat into spring wheat depended on just two plants, one of which died during the year-long experiment.[55] Many of Lysenko's techniques assumed some form of acquired characteristics: wheat, for example, could be induced to pass on resistance to cold after experiencing cold and wet conditions. Lysenko therefore had a dual incentive to denounce Mendelian geneticists: he gained by eliminating their rival claims on both patronage and the authority of science. His promise to solve a major problem – food production – convinced his patrons among the agricultural bureaucrats. It is this promised utility, rather than alternative explanations (derivation from Marxism–Leninism or intellectual support for creating a 'New Soviet Man'), that accounts, argues historian David Joravsky, for Lysenko's rise.

In 1935, on the eve of the Great Terror, Lysenko spoke at a conference. 'You know Comrades', he announced, 'wreckers and kulaks are located not only in your collective farms . . . They are just as dangerous, just as resolute, in science . . . And whether he is in the academic world or not in the academic world, a class enemy is always a class enemy.' Stalin, presiding, applauded: 'Bravo, comrade Lysenko, bravo.'[56] Later in the year, Lysenko denounced specified academic 'wreckers' to Stalin's police. Vavilov was named. The geneticist's friends and colleagues began to melt away, some adding denunciations in attempts to save their own skins. Vavilov, his other options closed, went on the offensive, attacking Lysenko's science. In August 1940, Vavilov was arrested and imprisoned. He died of malnutrition three years later.

The period of purges was terrifying but also arbitrary: Lysenko's supporters as well as geneticists could fall victim as denunciations and counter-denunciations flew.[57] Nevertheless, four fates were possible. First, some scientists and engineers, those who were fortunate not to attract the attention of the authorities, or who collaborated, were able to keep their jobs. Second, a select few, with expertise that was crucial to the survival of the Soviet Union in peace and war, received lavish patronage, albeit with the implied threat of punishment if success was not attained. The nuclear scientists and rocket engineers of the 1940s and 1950s are examples of such groups and are examined later. The long career of Lysenko, however, shows that punishment could be deferred almost indefinitely. Third, less useful and less lucky scientists and engineers lost their lives in the gulags. Several cases have been witnessed, but there were many more. Finally, there was the possibility of escape and exile. For the history of twentieth-century science this group is perhaps most interesting. In particular, the achievements and peculiar insights of early Soviet genetics were carried westwards, where they would form the heart of a consolidated revolution in the life sciences.

The fortunes of Soviet genetics

To see this significance, we need to rewind slightly and follow the fortunes of the schools of Soviet genetics. Historian Mark B. Adams notes four factors that shaped their direction.[58] First, the Bolshevik government, for ideological and pragmatic reasons, supported science. It saved from execution, for example, biologist Nikolai Konstantinovich Kolt'sov, who had helped organize White Russian resistance in Moscow. Kolt'sov ran an Institute of Experimental Biology, funded before the revolution mostly by private philanthropy. After the Bolshevik revolution he was funded by the state, including the Academy of Sciences Commission for the Study of Natural Productive Forces of the USSR, the Commissariat of Agriculture and the Commissariat of Public Health. Second, notes Adams, Darwinism was encouraged because it chimed with the prevailing ideology: 'following the revolution, with the advent of an avowedly atheistic, materialistic ideology, religious opposition was largely silenced and Darwinism enjoyed an unparalleled heyday.'[59] Nevertheless, the Darwinism favoured had had a particular Russian spin, with influential pre-revolutionary interpreters such as K. A. Timiriazev rejecting Darwin's stress on the 'struggle for existence'.[60]

In general, though, as Adams neatly contrasts: 'Darwinism tended to be selectively favored as a "materialist" theory of the living world', while 'non-Darwinian or anti-Darwinian tracts were for the first time strongly selected against'.[61] Or rather, as Adams also argues, the third factor, the influence of dialectical materialism as a philosophy of science, encouraged a 'synthesis', for example, as geneticist Kolt'sov's student Alexander Serebrovsky noted, in defence of his subject from ideological attacks, a 'synthesis' of the 'thesis' of Weismann (hard heredity) and the 'antithesis' of Lamarck; in Adams's view, 'the emphasis in the 1920s on dialectics did tend to encourage synthetic enterprises . . . and also the formulation of scientific results in "synthetic" language.'[62]

Finally, in the tough conditions of the 1920s this synthetic work could only be done where there were resources to build research programmes. And genetics was especially well placed: it did not require expensive or complex equipment and it held the promise, through improved plant and animal breeding, of direct application.

Three genetics schools flourished. Vavilov and his Institute of Applied Biology and All-Union Institute of Agricultural Sciences in Leningrad (examined above). Iurii Filipchenko built the second school, also in Leningrad. Finally, Kolt'sov's Institute of Experimental Biology expanded, mostly at its Anikovo genetics station, north-east of Moscow, with a hydrobiological offshoot nearby at Zvenigorod. One cheap, promising line of research was on *Drosophila*. News of the achievements of the Columbian fly school had begun to reach Russia in the late 1910s. Then, in 1922, Hermann Muller visited, bringing with him an 'invaluable gift': bottles containing fruit flies, including thirty-two live *Drosophila melanogaster* mutants.[63] To work with them, Kolt'sov passed the material on to a group that included a younger acquaintance whose enthusiasm – butterfly collecting – might have marked him out as an out-of-fashion naturalist if he had not also taught a spot of biometrics: Sergei Chetverikov.

In fact, Chetverikov's vast experience as a naturalist, when combined with the rigours of biometry and laboratory genetics, was the model for a new synthesis. Chetverikov argued that recessive genes would be widespread in a wild community, and that this hidden genetic variation was what natural selection acted on to produce evolution.[64] In the warmer months of 1925, Chetverikov's group set out to study the wild populations of *Drosophila* – such as *D. phalerata*, *D. transversa*, *D. vibrissina* and *D. obscura* – that could be found in locations around Moscow. 'Unlike geneticists or experimental biologists, who had to find their way back to Darwinism and natural populations', concludes Adams, 'Chetverikov and his group were Darwinian naturalists who had to learn their genetics.'[65]

In 1929 Chetverikov was arrested and exiled to Sverdlovsk. Adams gives three scenarios for the immediate cause of the arrest: he might have been fingered in the suicide note of the neo-Lamarckian Paul Kammerer, it might have been an expression of student resentment, or it might have been the consequence of officials' suspicion that the closed *Drosophila* seminars were a space for dissent and conspiracy.[66] Almost all the group dispersed, although Kolt'sov managed to relaunch genetics under the orphan Nikolai Dubinin. While starting in chicken genetics, Dubinin would develop ideas about genetic drift in parallel with Sewall Wright's work in the United States, and he would explore genetic drift in wild *Drosophila*.

However, Soviet genetics was exported out of Russia along three paths.[67] The husband and wife Nikolay Vladimirovich and Helene Aleksandrovana Timofeeff-Ressovsky, both members of Chetverikov's group, had left the Soviet Union in 1925 for the Kaiser

Wilhelm Institute in Brain Research, Berlin, as part of an official exchange programme. Second, Chetverikov met J. B. S. Haldane in Berlin in 1927. The British biologist arranged for Chetverikov's work to be translated. Finally, in the same year, Theodosius Dobzhansky, a member of Filipchenko's school, left Leningrad to work with T. H. Morgan in the United States. Dobzhansky records that he had had 'the privilege of several visits' to Chetverikov, who 'was unstinting with his time and inspiring with his advice'.[68] Crucially, Soviet genetics was the study of wild populations, informed by natural history, and was thus able to be a 'bridge builder' in the Evolutionary Synthesis.[69]

The Evolutionary Synthesis

Against a background of increasing disciplinary specialization, the life sciences in the late nineteenth and early twentieth centuries were also marked by a widening divergence between two research traditions.[70] On the one side there were those who studied the evolutionary biology of whole organisms, among them many 'naturalists', paleontologists and experts in systematics. On the other side were those who championed 'experimental biology', and included the laboratory investigations of *Drosophila* chromosomes as well as the Mendelian believers in de Vriesian saltation. A 'communication gap' separated the two sides.[71] They spoke different languages, and when they did read each others' work they found prominent misunderstandings that discouraged further engagement. So, for example, the evolutionary biologists found the experimenters holding on to a naïve concept of species – species as a type rather than a population. Conversely, the experimental biologists regarded evolutionary biology as 'speculative and unreliable'.[72] Experimental biologists shared these values with physical scientists.[73] The Mendelians who wrote about evolution held the view that discontinuity in inheritance of factors implied discontinuous evolution. In a way each side saw the other as it had been decades ago: recent work went unread. When they did get together, such as at a conference at Tübingen in 1929, 'there was no dialogue'.[74]

Nevertheless a synthesis of these traditions did take place in the 1930s and 1940s. The Evolutionary Synthesis was the widespread acceptance of Darwinian natural selection as the mechanism of evolution, an acceptance which in turn depended on the widespread adoption of interpreting taxa – especially species but also higher taxa – as populations of varying individuals. The Darwinian mechanism –

natural selection – had not been at all popular in the first few decades of the twentieth century. Paleontologists, for example, preferred either orthogenetic mechanisms (the unfolding of evolution along internally directed paths) or saltations (the giant mutations by which an individual would leap to become a new species). Darwinism, for them, was a refuted theory. Thomas Morgan, as late as 1932, wrote that 'natural selection does not play the role of a creative principle in evolution.'[75] Yet by 1947, the date of the Princeton conference, evolution by natural selection explained, to almost everybody's satisfaction, the shape of life on earth.

Usually the Evolutionary Synthesis is presented as a march of books: starting with Dobzhansky's *Genetics and the Origin of Species* (1937) and passing through Julian Huxley's *Evolution: The Modern Synthesis* (1942), Ernst Mayr's *Systematics and the Origin of Species from the Viewpoint of a Zoologist* (1942), George Gaylord Simpson's *Tempo and Mode in Evolution* (1944), Bernhard Rensch's *Neuere Probleme der Abstammungslehre* (1947, translated as *Evolution Above the Species Level*, 1959) and G. Ledyard Stebbins's *Variation and Evolution in Plants* (1950). Each text was a major synthesis in its own right. Dobzhansky, for example, had studied the experimental genetics of wild populations in Russia before joining Morgan's school of cytological genetics in 1927, courtesy of a scholarship from the International Education Board of the Rockefeller Foundation. Dobzhansky brought together laboratory-based insights with such studies of the variability of wild populations as Heincke's turn-of-the-century measurements of herring, Henry Crampton's Hawaiian *Partula* snails, Lee Dice's *Peromyscus* deer mice, Kinsey's gall wasps, Chetverikov's and Timofeeff-Ressovsky's wild *Drosophila*, and his own Russian ladybird beetles. But he also wove third and fourth strands into the synthesis in the forms of the theoretical arguments of Sewall Wright concerning 'genetic drift' within 'adaptive landscapes' and the mathematical population genetics of Ronald Fisher.[76] He 'treated [natural] selection not merely as a theory but as a process that can be substantiated experimentally.'[77] It amounted, argues the historian Jonathan Hodge, to 'the single most influential text in its generation, perhaps in the century'.[78] *Genetics and the Origin of Species* passed through many editions, soon reaching canonical status.

Julian Huxley's *Evolution: The Modern Synthesis* (1942) identified the synthetic movement as a distinct project and gave it a title:

Evolution may lay claim to be considered the most central and the most important of the problems of biology. For an attack upon it we need facts and methods from every branch of the science – ecology, genetics, paleontology, geographical distribution, embryology, systematics, comparative anatomy – not to mention reinforcements from other disciplines such as geology, geography, and mathematics.

Biology at the present time is embarking upon a phase of synthesis after a period in which new disciplines were taken up in turn and worked out in comparative isolation. Nowhere is this movement toward unification more likely to be valuable than in this many-sided topic of evolution; and already we are seeing the first-fruits in the reanimation of Darwinism.[79]

'My debt to Fisher's work is obvious', wrote Huxley. 'Fisher has radically transformed our outlook . . . by pointing out how the effects of a mutation can be altered by new combinations and mutations of other genes', adding that 'Any originality which this book may possess lies partly in its attempting to generalize this idea further.'[80] Like Dobzhansky, Huxley showed that the insights of laboratory genetics and Fisher's mathematical population genetics were fully compatible with the diversity described by natural history. *Drosophila* was joined by a host of fauna and flora: Adriatic lizards and koalas, gyrfalcons and hawkweeds, fox sparrows and chalk-hill blues. Birds were especially prominent, reflecting Huxley's passion.

Huxley was able to be synthetic because of where he came from, in two senses. First, he was extraordinarily well connected within elite science. Not only was he the grandson of T. H. Huxley, but his early career took him through the networking hotspots of global science, such as marine embryology in Naples and *Drosophila* studies in Morgan's fly school. Colleagues included E. B. Ford, whose 'ecological genetics' began under Huxley's supervision at Oxford, and J. B. S. Haldane, whose *Animal Biology* was co-written with Julian and published by Oxford's Clarendon Press in 1927, and whose work of proto-synthesis, *The Causes of Evolution* (1932), influenced Huxley's own. The historian of Oxford science Jack Morrell argues that the tutorial system encouraged Oxford academics to be generalists (to be taught by specialists was regarded as 'dreadfully provincial'), an intellectual style which in turn Julian Huxley expressed in his synthesis.[81]

But Huxley also had a lifelong interest in the planned organization and channelling of popular science. He had given up academic work in 1927 to collaborate with H. G. Wells (and son G. P. Wells) on the encyclopaedic *The Science of Life* volumes. Historian of biology Robert Olby argues that first drafts of Huxley's synthetic argument can be found in the entries on 'Evolution' that Huxley penned for Wells's popularization.[82] He interpreted bird behaviour in Oscar-winning nature documentaries such as *The Private Life of Gannets* (1934). When he wrote *Evolution: The Modern Synthesis* he was, as secretary of the Zoological Society of London, director of London Zoo. Encyclopaedia articles, zoo design and films were all ways to channel scientific knowledge to a public audience. But the flow of information was two-way: Julian Huxley harnessed the resurgent growth in popular naturalist and amateur science reporting, exemplified, for example, by the British Trust for Ornithology (founded 1932). From the 1910s, in the United States and Britain, Huxley had been encouraging amateur ornithologists to 'direct their emerging observational networks at problems of scientific moment'.[83] His synthesis benefited from this burgeoning of citizen science.

Julian Huxley also tried to organize and channel the productivity of human populations in another, not disconnected sense: he was, like Ronald Fisher, a staunch proponent of eugenics. *Evolution: The Modern Synthesis* culminates with a discussion of 'evolutionary progress'. J. B. S. Haldane (1932) had written that talk of 'progress', especially evolutionary progress that reached a pinnacle in humankind, represented the 'tendency of man to pat himself on the back . . . We must remember that when we speak of progress in Evolution we are already leaving the relatively firm ground of scientific objectivity for the shifting morass of human values.' 'This I would deny', responded Huxley:

> Haldane has neglected to observe that man possesses greater power of control over nature, and lives in greater independence of his environment than any monkey . . . The definitions of progress that we were able to name as a result of a survey of evolutionary facts, though admittedly very general, are not subjective but objective in their character. That the idea of progress is not an anthropomorphism can immediately be seen if we consider what views would be taken by a philosophic tapeworm or jelly fish . . . Man is the latest dominant type to be evolved, and this being so, we are justified in calling the trends which we have led to his development progressive.[84]

And, having defended progress of humankind so far, Huxley turned to look ahead. 'The future of progressive evolution is the future of man', he concluded, and 'The future of man, if it is to be progress and not merely standstill or a degeneration, must be guided by deliberate purpose.' As the privileged interpreter of a prodigious 'survey of evolutionary facts', the author of *Evolution: The Modern Synthesis* defined the 'progress of man' in theory. In practice this guidance meant, as Huxley urged in such eugenic films as *Heredity in Man* (1937), 'voluntary' sterilization of the unfit.

Each of the other synthetic texts brought in further fields of the life sciences. Ernst Mayr (1942) demonstrated how the science of biological classification, systematics, could be refounded in the light of synthetic argument.[85] Mayr had been hired by the American Museum of Natural History in 1931, and his contribution and that of systematists more generally to the synthesis testifies to the continued importance of museum-based collections.[86] George Gaylord Simpson (1944) brought paleontology into the Modern Evolutionary Synthesis by a tour-de-force account of how the macro-evolution of the horse from *Eohippus* to *Equus*, previously an exemplar of orthogenesis, occurred by the same processes as the micro-evolution by natural and artificial selection observed in *Drosophila*. 'Paleontology's status as an evolutionary discipline has been upgraded to respectability and far beyond', wrote Stephen Jay Gould in his assessment; 'we must thank Simpson'.[87] The combination of statistical analysis and an understanding of population genetics meant that *Tempo and Mode in Evolution* won over biologists, while leaving an older generation of paleontologists frozen out.[88]

Rensch (1947), writing in German and on the basis of fieldwork in Indonesia, reaffirmed the view that macro- and micro-evolution were consistent. (Nevertheless, in Germany, Otto Schindewolf's challenge to the Evolutionary Synthesis, his 'typostrophe' theory of sudden evolutionary change, was influential until the 1970s.)[89] Stebbins (1950) surveyed botany and showed that natural selection accounted for the evolution of plants. (Although Dobzhansky had already devoted a chapter to polyploidy, the phenomenon, commonplace in plants, of the multiplication of chromosomes.) Some of these books – Rensch, Simpson, Mayr – were written at the same time as Huxley's but were delayed by world conflict. Gradually the 'communicative gap' was filled in. One successful strategy was to devise new jargon, avoiding old pitfalls and knee-jerk reactions. 'Nearly all the architects of the new synthesis', notes Mayr, 'contributed terminological innovations.'[90]

However, if we understand the Evolutionary Synthesis to be just the publication and reception of texts, then we miss something very important. Indeed historian of biology Joe Cain suggests we drop the concept altogether:

> There was more to evolutionary studies in the synthesis period than the synthetic theory. There was more to the synthesis project than work done by the so-called architects of the supposed merger of Mendelian genetics and selection theory via mathematical theory. The evolution

books of the revived Columbia Biological Series have been far too dominant in synthesis historiography.[91]

The books, enlightening as they are, are like Christmas lights.
Holding them up is a Christmas tree of scientific organization, fieldwork and, especially, a working world of breeding. Agricultural
research stations supported the experimental study of genetics,[92] and
abstracted, mathematical representations of this working world fed
directly into the synthesis. Key figures, for example, were Ronald
Fisher, resident statistician at the British experimental agricultural
station at Rothamsted, Sewall Wright at the agricultural station in
Madison, and J. B. S. Haldane, who combined an academic career
at Cambridge and University College London with time at the John
Innes Horticultural Institution.[93] An experiment started in 1896 in
Illinois, selecting each year maize plants high and low in oil and
protein, generated measurements that were in turn analysed by
Fisher and reported by Huxley as 'particularly impressive' evidence
for selection.[94] (The Illinois Long-Term Selection Experiment is still
going, the longest-running plant genetics experiment in the world,
although Rothamsted has an older crop science experiment.) In many
ways this was a continuation of the debt that Darwin owed to breeders in his day.

And there were other working world contexts. Fisher had first
worked on the kinetic theory of gases, a task generated by the
requirements of industry.[95] He transferred the statistical analysis of
populations of gas molecules to genetics, and he promoted eugenic
programmes based on analyses of human genetics. His position, I
suggest, is typical. His science related to a working world – not reduc-
ible to politics, or economy, or industry or commerce alone; it was a
science of all, together.

The second structure holding up the Christmas lights was the social
organization of scientific work, in particular those that encouraged
interdisciplinarity. The early phase of synthesis was within disciplines: the argument that natural selection and Mendelism were not
incompatible was first made within genetics, while the articulation of
the biological species concept was within systematics.[96] Both discipli-
narity and interdisciplinarity are social institutions. But what facilitated the movement of the synthesis outside disciplines, to become
indeed a full-scale synthesis, were bridge-building institutions, such
as the interdisciplinary communities sustained in the Russian school
of 'naturalist-geneticists'[97] or the elite English tradition of general-
ism.[98] (A note of cautionary reflection might be warranted here:

interdisciplinarity was highly valued in the late twentieth century because of profound shifts in the working world. The histories of the Evolutionary Synthesis date from this time and have a tendency to find contemporary values in the past.)

The Evolutionary Synthesis, at its centre, was a merging of research traditions around the common claim that the Darwinian theory of evolution by natural selection was correct and broadly applicable across disciplines. In this chapter on revolution and materialism I began by tracing developments in Russia. However, the most extraordinary advance of materialism in the twentieth century, the settlement of a revolution in how we understand our world, was the triumph of a unified biology.

— 10 —

NAZI SCIENCE

Histories of science written in the first three decades after the Second World War presented Nazism as catastrophic for good science. For example, René Taton, in a 1964 French encyclopaedic survey of the history of modern science, translated as *Science in the Twentieth Century*, wrote that the brilliant revival of Weimar science

> was completely reversed in 1933, when victorious Hitlerism, with its attacks upon academic freedom, led to the emigration, within a few months, of more than a third of Germany's leading physicists and a great drop in the number of foreign students. Soon afterwards, numerous avenues of research were officially closed, and original work began seriously to decline.
>
> Hitlerism culminated in the Second World War, at the end of which most German laboratories had been destroyed or severely damaged, most leading scientists had gone into forced or voluntary exile, and new frontiers had appeared that proved more and more unpropitious to intellectual exchanges. Yet despite all these crippling handicaps, German science revived quite quickly.[1]

The Nazi period in this historiography was regarded as an extraordinary episode. When contrasts were drawn between the modern and the anti-modern, the rational and the anti-rational, and the progressive or democratic and the conservative or reactionary, fascism in general and German National Socialism in particular was placed, firmly, in the second column. Nazism was seen also as a rejection of industrial values. So, for example, a historian such as Fritz Stern might link the trauma of very fast industrialization with a rejection of modern, industrial values in favour of the values of the '*Volk*'.[2] The *Volk*, in Nazi ideology, was a natural unit, defined by relationship of a people to blood and soil. The leader – the '*Führer*' – was an

211

embodiment of the *Volk*, a position from which both the authority of the *Führer* and the irrelevance of any alternative democratic legitimation followed.

However, historical research of the last few decades has problematized the use of sharp dichotomies and also sought to explain how a technologically prodigious culture, as Nazi Germany was in some sectors, could thrive alongside anti-modernism. Jeffrey Herf, in *Reactionary Modernism*, for example, argued that Nazi Germany exhibited an irrationalist embrace of technology; paradoxically, some Nazis were anti-Semitic, anti-rationalist and anti-modern yet pro-machine.[3] Experience of the front during the First World War accounts for some of this attitude. Reactionary modernism was expressed in technological projects such as the Autobahns, rocketry, the technologies of mass killing, and the sophisticated propaganda. It helped of course where the development of a technology followed particular Nazi interests: Autobahns, for example, not only united Germany but also, by following contour lines, were a direct impression of German soil.

The new historiography has also told a much more complex, morally ambivalent story of the relationship of science and Nazism. The title of the English translation of Benno Müller-Hill's pioneering *Murderous Science*, a study of the science used to 'select' people to be eliminated, published in German in 1984, suggests little room for ambivalence. Robert Proctor, perhaps our foremost historian on the subject, has demonstrated three more subtle, general points.[4] First, Nazism cannot be seen as merely destructive of science. Not only did much science continue under Nazism, but science contributed to how Nazism was conceived. Second, while the abhorrent atrocities of concentration camp science are familiar to us, Nazi values also supported less well-known scientific developments which many associate now with progressive values. Examples include the Nazi investigations into the link between smoking and lung cancer and the resultant environmental regulations.[5]

Third, Proctor notes that the reluctance to talk about how Nazism and science could work together, and therefore making of the 'myth' of 'Nazi science' as an 'oxymoron', was in the interests of four groups.[6] German scientists who lived through the Nazi era used the 'myth of suppressed science' as a way 'to distance themselves from their Nazi past', discouraging moves to investigate their activities in the 1930s and 1940s. Jewish émigré scholars were unwilling 'to believe that the system that treated them so shoddily had continued to produce good science'. Meanwhile political authorities responsible for post-war Germany, in particular American authorities, 'were busily trying to recruit Nazi talent' for use in their own projects. Indeed, over 1,600 scientists moved from Germany to the United States under Project Paperclip in the late 1940s. Finally, 'the myth', writes Proctor, 'served to reassure the American public', as perhaps it did other Western publics, 'that abuses like those of the Nazi era could never occur in a liberal democracy.'

In the following I will draw on the new and old histories to survey the relationships between Nazism and science. In some cases the impact appears to be very slight. For example, Thomas Junker and Uwe Hossfeld have argued that, among the German biologists who contributed to the Evolutionary Synthesis, including Hans Bauer, Timoféef-Ressovsky and Bernhard Rensch, while many articulated eugenic positions, there was no particular correlation between 'scientific Darwinism and national socialist ideology'.[7] In other cases, such as those described in Alan Beyerchen's *Scientists under Hitler*, a split developed between ideologically inflected Nazi science – the project, discussed below, for an Aryan physics – and the science of its opponents.[8] However, the relationship between science and Nazism could also be strong and mutually constitutive. Human genetics, anthropology and medicine, for example, not only provide plenty of cases where Nazism furthered the developments of the sciences in certain directions but also illustrate that the sciences created languages, illustrations, evidence and theories that helped create Nazi ideologies and guide Nazi actions.

Nazi law and the employment of scientists

In April 1933 the new Nazi government passed the Law for the Re-establishment of the Professional Civil Service, which ruled that 'non-Aryan', 'unreliable' state employees should be removed from office. Since university professors in Germany were civil servants, the law immediately and profoundly affected academic science. Some specialties were disrupted almost completely, although the earlier view that some disciplines were destroyed has been shown to be too simplistic. Gestalt psychology provides an illustration. Of the three founding figures, Max Wertheimer, a liberal-leftist Jew, taught at Frankfurt, Wolfgang Köhler, not Jewish, led research in Berlin, while Kurt Koffka had emigrated to the United States in 1927. Wertheimer, undergoing medical treatment for a heart condition in Marienbad at the time of the passing of the civil service law, was suspended by the

213

University of Frankfurt; he was just one of fifty-two faculty members to lose their posts in 1933 alone.[9]

Köhler vacillated. He saw his colleague Kurt Lewin forced out in 1933, despite private protests from Köhler to the authorities. Lewin returned to Berlin, picked up his family, and left for Cornell University. The resignation of the physicist James Franck, made in protest against the dismissal of Jewish colleagues, prompted Köhler to act publicly, albeit in a compromised fashion.[10] He published an article in a prominent newspaper complaining that the dismissal of German non-Aryan patriots of irreplaceable 'expertise' and 'character' was impractical; while arguing that 'Germans have the right to control the composition of their population and to reduce the proportion of Jews in the leadership of all the essential affairs of the people', the measures that should be taken were ones

> which do not damage Germany indirectly, which do not suddenly destroy the existence of innocent people, and which do not sorely injure the significant, superior human beings among the German Jews. For my friends [on whose behalf Köhler wrote] do not want to agree to the thesis that every Jew, as a Jew, represents a lower, inferior form of humanity.[11]

Köhler's attempt to 'take a middle position' in order to defend his institute continued for two years, under increasing harassment from Nazi students. Eventually, in 1935, he resigned and took up a teaching position at Swarthmore College in the United States, helped by funds from the Rockefeller Foundation and the Emergency Committee in Aid of Displaced Foreign Scholars. Psychology, even the Gestalt strand, nevertheless survived in Germany. Psychologists shifted research interests from the experimental study of cognition to 'characterology', building personality profiles for the practical benefit of Nazi administrators.[12] Gestalt – a whole, like the *Volk* – was claimed for Nazism, as Friedrich Sander, professor of psychology at Jena, celebrated in a public lecture in 1937:

> He who, with believing heart and thoughtful sense, intuits the driving idea of National Socialism back to its source, will everywhere rediscover two basic motives standing behind the German movement's colossal struggles: the longing for wholeness and the will towards Gestalt . . . . Wholeness and Gestalt, the ruling ideas of the German movement, have become central concepts of German psychology . . . scientific psychology is on the brink of simultaneously becoming a useful tool for actualizing the aims of National Socialism.[13]

214

Köhler's failed attempt to appeal to political authority on the practical grounds that it was both possible and essential to distinguish between 'significant, superior' Jews and others was also a tactic attempted within physics by Max Planck. On 16 May 1933, the elderly physicist secured an audience with the *Führer* himself. To Planck's urging that 'there are different kinds of Jews, worthy and worthless to humankind, and among the former old families with the best German culture, and that one must make distinctions', Hitler's response was simple: 'That is not correct. A Jew is a Jew – all Jews hang together like chains . . . I must therefore act against all Jews in the same way.'[14]

Aryan science

While the 'majority of German physicists, like biologists, welcomed Nazi rule', as historian Paul Josephson observes, 'perhaps one-quarter of German physicists [were] forced from their jobs, mostly by the laws excluding Jews from the civil service.'[15] New fields provide opportunities for members of relatively disadvantaged social groups to enter into science. This social phenomenon provides a partial explanation for the relative prominence of Jews in physics and in theoretical physics in particular.[16] Conversely, the older research traditions, especially experimental physics of a classical nature, housed German scientists who were bitter at the revolutionary success of the theoretical physics of quantum mechanics and relativity. Men such as Philipp Lenard and Johannes Stark, both Nobel Prize winners in the first decade of the twentieth century, honouring meticulous experimental work in the Newtonian tradition, loathed the new physics.[17] They both thought theoreticians of the new physics had blocked their careers; both were vehemently anti-Semitic. They hated Einstein as an internationalist, as a theorist, as destroyer of the ether and as a Jew.

A distinct Aryan science was developed in line with Nazi ideology: 'All science (and all morality and truth)', writes Josephson, was to be 'judged by its accordance with the interest and preservation of the Volk', the 'metaphysical belief in an essential German people of organic purity and their historic mission to control world civilization'.[18] Lenard and Stark touted an alternative physics: Aryan physics, which would be traditional and experimental, and would reject quantum and relativity theory on racial grounds. Both had welcomed the rise to power of Adolf Hitler. In 1933 Lenard attempted to corral his fellow Nobel Prize winners to join a public declaration of support. Stark agreed, but Heisenberg, Max von Laue and Walther Nernst ducked out. Lenard spelled out Aryan physics in four volumes, his *Deutsche Physik*, published in 1936 and 1937. By rejecting the abstract, Aryan physics promised to stay close to the concrete, be of greater benefit to industry, and therefore serve the Nazi aims of economic self-sufficiency.[19]

The Aryan physics project flourished in the mid-1930s, when it received patronage from pillars of the Nazi regime, including the Reich education minister Bernhard Rust, and the Teachers' League and the Students' League. In 1936, the Aryan physics movement attacked the remaining great theoretician, Werner Heisenberg, calling him, and others, 'white Jews' for continuing to defend Einstein's physics. But Heisenberg had even more powerful patrons than Lenard and Stark: an appeal to defend his honour to the SS leader Heinrich Himmler, in the summer of 1938, was eventually successful. Himmler was persuaded, partly by the advice of the Göttingen applied mathematician Ludwig Prandtl, that Einsteinian physics could be used with benefit without endorsing Einstein the non-Aryan.[20] In the preparation for war, applicability trumped racial purity. Fortunately for Heisenberg and the 'white Jews', the Nazi authorities regarded the controversy as an 'intramural dispute, not a political one, judging both groups to be loyal to the regime, and this proved to be what saved the new physics in Nazi Germany'.[21]

The mutual support of science and Nazi values

The story of Aryan physics, in the hands of Lenard and Stark, has been told, along with the career of Lysenko's biology, as the primary twentieth-century case studies of the disastrous results of mixing science with totalitarian ideology. Unfortunately, the relationship between science and politics was not so simple. Take the cases of aeronautics, rocketry (considered later) and the quantum theories of Pascual Jordan.

German aeronautics had become an institutionalized science earlier than the First World War at places such as Göttingen and Berlin before falling into a deep depression following the restrictions on military aviation introduced by the Treaty of Versailles. The Weimar Republic had little money to spare either, and had cut back spending to the bone, so 'the removal of the democratic parliamentary bodies by the National Socialists was largely well-received by those in aeronautical research'.[22] The trickle of funds for aeronautical research became a gushing torrent. 'Infinite possibilities seemed to present themselves', writes historian of technology Helmuth Trischler, research 'previously unthinkable due to the high costs involved, were suddenly approved without question'. The Experimental Centre for Aviation (DVL) at Berlin-Adlershof employed a workforce of 2,000 by 1939. Aeronuatics was also now introduced in schools, partly to cultivate martial values in children. An aeronautical academy – the first ever in Germany for a technical science – opened in 1936. The generosity of patronage led to advances in aeronautical science and the development of tools used to pursue it. Giant and innovative wind tunnels were built.[23] One of the first electromechanical computers, Konrad Zuse's Z1 machine, was built in the mid-1930s to aid the solution of simultaneous equations of many variables generated by analysis of the tensions of struts in airframes.[24]

Pascual Jordan was a Göttingen physicist who wrote, with Max Born and Werner Heisenberg, the famous '*Drei-männer Arbeit*' (Three men's work) papers of 1925 and 1926 on mathematics of quantum mechanics. In quantum theory, Jordan found both a justification for and a source of Nazi ideas. So, for example, the leading physicists – Bohr in particular, but Einstein also qualified – he called by the name '*Führer*', fitting their roles as revolutionary, brutal leaders. Physics, foresaw Jordan in 1935, would be the source of 'technical energy sources' which would make a 'Niagara power station appear trifling and explosive materials in comparison to which all present explosives' would be 'harmless toys', led into such a world by *Führer*s of physics,

> Only the force of the strongest powers will force the multiplicity of competing individual interests into a higher unity and protect against attempts at disruption. Therefore, in the wide spaces of the future, physics will also supply . . . the positive constructive means for great developments.[25]

In other words, such physics supported the *Führer* principle. Note that Jordan rejected Aryan physics, which was grounded in the other main ideological prop of Nazism, blood and soil, and the *Volk*. Jordan's quantum theory was elaborated alongside his Nazism and supported new mathematical physics; it also led him to propose quantum theories of biology and psychology. Historian Norton Wise concludes that Jordan's 'Nazi values' were 'important to his fundamental work in mathematical physics', adding, just in case the unpleasantness of the point had been missed:

Jordan's elegant transformation theory and his work on second quantization expressed directly his views on the hierarchical, organistic structure of the natural and social order. It is necessary to make this point explicitly because we live with the tenacious myth that the acquisition of fundamental knowledge had to cease when scientists embraced Hitler.[26]

A similar and in fact further-reaching contribution can be found in the life sciences. The theoretical biologist Ludwig von Bertalanffy, argues Ute Deichmann, sought a biology that supported Nazism, and vice versa: to match an 'organismic age' with an 'organismic biology', hoping that 'the atomizing concept of state and society will be followed by a biological one that recognizes the holistic nature of life and of the Volk'.[27] Robert Proctor, in his study of medicine under the Nazis, *Racial Hygiene*, insists that 'biomedical scientists played an active, even leading role in the initiation, administration, and execution of Nazi racial programs ... there is strong evidence that scientists actively designed and administered central aspects of National Socialist racial policy.'[28] How does he make his case? He argues that the way that groupings of life scientists and doctors interpreted Darwinism and genetics according to their social situations presented them with the tools to assist the creation of National Socialist policies and ideologies.

As we have seen, evolutionary theories in general, and Darwin's in particular, were read for messages about change in human societies, in all countries. In the United States, for example, social Darwinism justified capitalism, red in tooth and claw. But in Germany in the later nineteenth century, against a background of political instability, industrialization, and anxieties about failing to reach imperial ambitions, social Darwinism was associated with a rejection of laissez-faire free markets and legitimated state intervention to prevent degeneration of German stock.[29] This interpretation provided German eugenics with a distinctive inflection. Alfred Ploetz was a leader of the movement. Germany, as did other industrial nations, founded eugenic societies that attracted the middle-class professionals as members. The medical doctor Ploetz, along with a psychiatrist, a lawyer and an anthropologist, established the Gesellschaft für Rassenhygiene, or the Society for Racial Hygiene, in 1905. Branches were soon set up in Germany's larger cities; the Berlin branch of 'notable scholars, physicians, industrialists, and representatives of other professions' was the base for Ploetz and the geneticist Erwin Baur.[30]

The Society for Racial Hygiene was about the same size as equivalent national societies: 1,300 members in 1930[31] compares to 1,200 members of the Eugenics Society in Britain.[32] The topics of lectures, policy campaigns and anxieties expressed were also similar: degeneration of the race, and the higher rates of birth among the poor, the criminal and the unfit more generally. Proctor even suggests that the German racial hygiene movement before the mid-1920s was 'less concerned with the comparison of one race against another than with discovering principles of improving the human race in general – or at least the Western "cultured races"'.[33]

Furthermore, the Society for Racial Hygiene, before the mid-1920s, attracted members from across the political spectrum, which of course was gapingly wide in Weimar Germany: liberals and reactionaries, left and right.[34] Crucially, however, the right wing controlled the 'institutional centres' of the movement, including the medical publishers. The early recipient of the Nazi Gold Medal of Honour, the publisher Julius Friedrich Lehmann, for example, seized control of the racial hygiene movement's premier journal.[35] Interconnections grew between this right wing of the racial hygienists and the Nordic movement, a sort of whites-only socialism, which was one of the tributaries of Nazism. Biological theory and practice increasingly supported racist and anti-Semitic argument.

Raised on a groundswell of interest in racial hygiene, resources poured into institutions for the study of human genetics. A professorial chair in racial hygiene was created at Munich, and was filled by Fritz Lenz in 1923. A Kaiser Wilhelm Institute for Anthropology, Human Heredity and Eugenics opened in Berlin-Dahlem in 1927. The geneticist Eugen Fischer, famous for his supposed application of Mendelism to the offspring of black South-West Africans and German colonists and a former chair of the Freiburg chapter of the Society for Racial Hygiene, was appointed director.[36] Stung by some early criticism, Fischer directed the broad, deep and well-funded research programme of the institute into 'areas that fit well with Nazi biologistic goals'.[37] The institute was a pioneer of twin studies as a methodology in human genetics. Two camps near the North Sea were opened to hold twins for study. At the institute, 150 'criminal twins' were examined in a search for the heritable causes of criminality. Other research included the human genetics of Down's syndrome, diabetes, tuberculosis, rickets and muscular dystrophy and studies of racial miscegenation. By the mid-1930s more Reichsmarks were being channelled to the Kaiser Wilhelm Institute for Anthropology, Human Heredity and Eugenics than to the institutes for physics and chemistry combined.[38]

Influential texts were also institutions. Lenz and Fischer, with geneticist Erwin Baur, had been co-authors of the *Outline of Human Genetics and Racial Hygiene*, first published in two volumes in 1921. These institutions – professorial chairs, institutes, journals, books – channelled knowledge that fed the development of National Socialist ideology. In turn the Nazis, on seizing power in 1933, provided further support. For example, Fischer, Lenz and Baur's *Outline* offered a selectionist account of racial hierarchies (from Aborigine to 'Negro' to Mongol to Mediterranean to Alpine and, finally, Nordic types), of masculine and feminine roles ('Men are specially selected for the control of nature, for success in war and the chase and the winning of women; whereas women are specially selected as breeders and rearers of children . . .') and of the character of Jews as a 'race'.[39] As Robert Proctor summarizes Lenz's contribution: 'Jews are precocious and witty but lacking in genuinely creative talent'; 'the Jew is more successful as an interpreter than as a producer of knowledge'; the Jew 'has been selected not for the control and exploitation of nature, but for the control and exploitation of other men'. The Jew as a type was parasitic: 'it is true', Lenz wrote, 'that whereas the Teutons could get along fairly well without the Jews, the Jews could not get along without the Teutons.'[40]

While in Landsberg prison after the Munich Beer-Hall Putsch, Hitler read the *Outline* and wrote *Mein Kampf*.[41] After the Nazi seizure of power in 1933 it was books like that of Fischer, Lenz and Baur that 'made it possible for Nazi-minded biologists and biology-minded Nazis to claim that 'the National Socialist world view has conquered Germany, and a central part of this world view is biological science' (the biologist Ernst Lehmann writing in 1933).[42] 'National Socialism as the Political Expression of Our Biological Knowledge' had been the tag-line of a celebration of the life and work of Alfred Ploetz in the *National Socialist Monthly*, in which a distinction drawn between the 'hard genetics' associated with August Weismann and the environmental shaping associated with Lamarck was one of the ways that a clear line was drawn between National Socialism and Marxist socialism.[43] Whereas the latter assumed that humans were biologically equal, National Socialism started from the observation that the 'genetics of particular individuals, races, and race mixtures are different', and that genetic traits could not be changed. By the time that the Nazi Party published the *Handbook for the Hitler Youth* in 1937, this line of scientific justification for Nazism was a commonplace:

[What we] need to learn from these [meaning Weismann's] experiments is the following: Environmental influences have never been known to bring about the formation of a new race. That is one more reason for our belief that a Jew remains a Jew, in Germany or in any other country. He can never change his race, even by centuries of residence.[44]

Nor was linkage between studies of heredity and Nazi ideology achieved merely through human genetics. 'National Socialism without a scientific knowledge of genetics is like a house without an important part of its foundation', wrote Hermann Boehm, as he set up an Institute for Genetics at Alt-Rehse to instruct medical doctors in the science. 'Basic scientific knowledge is indispensable for anyone who wants to work in the arena of genetic and racial care, and experimental genetics must provide the foundations for that care.'[45] 'Many a budding young Nazi doctor', notes Proctor, 'first learned experimental genetics by crossing *Drosophila melanogaster* at the SS Doctors' Führer School in Alt-Rehse;[46] the mutant fruit flies were supplied by Timoféef-Ressovsky.

These arguments were not the imposition of Nazi ideology onto science, warping it, but rather they were articulated within German science, and in circulation before the rise of the Nazis to power. Versions of human genetics were therefore partly constitutive of National Socialism. What science brought were not only ideas – such as immutability of races due to hard heredity – but also a language of objectivity that made it nearly impossible for the victims of Nazism to refute the prejudices of the experts.[47]

Doctors, science and the 'Jewish Question'

A second strand through which the biomedical sciences shaped Nazi policy was through the beliefs and actions of medical doctors. German doctors saw themselves as under intolerable strains. Many felt caught in a vice, squeezed between the demands of insurance companies, charged with administering access to healthcare under the welfare system created by Bismarck, and calls for further socialization of medicine from the political left. They also complained of the scientization of medicine, as the medical insurance companies encouraged large, relatively impersonal clinics with laboratories.[48] They saw the quality of their working life being eroded, deskilled. Some doctors diagnosed a 'crisis in medicine', a response typical of Weimar culture, which we have seen before. Many doctors sought a scapegoat. Jews were traditionally prominent in the medical professions – over one-third of doctors in late nineteenth-century Berlin were Jewish – while, from the turn of the century, pogroms in Russia forced emigrants west, where they competed for employment. However surprising it may be to a superficial first glance, the fact is that many doctors were early and enthusiastic supporters of fascism, more so than members of other professions. Over 2,500 doctors were members of the Nazi Party before it came to power in 1933, and in total 45 per cent of the profession were party members.[49]

The previously diverse medical press was brought under Nazi control: Jews were purged from editorial boards, and ideological commentary accompanied the usual medical research papers.[50] New popular magazines, ideal for waiting rooms, spelled out how the Nazis and the doctors were complementary. A sample of this writing, from a Stuttgart doctor, illustrates just how scientific experts of the human interior could serve their political masters by providing a naturalistic basis for ideology: 'In describing the various races, we must not stop with the external shape of the body, nor even with mental characteristics . . . We must go beyond this, to explore equally important differences in the inner organs of the body, differences that may reflect deeper, physiological differences among the races.'[51] Anti-Jewish policies began to accelerate. Jews were expelled from medical practices.

Doctors did not wait to be asked before eugenic policies were proposed and even carried out. Physicians and racial hygienists sat on the new Expert Committee on Questions of Population and Racial Policy, set up in June 1933. The Sterilization Law followed the next month. It set out to 'eliminate an entire generation of what were considered genetic defectives'.[52] A patient would now be sterilized if judged unfit by a genetics health court. The techniques were vasectomy for men and tubal ligation for women. The congenitally feeble-minded, schizophrenics, manic-depressives, epileptics, the genetically blind or deaf, alcoholics or sufferers of Huntington's Chorea were highly likely to be judged guilty. Of 64,499 decisions made by 1934, 56,244 were in favour of sterilization, the majority for being 'feeble-minded'.[53] Historian Gisela Bock estimates that the total number sterilized was around 400,000. This huge number, notes Proctor, required the full and enthusiastic contribution of the German medical profession: 'Doctors competed to fulfil sterilization quotas; sterilization research and engineering rapidly became one of the largest medical industries', while medical students 'wrote at least 183 doctoral theses exploring the criteria, methods and consequences of sterilization'.[54] In general, 'every doctor had to become a genetic doctor'.

In autumn 1935 the Nuremberg Laws, including the Blood Protection Law, introduced strict definitions of degrees of Jewishness, based on ancestry, and regulated marriage. For example, a man with one Jewish grandparent could not marry a woman who also had one Jewish grandparent. Marriages between the 'genetically ill' were forbidden.[55] Such laws had been called for by the medical profession, and physicians participated in the design and administration of the law. Research programmes set out to discover new and effective tests for determining Jewishness from biometrical data such as earlobe shape or blood group.

The medical profession also led the way into ever more extreme eugenic and anti-Jewish policies. The Nazi euthanasia programmes were under development from 1935 but were introduced as part of the final mobilization for war in 1939. (The primary justification made appeal to warfare: why should the healthy lay down their lives if the unhealthy did not? The racial hygienist Alfred Ploetz had argued in the early twentieth century that war was anti-eugenic – since the vigorous Nordic male was supposedly more ready to fight – and Hitler had recycled Ploetz's argument in 1935.) Euthanasia was cast as scientific research: the detailed procedures were drawn up by a Committee for the Scientific Treatment of Severe, Genetically Determined Illnesses, which reported in August 1939. The paper trail was as follows: midwives and doctors reported potential cases to a local health office (midwives received a couple of Reichsmarks for each completed form), the forms were collected at a central Berlin office, and three professors then annotated the names with plus signs or minus signs. Children with plus signs were sent to institutions, including hospitals, where there were extermination facilities. Adult euthanasia soon followed.

'Doctors were never ordered to murder psychiatric patients and handicapped children', concludes Robert Proctor. 'They were empowered to do so, and fulfilled their task without protest, often on their own initiative.'[56] A physician, for example, designed and ran an early gas chamber. Indeed, the Nazi political authorities were often reduced to the roles of coordinators. Medical science contributed by framing the Jews as a diseased race – and therefore a proper subject for their expertise. 'Science thus conspired in solution to the Jewish Question', writes Proctor. 'By the late 1920s, German medical science had constructed an elaborate world view equating mental infirmity, moral depravity, criminality and racial impurity.'[57] The passage from eugenic sterilization through euthanasia to the Final Solution was continuous. The same combinations of doctors, equipment and technique, for example, moved from euthanasia hospitals to concentration camps. 'The decision to destroy Europe's Jews by gassing them in concentration camps emerged', argues Proctor, 'from the fact that the technical apparatus already existed for the destruction of the mentally ill.'[58]

Historians have therefore shown that the claim that the Nazis simply corrupted or abused science is false. The actions of medical scientists illustrates, notes Proctor, that the 'scientists themselves participated in the construction of Nazi racial policy'.[59] Cases such as those of Pascual Jordan, as interpreted by Norton Wise, or Ludwig von Bertalanffy, as interpreted by Ute Deichmann, reveal that scientists could find inspiration in Nazism and return the favour by providing naturalistic grounds for ideology. Such science reflected on and built from the working world of Nazism. Scientists could also contribute in a more direct utilitarian fashion to the advance of National Socialist aims. Typically the picture was of a meeting of interests. The historian of technology Thomas P. Hughes argues that the continuation of the project of synthetic petroleum production through hydrogenation provides an example of this contribution.[60]

In the 1920s, the chemicals firm BASF had applied its knowledge of high-pressure, high-temperature catalytic chemical engineering – learned through its scaling up of the Haber–Bosch process – to the large-scale production of first synthetic methanol and then synthetic petroleum. Driving the decision in the mid-1920s was international competition in new technologies and contingencies of sources of natural raw materials: industrialists and politicians were alarmed that Germany was behind in the 'internal combustion revolution' but were also reluctant to spend reserves on importing petroleum. As part of the new IG Farben combine, 100 million Marks had been spent before Hitler came to power. With the Great Depression, world petroleum prices plummeted, leaving IG Farben with a white elephant in its synthetic petroleum plants. Negotiations with the future *Führer* began as early as 1932, and an agreement hammered out in December 1933 guaranteed production targets, prices and a market for IG Farben's product. The chemical combine's interests met Hitler's own interest in economic self-sufficiency, autarky.

The chemistry of synthetic petroleum cannot be said to have had a particular Nazi stamp. In the sciences of racial hygiene, anthropology and physics, however, as we saw above, Nazi ideology provided a resource used in the creation of new scientific content. Some of this science would be judged (both by contemporaries and by later commentators) as prejudiced and biased. But it was not merely the use

224

of Nazi ideology as a resource that made Aryan physics, say, poor science – after all Pascual Jordan's physics did the same. Rather, Aryan physics was poor science because it was wrong in the assumption that abstract science – theory – could not be a science of the working world.

Finally, recent historians have stressed that the relationship between National Socialism and science should not be seen primarily as one of political power pressurizing science. Rather, as Proctor argues, coercion was in 'the form of one part of the scientific community coercing another, rather than a non-scientific political force imposing its will on an apolitical scientific community'.[61] And this coercion could be local to an institute, a university, a discipline or a professional grouping. As the historian William S. Allen made clear in his micro-study of a single town, the Nazi seizure of power was achieved as much from below as from above, and depended upon Nazification of many of the civil society bodies active in a community.[62] A parallel story has been shown to be true for the scientific community. Furthermore, just as Allen shows for his town, Nazification was never complete. The spectrum of scientists' responses to Nazism ran the gamut from intense enthusiasm, through professional opportunism or feelings that politics was a world beneath them, to dislike and contempt.

Émigrés

Fascism in Europe forced many people to migrate. Elite scientists were often in the relatively lucky position of possessing the resources – money, international contacts and reputation – to be able to move. The diaspora of scientists from Germany, Austria, Italy and Hungary had a profound influence on the course of science in the twentieth century. The pattern of displacement varied enormously according to cause, the place of departure, discipline, social position, and path and terminus of emigration. The timing of emigration reflected the particular histories of fascist oppression. For Germany the key dates were the passing of the civil service (1933) and Nuremberg (1935) laws, and the pogroms and Kristallnacht of 1938. Italian fascist government dates from 1922. Italy introduced anti-Semitic laws later (1938) than Germany, but also had experienced a century of emigration dating back to the conflicts of unification. A naval admiral turned regent, Miklós Horthy, governed Hungary for over two decades, harassing the political left and limiting Jews in academic faculties. A fully fascist Hungarian regime briefly followed in 1944.

From Germany, out of half a million refugees, historians Mitchell Ash and Alfons Söllner report that around 2,000 were scientists or other scholars, a figure made up of 1,100 to 1,500 university professors with the remainder an estimate of non-university research scientists (such as those employed at the Kaiser Wilhelm Institutes) and younger scholars.[63] Some disciplines were hit hard: 15 per cent of German physicists and psychologists left. Among prominent physicist refugees were Max Born, Hans Bethe, James Franck, the Swiss-born Felix Bloch, Lise Meitner, Wolfgang Pauli and Rudolf Peierls. Albert Einstein, the most famous physicist of them all, was in the United States in 1933 and decided not to return to Germany. Italian physics, undergoing a minor renaissance in the 1920s, was severely disrupted by emigration. Senior physicists who left on account of the 1938 anti-Semitic laws included Bruno Rossi, Emilio Segrè, Giulio Racah and Enrico Fermi, whose wife was Jewish. They were joined by some younger scholars such as Bruno Pontecorvo. Hungarian mathematicians and theoretical physicists who left their homeland (often before fascism arrived) formed a particularly bright constellation of talent: the mathematician John von Neumann, the physicists Theodore von Kármán, Leo Szilard and Eugene Wigner, and the chemist Michael Polanyi, among others. More than one hundred physicists from Central Europe found academic posts in the United States before 1941.[64] The argument of Donald Fleming that four particular émigré scientists – three physicists (the Germans Max Delbrück and Erwin Schrödinger and the Hungarian Leo Szilard) and one biologist (the Italian Salvador Luria) – decisively transformed biology will be assessed later.

Fritz Haber, who had converted to Christianity, and had regarded science in the service of the German state as his vocation, was not spared. He had developed poison gas for the German military and served loyally in command of the gas troops. He had even, after the Treaty of Versailles saddled Weimar Germany with crippling reparations, sought to extract gold from seawater to settle the debts. Nazi law defined a Jew by ancestry, which overrode Haber's service and Christian conversion. In the spring of 1933, Haber was told to dismiss junior Jewish scientists in his institute. He chose instead to fire two senior scientists, Herbert Freundlich and Michael Polanyi, assuming that they would have a greater chance of securing safe havens abroad.[65] Haber resigned soon after. Max Planck's attempt to intervene on behalf of senior Jewish scientists, first with the Prussian minister of science, art and education Bernhard Rust and then with Hitler himself, failed. Haber journeyed through Holland, France and

England. 'I am especially glad', wrote Einstein to Haber, 'that your earlier love for the blond beast has cooled off a bit.'[66] In 1934, ailing, he left for a sanatorium in Switzerland. He died two days after arriving in Basel.

Individual lives were deeply scarred, and worse. Some particular institutions were also singled out: the Göttingen institutes of physics and mathematics were nearly emptied. Richard Courant subsequently built up a whole new institute of applied mathematics at New York University. In Austria, the Vienna Circle was broken. Moritz Schlick was shot by an anti-Semitic student in 1936. Rudolf Carnap had left for the United States the year before. Otto Neurath, a socialist with Jewish ancestors, was in immediate danger following the Anschluss; he left first for the Netherlands and then Britain.

Other disciplines were relatively unaffected. Dismissal and exile for some presented individual and professional opportunities for others: in 1942 there were more professors of psychology than there had been in the year before the Nazis seized power.[67] Deichmann concludes that 13 per cent of biologists active in 1932 were dismissed and 10 per cent emigrated.[68] These émigrés were much more productive than those biologists left behind, as measured by citation counts. Nevertheless, says Deichmann, despite such losses the forced emigration of scientists 'did not lead to the disappearance of entire fields of research'; indeed, 'funding for basic biological research increased steadily from 1933 to the end of the war.'[69] The Kaiser Wilhelm Societies benefited particularly. Biologists did not even find it a particular advantage to be a member of the Nazi Party. Deichmann finds the big picture of biology in Nazi Germany to be one of generous funding and 'continuity' in research, of universities and institutes where 'good basic research' was done at the same time as biologists legitimated racial hygienist policies. 'Compromise and collaboration' was found in both physics and biology.[70]

In general, while historians have understandably been drawn to study the 'emigration of scientists and scholars [that was] a mass phenomenon unprecedented in the modern history of academic life', and while we will see the impacts that the refugees made on science outside Central Europe, Mitchell Ash and Alfons Söllner warn against treating the subject too simplistically. In particular, the history of forced migration should not be seen just in terms of loss and gain, as if the émigrés were 'a sort of human or intellectual capital, or . . . prestige objects'.[71] The scientists who travelled did indeed bring new ideas and new skills to new lands. An émigré, Michael Polanyi, provided a term of analysis – 'tacit knowledge' – that named the bodily skills, knowledge and practices that would otherwise have not trav-elled well. The émigrés took with them their tacit knowledge.[72] The export of German research styles may even have lessened differences between national styles – a process of 'denationalization' of science that may have been encouraged by forced migration.[73] But the knowl-edge often changed – was not just lost or gained – as it was replanted in new contexts. Forced migration 'made possible careers that could not have happened in the smaller, more restrictive university and science systems of Central Europe', note Ash and Söllner, and 'the pressure to respond to new circumstances may have led to innova-tions that might not have occurred in the same way otherwise.' Or the new contexts could prove barren too. Even when they had escaped Germany, Jewish scholars confronted anti-Semitism. The physicist James Franck, a Nobel laureate, who had arrived at Johns Hopkins in 1933, was soon forced out by the prejudices of the university presi-dent, Isaiah Bowman.[74]

— 11 —

SCALING UP, SCALING DOWN

This chapter concerns mostly, but not exclusively, American science of the 1920s and 1930s. There are seven interlinking themes. The first theme is the increasing size of scientific instruments. For example, take telescopes. A measure of the power of a reflecting telescope is the diameter of the primary mirror: Mount Wilson's 60-inch instrument was completed in 1908 and a 100 inch followed in 1917. In 1928, the International Education Board of the Rockefeller Foundation awarded the funds to build a telescope of twice the latter's size. It took the good part of two decades to build, but the Mount Palomar 200-inch instrument, which would be the dominant optical telescope until the launch of the Hubble Space Telescope, saw first light in 1949. Cyclotrons, which would become the central instrument of nuclear physics, use powerful magnets to hold charged particles to circular tracks. The larger a cyclotron was, the higher energies it could reach and the more powerful it became. In the 1930s cyclotrons grew from a few inches in diameter to many metres. Cyclotrons joined other large-scale apparatuses, such as van de Graaf generators, mass spectrometers and linear accelerators, in the physics laboratories of the decade. Nor was the trend restricted to the physical sciences. Large, expensive instruments, such as ultracentrifuges, which used incredibly fast spin speeds to separate out different-sized objects of biological interest, and electron microscopes, which drew on the quantum understanding of matter as waves to probe and picture the world at tiny scales, were to transform investigation in the life sciences from mid-century.

Large instruments were expensive. They also required different sorts of skills to build than might have been possessed by the technicians found in the backrooms of turn-of-the-century laboratories.

Specifically, skills, such as would be needed to cast a giant electro-magnet for a building-sized cyclotron, were to be found in industry. So the second and third themes are increased roles of patronage and industry. How was the money secured to build the large-scale scientific instruments of the 1930s? What sort of industrial collaboration typified the sciences performed with such instruments? Crucially, the context of much 1930s science, especially the physical sciences, was the growth of industries: the radio industry, electronics and radio physics came as a package, as did the nuclear isotopes industry and nuclear physics.

There were also interconnections: building bigger accelerators required industrial electrical and radio skills. The resolution of controversies such as that between Cambridge Cavendish Laboratory physicists and their rivals at Vienna on the nature of radioactive decay required new electrical instruments for detecting and counting particles. These instruments in turn could only be built with the availability of the skills and the 'reliable and inexpensive electronic components of a booming radio industry'.[1] The radio industry, as it spun off from telephony, founded its own research laboratories. AT&T's Bell Labs, for example, was opened in 1925. Governments and radio companies shared an interest in how radio waves were propagated. Navies needed to communicate, and radio networks needed to broadcast. The shared interest saw money channelled into investigations of the upper atmosphere. An 'ionosphere', far above the clouds, was named and explored. From 1924, Edward Appleton, a British scientist, used BBC transmissions to measure the heights of reflecting layers in the ionosphere. He received a Nobel Prize in 1947, partly in recognition of the fact that the development of interconnecting techniques in Britain had been crucial to the early development of radar.

Nevertheless, the history discussed in this chapter is overwhelmingly American and, indeed, predominantly Californian. How do themes of instrumental scale, economies of patronage, industrial skills and knowledge relate to this fourth theme of location, of the geography of scientific leadership? The great scientific institutions of the east coast (and mid-west) of the early twentieth century – the research universities of Johns Hopkins and Chicago, the continuing traditional strengths of Harvard and MIT, the institutes supported by Rockefeller and Carnegie – had been on a par with their analogues in the old world. The mushrooming sites of science of the west coast – Mount Wilson and Mount Palomar observatories, Lawrence's cyclotron laboratories, the rapidly growing California Institute of Technology (Caltech), home to Linus Pauling, Theodor von Karman and T. H. Morgan, or Berkeley, home to Robert Oppenheimer – would influence the world. The 1930s was the decade when American leadership in science was discernible. 'The United States leads the world in physics', ran a *Newsweek* article in 1936; historian Daniel Kevles adds his assessment: 'so it did'.[2] The same case could be made for the life sciences. Furthermore, the leadership was western in emphasis. Does it make sense, and, if so, how and why, for historians to speak of the Americanization, or even the Californification, of science by the mid-twentieth century?

In the early 1930s it was not just science that was on the move: streams of humanity were leaving the Dust Bowl of Texas and Oklahoma to find a new beginning by the Pacific. The Wall Street Crash of 1929 and the Depression that followed it provide the inescapable economic context for work in the 1930s, including scientific work. Indeed, science was seen by some as a contributory cause of the crash. In the 1920s there had been a small, countervailing tradition of humanist dissent with the apparent successes of science. In 1927, the bishop of Ripon had called, apparently partly in jest, for a ten-year moratorium on scientific research; but he struck a chord with 'humanists' – a handful of professors of literature, writers and poets.[3] After the crash, 'going beyond the earlier humanist critique', Kevles notes, 'thoughtful Americans naturally asked whether science was not responsible, at least in part, for the failure of machine civilization.'[4]

Others saw science as part of the solution. The economical framework outlined by the British economist John Maynard Keynes in his 1936 magnum opus *The General Theory of Employment, Interest and Money* was an analysis and remedy for the working world of the Depression. Some American scientists even looked to the Soviet experiment as an example to follow.[5] Awakened to radical politics by the Depression, the parallel cause of anti-Nazism encouraged a small number of American scientists, such as the physiologist Walter B. Cannon and the anthropologist Franz Boas, to campaign and organize for causes of the left. Hermann Muller, who had worked in Texas in the 1920s, having published his ground-breaking experiments on the production of mutations in *Drosophila* by X-rays in 1926, even moved to the Soviet Union in 1933.

In general, the response to the Depression was to encourage and deepen pre-existing trends towards large-scale organization and expert planning as paths out of the crisis. This emphasis is the fifth theme. Progressive engineers in the first quarter of the twentieth century had hoped that their expertise could be deployed to cure America's ills; they had pinned their hopes on Herbert Hoover, but the movement had run out of steam.[6] After 1929 the technocratic idea returned, with force. Indeed, in the hands of Howard Scott, a Technocracy Party called for an economy planned centrally by experts, with the value of goods measured not by the price obtained at market but by the amount of energy used in production.[7] The Century of Progress International Exposition in Chicago during 1933 and 1934 was held under the banner 'Science Finds, Industry Applies, Man Conforms'.

Planning experienced a true golden age in the bodies set up under Roosevelt's New Deal. Social science thrived. Apart from a flurry of industrial planning in the First World War, it was not until the Depression that the federal government was regarded as a legitimate agent of technological transformation, outside of the working world of the military.[8] In the 1920s, enthusiasm for regional planning began to grow, encouraged by the commentator of technology Lewis Mumford, who had read the ideas of an Edinburgh biologist, Patrick Geddes. With Roosevelt, the regional planners' chance had come. The Tennessee Valley Authority (TVA), set up in 1933, planned the economic development of a whole region through the re-engineering of river systems: distributing of electricity from hydroelectric power stations, controlling floods, and conserving soil. 'For the first time in history', celebrated one of the TVA leaders, David Lilienthal, 'the resources of a river were not only to be "envisioned in their entirety"; they were to be developed in that unity with which nature herself regards her resources – the waters, the land, and the forest together, a "seamless web".'[9] The TVA was a high-water mark for pre-war technocracy. Its model of planned governmental action, as well as its prodigious electricity, would, as we see later, be critical for the watershed scientific and technological project of the twentieth century.

The TVA was an example of the intention to organize a way out of the Depression. Planning and organization on an ambitious scale also marks other American projects of the 1930s, including philanthropic ones. Before the 1930s, the American Society for the Control of Cancer had worked quietly to improve cancer education and encourage early diagnosis through self-examination. In the New Deal period, the organization rebranded itself as the American Cancer Society, set up a Women's Field Army, numbering 100,000 by 1939, to collect money, and eventually, led by Mary Lasker, became the 'leading medical research entrepreneur' in the United States.[10] Likewise, the March of the Dimes (the National Foundation for Infantile Paralysis, launched 1938) tapped large-scale organization to raise funds for research into polio, the disease which ravaged many, not least the president himself. Finally, in a case investigated at length below, the Rockefeller Foundation launched in 1933 a major programme, the Science of Man, which would coordinate and reshape key components of the life sciences.

If the fifth theme was large-scale organization and expert planning, then the sixth is a rider: scaling up was not trivial, it was an achievement which could only be reached after considerable further experiment and development. The skills learned about scaling up would not be lost and would be critical to the success of large-scale scientific projects, especially American projects, after the 1930s. Knowledge of management and production were brought together with research disciplines. The problems of scaling up present one set of examples of the seventh and final theme, interdisciplinarity. Other sets will be found in the problems of designing, building and using large instruments (theme 1) and the coordinated realignment of disciplines found in projects such as the Rockefeller Foundation's Science of Man (theme 5). Interdisciplinarity thrived in American universities of the 1930s, especially in Californian universities such as Caltech and Berkeley, at least when compared to, say, German institutions.[11] A new question becomes pertinent: was a distinctive interdisciplinarity a crucial component of American leadership in the sciences from the mid-twentieth century?

Having introduced the themes, here is the order of play. We start with three Californian scientists: Ernest O. Lawrence, J. Robert Oppenheimer and Linus Pauling, discussed in relation to their respective sciences – particle and nuclear physics, theoretical physics, and chemistry. Pauling was a key player in the next topic discussed: the science of macromolecules and the emergence of a 'molecular biology', as coordinated by the Rockefeller Foundation and given shape by the ideas of émigré physicists. Instrumentation, too, was crucial to the form and direction taken by these life sciences. Three cases will be examined that reveal what needed to be learned in order to scale up laboratory science to industrial production: radium, insulin and nylon. Finally, we end a chapter that covers many different, and some novel, ways of combining public and private enterprise to support science with a brief discussion of three contrasting cases that proposed models for how science should relate to the working world: the macro-planning of John Maynard Keynes and J. D. Bernal, the early experiments with venture capital in Boston, and the corporate vision expressed at the New York World's Fair of 1939.

New stylists: Californians and the Cavendish

We have already seen how the California Institute of Technology, Caltech, was built from patronage from Rockefeller, Carnegie and local business in return for a civic contribution, providing expertise for Californian projects of electrification. Robert Millikan, the Caltech physicist, had benefited from a large-scale million-volt laboratory built for him by Southern California Edison; the transformer alone filled a building 50 feet high and with a footprint of 300 square feet.[12] Such huge voltages could be used to accelerate particles which could then in turn be smashed into other objects, such as atomic nuclei or each other. The higher the energies involved, the more physics was revealed. One approach to increase the energy was to use very high potential differences in a one-off acceleration. Merle Tuve at the Carnegie Institute of Washington and Ernest Walton at the Cavendish Laboratory, Cambridge, used the new high-voltage generators developed by Robert van de Graaf in this manner. Another approach was to push the particle a little faster many times. If this was done in a straight line, as the Norwegian Rolf Wideröe suggested, one could build a linear accelerator. The trouble was that one ran out of space. Ernest Lawrence, of the University of California, Berkeley, developed the idea of constraining charged particles to a spiral path, using magnets to bend the paths, and accelerating them bit by bit: the principle of cyclotrons. The techniques of electrical field pulsations were radio engineering ones. Indeed, an 84-ton magnet used by Lawrence was also scavenged from the radio industry: it had been part of an old Palo Alto arc generator used for intercontinental radio communication.[13]

Lawrence's skills and contacts, necessary to build cyclotrons, were therefore found in the working world of Californian electrical and radio engineering. Starting in 1930, Lawrence and his student, Stanley Livingston, built a proof of concept device, 4½ inches in diameter, that accelerated hydrogen ions to energies of 80,000 electron volts. Later, in 1931, they had built an 11-inch cyclotron that reached 1 million electron volts. As his cyclotrons grew in size, so they required larger facilities to house them and new ways of funding and managing the work. The next cyclotron, 27 inches in diameter, was the one that required the 84-ton magnet. Its building was called the Radiation Laboratory. A 37-inch cyclotron was complete by 1937, a 60 inch by 1939, and a 184 inch during the war. Funds were found from the state of California as well as from philanthropic sources: the university's own spin-off foundation, the Research Corporation, which channelled revenues from the chemist Frederick Gardner Cottrell's patent on his electrostatic precipitator, funded the 27-inch cyclotron, while the Rockefeller Foundation would later donate $1.15 million.[14] Lawrence reinterpreted the cyclotron as a medical instrument to encourage these donations, offering high-energy X-ray and neutron sources as cancer therapies, despite doubts about their effectiveness. As Lawrence admitted to Bohr in 1935: 'I must confess that one reason we have undertaken this biological work is that we thereby have been able to get financial support for all of the work in the laboratory. As you know, it is much easier to get funds for medical research.'[15]

Nevertheless, this therapeutic role was enough to persuade the federal government, through the National Advisory Cancer Council, to support Lawrence's work as well.[16] The New Deal also contributed: the Radiation Laboratory, through the Works Progress Administration and the National Youth Authority, employed electricians, machinists, carpenters and other technicians, all on the government's payroll.[17] The total funds supplied to Lawrence's Radiation Laboratory, calculates Robert Seidel, equalled 'the entire sum going to the support of all academic physics, exclusive of new plant, in 1900'.[18] And all this in the Depression.

In addition to the increasing scale of instruments and the diversion of unprecedented streams of patronage, historian Robert Seidel outlines other ways that Lawrence's Radiation Laboratory prefigured twentieth-century shifts in the style and leadership of science. First, he says, distinctions between science and technology were hard to draw. Was building and using cyclotrons physics or engineering, or both? Certainly, mechanical and electrical engineering skills were necessary, not just to set up but also to operate, maintain and interpret the output of the instrument. However, if we remember that much of nineteenth- and twentieth-century physics, even the theoretical physics, must be considered as part of working worlds, then we realize that this technological aspect was not new – rather, perhaps, it was brought to the surface, in Lawrence's laboratory. Second, cyclotron science was interdisciplinary science, whether it was combinations of nuclear physicists, radiochemists, radiobiologists and medics for the therapy work or the teams of engineers and physicists for the instrument's experiments. Furthermore, a hierarchy of management was installed to manage this organizational complexity.

Putting this all together, the historians John Krige and Dominique Pestre identify the emergence of a new style of science:

a profound symbiosis previously unknown in basic science, a fusion of 'pure' science, technology and engineering. It was the emergence of a new practice, a new way of doing physics, the emergence of a new kind of researcher who can be described at once as physicist, in touch with the evolution of the discipline and its key theoretical and experimental issues, as conceiver of apparatus and engineer, knowledgeable and innovative in the most advanced techniques (like electronics at the time) and able to put them to good use, and entrepreneur, capable of raising large sums of money, of getting people with different expertise together, of mobilizing technical resources.[19]

Lawrence, they suggest, was the first example of this type. Others like him – Luis W. Alvarez, Edward Lofgren, Edwin McMillan, Wolfgang Panofsky and Robert Wilson – 'became the masters of the new physics and who imposed their rhythm on world science'.[20] 'The fascination with hardware', writes Seidel, 'like the subordination of individual to group science, was to spread from accelerator laboratories to other parts of physics and from the United States to the rest of the world, and to make science and technology one.'[21]

Significantly, the most productive physical laboratories in the interwar years were those, such as the Cavendish Laboratory at Cambridge, which also scaled up their research. We have seen in a previous chapter how Ernest Rutherford brought his experimental physics to Cambridge and how a model of the nuclear atom was articulated in conversation with theorists such as Niels Bohr. After the First World War, Rutherford's Cavendish had become one of the essential places to pass through for aspiring physicists, along with Göttingen and Copenhagen. As numbers of researchers grew, so the Cavendish began to organize into research groups. Nuclear research continued. The Russian Peter Kapitza arrived in 1921, and would become in effect a deputy head to Rutherford, as well as spinning out new lines of research of his own, such as low-temperature studies in the early 1930s. Kapitza, having returned briefly to the motherland, would be ordered by Stalin to remain in the Soviet Union in 1934.

The articulation of nuclear theory and experiment continued alongside the developments in instrumentation. Francis Aston, a chemist, had looked for ways to settle the question of isotopes. After the interruption of the First World War, Aston built, in 1919, a mass spectrograph, a means of determining the mass of ions by measuring how much their paths were deflected by a magnetic field, heavier ions being deflected less than lighter ones. The mass spectrograph would become a vital instrument of twentieth-century physics. Other instruments were devised for imaging and measuring: C. T. R. Wilson's cloud chamber, devised in the 1900s and 1910s, was adapted by other Cavendish physicists, such as Takeo Shimizu and Patrick M. S. Blackett, an ex-naval cadet.

Blackett's Cavendish research provides some telling indicators of scale. Rutherford had bombarded nitrogen nuclei with alpha particles, observing what he thought was the disintegration of the nitrogen as revealed by the dimmest of green fluorescent flashes on a zinc sulphide screen. Each flash had to be registered by a human eye and recorded with a human hand. Rutherford asked Blackett to pick up Schmizu's project to generate records of these events. He took the cloud chamber and added a spring action that triggered both the expansion of the chamber and the shutter of a camera.[22] This improved cloud chamber was ready in 1922. By the summer of 1924, Blackett had collected 415,000 tracks of ionized particles on 23,000 photographs. Among them were just eight tracks of primary interest: they showed an alpha particle hitting a nitrogen nucleus, producing a proton and an oxygen ion. Against Rutherford's hypothesis, it was not disintegration but the incorporation of an alpha particle – perhaps due to some kind of 'nuclear' force?[23] This raised important questions for physics, but, crucially, the collection of evidence was practically achievable only through the automatic registry and photography of particles. Automation was another lesson learned for dealing with the increasing scale of physical experiment.

Some significant Cavendish work remained relatively small-scale. James Chadwick, Rutherford's right-hand man on the nitrogen disintegration experiments, continued the studies of the transformation of elements. He knew from work by Walther Bothe and others in Germany that, when high-energy alpha particles bombarded certain light elements, a penetrating radiation, presumed to be gamma rays, was produced. The Parisians Irène Joliot-Curie and Frédéric Joliot had shown that these supposed gamma rays in turn produced protons when the target was paraffin. Chadwick ruled out the gamma ray suggestion, instead proposing in *Nature* in 1932 that the penetrating radiation was a new particle, the neutron. This was novel, but depended on familiar and traditional methods. However, most of the Cavendish discoveries that depended on relatively small-scale apparatus directly made use of these scaling techniques indirectly. In 1933, working with Giuseppe Occhialini, Blackett, deploying automation techniques – using Geiger counters to trigger cloud chamber photographs on the passage of cosmic rays – found tracks that he attributed to positive electrons. This finding confirmed the observations made the previous year by Carl D. Anderson, a protégé of

Millikan, at Caltech. The positron had been described as a theoretical possibility by the Cambridge mathematical physicist Paul Dirac in 1928, in his relativistic theory of electrodynamics. Notice, however, that Californian science was nosing ahead, at least in experimental physics.

A final Cavendish example illustrates the phenomenon of scaling up. Two researchers who joined the Cavendish in the 1920s were the Irishman Ernest Walton and the Lancastrian John Cockcroft. The task they were set was the bombardment of lithium nuclei with protons. This project had been prompted by calculations made by George Gamow, an Odessa-born theorist who had passed from Leningrad through Göttingen and Copenhagen and was pausing in Cambridge before returning east. (He would defect to the United States in 1934.) But the bombardment of lithium was also an entirely mainstream project within Rutherford's Cavendish research tradition. However, to succeed in their task Walton and Cockcroft began to build on an industrial scale. Cockcroft had been previously employed at the specialist engineering firm Metropolitan-Vickers of Manchester, and he brought these industrial machine-building skills, contacts and products to the Cavendish.[24] Their room-sized apparatus generated immensely high potential differences. Using some 700,000 volts to accelerate the protons, Cockcroft and Walton broke apart the lithium nuclei, an artificial disintegration. A good Cambridge journalist contact with the national press, J. G. Crowther, made sure that the event was splashed in the media. 1932 became publicly celebrated as an 'annus mirabilis' for the Cavendish, and newspapers presented nuclear science as 'the most exciting branch of contemporary science'.[25] Like Lawrence's cyclotrons, the scale and expense of these machines was encouraging not only the industrialization of research but also management of the public image of science.

The discoveries made at the Cavendish – along with others such as Frédéric Joliot and Irène Joliot-Curie's creation of artificially radioactive isotopes in 1934 – were regarded by Lawrence as missed opportunities. An 'obsession' with scaling up the Radiation Laboratory machines, notes Peter Westwick, 'could come at the cost of research results'; some of the 'discoveries could have been made in Berkeley, if Lawrence had diverted some attention from improving accelerators to the particles produced and the means of detecting them.'[26] Nevertheless, the strategy did begin to pay off. The Radiation Laboratory became a factory for new isotopes. The Italian Emilio Segrè found a new artificial element, the first, technetium (atomic number 43), in a sample of molybdenum used as a target in one of

Lawrence's cyclotrons. (Segrè would emigrate to join Lawrence when Mussolini approved anti-Semitic laws in 1938.) Edwin McMillan and Philip Abelson made the first transuranic element, neptunium (atomic number 93), in 1940. An even more significant element, plutonium, was soon added by Glenn T. Seaborg, and will be discussed later. The young Luis Alvarez, with the older Felix Bloch, carefully measured the magnetic moment of the neutron. This fundamental study would later ground developments in nuclear magnetic resonance.

However, one set of experiments in particular illustrates the inter-disciplinary promise of work at Lawrence's laboratory. Given the centrality of carbon chemistry to life, in the late 1930s both Harold Urey's team at Columbia University and Lawrence's group developed an interest in isotopes of carbon as biological tracers. Carbon-12 was the familiar, stable, 'normal' isotope. Another isotope was known, carbon-11, but it had a half-life – just twenty-one minutes – that was too short to be useful. In late 1939, Lawrence set two young research-ers, the Chicago-trained Martin D. Kamen and the chemist Samuel Ruben, the task of finding a heavier, long-lived isotope. In February 1940, Kamen ran a three-day experiment, bombarding graphite with deuterons from the 37-inch cyclotron. (Deuterons had been identified as combinations of protons and neutrons by Urey in 1931.) George Kauffman picks up the story:

> For three nights in a row Kamen did not sleep so that the target probe could accumulate a [long exposure] . . . Just before dawn on February 19, 1940 he withdrew the probe, scraped off the graphite, and left it on Ruben's desk in a weighing bottle. On his return home, during a thunderstorm, 'with eyes red-rimmed from lack of sleep, unsteady gait from weariness, and a three-day growth of beard', he 'was picked up by the police as a likely suspect for a mass murder perpetrated a few hours earlier somewhere in the East Bay'. When the victim of the crime failed to recognize him, he was released. On returning to his one-room apart-ment he collapsed into a twelve-hour sleep.[27]

Ruben, carefully ruling out contaminants, determined that some of the carbon-12 had absorbed a neutron from the deuterons to form a new isotope, carbon-14, with an estimated half-life of over a thousand years. Willard Libby would turn measurements of this isotope into a revolutionary tool for fields as diverse as archaeol-ogy and anthropology – radiocarbon dating, discussed later. Kamen himself, and others in a story also discussed later in chapter 15, would use carbon-12 as a biological tracer to unpick a biochemical process central to life on earth: photosynthesis. Knowledge of other metabolic pathways would follow. While being only one of many projects for

Lawrence's laboratory, carbon-12 would be at the heart of important and novel interdisciplinary science.

Viewed from the standpoint of disciplinary histories, a seeming paradox emerges: at the same time as setting off interdisciplinary lines of research, nuclear physics was emerging as a field with its own stability and identity. Hughes dates this emergence to the period 1932 to 1940, with Enrico Fermi's Rome conference of 1931, 'Congresso Internazionale di Fisica Nucleare', an early marker.[28] Lawrence's Radiation Laboratory showed that nuclear physics could flourish on the west coast of the United States. Oppenheimer's career shows that theoretical physics, too, could take root.

J. Robert Oppenheimer was born in 1904 in a family that was 'ethnically and culturally Jewish', but which 'belonged to no synagogue'; his parents instead sent him to the schools of the Ethical Culture Society, which inculcated rationalism, secular humanism, social justice and faith in progress.[29] This upbringing was pressurized and drove Oppenheimer to displays of prodigality: 'Ask me a question in Latin', he would boast, 'I will answer you in Greek.'[30] He enrolled at Harvard, where he was taught physics by Percy Bridgman. On graduation, Oppenheimer aimed, as all aspiring physicists did, to complete further study at the great European centres. To Rutherford at Cambridge, Bridgman wrote that, despite Oppenheimer's

> perfectly prodigious power of assimilation ... his weakness is on the experimental side. His type of mind is analytical, rather than physical, and he is not at home in the manipulations of the laboratory ... It appears to me that it is a bit of a gamble as to whether Oppenheimer will ever make any real contributions of an important character, but if he does make good at all, I believe he will be an unusual success.

He added: 'As appears from his name, Oppenheimer is a Jew, but entirely without the usual qualifications of his race.'[31] Despite such warm endorsement, Oppenheimer was accepted into the Cavendish in 1925.

At Cambridge, Oppenheimer met Blackett, consummate experimenter and socialist. The confrontation seems to have triggered a breakdown in the emotionally brittle young theorist. 'Consumed by feelings of inadequacy and intense jealousy', Oppenheimer announced that he had left a poisoned apple on Blackett's bench.[32] Blackett was unharmed. Oppenheimer, however, rode an emotional crisis that peaked, abruptly, and dissipated in 1926. He moved to the next station on his European tour: Göttingen, where his theoretical talents shone and his American wealth made him a social hub. His confidence returning, he composed a string of influential papers on topics including the quantum mechanics of molecules (the 'Born–Oppenheimer approximation') and, in 1927, the phenomenon of quantum tunnelling.

Before taking up a highly unusual double Californian appointment, as professor of physics at both Caltech and Berkeley, Oppenheimer completed postgraduate finishing school with stints with Paul Ehrenfest at Leiden and – travelling with a new friend, Isidor Rabi – Wolfgang Pauli at Zurich (in preference to Bohr at Copenhagen). Once settled on the west coast, Oppenheimer at Berkeley became a magnet for American theoretical physicists, attracted too by the data being generated by Lawrence's machines. He was charismatic: the students aped his style, his clothes, his posture and his gestures and adopted his brand of cigarettes (Chesterfields, say biographers Kai Bird and Martin Sherwin).[33] Oppenheimer and Lawrence were great friends but diverged politically in the 1930s: the theorist, perhaps in belated accordance with his Ethical Culture Society schooling, swung to the liberal left in the mid-1930s, while the cyclotron-builder adopted the conservative, anti-New Deal politics of the patrons he schmoozed.[34] Just as significantly, Oppenheimer and Lawrence shared a style of work. Just as Lawrence's aim was a working cyclotron rather than elegant design, so, as colleague Robert Serber told Sherwin:

> Oppie was extremely good at seeing the physics and doing the calculation on the back of the envelope and getting all the main factors ... As far as finishing and doing an elegant job like Dirac would do, that wasn't Oppie's style. [He worked] fast and dirty, like the American way of building a machine.[35]

In the late 1930s Oppenheimer, with his students, dashed off extraordinary physics papers, on cosmic rays, gamma rays, electrodynamics, nuclear physics (with Melba Phillips), and astrophysical topics such as neutron stars (with Serber and George Volkoff) and gravitational collapse to form black holes (with Hartland Snyder). These last two objects would be at the centre of whole observational programmes for astronomy in the later twentieth century.

Oppenheimer's late 1930s work was distinctive: it drew on a massive range of physics, it was synthetic, and it cut to the chase, following theory through to illuminate astrophysical or experimental topics. It drew on European achievements. Wolfgang Pauli in 1930 had proposed a radical step to preserve energy conservation in beta decay: a 'desperate remedy' in which a new neutral particle – what we now call the neutrino – was predicted and described. Enrico Fermi, in 1933, offered a quantum field theory in which electrons 'acquire their existence at the very moment when they are emitted; in the same manner as a quantum of light. In this theory, the total number of electrons and of the neutrinos . . . will not be constant, since there might be processes of creation and destruction of these light particles.'[36] Here was a new model for describing the electromagnetic force, in terms of the creation and destruction of particles. Quantum electrodynamics (QED), as we shall see, would be articulated in its classic form after the Second World War.

A new picture of nuclear forces was also emerging in the late 1930s, one that illustrates both the interplay between theory, observation and experiment and the changing global geography of physics. In 1932, Werner Heisenberg attributed a new property, 'isotopic spin', to neutrons and protons. This was, notes historian Sam Schweber, 'the first example of the two kinds of internal quantum numbers eventually used to classify particles, namely . . . [, first, conserved] additive quantum numbers, like electric charge, strangeness, baryon and lepton numbers; and [, second,] "non-abelian" quantum numbers, such as isotopic spin, that label families of particles.'[37] In other words, Heisenberg's move, like Fermi's treatment of electrons, was a model for post-war physics. Heisenberg had offered an explanation for nuclear force in terms of the exchange of an electron (later an electron–neutrino pair, to conserve energy) between a proton and neutron. In 1935, the Japanese physicist Hideki Yukawa, working at Osaka University, published a paper, 'On the interaction of elementary particles, I', in English, in the *Proceedings of the Physico-Mathematical Society of Japan*.[38] Yukawa proposed a field theory of nuclear forces and predicted the existence of a new particle, the 'U-quantum' (what would later be called a 'meson'). The short range of the nuclear forces would be explained by the fact that the meson had mass.

The 'initial reaction', report historians Helmut Rechenberg and Laurie Brown, 'was practically nil even in Japan, and although the paper was in English and in a well-recognized journal, no one abroad mentioned it for nearly two years'.[39] Theoretical and experimental physicists were relatively separated in Japan (except at Yoshio Nishina's Riken laboratories), and anyway Yukawa's theory depended upon a particle no one had seen. However, studies of cosmic rays using counter-controlled cloud chambers were under way in London (by Blackett, located on a disused platform of Holborn underground station) and California.[40] Carl Anderson and Seth

Neddermeyer, at Pike's Peak and Pasadena, California, captured cosmic ray tracks, some of which fell within quantum electrodynamics' predictions for electrons, while others looked peculiar (indeed, anomalous cosmic ray tracks had been picked up since 1933, without prompting any discovery claims). Anderson discussed his results in-house at a Caltech colloquium and also, at the end of his Nobel Prize address of December 1936 (the prize had been awarded for his finding the positron in earlier cosmic ray tracks), announced this recent discovery of new 'highly penetrating particles'.

In the same month, a few hundred miles north, at Berkeley, Oppenheimer and his student Franklin Carson wrote a paper for the *Physical Review* that proposed a 'radical alternative' to identifying the cosmic ray tracks as electrons: 'If they are not electrons, they are particles not previously known to physics.'[41] Yukawa, following Anderson's announcements in Japan, fired off a letter to *Nature*, claiming that his U-quanta could be the penetrating particles. *Nature* rejected it.[42] Meanwhile Japanese experimenters, part of Nishina's team, temporarily transposed to Yokusuka Naval Arsenal to borrow a generator normally used to charge up submarine batteries, ran cloud chamber experiments that led to estimates of the particle's mass – roughly a tenth of a proton. (A delay in publication allowed an American team to claim credit for this first determination.) In mid-1937, letters to physics journals, especially Oppenheimer and Serber's letter to the *Physical Review*, identified the middle-weight, penetrating cosmic ray particle with Yukawa's U-quantum, and, although they were dismissive of Yukawa's theory,[43] the link to the Japanese theorist had been forged.

Yet, ultimately, the broader significance of the Yukawa story is not just to illustrate the difficulties and contingencies that prevented or allowed Japanese science to claim credit for its achievements in the twentieth century, but rather to emphasize that leadership in physics by the 1930s depended on the possession of both thriving and interconnecting centres of theoretical and experimental physics. Japanese physics was building these links, as the Nishina group illustrates. But California, at Caltech and Berkeley, was where the growth and linkages were strongest.

A third major figure of twentieth-century Californian science, the chemist Linus Pauling, also fits this description. Pauling's life spanned nearly the whole century: born in 1901 in Portland, Oregon, he died in 1994, having collected two Nobel prizes along the way: for chemistry in 1954 and peace in 1962. The early death of his father, in 1910, was traumatic for the young Pauling, and the financial difficulties that the family subsequently faced pushed him to make a living to contribute.[44] Nevertheless, he graduated from Oregon Agricultural College in chemical engineering in 1922, and moved south to Caltech, just as Hale, Millikan and chemist A. A. Noyes were reinventing and expanding the university as an interdisciplinary powerhouse serving both the ideals of research and the working world of Southern Californian industry. Pauling quickly found his feet, finishing a doctoral dissertation on 'The determination with X-rays of the structure of crystals' in 1925. (He also married: Ava Helen Miller had been a fellow student at Oregon Agricultural College. A committed pacifist, the daughter of a left-wing father and suffragist mother, she greatly influenced Pauling's political campaigns.) Noyes secured his 'favourite student' a new Guggenheim fellowship to take him to Europe to study with Arnold Sommerfeld in Munich.[45] It was a bit of a honeymoon. He was also in the right place (Germany) at the right time (1926) to witness the great intellectual three-way debate over quantum mechanics: Heisenberg and Pauli wrestled with the old Bohr–Sommerfeld model of the quantum atom, while Schrödinger offered wave mechanics as a brilliant alternative.

Pauling began to write papers that used quantum mechanics to explain the shape and size of atoms, by accounting for the behaviour of an atom's outermost electrons. He was also reading proofs of Heisenberg's 1927 paper on the uncertainty principle. Significantly, he was not distracted at all by the philosophical questions raised, as he later recalled: 'Even if this were a classical world, it would be impossible for us to determine the positions and momenta of all the particles of the universe by experiment. Even if we did know all of them, how could we carry out the computations? . . . I think it is quite meaningless to argue about determination versus free will, quite independent of the uncertainty principle', adding, 'I have never been bothered by the detailed or penetrating questions about interpretation of quantum mechanics.'[46] Pragmatic success was far more important than philosophical niceties for Pauling.

Pauling passed through Copenhagen (to learn from Bohr), via Göttingen (meeting Dirac, Heisenberg again, Jordan and Born, and crossing paths with Oppenheimer) and on to Zurich (for Schrödinger). In Zurich, he heard from two young researchers, Walter Heitler and Fritz London, that they had applied wave mechanics, as well as Heisenberg's concept of 'resonance', to give a radical new account of how atoms could form bonds.[47] Two hydrogen atoms could exchange electrons, with the effect of drawing the atoms together, creating a bond. It had a basis in quantum theory for the shared-electron bond proposed by the American chemist Gilbert N. Lewis. Pauling immediately grasped that Heitler and London had shown, in one case, how new physics – quantum mechanics – could provide a new foundation for chemistry.

On his return to Caltech in late 1927, Pauling had the knowledge and skills and vision for a lifetime's research programme in chemistry. By 1928 he had worked out a suggestive but not watertight account of the tetrahedral bonds of carbon, applying the resonance idea. He also developed effective pragmatic rules for suggesting solutions for crystal structures given the clues of X-ray diffraction patterns. The general problem was that the computation work to move backwards from X-ray diffraction patterns to determine the shape of the crystal that formed them was immense, and grew with the size and complexity of the crystal. The choice was either hours, days, even months of calculation which might deliver an exact solution or, alternatively, to find and justify heuristics that cut out the improbable solutions, simplifying and speeding up the calculations. Just as Lawrence cut to the chase to make cyclotrons that worked, so Pauling wanted to find the quick, effective solutions. An engineering feel was required. 'Mustering everything he knew about chemistry and physics', writes his biographer Thomas Hager, 'Pauling was able to leap to a solution where others were left mired in a swamp of confusing x-ray data.'[48] Rules in place, Pauling churned out insight after insight into chemical structure: mica, talc, zeolites. In the early 1930s he was publishing significant papers at the rate of one every five weeks.[49]

Pauling summarized his quantum mechanical rules for chemistry in the *Journal of the American Chemical Society* in 1931 as 'The nature of the chemical bond', beginning a series of articles of the same name. One of the ideas that he developed was the notion that bonds could be placed on a spectrum from ionic and covalent extremes or, rather, resonant forms combining the two. Pauling's approach, while grounded in the techniques, both mathematical (quantum mechanics) and instrumental (X-ray crystallography), that were far from the experience of most other chemists of the first half of the twentieth century, could be expressed in a form that was nevertheless graspable. His style, for example, was in complete contrast to the obscure 'anti-intuitive' approach of the Chicago chemist Robert Mulliken, creator of a rival theory of bonds, the molecular orbital theory.[50] 'I felt', recalled Pauling of his state in 1935, 'that I had an essentially complete understanding of the chemical bond.'[51] This understanding was distilled in *The Nature of the Chemical Bond and the Structure of Molecules and Crystals: An Introduction to Modern Structural*

Chemistry, the profoundly influential textbook published in 1939. Here was chemistry as physics: 'chemistry that', noted Max Perutz, 'could be understood rather than memorized'.[52]

From large molecules to molecular biology

Yet present too was a potential to reform the life sciences. Triumphantly in 1930, Pauling had finally nailed the account of the tetrahedral bonds of carbon. In Pauling's hands, quantum mechanics had made sense of the chemistry of the basic element of life, a connection from physics, through chemistry, to biology. In 1932 and 1933, Pauling, working with his student George Wheland, analysed the structure of the benzene ring in terms of resonance. Organic chemistry, too, had fallen. Larger molecules beckoned.

To understand what happened next we need to step back and consider, first, the working world that generated studies of large molecules and, second, changing forces of philanthropic funding of science in the context of the gathering economic depression of the early 1930s. It is in the confluence of the two that we will find something beyond a mere potential: a plan to reform the life sciences.

First, large molecules: an early twentieth-century debate centred on the molecular interpretation of substances such as cellulose, rubber, starch, proteins and resins. The consensus, albeit supported by different and often opposing branches of chemistry, was that the molecules involved were small.[53] The branch called aggregate theory, for example, held that the substances formed by physically clumping the molecules together. This was not high-prestige science; rather it was what Adolf von Baeyer had dismissively labelled 'the chemistry of grease'. Hermann Staudinger, working at technical universities in Karlsruhe and Zurich, rejected this small molecule theory. Instead, he held that molecules larger than other chemists had dared to imagine not only existed but could explain the unusual properties of substances such as rubber and cellulose. He invented the term 'macromolecules' in 1922.[54] Physical methods, such as the X-ray crystallographic investigations of silk and cellulose at the Kaiser Wilhelm Instiutute for Fibre Chemistry in Berlin-Dahlem, at first supported the small molecule hypothesis. Staudinger, a 'conservative organic chemist' who, notes historian Yasu Furukawa, 'disdained physical chemistry and distrusted physical methods', rejected such evidence.[55]

However, in the 1920s, new instruments came on the scene. Theodor Svedberg was a chemist at Uppsala University who investigated the nature of colloids – suspensions of fine particles in fluids. By using the ultramicroscope, invented in 1903, colloid chemists had observed the particles being buffeted, displaying Brownian motion. Careful but tedious measurement of these motions could be used to deduce the range of particle size. Svedberg was looking for a faster technique.[56] At first he allowed gravity slowly to settle the colloid material – finer particles took longer to fall, and so size could be correlated to layers of sediment. Then, in 1922, Svedberg saw that he could replace gravity with the force of spin. Placing colloids in a centrifuge and spinning them as fast as possible resulted in sedimentation patterns that formed far more quickly than under gravity. This was the principle behind Svedberg's ultracentrifuge, operational from 1924. Svedberg tested his machine on a suspension of fine gold particles, and later in the year started investigating proteins. On the night of 16 October 1924, Svedberg was woken up by a telephone call from his assistant. Horse haemoglobin had quickly formed a single band in the ultracentrifuge; horse haemoglobin therefore seemed to be a molecule of a single, large size – they measured a molecular mass of 66,800 units.[57]

Here was evidence that supported Staudinger's claim of the existence of macromolecules. Other lines of evidence seemed to support the theory too: Wallace H. Carothers at Du Pont, discussed further below, showed that molecules made of repeating simple molecular forms could be made – synthetic polymers. With the upper molecular size limit removed, the architectural possibilities of chemistry seemed boundless, as Staudinger recalled in his autobiography:

> Molecules as well as macromolecules can be compared to buildings which are built essentially from a few types of building stones ... If only 12 or 100 building units are available, then only small relatively primitive buildings can be constructed. With 10,000 or 100,000 building units an infinite variety of buildings can be made: apartment houses, factories, skyscrapers, palaces, and so on.[58]

The 'grease' chemistry of synthetic polymers, rubber, cellulose, proteins, silk and starch was firmly a science that served the working world. For example, Emil Fischer worked with materials of direct interest to industry – such as lichen-derived substances for tanning. In the 1900s and 1910s he synthesized ever larger polypeptides, and interpreted proteins as such. The macromolecular hypothesis was developed critically by scientists such as Herman F. Mark and Kurt H. Meyer of IG Farben as part of industrial projects to work with rubber and to mass-produce synthetic rubber and new plastics such as

247

polystyrene. Understanding fibrous protein – the keratin of horn and wool – mattered to the textile industry and supported work at places such as the Kaiser Wilhelm Society for Fibre Chemistry and people such as William Astbury at Leeds University. This working world link provided the funds for expensive instruments such as X-ray crystallography apparatus and ultracentrifuges and the motivation for the science.

And it pushed scientists towards an instrumental view – a building blocks view – of life's processes. In 1930, Linus Pauling had visited Herman Mark at IG Farben. Mark demonstrated the X-ray crystallographic techniques and taught the American to relate the large-scale 'springy' character of rubber to its spiral-shaped macromolecular structure; the same techniques, suggested Mark to Pauling, were applicable to proteins.[59] A research programme here passed from Germany to the United States.

In 1932, Linus Pauling sat at his desk and wrote out a duplicate application for funds from the Rockefeller Foundation and the Carnegie Institute. He wanted $15,000 a year, over five years, for 'a unified series of investigations on the structure of inorganic and organic substances, involving both theoretical and experimental work', specifically:

> I desire to solve the wave equation for simple organic crystals and molecules [and to] develop a set of atomic radii and of structural principles enabling one to predict with confidence the atomic arrangement, including interatomic distances, of the normal electronic state of any molecule, and its stability relative to other molecules. This knowledge may be of great importance to biochemistry, resulting in the determination of the structure of proteins, haemoglobin, and other complicated organic substances.[60]

The mention of 'proteins' immediately interested the Rockefeller Foundation. Now, while it is easy to see why the textile industry might support keratin research, it is not immediately obvious why the Rockefeller Foundation should be interested in the shape of proteins. The answer lies in the interests of Rockefeller Foundation administrators and trustees in response to the social and economic storms of the late 1920s and the 1930s.

The two administrators who had most influence on the direction of the Rockefeller Foundation's research programme from the late 1920s were Max Mason and Warren Weaver. Mason was an American mathematician who had travelled to Germany to write a doctorate under Hilbert at Göttingen. After returning to the United States, he had taught at MIT, Yale and Wisconsin and, between 1925 and 1928, served as president of the University of Chicago. This post, at the research university created by philanthropic patronage, as well as his war work devising acoustic submarine detection devices,[61] which provided him with invaluable experience in working with scientists and engineers of diverse disciplines, led to Mason's being appointed director of natural sciences at the Rockefeller Foundation. The following year, 1929, he stepped up to become president of the foundation, and soon recruited Warren Weaver into his old post. Weaver shared Mason's Wisconsin background and, although the younger man by a couple of decades, had also served in the First World War, with the early United States Air Force. Weaver's academic career was built at Caltech in the 1920s. The pair had co-authored a book, *The Electromagnetic Field*, in 1929.

Mason recruited Weaver because he thought the younger man was an ideal leader of a new, large-scale research agenda. The Rockefeller Foundation's Science of Man programme, described by the historian Lily Kay as an attempt 'to develop the human sciences as a comprehensive explanatory and applied framework of social control grounded in the natural, medical and social sciences', was launched in 1933.[62] This technocratic vision was shared by others. For example, H. G. Wells, his son G. P. Wells and Julian Huxley in their popular encyclopaedia *The Science of Life: A Summary of Contemporary Knowledge about Life and its Possibilities*, published in 1929 and 1930, argued that 'biological science . . . equipped with a mass of proved and applicable knowledge beyond anything we can now imagine [would allow] the ultimate collective control of human destinies'. Similarly, in the Rockefeller Foundation's Science of Man vision:

> Science has made significant progress in the analysis and control of inanimate forces, but science had not made equal advances in the more delicate, more difficult and more important problem of the analysis and control of animate forces. This indicates the desirability of greatly increasing emphasis on biology and psychology, and upon those special developments in mathematics, physics and chemistry which are themselves fundamental to biology and psychology.[63]

The invitation was to use the power of direction given by Rockefeller grant allocation to reshape the human sciences by reference to the physical sciences. The pay-off from this reductive programme seemed glittering:

> The challenge of this situation is obvious. Can man gain an intelligent control of his own power? Can we develop so sound and extensive a

genetics that we can hope to breed, in the future, superior men? Can we obtain enough knowledge of the physiology and psychobiology of sex that man can bring this pervasive, highly important, and dangerous aspect of life under rational control? Can we unravel the tangled problem of the endocrine glands, and develop, before it is too late, a therapy for the whole range of mental and physical disorders which result from glandular disturbances? Can we solve the mysteries of the various vitamins so that we can nurture a race sufficiently healthy and resistant? Can we release psychology from its present confusion and ineffectiveness and shape it into a tool which every man can use every day? Can man acquire enough knowledge of his own vital processes so that we can hope to rationalize human behaviour? Can we, in short, create a new science of man?[64]

So now we can see why the Rockefeller Foundation would want to know the structure of proteins. It was part of an audacious hope to re-engineer society from the bottom up. The shape of proteins was a first impression of the shape of things to come.

Lily Kay regards the concern for social control as being a social interest – essentially the prejudices and worldview of a white, Anglo-Saxon, protestant industrial elite – held by the early trustees of the Rockefeller Foundation. Scientific programmes funded under the early Rockefeller Foundation followed this interest. So did eugenics. The Science of Man programme was a refined, late, very generous step down this path. It was also an alternative to the faltering scientific credibility of old-style eugenics.[65] A rational response to lack of control meant encouraging the relevant centres of science. Many universities benefited from the Rockefeller Foundation policy of 'making the peaks higher': Chicago, Stanford, Columbia, Harvard and Wisconsin. But none received the support lavished on Caltech. The values of Weaver and Mason were fully in accord with the style of Caltech: support for interdisciplinary research, teamwork, a style of managerial organization that mirrored those found in corporations. Indeed historian John Servos describes the expansion of Caltech as the invention of a 'knowledge corporation'.[66] Kay implicitly assumes an odd version of the frontier thesis to explain Caltech: Hale, Noyes and Millikan had been able to build a new kind of university because it was 'unimpeded by intellectual and political residues of prior scientific traditions'.[67]

Pauling got his money. For the first year or so Warren Weaver was anxious that Pauling's research, while admirable, was insufficiently aligned to the Science of Man project. He hinted that funds might dry up. This pricked Pauling: he started serious work on the structure of the blood protein haemoglobin. Later, in 1949, Pauling, with younger colleagues Harvey Itano and John Singer, would show that the haemoglobin of blood drawn from sufferers of sickle-cells was physically different from 'normal' haemoglobin; their paper 'Sickle-cell anemia, a molecular disease' was a bold statement that diseases could have simple, identifiable molecular causes.

Rockefeller money supported research across Caltech's Pasadena campus. Thomas Morgan moved his fly laboratories there in 1928. Local philanthropic eugenicists also welcomed his arrival.[68] (Remember that California was a world leader in eugenic sterilization.) Another local philanthropist, the widow of the hydroelectric baron William G. Kerckhoff, supported Morgan's laboratory so long as she could see a medical application. Morgan also spelled out other practical applications:

> First then, the physical and physiological process involved in the growth of genes and their duplication (or as we say 'division') are phenomena on which the whole process rests. Second: an interpretation in physical terms of the changes that take place during and after conjugation of the chromosomes. Third: the relation of genes to characters. Fourth: the nature of the mutation process – perhaps I may say the chemicophysical changes involved when a gene changes to a new one. Fifth: the application of genetics to horticulture and to animal husbandry.[69]

Morgan's new laboratory became an interdisciplinary workshop for geneticists, bioorganic chemists, biochemists, and physiologists. 'Caltech's biology was thus constrained and liberated by its ambiguous service role, navigating between the Scylla of medicine and the Chraybdis of eugenics [and good breeding more generally] until World War II', writes Kay. Its leaders 'managed to garner sufficient resources by holding out vague promises to both enterprises', but also avoided firm commitment: 'this open-endedness created an intellectual space for developing a distinctive biological identity. Under the auspices of the Rockefeller Foundation, the new biology could focus on long-range visions rather than immediate concerns'.[70]

Specifically, the philanthropic support cemented what historians such as Kay have called the 'protein paradigm': the view widely held by scientists from the 1910s to 1940s that proteins were the hereditary material. Morgan held this position, as did Svedberg, who received large Rockefeller grants to support his expensive instrumentation. In 1929–30, John Northrop, working at the Rockefeller Institute in New York, crystallized the enzymes pepsin and trysin and identified them as proteins, encouraging the view of the centrality of proteins to biochemistry. In 1935, Wendell M. Stanley, also working at the Rockefeller Institute, managed to form crystals of the Tobacco Mosaic Virus (TMV) and determined that it, too, was proteinaceous. A virus, in this view, was an 'autocatalytic protein', capable of reproducing itself.[71] 'This world picture', the protein paradigm, writes Kay, 'greatly stymied curiosity about the chemical structure and physiological role of nucleic acids.'[72]

Two of Morgan's younger researchers at the Caltech fly lab were George Beadle and Edward Tatum. The pair decided to focus on a new model organism, the bread mould *Neurospora* that promised to be more revealing about biochemistry than *Drosophila*. *Neurospora* would multiply quickly when placed in suitable mixtures of nutrients, such as sugars, and such simple nutrients were all that it needed to produce amino acids and other vital substances. While at Stanford from 1937, Beadle and Tatum exposed the *Neurospora* to X-rays, producing, as would have been readily expected from *Drosophila* studies, a range of mutants. These mutants varied in how they took up and used nutrients. Some of the mutants, for example, could not produce one of the twenty amino acids; Beadle and Tatum were able to demonstrate this by comparing the growth of such mutants in a medium containing just the basic nutrients (no growth) with one containing the basic nutrients plus the missing substance of interest (growth). In 1941 they reported in the *Proceedings of the National Academy of Sciences* that they had found three mutants unable to synthesize pyridoxine (Vitamin B_6), thiazole (part of Vitamin B_1) and para-aminobenzoic acid. They argued that the mutations were of a single gene. This finding is remembered in biology textbooks as the origin of the one-gene-one-enzyme hypothesis (named as such in 1948) and as the identification of metabolic pathways. However, within the protein paradigm this is better expressed by the idea that one gene ≡ one enzyme. Beadle and Tatum hedged their bet:

> From the standpoint of physiological genetics the development and functioning of an organism consist essentially of an integrated system of chemical reactions controlled in some manner by genes. It is entirely tenable to suppose that these genes which are themselves a part of the system, control or regulate specific reactions in the system either by acting directly as enzymes or by determining the specificities of enzymes.[73]

Persuaded by Pauling, George Beadle would return to Caltech after the Second World War to lead a biological division committed to a molecular vision of biology.[74] This sort of science had been given a name – 'molecular biology' – in 1938 by the person who held the purse strings: Warren Weaver of the Rockefeller Foundation.

Molecular biology was interdisciplinary by design – a 'grand fusion', writes Yasu Furukawa, 'of the methods, techniques and concepts of organic chemistry, polymer chemistry, biochemistry, physical chemistry, X-ray crystallography, genetics, and bacteriology'.[75] But molecular biology 'did not just evolve by the natural selection of randomly distributed disciplinary variants, nor did it ascend solely through the compelling power of its ideas and its leaders', writes Lily Kay. 'Rather the rise of the new biology was an expression of the systematic cooperative efforts of America's scientific establishment – scientists and their patrons – to direct the study of animate phenomena along selected paths toward a shared vision of science and society.'[76]

Molecular biology was largely an American creation – arguably indeed a Californian creation – but it depended on the import of European, largely German, ideas and practices. Some of this knowledge was acquired by travelling encounters, such as Pauling and Mark at the IG Farben laboratory at Ludwigshafen. Another major route was the tracks of émigrés from fascism. Historian Donald Fleming argues that the reformulation of biology as a molecular science was made possible by the fresh, outsider vision provided by émigrés, especially émigré physicists, and four men in particular: Erwin Schrödinger, Leo Szilard, Max Delbrück and Salvador Luria.[77] Fleming's argument partly revolves around the DNA story, and that part will be left to later. However, his account of Delbrück's pre-war career is highly relevant to this chapter.

Max Delbrück was born in 1906 in Berlin. He was scientific aristocracy: his mother was a great-grand-daughter of Justus von Liebig. He studied theoretical physics at Göttingen and pursued post-doctoral work in England, as well as in Zurich and Copenhagen. He arrived in the Danish capital just in time to hear Niels Bohr's famous address on 'Light and life' in 1932, in which the great Dane spoke of the implications for, and parallels in, biology of his notion of complementarity. The speech seems to have turned Delbrück towards the sciences of life. He took a physics job under Lisa Meitner in Berlin, but sought ways to talk about biology. An unofficial seminar with like-minded Berlin scientists led to a 1935 paper, authored by N. V. Timoféeff-Ressovsky, Karl Zimmer and Delbrück, on gene mutation. (This paper is also sometimes nicknamed the *Dreimännerarbeit* – Three men's work – acknowledging a similar status to the great *Dreimännerarbeit* of quantum mechanics, the Born–Heisenberg–Jordan paper.) In 1937, Delbrück was able to leave Nazi Germany

253

on a Rockefeller fellowship. He went straight to Caltech. He spent his first year looking for the 'atoms of biology', basic organisms, and settled on Emory Ellis's bacteriophages – viruses that infect bacteria, 'phages' for short.[78]

Phages had first been recognized during the First World War as those agents that passed through the finest filters and destroyed bacteria (one scientist, the Frenchman Felix d'Hérelle, was investigating dysentery in cavalry units when he had made this observation). The potential of phage as a focus of research had been pinpointed in 1921 by Hermann Muller: if 'd'Hérelle substances' (phages) were genes, and, since phages were 'filterable, to some extent isolable, can be handled in test-tubes, and their properties, as shown by their effect on the bacteria, can then be studied after treatment'; perhaps then 'we may be able to grind genes in a mortar and cook them in a beaker after all. Must we geneticists become bacteriologists, physiological chemists and physicists, simultaneously with being zoologists and botanists? Let us hope so.'[79]

Delbrück would continue his phage work at Vanderbilt University and, most influentially, with émigré Italian bacteriologist Salvador Luria at 1940s summer schools at Cold Spring Harbor Laboratory on Long Island, New York. Students there were taught relatively sophisticated statistical techniques – sampling, Poisson and binomial distributions – as well as how to handle phage. These students then spread the techniques and enthusiasm for starting with 'atoms of biology' with them as they travelled. The symposia, wrote Fleming of the émigré's haven, were 'one of the little communities of intellectual purpose and excitement that constitute the only genuine utopias of the twentieth century'.[80] As the students travelled, so news of the utopia spread.

In general, notes Kay, molecular biology in the 1930s displayed eight features.[81] First, molecular biology placed a stress on the unity of life rather than on its diversity, and therefore followed a preference to study phenomena, such as respiration and reproduction, common to all organisms. Second, based on this rationale, 'it became far more convenient to study fundamental vital phenomena on their minimalist levels'. Delbrück searched for his 'atoms of biology' and found phage; others studied viruses or bacteria such as *Escherichia coli*. Third, the aim was to discover 'general physicochemical laws governing vital phenomena', which bracketed out higher-level, interactive processes or historical explanations and encouraged only explanation by upward causation. Fourth, molecular biology borrowed methods from all other sciences: it was interdisciplinary by design. Fifth, macromolecules, especially proteins, became the 'principal focus', and, sixth, the scale of interest was of the order of size of these objects, roughly 10^{-8} and 10^{-9} metres. (Molecular biology was, if you will, the first nanoscience.) Seventh, molecular biology depended upon the design, provision and maintenance of complex and expensive apparatus: not only X-ray diffraction equipment, ultracentrifuges and new microscopes but also electron microscopes, electrophoresis machines, spectroscopes, and materials such as radioisotopes and scintillation counters. Finally, the 'cognitive focus on the molecular level also shaped the social structure of research': interdisciplinary science meant teamwork, and effective teamwork required the prizing of cooperation.

Making the small real: virology, cytology and organic chemistry

A combination of working world relevance, philanthropic visions and a trend for expensive instrumentation also shaped disciplines close to the emerging molecular biology. Virology and cytology are good examples. Meanwhile there was a mid-century instrumental revolution in chemistry, which will be summarized below.

In 1911 Peyton Rous had observed that leukaemia in chickens could be passed on through filtered extracts of chicken blood: whatever the agent was, it must be small, a virus by definition. In the 1930s, the idea that cancer could be caused by a virus went from being an 'oddity' to mainstream science.[82] The theory accorded with the Rockefeller Foundation's vision of controlling the devastating wasteful scourges of the age by winning understanding and intervening at the smallest scales. The rabbit papilloma virus soon became an object of intense study for the Rockefeller Institute. And small objects required large instruments. From 1929 at the Rockefeller Institute Albert Claude broke down cells with centrifuges and in the 1930s studied the contents with electron microscopes. Claude's project had sought to identify the bodies that caused chicken tumours. However, it found particles that were common throughout the chick embryo cells but were not associated with disease: microsomes. Deploying cell-fractionation, electron microscopy and biochemical characterisation, Claude isolated many parts of the cell's interior architecture: mitochondria, chromatin threads, the reticulum.[83] Scaled up, Claude provided much of the detail of the twentieth-century picture of the cell. But behind it are philanthropic visions of desirable science, and behind those are a working world of medicine, of cancer and other diseases.

Between 1920 and 1950, chemistry laboratories, of the kinds devoted to analytical or organic chemistry, were transformed.[84] The process began in the 1920s, was accelerated by wartime projects, such as synthetic rubber production, and matured by the 1950s. In came a host of instruments – infrared spectrophotometers, polarographs, mass spectrometers – instruments that may have been smaller scale than those in physics but which, given the sheer number of chemical laboratories around the world, marked a similar level of capital investment. 'Prior to 1920 analytical chemists determined the chemical constitution of some unknown by treating it with a series of known compounds and observing the kind of reactions it underwent', philosopher Davis Baird explains of how what it meant to be an analytical chemist changed. 'After 1950, analytical chemists determined the chemical constitution of an unknown by using a variety of instruments' to distinguish between chemicals.[85] It was also likely to be teamwork, with each 'analyst' a technical specialist, rather than a trained general chemist.[86]

Scaling up from laboratory to mass production: radium, insulin, nylon

Claude or the 'analysts' identifying specific chemicals with their instruments were producing large representations of small objects. This was another kind of scaling up to be added to those already identified. The production of large quantities of scarce substances, and the development of transferable skills needed to move from laboratory preparation to mass production, was a feature of the interwar biomedical and synthetic textile industries. The Curie Foundation and Radium Institute, for example, invented a kind of 'big medicine' that paralleled Lawrence's Radiation Laboratory. The Paris 'radium bomb' contained immense amounts of the expensive element, creating a radioactive beam used to treat deep tumours. Only expanded uranium mining in the Congo could meet demand. The 'existence of the bomb changed local views about cancer, linking therapeutic efficiency with machines of escalating proportions', writes historian Jean-Paul Gaudillière of the 1920s Parisian cancer clinic, adding that thus was invented 'a form of "big medicine" which was soon to affect other areas of medical care'.[87] In the process, radium had changed from being a substance manipulated in microscopic amounts in Curie's laboratory to something made on an industrial scale.

The story of insulin also illustrates this process of scaling up from the lab to mass production. First, here is the story told in heroic narrative mode. Frederick Banting was a Canadian surgeon with a strong interest in diabetes. In Toronto in 1921, with his student Charles Best, Banting experimented with dogs, tying the pancreatic duct and extracting a secretion. This secretion kept other diabetic dogs alive. His team found that a purified extract – insulin – saved human victims of diabetes. A disease that caused blindness and early death for hundreds of thousands was now made manageable. Banting won the Nobel Prize in 1923. Physiology had often been used to give medicine a scientific gloss, but here was clinical physiology providing a treatment for a major disease.[88] Insulin imagery was also used to fight back against the anti-vivisection movement[89] – after all, the death of a handful of dogs had led to science that saved thousands of human lives. However, in contrast to this narrative, as crucial to the picture as was Banting's small team of physiologists (who took out a patent and assigned it to the University of Toronto), equally important was the contribution of the company Eli Lilly, which scaled up insulin production in return for one year's monopoly in the United States market.[90]

Likewise the case of nylon, the wonder synthetic material of the age, used for tyres, toothbrushes, carpets, and many other applications, most famously stockings.[91] Du Pont, an explosives company, had decided to diversify into dyestuffs as the First World War drew to a close, and built new laboratories to help figure out German patents in the area.[92] By 1921, $21 million had been spent, and a similar amount was forecast as necessary before any progress to be made. In the future, company executives decided, research and manufacturing expertise would be bought in rather than developed from scratch. In the 1920s, Du Pont bought into rayon fibre, cellophane, synthetic ammonia, tetraethyl lead and Freon refrigerants.[93] The result was a diversified company, with ten industrial departments, each with laboratories.

In 1926, Du Pont decided to build up 'fundamental' research in a central laboratory as a means of improving efficiency in the long term across common areas of the research interests of the company.[94] Such areas were catalysis, high-pressure chemical engineering and polymerization.[95] The expansion, and the apparent freedom to research in 'fundamental', perhaps even 'pure', areas, attracted in a Harvard chemist, Wallace H. Carothers. Carothers, munificently supported by Du Pont, turned out both theoretical interventions – arguing, for example, for the reality of polymers as large molecules, not mere aggregates – and practical substances, such as neoprene, a synthetic rubber, in 1930. In 1934 and 1935, a polyamide was found by Carothers that could be spun into a fibre – nylon – which possessed alluring properties.

But there was a big difference between making polyamide polymers in the laboratory and mass-producing nylon fibres. Carothers's research cost just $75,000, reckons historian David Hounshell.[96] But many millions of dollars would be spent by the company on research and development before successful mass production was possible. Scaling up was non-trivial. What was invented was how to manage scaling up effectively: freezing research once a good polyamide had been found rather than wasting time determining the best, developing parts of the process in parallel rather than in series (risky, but wonderfully time-efficient if successful), and intensely focused on a commercial target – women's stockings.[97] As nylon moved from laboratory, to 'semi-works', to pilot plant, to commercial production, the process itself was studied and the data used to guide improvements.[98] Target-driven, parallel-organized, scientifically managed research would be the secret of future American success, not least with the atomic bomb.

Three places, four ways of organizing science

Let us end this review of science in the hungry 1930s by visiting four places to see how it might be organized. We are already in Du Pont's 'Purity Hall', the company's central research laboratories near Wilmington, Delaware. The lesson about organizing science is to treat the process of moving from laboratory to production as its own working world and develop a management science of science.

North to Flushing Meadows, New York, 1939: the New York World's Fair is under way. There are displays from all of the great corporations of the United States. Inside the Perisphere was a futuristic townscape, called Democracity. Lewis Mumford called the Perisphere 'the great egg out of which civilization is to be born'.[99] Nearby was the huge Futurama display from General Motors, designed by Norman Bel Geddes as a 'scientific pilgrimage' that also led to a climactic creation of a new world. The promise made to visitors – who were given a pin to wear proclaiming 'I've seen the future' – was that the products of corporate research laboratories would indeed shape a new world, so long as government interference was successfully resisted.[100]

If we move further north, to Boston, Massachusetts, we find the inkling of a new way of organizing science. 'A form of partnership was proposed between the scientific expertise of the university', writes Terry Shinn of MIT in the 1930s, 'and the entrepreneurial expertise of local businessmen': the birth of venture capital support for science.[101] The American Research and Development Corporation would be founded in Cambridge, Massachusetts, in 1946. It would be especially important for the development of new biotechnology from the 1970s.

Finally, if we fly across the Atlantic (a regular airmail route from the 1930s) and drive to Cambridge, England, we find rival visions of organization. At King's College, the bursar, an aesthete economist called John Maynard Keynes, is extracting from the (un)working world of the Depression a model and theory of the economy.[102] In his *General Theory of Employment, Interest and Money* of 1936, Keynes argues that the economy as a whole must be understood in a different fashion from simple market exchanges. A macroeconomy was therefore distinguished from a microeconomy. The macroeconomy could be managed for full employment, argued Keynes, by spending during economic downturns. Roosevelt adopted Keynesian spending in the late 1930s – a sign of the willingness of the government to intervene that the World's Fair set out to head off.

Also in Cambridge in the 1930s was a firebrand Irish crystallographer: John Desmond Bernal.[103] His group worked on the structure of large molecules of relevance to living processes, including vitamins (B_1, D_2), enzymes (such as pepsin) and other proteins, and the tobacco mosaic virus. He was a socialist of deepest red. For Bernal, science was not an enterprise that could be left to corporations or the private sector more generally if it was to provide the solutions to all society's problems. Instead, it must be planned. His arguments were set out in *The Social Function of Science*, published in 1939, and offered, on the eve of war, an alternative vision for the organization of science.

Part III

Second World War and Cold War

— 12 —

SCIENCE AND THE SECOND WORLD WAR

Did the Second World War mark continuity or a discontinuity in the story of science in the twentieth century? Was it a turning point? A break? Or business as usual? Certainly it was a war fought largely with old chemistry rather than new physics. 'Physics may have been the pinup science of World War II', writes historian Edmund Russell, 'but chemistry was the workhorse science of urban destruction.'[1] Nevertheless, the Second World War, notes Alex Roland, when compared to previous wars, does seem to be different from all that had gone on before, going

> beyond mere production in quantity; it systematically improved the quality of weapons. Indeed, it introduced whole new categories of weapons. Jet engines, liquid-fuel rockets, the proximity fuse, and the atomic bomb moved from concept to application . . . [For] the first time in history, a substantial rearming of combatants [took place during the time-span of a conflict]; the victors emerged from the war with a different arsenal than they had at the outset.[2]

Thus from the point of view of military technology the Second World War seems to mark an abrupt, qualitative change. However, Roland's characterization stresses novelty, whereas to understand science's roles in the Second World War the eye should be kept on older developments and trends. In the last chapter I argued that the 1920s and 1930s were notable for the ways that lessons about how to scale up science were learned in various settings. The same question – how can the scaling up of science be managed and planned? – was asked in different contexts for different reasons. In this chapter I argue that much of the triumphant uses of science in the Second World War make sense when seen as a continuation of this pre-war trend.

Indeed, the interwar scaling up of science (and numbers of scientists) shaped the character of the conflict. 'With the proliferation of corporate laboratories, science became part of the mass-production/mass-consumption culture', notes historian of science Michael Dennis, and in a thought-provoking aside he asks: 'Could we not suggest that the policy of "mass-producing" scientists between the wars had the unexpected effect of creating a supply that only the military could use?'[3]

Organizing science for war

In the First World War scientists had sought and organized ways to contribute to the war effort, deepening cooperation with government and industry. In the 1930s such military ties were either reinvented or deepened. In Germany, in addition to expanded military research and development, a new central organization for scientific research was established. 'The great tasks ahead in German science', proclaimed minister Bernhard Rust, announcing the new Reich Research Council, 'require that all resources in the area of research contributing to the fulfilment of these tasks be unified and brought into action.'[4] In France, preparation for war prompted the establishment in 1938 of the Centre national de la recherche scientifique appliquée. The 'appliquée' was dropped the following year, and the CNRS has been France's premier research organization ever since.[5] In Britain, research was coordinated primarily through the armed service ministries: the Admiralty, War Office and Air Ministry, as well as through the research councils, including the Department for Scientific and Industrial Research. These organizations were bolstered by the addition of advisory councils, which channelled and formalized routine scientific advice to government.[6]

In the United States, the prime movers of forging stronger links between scientists and the military were a cabal of east coast scientific brahmins: the MIT electrical engineer Vannevar Bush, the MIT president Karl Compton, the president of Bell Laboratories in New York City, Frank Jewett, and the chemist and president of Harvard University, James Bryant Conant. Bush had been a professor of electrical engineering at MIT since 1919 and was a rich man owing to income from patents used to establish the Raytheon Company. Bush can be compared and contrasted to George Ellery Hale; they both, notes historian Daniel Kevles, embraced the role of being an 'entrepreneur of organization';[7] the difference was that, where Hale distrusted and resisted federal involvement in science, Bush was far more flexible. Bush, reviewing the experience of science–military organizational relations in the United States, regarded Hale's hands-off National Research Council as a failure and the National Advisory Committee for Aeronautics (NACA) as a success and a model for the future.[8]

Bush, supported by the cabal, persuaded Roosevelt to establish a new organization, the National Defense Research Committee (NDRC), in June 1940, after the outbreak of the war in Europe but well in advance of direct American involvement in the conflict. 'There were those who protested that the action of setting up NDRC was an end run, a grab by which a small company of scientists and engineers, acting outside established channels, got hold of the authority and money for the program of developing weapons', recalled Bush. 'That, in fact, is exactly what it was.'[9] The NDRC 'agreed to conduct defense research primarily by letting contracts for military purposes to industrial corporations and universities . . . [and committed] the NDRC, like the NACA, to both the large-project and individual grant approaches, and extended the combination beyond aeronautics to federal sponsorship of all the militarily relevant sciences', writes Kevles – 'a decision with major implications for the future.'[10] NACA was, agrees historian of airpower Michael Sherry, 'the prototype for the mobilization of industrial and academic science' during the Second World War.[11] Historian Hunter Dupree has called this moment a second 'great instauration', explicitly comparing it in importance to Francis Bacon's plans for the scientific revolution of the seventeenth century.[12] Bush brought in administrators – in particular Warren Weaver of the Rockefeller Foundation – who knew how to connect lavish patronage to scientific teamwork to achieve directed results.[13]

In 1941, the system was strengthened even further. An Office of Scientific Research and Development (OSRD) was established which, while under civilian control, channelled immense funds to military research and development and had a direct line of communication to President Roosevelt: 'I knew that you couldn't get anything done in that damn town [Washington] unless you organized under the wing of the President', recalled Vannevar Bush.[14] The OSRD included a Committee on Medical Research, expanding its coordinating remit.

All combatant nations deepened the organizational ties between science and the military in preparation for, and during, the Second World War. Much of this effort centred on the scaling up of research projects, translating laboratory and demonstration projects into working systems. But was such organization equally effective? Certainly many of the celebrated cases of science at war are best understood as achievements in the scaling up of production, deploying skills learned during the preceding decades. This thesis best guides interpretation of penicillin, radar and the atomic bomb.

Penicillin

The story of Alexander Fleming's discovery of an antibacterial agent, 'penicillin', in 1928 has been retold many times as an example of serendipity in science. The Scottish medical scientist, working at St Mary's Hospital, London, had noticed that a mould that had accidentally contaminated a culture of *staphylococcus* was destroying it. Fleming soon found that his 'mould juice' could destroy other bacteria. But he also concluded that penicillin was an unstable agent, too easily broken down and too hard to produce in medically useful quantities.[15] In the late 1930s, a second group, led by an Australian, Howard Florey, and a refugee German biochemist, Ernst Chain, at the Sir William Dunn School of Pathology at Oxford University, re-examined the problem. Chain proposed possible models for the structure of penicillin; Dorothy Hodgkin, an X-ray crystallographer who had moved from Bernal's Cambridge team to set up in Oxford, would confirm Chain's structure in 1945. From the late 1930s, the Medical Research Council and the Rockefeller Foundation funded further tests conducted by Florey's team on animals and human patients. Historian Guy Hartcup describes the two methods used to cultivate the penicillin mould:

> In the one used by Florey, known as the surface or stationary culture, the fungus grew as a mat on the surface of a liquid medium . . . After ten days or so, the penicillin content of the clear liquid was poured off and its penicillin extracted and purified. In submerged or deep culture the fungus was cultivated in large fermenters, of up to 10,000 gallons or more, vigorously aerated by sterile air, and stirred with powerful stirrers operating at several hundred horse power.[16]

The successful operation of such giant fermenters depended on the skills and knowledge of the pharmaceutical, chemical engineering and biotechnological industries.[17] The task of scaling up was shared with the United States. After contact with Florey's team, A. J. Moyer of the Northern Regional Research Laboratory of Peoria, Illinois, devised ways of increasing yield. Nevertheless, mass production was a collective, organizational achievement of many bodies and companies, with each stage presenting new and difficult problems to solve. Business historian John Swann reports that 'over two dozen American and British pharmaceutical companies took part' in the development of penicillin manufacture, alongside philanthropic institutions and academic and government scientists; the whole required coordination, organizational entrepreneurship and the pooling of knowledge.[18] 'Large networks of academic and government laboratories and pharmaceutical manufacturers in Britain and the United States were coordinated by agencies of the two governments', writes historian of biotechnology Robert Bud, of the interdisciplinary achievement of coordination, with the OSRD's Committee on Medical Research and the Medical Research Council in mind; in all, an 'unanticipated combination of genetics, biochemistry, chemistry and chemical engineering skills had been required'.[19]

Mass-produced penicillin had immediate and long-term impacts. American output of penicillin – 150,000 million units per month in late 1943[20] – was available in time to prevent the Allied armies in North Africa and, after the Normandy landings, Europe succumbing more than they would have done to the great killer in war: infection of wounds. Penicillin replaced the sulphonamides, former wonder drugs developed in Germany in the 1930s. But the methods of production were as much a prize as the antibacterial agent itself: the skills of building and operating the giant fermenters were necessary for postwar industrial production of new antibiotics and, indeed, the steroids that would be used in contraceptive pills.[21] 'We all know how much the new drug, penicillin, has meant to our grievously wounded men on the grim battlefronts of this war – the countless lives it has saved – the incalculable suffering which its use has prevented', reported Vannevar Bush to his new president in 1945. 'Science and the practical genius of this nation made this achievement possible.'

Likewise, the antibiotic streptomycin, the first direct treatment for the great killer of the nineteenth century, tuberculosis, discovered in 1943 at Rutgers University, required the industrial resources and knowledge of Merck & Company. A parallel but much less well-known story can be told of the antibiotic gramicidin.[22] The Soviet ecologist Georgii Frantsevich Gause had argued that organisms that fitted too similar niches would compete to exclude each other. This was as true for moulds on Petri dishes as it was for large mammals on the steppes. Just as Lysenko attempted to win favour from Soviet political authorities by demonstrations of application of his research, so Gause aimed to show that his ecological theory had applications of use to Soviet aspirations. This orientation encouraged Gause to examine competing moulds for antibiotic effects, leading in 1942 to the identification of gramicidin S ('S' for Soviet), a chemical defence produced by a *Bacillus* bacterium. Again there were important and difficult stages of research (including samples sent to Western allies for X-ray crystallographic investigation) and development before mass production and use in military hospitals were possible.

Radar

Effective, interdisciplinary, cross-Atlantic cooperation in scaling up science was also a feature of radar.[23] In the First World War, Zeppelin airships had dropped bombs on London, and the capital had responded by organizing a system of reporting information, a blueprint for the national information reporting system of radar.[24] Nearer the front, aeroplanes were developed and found a number of niche roles, such as reconnaissance. However, the aeroplane as a bomber was developed in practice and theory during the 1920s and 1930s. There was a consensus around the theory of the Italian strategist Giulio Douhet, published in *The Command of the Air* (1921), that a quick victory could be achieved through air attack on vital centres, leaping over armies in the field. The 'man in the street [has] to realise that there is no power on earth that can protect him from being bombed', argued the British ex-prime minister Stanley Baldwin in 1932. 'The bomber will always get through.' The destruction of the Basque town of Guernica by the German Condor Legion in 1937 seemed to confirm such fatalism.

Early warning depended on visual spotting and acoustic detection. Large concrete parabolic mirrors, used to focus sound waves, can still be seen on the cliffs of England's south coast, relics of the 1930s. Yet the idea that radio waves could be used to detect distant objects by measuring the scattered reflections was appreciated and widespread in the first half of the twentieth century. A German, Christian Huelsmeyer, had patented it in 1904. Albert Hoy Taylor and Leo C. Young, electrical engineers working for the United States Navy, had experimented with continuous wave radio detection in the 1920s. The rapidly expanding wireless industry delivered cheap components and skilled technicians. The potential of radar was available to all. Yet radar as an effective tool was primarily a British invention. The reasons why are important, as are the consequences for science, technology and industry in the second half of the twentieth century.

At the Royal Air Force summer exercises of 1934, bombers managed to attack the target for half of the day without being intercepted by defending fighter aircraft. In urgent response, a Committee for Scientific Study of Air Defence was set up, chaired by Henry Tizard, a government chemist with knowledge and experience of aeronautics. Sitting around the committee table were the muscle physiologist A. V. Hill, the ex-naval lieutenant and Cavendish physicist Patrick Blackett, the Air Ministry's scientist H. E. Wimperis, and his junior colleague A. P. Rowe. Wimperis investigated proposals for new military technologies. The Air Ministry had offered a prize of £1,000 for the person who could build a 'death ray' that could kill a sheep at 200 yards. Wimperis asked Hill to calculate how much energy would be needed. Hill's calculation was passed on to Robert Watson-Watt, a Scottish meteorologist. Watson-Watt had been working on methods of locating thunderstorms by finding and triangulating the direction of bursts of radio static. He replied to Hill and Wimperis that his energy calculations confirmed that a death ray was impossible, but suggested the idea of using radio to detect, rather than destroy, aircraft. It was a simple transferral of methods from locating thunderstorms to locating bombers. By pulsing the radio waves and measuring the time between transmission and reflection of echoes, it was possible, argued Watson-Watt and Arnold F. Wilkins, to determine both direction and range of a target.

On 26 February 1935, Watson-Watt supervised an experiment: an airplane flew back and forth past the BBC radio transmitter at Daventry while a ground crew sought to find and measure scattered radio waves. It was a success: the aircraft was detected at 8 miles distance. The Air Ministry established an experimental station, at Bawdsey, Orfordness, a remote spot on the Suffolk coast. Over the next few years bright young scientists, many of them physicists, were brought to the Telecommunications Research Establishment (TRE) to work on 'Radio Direction Finding' ('radar' is the 1940 name, coined by the United States Navy). The first major system developed, in 1936–7, was Chain Home, a network of 100 metre tall wooden towers each holding a transmitter that floodlit the sky with pulsed radio waves with a receiver stations behind. By 1937 a Chain Home station could detect an approaching aircraft 100 miles distant. This was far enough away for fighters to be scrambled to intercept.

But the transmitters and receivers were mere components of radar. The truly important innovation was the handling of 'information' as part of a technological system.[25] To be effective, the incoming intelligence had to be gathered, assessed for its quality, brought together to a central location, and presented in a format that allowed quick and accurate tactical decisions to be made. Then the decisions had to be broken down into specific tactical messages and passed out to the various defensive sub-systems: the fighter aircraft, the anti-aircraft batteries, the barrage balloon operators, and so on. Only then could the scarce defensive forces be marshalled effectively to respond to an otherwise overwhelming bomber threat. As important an invention as pulsed radio transmitter/receivers were 'filter rooms', places that reduced the immense quantities of incoming data into simpler, clearer forms to be passed to the central 'operations rooms', where the information was arranged and decisions made. It was this total 'information system' – indeed, the term has its origins here – that won the Battle of Britain, the aerial combat for superiority over southern Britain of summer 1940.

The radar system was constantly being developed in response to emerging threats and the availability of new techniques. One of the problems was that Chain Home was inaccurate, partly because of the long wavelengths it used. Chain Home could guide fighters near but not directly to incoming bombers. From mid-1940 the Plan Position Indicator – the familiar dial face of radar systems – improved the visual representation of radar information. Now a radar operator on the ground could watch the position of bombers and fighters and guide the latter to the former. This technique was called Ground Controlled Interception, in use from early 1941. In effect, it made the pilot and his aircraft part of the wider system. There was also an attempt to build Air Interception (AI) radar into the fighter so that the last step of location, which was where Chain Home failed, could be completed by the pilot. This failed because the AI radar apparatus was too large.

The technical obstacles to improved radar in 1940–1, therefore, were the interconnected ones of size and weight of equipment (too heavy for most aircraft) and the long radio wavelengths used. Shorter wavelengths would mean more accurate radar and lighter equipment, but designers were already working at the edge of the state of the art of 1930s electrical engineering. Chain Home wavelengths were around 10 metres; an ideal wavelength would be around 10 centimetres. Finding ways to produce and manipulate short wavelength radio waves were the crucial contribution of mobilized scientists, often from academia, working closely with industrial colleagues and military patrons. It was also where the division of labour, tasks and production between Britain and the United States were most important.

The Admiralty, consumed by its need to detect submarines, was as interested in short wavelength radar as the Air Ministry: the brief but essential moments when a U-boat surfaced presented a weakness that could be exploited if the conning tower or even the periscope could be detected. The Admiralty therefore sponsored research into the production of high-frequency radio waves. John Randall, an industrial physicist seconded to the University of Birmingham, and his colleague Harry Boot were recipients of this sponsorship. In spring 1940, Randall and Boot developed a radically new method of generating short wavelength radio waves. Their 'cavity magnetron' consisted of a simple pattern of holes bored into steel. Resonating fields generated short wavelength – what we would call 'microwave' – electromagnetic waves of unprecedented intensity: one hundred times the power at 9.8 centimetres of its nearest vacuum-tube competitor, Stanford University's klystron. At the Telecommunications Research Establishment a magnetron-powered radar was used in August 1940 to track a periscope-sized object – a bicycle. Scaling up the cavity magnetron from laboratory bench to mass production was the work of Randall's employer, the General Electric Company. Strong and intimate academic–industrial–military links were crucial to the success of the cavity magnetron.

The development of centimetre radar, which held so much potential, required time, money and engineering skills, all of which were in desperately short supply in Britain, threatened in 1940 with imminent invasion. In September 1940, Henry Tizard led a mission across the Atlantic. He was accompanied by the Cavendish nuclear scientist John Cockcroft and the radar scientist Edward G. Bowen. Avoiding U-boats on the way, Tizard's mission to the United States carried the most valuable cargo ever to pass between the two countries. In Tizard's possession – all in one trunk – were designs for an anti-aircraft predictor, crucial nuclear reaction calculations, Frank Whittle's plans for a jet engine, and a cavity magnetron.[26] Cockcroft, who knew the Americans, passed on the magnetron and discussed other aspects of British radar with Vannevar Bush and the chair of the NDRC's microwave committee, a member of the cabal, Alfred Loomis. Swiftly, the Radiation Laboratory, or 'Rad Lab', was established at MIT to develop short wavelength radar models for mass production by American industry. Growing from tens to thousands of employees before the war's end, the Rad Lab pushed back the boundaries of radar science and technology, developing methods for even shorter wavelengths (3 centimetres), as well as long-range radio navigation systems (Gee) to guide strategic bombing and gun-laying radars to improve anti-aircraft accuracy.

As Allied defence turned to offence in Europe, after the entry of the United States into war (Pearl Harbor was attacked on 7 December

1941) and the German army was turned back outside Stalingrad in the winter of 1942, the strategic bombing of German targets became ever more intense. Radio navigation systems, such as Gee, Oboe and Rebecca, brought the bombers over German cities. By 1943, centimetre radar allowed Allied pilots to 'see' the outline of the cities they were attacking: Home Sweet Home (H2S) radar put a picture in front of the navigators' eyes. H2S was the proposal of an academic physicist seconded to TRE, Bernard Lovell, and was developed with Alan Blumlein, a talented specialist from the radio company Electrical and Musical Industries (EMI). Again, strong and intimate academic–industrial–military links were crucial to the success of war innovation. Translated to naval applications as Air to Surface Vessel (ASV), radar was crucial to the Allied winning of the Battle of the Atlantic, the war against the U-boats. So much shipping was being lost to torpedoes that this particular use of radar has the strongest claim to the technology being a decisive tool of the Second World War.

Germany also had radar during the Second World War (as did all combatant nations). The Gesellschaft für electroakustische und mechanische Apparate (GEMA) had developed an early warning radar, Freya, for the German navy in the late 1930s. It was like a smaller mobile version of the Chain Home transmitter/receivers. Even more flexible Würzburg radar dishes were introduced in 1940. Freya could detect incoming aircraft at 75 miles, while the Würzburg was shorter range, more accurate and elegantly designed. German equivalents of gun-laying radar, air interception and almost all other types were also built. The difference between British and German radar was not the apparatus to transmit and receive – if anything the German engineering of these was better; rather, it lay in the system to collect, filter and represent information.[27] The effectiveness of filter and operations rooms was the secret. E. C. Williams of TRE, just twenty-one years old, had led the Chain Home filter room development.

Research establishments such as TRE and the Rad Lab employed thousands of people, including many hundred scientists drawn from many disciplines: biologists, mathematicians, electrical engineers, physiologists – but especially physicists. Many were young: fresh graduates. They were organized into teams. They were set targets. Success meant a working apparatus. Teamwork, target-oriented research, and an instrumental definition of success were all experiences, with associated values, that would be carried into, and indeed define much of, post-war prestige science.

The young generation also learned how to work with – and on – a patron.[28] The academic scientists were learning what industrial and military science was like. They learned how to sell an idea: how to pitch a complicated, expensive system to a patron, a user who had little time to spare. Lovell, for example, successfully persuaded military patrons, such as Lord Portal, of the potential of H2S through sophisticated and dramatic simulations. A. P. Rowe, superintendent of TRE, recalled in his history of radar, published in 1948, that important visitors were taken to 'our Hall of Magic', and

> we usually wanted something. We might need more staff or more buildings, or we might want to tell a high-ranking RAF officer our views on the need for some new radar device or how an existing device might be more profitably used. It was once said in London that 'If you let an Air Marshall get near TRE they will sell him anything'.[29]

In summary, then, the generation to emerge from such war work had learned the art of how to get things done.

Radar sciences

The working world of military radar also generated its own sciences. Operational research and communication theory (and, from the late 1940s, cybernetics) were seen as new sciences, and they are best understood as sciences of the working world of Second World War organization.

The idea of a science of war is a very old one. In the 1920s the strategic thinker Basil Liddell Hart had called for the systematic application of scientific techniques, such as statistical analysis, to aid the British Army.[30] In June 1940, a group of mostly prominent scientists (Julian Huxley, J. D. Bernal, Solly Zuckerman, C. H. Waddington), who formed an informal dining club, the Tots and Quots, published a Penguin paperback, *Science in War*, to shame the British government into making more use of scientists. (It sold well: copies of *Science in War*, printed on tissue-thin rationed wartime paper, can still be found in most good second-hand bookshops.) But the very title of their intervention slightly misses the novel developments already under way: not science in war but a science of war, specifically a science that took its ontology from the bureaucracy of war.

Operational research examined the informational products of the operations room: the location, pattern and effectiveness of engagements with the enemy, the information extracted from radar and from aerial reconnaissance as filtered through the system. Patrick Blackett identified the sources in his memorandum 'Scientists at the operational level', circulated to political and military masters in December 1941:

> Operational staffs provide the scientists with the operational outlook and data. The scientists apply scientific methods of analysis to these data, and are thus able to give useful advice.
>
> The main field of this activity is clearly the analysis of actual operations, using as data the material to be found in an operations room, e.g. signals, track charts, combat reports, meteorological information, etc.
>
> It will be noted that these data are not, and on secrecy grounds cannot, in general, be made available to the technical establishments. Thus such scientific analysis, if done at all, must be done in or near operations rooms.[31]

Operational research was a science of the reduced representations of the working world of war – 'signals, track charts, combat reports, meteorological information' – that allowed its practitioners to speak authoritatively at the top of the hierarchy. Blackett, for example, had the ear of commanders-in-chief. Formal scientific advisory positions multiplied. Scientists made knowledge and gained power.

Operational research staff teams were attached to all of the British military commands. Coastal Command's Operational Research Section, assessing anti-U-boat strategies, was typically interdisciplinary: three physicists, one physical chemist, three radar engineers, four mathematicians, two astronomers, and eight physiologists and biologists.[32] By tweaking such parameters as the size of convoys, the depth at which depth charges detonated, or the relative number of days spent maintaining compared to flying anti-U-boat aircraft, the effectiveness of operations was ratcheted up. Cecil Gordon, treating Coastal Command 'as though it were a colony of ... *Drosophila*', showed that the shortage of skilled maintenance labour was the bottleneck in the war against the U-boat, while Bernal analysed the effectiveness of air raid shelters.[33] Improvements were passed on. Operational research spread through the United States military, where it was called, even more accurately, operations research.[34]

Nevertheless, this apparent evidence-led war-making could run into serious problems when both sides possessed scientific advisors. The most contentious aspect of the Allied conduct of the European war has been the morality and practicality of strategic bombing. Carpet bombing of cities containing concentrations of civilians was justified at the time as an attack on the morale of the enemy workforce. In practice, carpet bombing was done because it was possible. Lübeck, for example, flattened on 18 March 1942, had almost no military or industrial value but was within reach of the bombers. In its aftermath, in April 1942, Bernal and Zuckerman analysed the German bombing of the industrial city of Birmingham and of the port of Hull and calculated that 1 ton of bombs had killed a mere four people, that high-explosive munitions were not as effective as incendiaries, and that there was no evidence of a collapse in morale.[35] However, Prime Minister Churchill had his own scientific advisor. Frederick Lindemann, Lord Cherwell, an Oxford professor of physics, argued that morale had fallen, and therefore that strategic bombing was effective. Blackett and Tizard joined the debate on the side of Bernal and Zuckerman. The debate has raged ever since, with the current balance tipping against Cherwell. But who was 'right' is not the issue. Rather it is this: while operational research was a science of the working world of operations, it could not resolve disputes where there were entrenched opposing strategic views.

Communication theory was another science of the working world of war that had a more refined, mathematical, distanced character when compared to operational research, as befits the more distanced location of its invention. American development involved a network of laboratories and factories, from the Rad Lab to equipment manufacturers such as Western Electric. A crucial node in this network was Bell Laboratories, the research end of AT&T's Bell System. War work would pump-prime Bell Labs to be a leader of post-war electronic research and development, not least in creating the conditions for its rapid development of the transistor. However, other work was more theoretical. Claude Shannon was an MIT-trained mathematician. At Bell Labs his task to improve a speech enciphering system, Project X, was new because it was digital: rather than merely distorting a voice message to thwart eavesdroppers, this system would digitize it and mix it with a digital key.[36] He wrote up the improvements in a confidential report, 'Communication theory of secrecy systems' (1945); for the audience outside the secret realm, these ideas, plainly the analysis of a working world, became the more abstract 'A mathematical theory of communication', the foundation of a post-war field, information science.

Chemical and biological weapons

While there may have been intense arguments about the morality and effectiveness of high explosives as delivered by strategic bombers, millions of tons of such weapons were of course used. More curious is the non-use of weapons, including those such as chemical and biological weapons, that absorbed significant quantities of time, money and expertise during the Second World War. In June 1918, the United States had gathered together its gas warfare branches under a single Chemical Warfare Service (CWS). It had seven divisions: Gas Defense, which made gas masks in New York; Gas Offense, which produced chemical agents and weapons at Edgewood Arsenal, Maryland; Proving Ground and Training, both in Lakehurst, New Jersey; Development, which made mustard gas; Medical; and Research, which was located at the American University, Washington.[37] After the war, the head of the CWS, General William L. Sibert, had to make a case for the continuation of his organization:

> Based on its effectiveness and humaneness, [chemical warfare] certainly will be an important element in any future war unless the use of it should be prohibited by international agreement. As to the probability of such action, I cannot venture an opinion.[38]

The CWS was indeed made permanent in 1920, and was not disbanded despite the Treaty of Washington (1922) and the Geneva Protocol (1925). The United States Senate had agreed to ratify the Treaty of Washington, which banned the use in war of 'asphyxiating, poisonous, and all analogous liquids, materials, or devices', but the French objected to how the treaty handled submarines, stalling the agreement. The Geneva Protocol used the treaty as a template and prohibited chemical weapons, as well as extending the range, explicitly ruling out 'bacteriological methods of warfare'. The protocol was signed on 17 June 1925 and entered into force on 8 February 1928. The United States signed but did not ratify it, responding to opposition from three sources: isolationists (who rejected international agreements), the American Legion (who feared legal reprisals against veterans) and the American chemical industry lobby.[39]

In the 1920s and 1930s, the CWS had maintained stockpiles of First World War agents, developed new gas masks, and undertook some special projects, such as fighting boll weevils, deemed helpful in peacetime.[40] Mobilization began in earnest in 1939, and by 1941, partly in response to Japanese developments, discussed below, the CWS was rapidly growing to meet the demands of the United States Army for its products. About 146,000 tons of chemical agents were made in the United States between 1940 and 1945 – mostly mustard, with some phosgene, Lewisite and cyanogen chloride – at both government factories (Edgewood Arsenal, Huntsville Arsenal, Pine Bluff Arsenal, Rocky Mountain Arsenal) and the factories of chemical companies (Monsanto's Duck River plant in Tennessee, American

Cyanamid's facilities in New York State and California, and Du Pont's various sites).[41] As well as the agents, factories produced mortars, bombs, masks and detectors. Thousands of scientists were mobilized – and some lost their lives: Samuel Ruben, the co-discoverer of carbon-14, introduced in the last chapter, for example, was killed in a phosgene accident in 1943.

Other countries were just as well prepared. In Germany, the chemical warfare programme took in not just First World War agents, but devastating new agents too.[42] Gerhard Schrader, a chemist at IG Farben developing organophosphates as possible insecticides, synthesized a new substance. In 1936 he reported 'tabun' to German military authorities, who noted its extreme toxicity. In 1938 Schrader found, and reported, a second such agent, 'sarin', five times more powerful. Developing tabun and sarin as nerve-gas chemical weapons became a large development programme for the German chemical industry. During the war, 78,000 tons of chemical warfare agents were produced, including 12,000 tons of tabun and a much smaller quantity of sarin.[43]

A similar state of preparation can be found in the case of biological weapons. Against a backdrop of fears of German and Russian biological armament, France had launched a biological warfare research programme in 1921.[44] Led by the director of the Naval Chemical Research Laboratory, Auguste Trillat, the notion that microbes were too fragile to survive explosive impact followed by exposure outside their animal host bodies to dry air was overturned. This finding meant that biological weapons were practical. Trillat favoured liquid cultures of bacterial agents detonated to form 'microbial clouds', in the form of bombs dropped by aeroplanes. Fearing Nazi developments, the French biological weapon programme picked up speed between 1935 and the German invasion of 1940. Agents studied included the microbes responsible for tularemia (rabbit rever), brucellosis (which strikes down livestock but can incapacitate humans), melioidosis (a disease similar to glanders), plague, anthrax and the bolulinum toxin. In selecting these agents, notes historian Jeanne Guillemin, the French 'prefigured the approaches of scientists in later state programs'.[45] Smallpox was rejected, since a vaccine was available, while diphtheria, poliomyelitis and tuberculosis infected their victims too slowly. The French biological weapon programme was dismantled, and documents burned, to prevent German capture.

In 1934, journalist Henry Wickham Steed reported, on the basis of dubious evidence, that German authorities were actively investigating the delivery of biological weapons by aeroplanes and had experimented with releasing bacteria on the London Underground and Paris Metro.[46] The Soviet Union was alarmed enough by Wickham Steed's reporting to launch a biological weapons programme of its own, although it would later be purged. In Britain, a powerful civil servant, Maurice Hankey, sought out senior British scientists' comments on Wickham Steed's allegations in particular and the likelihood of effective biological weapons in general. John Ledington, of the Lister Institute, William Topley, of the London School of Hygiene and Tropical Medicine, and Captain Stewart Douglas, of the National Institute of Medical Research, were sceptical, as was the secretary of the Medical Research Committee, Edward Mellanby. Routine public health measures were the best response: biological warfare would be treated like any normal epidemic. Hankey was not to be rebuffed.

Hankey received support from an unexpected quarter. In 1938, Frederick Banting, the Canadian clinical physiologist famous for insulin, had written a paper arguing that the British Empire should be equipped with an enormous retaliatory capability for biological warfare. He established ranges in Suffield, Alberta, to test biological agents. Canadian industrial philanthropists helped support his work. In 1939, Banting arrived in London to argue his case and offer his organization. His arguments persuaded Hankey to act. Now a minister without portfolio, which gave him political power combined with flexibility, Hankey started Britain's biological warfare programme, notes Guillemin, without 'consulting with either the War Cabinet, which [Hankey] considered far too busy to consider germ warfare, or Prime Minister Churchill'.[47] 'In view . . . of the extreme secrecy of the matter', wrote Hankey, 'I am inclined to go a certain distance without a Cabinet decision.'[48] A research station at Porton Down on Salisbury plain was established in 1940, under microbiologist Paul Fildes, who believed 'his mission was to achieve high-volume offensive capacity available at short notice'. Anthrax cakes were soon being produced. Formal approval did not arrive until 1942. Thus did Britain (and the world) sleepwalk into a state of 'inadvertent escalation', 'with high officials unaware that research on a new type of weapon had been authorized'.[49] Biological warfare programmes were largely driven by keen scientists and civil servants.

Despite the death of Banting, in an air crash in February 1941, the British-Canadian biological weapon programme continued to expand. Links were forged, via Vannevar Bush, with the United States, which channelled some funds to support the Canadian research. In Britain, animal testing began. In July 1943 a large-scale trial of the anthrax weapon was conducted by a team led by Porton

Down scientist Oliver Graham Sutton.[50] Sheep were brought to the uninhabited Scottish island of Gruinard and tethered. In their midst stood a gallows with a bomb containing 3 litres of anthrax in suspension. Detonated by remote control, a cloud of anthrax spores infected the sheep, killing those within 400 yards within the week. Further trials dropped anthrax weapons from Wellington bombers.

The Gruinard and other trials suggested that biological weapons were effective. Like other aspects of the war, the efficacy of bombers delivering biological weapons was itself the subject of analysis and argument. In early 1944, a fierce dispute broke out: those sceptical of the worth of strategic bombing likewise doubted the anthrax bomb, and vice versa. Just as with the strategic bombing debate, Churchill's personal scientific advisor, Lord Cherwell, possessed the most influential voice. Fildes convinced Cherwell, who in turn persuaded the prime minister:

Half a dozen Lancasters [bombers] could apparently carry enough [anthrax], if spread evenly, to kill anyone found within a square mile and to render in uninhabitable thereafter ... This appears to be a weapon of appalling potentiality; almost more formidable, because infinitely easier to make, than tube alloy [i.e., the atomic bomb]. It seems most urgent to explore and even prepare the counter-measures, if any there be, but in the meantime it seems to me we cannot afford not to have N [anthrax] bombs in our armoury.[51]

Fildes drew up plans: targeting six German cities (Berlin, Hamburg, Stuttgart, Frankfurt, Wilhelmshafen and Aachen) would kill 3 million civilians. For Fildes, biological warfare would be like any other form of warfare, but 'without the distressing preliminaries'.[52] Others were alarmed:

I cannot resist the feeling [wrote an Air Marshal] that the enthusiasm of the scientists, in particular, of course, but not solely, Dr Fildes, has tended to break loose from policy control and that decisions have been obtained and acted upon beyond the knowledge of the matter possessed by the War Cabinet as a whole. I am not sure that their scientific enthusiasm as inventors is not prejudicing their sense of responsibility; and I fear that there is a tendency not to weigh carefully enough the consequences of loosing these inventions upon mankind.[53]

Nevertheless, Churchill and the War Cabinet, once fully informed, took the view that mass production was not possible in war-stretched Britain. The problem was passed across the Atlantic. A Special Projects Division within the CWS at Camp Detrick, Maryland, repeated and extended British trials and developed a mass-producible anthrax bomb. Biological warfare research began on American campuses, such as Harvard's. A factory was built in Vigo, Indiana, primed to mass-produce anthrax. By the end of the war, $400 million had been spent, 'historically unprecedented', and a fifth of the cost to produce the atom bomb.[54]

Chemical and biological weapons therefore constituted a great investment of scientific and industrial resources during the Second World War. They were no side-show. There were opportunity costs too: the funds needed to conduct trials and to build factories, and the scientists deployed, were all in demand elsewhere in the war economies. Given this investment, the non-use of the weapons – by the countries that invested most – needs to be explained. Chemical and biological weapons were started because of fears that other countries were already further along in development. Events in Ethiopia and China justified some, but not all, of these fears. In 1935, Benito Mussolini's fascist Italy attacked Emperor Haile Selassie's Ethiopia. When the initial assault foundered, and under pressure from Mussolini for a swift demonstration of fascist superiority, the Italian commander, Pietro Badoglio, ordered the use of mustard agent, dropped in bombs and sprayed from aeroplanes. (Both Italy and Ethiopia had ratified the Geneva Protocol.) 'It was thus that, as from the end of January 1936', reported Selassie to the League of Nations,

> soldiers, women, children, cattle, rivers, lakes, and pastures were drenched continuously with this deadly rain . . . [The] Italian command made its aircraft pass over and over again. These fearful tactics succeeded. Men and animals succumbed. The deadly rain that fell from the aircraft made all this whom it touched fly shrieking with pain. All those who drank poisoned water or ate infected food also succumbed in dreadful suffering. In tens of thousands the victims of Italian mustard gas fell.[55]

Given such incidents, simple arguments that a taboo against using poisons in war are inadequate explanations. Certainly colonial ideologies of racial inferiority, already a pillar of fascism, contributed to allowing the use of mustard gas on bare-footed Ethiopians to be imagined. Another approach is to follow the historian Richard Price, who argues that the prohibition against chemical weapons was a political construct built in the twentieth century.[56] Any country breaking the taboo had to be convinced that the weapon would be decisive, not merely useful. By the Second World War this costly burden of proving decisiveness – not shared by new conventional weapons – was sufficient to prevent Germany, France, the United States, Russia and Britain from use. Note, though, that the supposed taboo did not prevent widespread and substantial preparation. Nor did it prevent the use of chemicals in the mass murder of the concentration camps; Zyklon B was a marketed form of hydrogen cyanide supplied by a company called Degesch (translated, the German Company for Pest Control), which was 42.5 per cent owned by IG Farben, the epitome of German science-based industry.[57]

Nor did prohibition prevent the Japanese from deploying chemical weapons in its war against China. Japan had invaded Manchuria in 1931 intent on colonial subjugation. Japan's war for regional supremacy was launched in parallel with Germany's war in Europe. On 7 December 1941, the surprise attack on the American naval base of Pearl Harbor in the Hawaiian Islands brought the United States into the global conflict. The fierce fighting through the southern Pacific pushed the Japanese forces back. A resurgent Chinese army fought on the mainland. In the spring of 1943, a desperate Japanese command approved the use of chemical weapons in the Battle of Changde, in Hunan province.

Japanese biological weapon research, development and use were even more extensive.[58] An officer, Ishii Shiro, had argued in 1927 that, since Japan had signed but not yet become a full party to the Geneva Protocol, signed two years previously, she was at an advantage. The banning of a class of weapons by most countries therefore was seen as providing an opportunity for others. (This entirely rational, but unwelcome, response to global regulation will be witnessed again later in the case of post-war Soviet chemical and biological warfare.) 'Western biological science, already promoted in Japanese universities and medical schools', notes Guillemin, 'promised to bring modern efficiency to the Japanese arsenal.'[59] Shiro exploited the fluid, brutal situation in Manchuria, setting up a biological warfare research complex at Ping Fan, later called Unit 731. Another research centre, Unit 100, was set up under Wakamatsu Yujiro. Units 731 and 100 included laboratories, production plant and crematoria. Shiro conducted human experiments, for example the deliberate infection of Chinese prisoners with anthrax spores and the repeated freezing and thawing of live human limbs to investigate frostbite.[60] During the Second World War, General Shiro led the use of biological weapons, first against Soviet troops and then against Chinese soldiers and civilians, ordering, for example, soldiers to dally behind enemy lines to spread plague fleas, and the contamination of water supplies with cholera and typhoid.

Shiro claimed that the deployment of cholera, dysentery, anthrax, plague and other biological agents had been a great success in the 1942 campaign in Chekiang province. In all likelihood, notes Guillemin, it was a 'disaster', with perhaps Japanese troops dying through self-infection, biological friendly fire.[61] Biological weapons were certainly not decisive. The Japanese authorities curtailed Shiro's programme thereafter. This seems to support Price's argument that assessments of decisiveness were critical in the decisions to use, or not to use, supposedly tabooed weapons. Certainly, there were plenty of scientists on both sides who shared Shiro's view of such weapons as modern and potentially effective in an era of total war where civilian populations, and specifically their morale, were valid targets for military aggression. To quote just one such advocate at length, Frederick Banting argued:

> In the past, war was confined for the most part to men in uniform, but with increased mechanization of armies and the introduction of air forces, there is an increased dependence on the home country, and eight to ten people working at home are now required to keep one man in the fighting line. This state of affairs alters the complexion of war. It really amounts to one nation fighting another nation. This being so, it is just as effective to kill or disable ten unarmed workers at home as to put a soldier out of action, and if this can be done with less risk, then it would be advantageous to employ any mode of warfare to accomplish this.[62]

The twentieth-century doctrines of total war created the object of the 'civilian population' as a subject of proliferating and expanding working world sciences. Advocates of chemical and biological weapons, as well as, of course, strategic bombing, made this subject their object, their target. Operational research measured the accuracy with which it was targeted with bombs, correlating the figures with measures of 'morale'. Nutritionists targeted the national diets. The immense armies were often proxies for the wider civilian population. George Beadle, for example, developed some of his mutant strains of *Neurospora* partly to find ways of making contributions to army rations.[63] Rationing in general presupposed a manageable, manipulable, measurable population. This population science, as much as penicillin development or the Manhattan Project,[64] was a distinctive science of the Second World War.

The atomic bomb

The project to build an atomic bomb displays many parallels with the mobilization of other scientific and technical talent during the Second

World War. In particular, the movement from laboratory to bomb needed the application of the lessons in scaling up as well as an appreciation of the fine arts of management learned in pre-war decades. The conventional history of the atomic bomb treats the weapon as a devastating application of the stunning advances in physical science of the first half of the twentieth century. Of course the new physics was a crucial part of the story. We have seen in earlier chapters how physics laboratories drew on an increasingly industrial scale of instrumentation to identify and investigate the particles and phenomena of the nuclear atom. In 1932, John Cockcroft and Ernest Walton had artificially disintegrated light nuclei. In 1934, Enrico Fermi in Italy had bombarded uranium atoms with slow neutrons and noted an artificially created radioactivity. Fermi thought he was producing trans-uranic elements.

In December 1938, Otto Hahn (who, long ago, had worked with Haber on poison gas) and Fritz Strassman in Berlin conducted a variant of these fascinating uranium experiments. Surprisingly, energy was released. Hahn's curious results were interpreted by Lise Meitner, a Jewish Austrian ex-colleague of Hahn's who had left Germany for Stockholm (where she was close enough to Bohr in Copenhagen to be part of the physicist network). Meitner and her nephew Otto Frisch argued that Hahn and Strassman had achieved 'fission': the splitting of uranium atoms into two roughly equally heavy parts. In 1938, nuclear science was open, academic science. Hahn and Strassman's experiment was written up and published in a German scholarly journal. It was widely circulated and read. Replication of their experiment was achieved in many laboratories in January 1939. The idea of a 'chain reaction', the use of neutrons released by the fission of uranium to trigger further fission, thereby greatly increasing the energy released, was also already widespread.

In Britain, two scientists who followed this open academic discussion with mounting interest and concern were Rudolph Peierls, a German Jewish physicist who had chosen to build a career in Britain after Hitler seized power, and Otto Frisch, who had been stranded by the outbreak of war while on a short trip. As a German and an Austrian, Peierls and Frisch were kept well away from the project that was absorbing most physicists in Britain – radar. Ironically, this ban freed Peierls and Frisch to instigate a much more secret project. They calculated the 'critical size' – 'about one pound' of Uranium-235 – the critical mass needed for the explosive chain reaction that would produce a 'super-bomb'. 'As a weapon, the super-bomb would be practically irresistible', they wrote. 'There is no material or structure that could be expected to resist the force of the explosion', while the radioactivity released would continue to kill for days after.

'We have no information', they continued, 'that the same idea has also occurred to other scientists but since all the theoretical data bearing on this problem are published it is quite conceivable that Germany is, in fact, developing this weapon.' Thus prompted by fears of German developments, felt acutely by émigrés who had already had to flee Nazi aggression, Frisch and Peierls in 1940 wrote an urgent memorandum to warn the British government.[65] 'The attached detailed report', they began,

> concerns the possibility of constructing a 'super-bomb' which utilizes the energy stored in atomic nuclei as a source of energy. The energy liberated in the explosion of such a super-bomb is about the same as that produced by the explosion of 1000 tons of dynamite. This energy is liberated in a small volume, in which it will, for an instant, produce a temperature comparable to that in the interior of the sun. The blast from such an explosion would destroy life in a wide area. The size of this area is difficult to estimate, but it will probably cover the centre of a big city.

The memorandum spelled out how much uranium-235 isotope would have to be separated to make an atomic bomb, and recommended that research on separation be immediately undertaken. Frisch and Peierls noted that separation by thermal diffusion was already being achieved by a Professor Clusius of Munich University. They even speculated on possible bomb designs, such as rapidly bringing together hemispheres of uranium by springs.

The British government responded to the Frisch–Peierls memorandum in a traditional way: it set up a committee. The MAUD Committee sat between April 1940 and July 1941, reviewing evidence, testing each stage in the logic of Frisch and Peierls's argument, reviewing likely separation strategies, and composing a report. The committee was put together by Tizard, and the influence of the committee-led development of radar can be seen in its form. Members were the Imperial College nuclear physicist George Paget Thomson, the Birmingham University physicist Mark Oliphant (in whose lab Frisch and Peierls were working), Oliphant's colleague Philip Moon, neutron expert James Chadwick, his Cambridge colleague and expert in the intersection of laboratory and industrial physics John Cockcroft, and the rising star of war science Patrick Blackett. The MAUD Committee reported that Frisch and Peierls's super-bomb was indeed feasible.

In support and in response to these calculations, Britain began a project to build an atomic bomb. The project was dubbed 'tube alloys', a deliberately dull-sounding title with a 'specious air of probability' about it.[66] Tube alloys focused on methods of separation. Its favoured candidate was gaseous diffusion: heavier molecules take longer to diffuse than lighter ones, so, if a uranium gas could be made, then the lighter uranium-235 that made up less than 1 per cent of a typical sample could be separated from the heavier, useless uranium-238. This was a problem both for laboratory chemistry and for chemical engineering. A gas, uranium hexafluoride, was made in substantial quantities by the British chemical combine ICI. The MAUD report of March 1941 concluded, first, that 'the scheme for a uranium bomb is practicable and likely to lead to decisive results in the war', second, that 'this work be continued on the highest priority and on the increasing scale necessary to obtain the weapon in the shortest possible time', and, lastly, that 'the present collaboration with America should be continued and extended especially in the region of experimental work'.

The reports of the MAUD Committee were sent to the United States. The Tizard mission had already exchanged information about nuclear research. American interest in the atomic bomb was minimal, initially. In August 1939, Leo Szilard had urged his fellow émigré Albert Einstein to put his signature to a letter, drafted by Szilard, to be sent to President Roosevelt. The letter spelled out the implications of the discovery of fission. After reporting that uranium 'may be turned into a new an important source of energy in the immediate future', Einstein wrote:

> This new phenomenon would also lead to the construction of bombs and it is conceivable – though much less certain – that extremely powerful bombs of a new type may thus be constructed. A single bomb of this type, carried by boat and exploded in a port, might very well destroy the whole port together with some of the surrounding territory. However, such bombs might very well prove to be too heavy for transportation by air.

Einstein urged 'watchfulness', but also immediate actions to secure a supply of uranium ore (only a small quantity lay within United States borders; most came from Bohemia in Czechoslovakia, soon under German control, and the Belgian Congo) and to fund an expansion of academic research and the 'co-operation of industrial laboratories, which have the necessary equipment'. Einstein ended on a dark hint, noting that Germany had stopped the sale of uranium from Czechoslovakian mines and that Carl Friedrich von Weizsäcker, expert in nucleosynthesis, was the son of Nazi politician Ernst von Weizsäcker.

In response to Einstein and Szilard's warning, a uranium committee was established and a small sum, $8,000, set aside for research. The major project supported was under Enrico Fermi in Chicago. Fermi had built up a reputation as the leader of a creative group of physicists in Rome in the 1920s and 1930s. He received a Nobel Prize in 1938, and, with his status more than confirmed, emigrated that year to the United States – his wife, Laura, being Jewish and targeted by fascist anti-Semitic laws. Fermi arrived in time to join the American replication of the fission experiments at Columbia University. In 1940, he was in charge of the project to scale up: to build a working nuclear pile, a controllable fission chain reaction.

In October 1941, the MAUD reports had fallen into the hands of the cabal of OSRD science administrators centred on James Conant and Vannevar Bush. They pushed hard for the implications of the MAUD Committee's conclusion – that atomic bombs were feasible and decisive weapons – to be fully appreciated. From October 1941, full-scale research into an atomic bomb was under way in the United States. By 1942, Fermi's team, which included Szilard, had completed his nuclear pile. It was a pile of uranium bricks, kept apart by blocks of graphite, which absorbed the neutrons and, while in place, prevented runaway fission. It was built under the wing of the university's football stadium, in the midst of suburban Chicago. On 2 December, the graphite was withdrawn, piece by piece. Energy began to surge. When it was clear that the first controlled nuclear chain reaction had been achieved, the graphite was replaced, with celebration and more than some relief.

In 1942, the escalating problem was placed under the Army Corps of Engineers, with a division name of the 'Manhattan Engineer District' – hence 'Manhattan Project'. Its leader was Colonel (soon General) Leslie Groves, a military organizer who was coming fresh from overseeing the construction of the Pentagon, one of the largest and most complex civil engineering tasks of the age. Bush and Conant were now key members of a highest-level committee, the S-1, with the war secretary, Henry Stimson, Roosevelt's chief of staff, George C. Marshall, and the vice president, Henry Wallace. In January 1942, on Ernest Lawrence's advice, the S-1 group recruited Robert Oppenheimer, who immediately began to put together a team of top physicist talent that would review and develop the science needed to build atomic bombs.[67]

This unusual summer school, arguably the most important meeting of scientists in the twentieth century, included several émigrés, among them the German Hans Bethe, now at Cornell, the Hungarian Edward Teller, now at George Washington University, and the Swiss physicist Felix Bloch. The group was united in the belief that possession of the atomic bomb was the surest means of defeating Hitler. They sketched designs for bombs. Edward Teller, starting down a lonesome path of profound implications, spent much of July 1942 arguing that a bomb that fused hydrogen together, rather than splitting heavy atoms, would be the better bet. Teller lost the argument, for now. Bethe calculated that the chance of accidentally igniting the atmosphere was 'near zero'.[68] In general, Oppenheimer's physicists confirmed, to their own satisfaction, that a few kilograms of uranium-235 would work, identified what further research was needed, and concluded that an expensive, industrial-scale plant was needed to separate the uranium-235 from the uranium-238.

The Manhattan Project therefore had a junior, civil, scientific side, increasingly led by Robert Oppenheimer, and a senior, military, organizational side, led by General Groves. Both sides had to manage the central task of the project: scaling up the processes, found in chemical and nuclear physics laboratories, for separating isotopes. On top of that task, a working design had to be drawn up, which in turn depended on further research (such as the measuring of precisely how neutrons scattered through uranium) and difficult, time-consuming calculation. The Manhattan Project developed on three main sites, each one a small science-engineering city built from scratch: Oak Ridge in Tennessee, Hanford in Washington state, and Los Alamos in New Mexico. Two connected sites were also relevant. One facility was built at Chalk River, Ontario, Canada. British and Canadian atomic bomb work would eventually be subsumed totally into the American project. A second would open at the University of Chicago. When Niels Bohr had told American physicists that 'to get the fissionable material necessary to make a bomb' it would be necessary to 'turn the whole country into a factory' he was exaggerating, but not by much.[69]

The central task of the Manhattan Project was to separate uranium isotopes, and various approaches were considered: gaseous diffusion has already been mentioned, while thermal diffusion relied on a similar principle that heavier isotopes moved slower than lighter ones. Another method considered was using centrifuges to separate light from heavy. Finally there was electromagnetic separation: to build what were, in effect, industrial-scale mass spectrometers that would separate and sort the ionized uranium, atom by atom. The experts in electromagnetic methods of separation were to be found in Lawrence's laboratory. Oak Ridge should be interpreted, therefore, as the place where Lawrence's scaling up of research continued on an even greater scale.[70] Oak Ridge covered 54,000 acres of land. Calutrons, modified cyclotrons, as well as gaseous diffusion plant, drew on the immense hydroelectric power available in Tennessee, thanks to the Tennessee Valley Authority, to separate out uranium-235.[71] The target of improving the output of calutrons generated new physics, such as the phenomenon of Bohm diffusion,[72] but, more importantly, issues of engineering. When electrostatic and gaseous diffusion faltered at a critical stage in 1944, thermal diffusion was tried. Indeed, in a triumph of engineering, all three were hitched together in a production line of uranium enrichment.[73]

Hanford is an even more remarkable story of scaling up. Between late 1940 and early 1941, a team led by Glenn T. Seaborg, a scientist at Lawrence's laboratory, had described a new transuranic element, 'plutonium', found in microscopic quantities after bombarding uranium in the 60-inch cyclotron. Almost nothing was known about the chemical or other properties of this artificial element. At Chicago University, Arthur Compton's Metallurgical Laboratory (another deliberately bland name), which would eventually have a staff of 2,000 scientists, engineers and administrators, investigated the properties of plutonium and its potential role in a second type of fission bomb.

Very soon tensions emerged between the academic, democratic culture of the Met Lab scientists and that of the military, engineering, industrial management under Groves.[74] The scientists, such as Szilard and Eugene Wigner, regarded Groves's organization as authoritarian; Groves considered the scientists to be stuck in an ivory tower, even trouble-makers. At root, though, the difference was an appreciation of what was necessary to scale up. Groves brought in Du Pont, the firm which possessed crucial experience of scaling up laboratory work – for nylon, now read plutonium – and also had adopted, in the nylon production, an approach now favoured by Groves as a risky necessity.[75] Groves argued that the United States could not afford to wait for early stages of production to be figured out before later stages were designed. Du Pont had followed this strategy for nylon, successfully. Du Pont, working with designs from Seaborg, and drawing on, after much agitation, the expertise of the Met Lab, oversaw the construction of giant reactors at Hanford, cooled by drawing on the great Columbia River at a rate of 30,000 gallons per minute for each pile.[76] Again, scaling up revealed new problems that needed considerable ingenuity to resolve. For example, the reactors at first seemed unaccountably to slow grind to a halt. Met Lab scientist John Wheeler analysed the problem as being due to the presence of xenon-135.[77] Soon the Hanford Engineer District was producing enough plutonium to be seen with the naked eye.

The third, and probably most famous, site was Los Alamos.[78] Robert Oppenheimer had ridden around this remote mesa in New Mexico with a friend, Katherine Page, in 1922. Now, in November 1942, Oppenheimer, Edwin McMillan, a member of the plutonium team, and General Groves were scouting nearby land for a new laboratory. The area they had come to see was unsuitable. 'If you go on up the canyon you come out on top of the mesa and there's a boys' school there which might be a usable site', suggested Oppenheimer; once there Groves made his mind up in an instant: 'This is the place'.[79] By March 1943, when Los Alamos officially opened, over a hundred scientists, engineers and other staff were already on site. Soon there were thousands, including luminaries such as Hans Bethe and Richard Feynman. A few others, in particular Isidor Rabi and Niels Bohr, took up distanced, more consultative roles. Very few, such as Lise Meitner, refused to join.[80] Los Alamos was many places at once: a weapons laboratory, an 'army camp', an 'ideal republic' of science, a Magic Mountain, a Shangri-La.[81] Newcomers went to school immediately to learn from Oppenheimer's protégé Robert Serber the nature of their task: to design and build a 'gadget', a working atomic bomb.

In 1943 and early 1944, the Manhattan Project was working at full tilt, producing macroscopic quantities of two kinds fissionable material – uranium-235 and plutonium – to be used in a gun design of bomb: firing one lump into another to create a critical mass. However, in spring 1944, research by Seth Neddermeyer revealed a flaw in the plutonium gadget: the gun would fizzle if fired. By July this problem had become a crisis, and an alternative design was improvised: if a hollow sphere of plutonium could be imploded fast enough, the device could detonate before fizzling. But the implosion design presented almost intractable unknowns: how could the plutonium be machined to the right shape? How would the chain reaction proceed in detail? The Hungarian prodigy John von Neumann was brought in to manage the fiendishly complex calculations.

On 7 May 1945 Germany surrendered. Over a year before this date the atomic bomb project had subtly changed from a means to defeat Nazism, the aim that especially united the émigré scientists, to something else, including, but not restricted to, a means to defeat Japan. Only one man left the Manhattan Project because of this change in purpose. Joseph Rotblat, a Polish émigré who was part of the British bomb effort, was at James Chadwick's house for dinner with General Groves in March 1944. 'You realize of course', Groves remarked in passing, 'that the main purpose of this project is to subdue the Russians.'[82] The remark festered in Rotblat's mind. In December 1944, when it was clear that Germany was heading for defeat, Rotblat walked out. The scientists that remained had become artfully managed and focused on their military mission, as is revealed by this comment from Oppenheimer to Groves:

> For the most part these men regard their work here not as a scientific adventure, but as a responsible mission which will have failed if it is let drop at the laboratory phase. I therefore do not have to take heroic measures to insure something which I know to be the common desire of the overwhelming majority of our personnel.
>
> [It would be a] fallacy [to regard] a controlled test as the culmination of the work of this laboratory. The laboratory is operating under a directive to produce weapons; this directive has been and will be rigorously adhered to.[83]

On 16 July 1945, the first Los Alamos gadget was ready to test. An implosion device was held by a scaffold on the Jornada del Muerto ('Dead man's journey'), near Alamogordo, New Mexico. At its core was a finely engineered shell of plutonium, an element not even discovered when Germany had invaded Poland in 1939. Physicist Samuel Allison counted down. At zero, the Trinity test, the first atomic detonation, lit up the sky. The immediate recorded responses are revealing: 'Still alive, no atmospheric ignition' (Allison), 'Now we're all sons of bitches' (physicist Kenneth Bainbridge), 'The war's over' (Groves) and, most typically, 'It worked' (Oppenheimer, in the recollection of his brother, Frank).[84] Years of targeted, finely managed work had reduced the most devastating weapon devised by humankind to a technically sweet problem, instrumentally assessed. Moral doubts had been temporarily erased, although, as we shall see, they returned in memory and in practice.

On 6 August 1945, a uranium gun-type atomic bomb, 'Little Boy', was dropped on the port of Hiroshima: 78,000 Japanese died immediately. Three days later, a plutonium implosion atomic bomb, 'Fat Man', devastated Nagasaki. Japan surrendered. The debate about why and on whom the atomic bombs were used began in earnest in early 1945 and has continued ever since. In June 1945, scientists at the Chicago Met Lab had composed the so-called Franck Report, named after the chairman of an impromptu committee, James Franck, which urged the demonstration of the device in front of international witnesses. The report was interrupted en route to Washington and was not read by the senior political authorities.[85] The Met Lab scientists had a misplaced trust that Oppenheimer would think along similar lines. But the Los Alamos scientists, heads down and working towards their target, saw it now as a weapon of war. A demonstration test, after all, might fizzle.

Certainly the destruction of cities concentrated the minds of at least three crucial audiences.[86] First, Japanese leaders, whom Allied leaders agreed were otherwise culturally indisposed to surrender, would order the resistance of a land invasion in which many American lives would be lost. Second, Stalin, whose Red Army was advancing on Japan through continental Asia. The atomic bombs presented a way of ending the Pacific war before Soviet troops could land, and partition, Japan. Hiroshima, according to historians who take this view, was an early shot in the Cold War. Finally, and most cynically, the United States Congress would find it far easier to accept an immensely expensive, secretive project if it ended the war decisively rather than seeming a costly distraction.

Lessons learned

The Manhattan Project cost just over $2 billion in the 1940s. One lesson not forgotten was that government-funded, goal-oriented, science-based missions could not only succeed but decide matters of global influence. Other matters of the scale and organization of the Manhattan Project are worth reviewing. The bomb project, like radar and conventional weapon development, depended on thorough links with industry, with firms such as Du Pont, Stone & Webster, Monsanto, and a host of smaller contractors. Staff numbers were high, filling small cities: 75,000 at Oak Ridge, 45,000 workers at Hanford. The three major sites – Oak Ridge, Hanford and Los Alamos – contained not only laboratories and factories but cinemas, churches, hospitals and schools. The scale brought a fine and complex division of labour, the management of which was another area brought to the project by the industrial and military experts. The style was of teamwork and managed cooperation, familiar to industrial researchers or those funded by the philanthropic foundations before the war, and now generalized.

The Manhattan Project was improvised in short order, but it drew on lessons learned earlier at four key sites: scaling up at Lawrence's laboratory[87] and Du Pont,[88] and the large-scale systems management needed to build the Pentagon[89] and, on a regional scale, the Tennessee Valley Authority.[90] Groves had learned how to build secrecy into organizations and architecture. The Manhattan Project brought this lesson to scientists. Under the militarization of nuclear science, the labour to design and build an atomic bomb was not only divided but compartmentalized: a team would be set an achievable task but would not necessarily know how it fitted in the overall plan. Compartmentalization was a security measure, under which scientists chafed, very occasionally rebelled (Richard Feynman), but ultimately accepted to such an extent that it encouraged the instrumental ethos revealed at Trinity. Compartmentalization therefore dampened moral concerns as well as serving to prevent careless talk.[91]

Teamwork was also encouraged by the interdisciplinary character of the Manhattan Project. Physicists were joined by chemists, chemical engineers, mathematicians, medical doctors, human computers, civil engineers, electrical engineers, experts in fuses and photography, industrial research managers, civil and military administrators, security chiefs, and many other stripes of expertise, authority and service staff. Yet the atomic bomb has been presented from 1945 onwards as a triumph of physics, as the epitome of the Physicists' War. This attribution is a construct, shaped by security concerns. In 1945, in response to incessant demands for more information, an official report on the development of the atomic bomb was commissioned by Groves from Henry DeWolf Smyth, physicist and NDRC Uranium Committee member. The Smyth Report, *Atomic Energy for Military Purposes*, released three days after Nagasaki, emphasized the physics, which was open and public, and downplayed the engineering and other expertise equally essential to producing a working bomb. Einstein's formula '$E = mc^2$', for example, was publicized, decisively, by the Smyth Report. Ever since, the atomic bomb has been seen as an achievement of physics.

If the historical lens is widened to consider atomic weapons alongside the chemical and biological warfare projects of the Second World War, then several other trends become prominent. First, the transfer of projects from Britain to the United States exemplifies a westward shift in leadership and power. The story of radar also followed this narrative. And, as it moved, science changed. Science as scaled-up, goal-driven, manager-organized, American-led projects would be a prominent, but by no means hegemonic, feature of the post-war world.

292

Second, almost all the science-based military projects of the twentieth century were intimately connected to the development of airpower. For radar the case is clear, but airpower also powerfully shaped thinking about chemical and biological weapons. In 1942, two American medical scientists, Theodor Rosebury and Elvin Kabat, of Columbia University's College of Physicians and Surgeons, wrote a comprehensive review of the state and scope of biological weapons, which was published in a brief window of openness on the subject in 1947. They concluded that the 'airplane is clearly the most useful means for the dissemination of infective agents'.[92] It was the combination of chemical weapons and airpower that had made the Italian war so devastating in Ethiopia. Most importantly, the establishment of strategic bombing as an acceptable form of warfare cleared the path for imagining and justifying atomic bombing. The two are continuous.

Historians have raked over the coals of Lübeck, Hamburg, Dresden and Tokyo. Richard Overy regards strategic bombing as decisive in the Allied victory.[93] Michael Sherry takes a different position, arguing that the cities did not become targets because they were important but were important targets because they could be bombed.[94] Sherry's analysis of the rise of American airpower can frame our understanding of the twentieth century; he argues that the 'casual' transition from mere coastal defence, to strategic bombing, to the capability and will to create fire-storms out of Tokyo (and Hiroshima and Nagasaki) was driven by the organizational momentum acquired as the newly established United States Air Force sought a role of its own. It's a frightening thesis of organizational and moral drift. Deeper forces supported this casualness: 'Air power appealed as well as to a deeper strain of anti-statism and anti-militarism in American culture', writes Sherry, 'because its reliance on a small, technically sophisticated elite apparently avoided the burdens of conscription, taxation and death.'[95] Furthermore, airpower distanced killer from killed in several mutually supporting ways – technological, rhetorical, moral. The technologies here were not merely the obvious ones of bombers operating far above human targets but also quantitative technologies[96] that could represent the devastating effects of, say, incendiary raids on Tokyo, in terms of a cost–benefit analysis (witness, for example, Robert McNamara discussing these raids in Errol Morris's 2004 film *The Fog of War*).

Airpower was a subset of the working world of the military, with its own sciences. Distancing technologies needed all kinds of experts – 'Air Prophets', operational scientists, statisticians – scientists as 'civilian militarists', to use Sherry's term.[97] Roles were created, and appeals to such roles justified expert authorities. When combined with secrecy, the effect was often anti-democratic: witness the start of large-scale Allied biological weapon production without formal political approval, the way that Roosevelt was barely informed about subsequent developments in this 'dirty business' by Henry Stimson,[98] or the funding of the Manhattan Project off the books.

The character of twentieth-century warfare, therefore, was powerfully shaped by civilian experts. Strategic bombing, for example, notes Overy, was 'chosen by civilians to be used against civilians, in the teeth of strong military opposition'.[99] Likewise, Allied biological warfare as a weapon and civilians as a legitimate target of such weapons were advocated by civilian militarist scientists such as – as we saw earlier – the 'hero of insulin', Frederick Banting. Japanese biological warfare in the Second World War stands out as being promoted from within the military, and also for not being motivated primarily by fears that rival nations were developing it (indeed, precisely the opposite). The French had started their biological warfare programme in the 1920s because an inspection of German pharmaceutical plants sparked concern.[100] Paul Fildes 'firmly believed' Germany possessed an 'identical' anthrax weapon to the one he was urging in the 1940s.[101] Fears, rumours, allegations and intelligence of dubious quality are a recurring motif of the history of weapons of mass destruction.

The German superweapons

Fear of a German atomic bomb project, of course, was the primary motive for starting the British and American atomic bomb projects. So what was the reality? Under the Nazis, despite the distraction of the Aryan science movement, German physics was still comparable in quality and quantity its rivals elsewhere in the world. In theoretical physics, talents such as Werner Heisenberg arguably gave it an edge. After the revelation of the Berlin fission experiments, a project to investigate and develop applications of nuclear energy started in 1939 under army control. The number of scientists involved (sixty to seventy), and the focus on techniques (centrifuge work to separate isotopes), paralleled British and American development between 1939 and 1942. Nevertheless, the self-imposed censorship by Allied physicists which led to a dearth of fission papers had one major effect: the Germans struggled to use heavy water to moderate chain reactions and did not find out about the much easier method deploying graphite.[102]

In September 1941, an exchange of views took place between Heisenberg and Niels Bohr in Copenhagen. The status and significance of this meeting has been disputed by its protagonists (Heisenberg, Bohr) as well as historians (Paul Lawrence Rose, Mark Walker, David Cassidy), and has even been addressed by playwrights (Michael Frayn). Heisenberg claimed he alluded to the German bomb project, saying he tried to pass a warning, through the Dane, to the rest of the world. Heisenberg hinted that he was using his pivotal position – as an expert authority in relation to his Nazi superiors – to delay progress.[103] Bohr says that Heisenberg passed him information but that Heisenberg remained pro-Nazi. Rose argues that Heisenberg was trying to recruit Bohr, or at least to extract useful knowledge.[104] Frayn concludes that the episode is irreducibly undecidable. Walker argues that the irreducible doubts have made the Copenhagen exchange seem more significant than it was and have contributed to its mythologization.[105] This is true, but myths tell us much about what groups want to believe, in this case about the responsibility, or otherwise, of scientists working in Nazi Germany.

More decisive was a review conducted by the German High Command in 1942 of the use of scarce resources in the light of progress of the war. As a result an entirely rational decision was taken to scale back dramatically investment in projects that were not deemed likely to be ready for use before the war ended and to allocate resources in favour of those weapons that would be ready. So atomic weapons projects were dropped, while other projects, in particular liquid-fuel rocketry and pilotless aircraft, were favoured. The Allied governments did not learn directly of this change of strategy, and so the Manhattan Project continued undisturbed. Operation Alsos, a military intelligence team that searched liberated Europe for evidence and interviewed (and wiretapped) captured nuclear scientists, concluded that Germany was not close to building an atomic bomb in 1945. Samuel Goudsmit, physicist and scientific leader of Alsos, concluded that this failure was due to the incompatibility of totalitarianism with science-based innovation.[106] His general conclusion is contradicted by my next example.

Amateur rocket societies had flourished in the United States, the Soviet Union and Germany from the 1920s. These societies developed the theory of rocketry, legitimated and publicized the idea of space flight, launched small rockets (both solid-fuel and liquid-fuel) and provided training in, and practical experience of, rocketry foundational to later twentieth-century space flight. But amateur groups did not have the resources to develop rockets very far. State patrons, however, did. The restrictions on the rearmament of existing military technologies under the Treaty of Versailles made rockets of particular interest to German military authorities. In 1932, senior officers of the German army had visited the German rocket society (the Verein für Raumschiffahrt, or the Society for Space Ship Travel) and recruited the engineer Wernher von Braun to work for them.

The 'German rocket programme was one of the first examples of state mobilization of massive engineering and scientific resources for the forced invention of a radical, new military technology', writes historian Michael Neufeld, adding that it 'preceded by a decade and was not dramatically smaller in scale than the even more revolutionary Manhattan Project'.[107] Von Braun became technical director of the Nazi rocket programme at Peenemünde, near the Baltic coast. He designed the A4 – later renamed the Vergeltungswaffe 2 (V-2) – a giant liquid-fuel rocket. Captain Walter Dornberger was the 'key system builder' who guided the V-2 from design to mass production – scaling up which required much investment in research and manufacturing methods.[108] Benefiting from the flow of resources after the 1942 change in strategy, the mass-produced rocket could be launched from the Netherlands and climb 100 kilometres, before plunging unstoppably to hit targets in the South of England.[109] It arrived too late to change the outcome of the Second World War in Europe.

V-2s were built partly with forced labour. Dora-Mittelbau, a satellite of the Buchenwald concentration camp system, provided inmates as a workforce. 'The V-2 was unique in the history of modern warfare', says Smithsonian curator Tom D. Crouch, 'in that more people [20,000] died building the rockets than lost their lives on the receiving end.'[110] The concentration camps also provided bodies – live and dead – for human experimentation. Germany in fact possessed the most stringent regulation of human experimentation, formed in response to medical controversies. The late nineteenth-century tests of syphilitic sera conducted by Albert Neisser on Breslau prostitutes had prompted Prussian authorities to outlaw experimentation on the young and the debilitated, while demanding consent from other potential subjects.[111] In 1939, doctors in Lübeck had injected a contaminated BCG vaccine, killing seventy-six children; the regulations that followed the Lübeck disaster included consent, the substitution of animal experimentation in place of some human experimentation, and further protection of the vulnerable.[112] Yet these strong regulations were powerless against Nazi abuse. During the Second

World War, concentration camps served up prisoners for a series of gruesome experiments, many designed to assist military planning and design. For example, Dachau inmates were immersed in freezing salt water to generate data relevant to determining how long ditched pilots could survive in the sea, while other prisoners were locked in low-pressure chambers to simulate the conditions of high-altitude pilots.[113] Josef Mengele dissected murdered twins in Auschwitz.[114] Experiment is about controlling and manipulating simplified representations of working worlds – and the almost total control granted the authorities of the concentration camps facilitated these dreadful experiments.

Aftermath: ethics and spoils

In 1946 and 1947, twenty-three Germans were put before a United States military tribunal in Nuremberg, the city which had given its name to the infamous race laws of the 1930s. Seven were hanged in 1948, including Hitler's personal doctor, Karl Brandt, Waldemar Hoven, the leading doctor at Buchenwald, and Wolfram Sievers, who had organized the collection of skulls of Jewish prisoners for anthropometric study. The verdict of the Doctors' Trial laid down the Nuremberg Code (1949), a guide to the ethics of human experimentation, consisting of ten points:

1 The voluntary consent of the human subject is absolutely essential. This means that the person involved should have legal capacity to give consent; should be so situated as to be able to exercise free power of choice, without the intervention of any element of force, fraud, deceit, duress, over-reaching, or other ulterior form of constraint or coercion; and should have sufficient knowledge and comprehension of the elements of the subject matter involved as to enable him to make an understanding and enlightened decision. This latter element requires that before the acceptance of an affirmative decision by the experimental subject there should be made known to him the nature, duration, and purpose of the experiment; the method and means by which it is to be conducted; all inconveniences and hazards reasonable to be expected; and the effects upon his health or person which may possibly come from his participation in the experiment.

The duty and responsibility for ascertaining the quality of the consent rests upon each individual who initiates, directs or engages in the experiment. It is a personal duty and responsibility which may not be delegated to another with impunity.

2 The experiment should be such as to yield fruitful results for the good of society, unprocurable by other methods or means of study, and not random and unnecessary in nature.

3 The experiment should be so designed and based on the results of animal experimentation and a knowledge of the natural history of the disease or other problem under study that the anticipated results will justify the performance of the experiment.

4 The experiment should be so conducted as to avoid all unnecessary physical and mental suffering and injury.

5 No experiment should be conducted where there is an a priori reason to believe that death or disabling injury will occur; except, perhaps, in those experiments where the experimental physicians also serve as subjects.

6 The degree of risk to be taken should never exceed that determined by the humanitarian importance of the problem to be solved by the experiment.

7 Proper preparations should be made and adequate facilities provided to protect the experimental subject against even remote possibilities of injury, disability, or death.

8 The experiment should be conducted only by scientifically qualified persons. The highest degree of skill and care should be required through all stages of the experiment of those who conduct or engage in the experiment.

9 During the course of the experiment the human subject should be at liberty to bring the experiment to an end if he has reached the physical or mental state where continuation of the experiment seems to him to be impossible.

10 During the course of the experiment the scientist in charge must be prepared to terminate the experiment at any stage, if he has probable cause to believe, in the exercise of the good faith, superior skill and careful judgment required of him that a continuation of the experiment is likely to result in injury, disability, or death to the experimental subject.

The Doctors' Trial and the Nuremberg Code were shaped by ethical advice from Andrew Ivy of the American Medical Association. Yet American medical research during the Second World War had not followed these principles: 'American investigators', concludes historian Susan Lederer, had 'routinely used populations unable to exercise voluntary consent, including children and the mentally-ill', while the OSRD had funded medical research in 'prisoners, inmates of mental institutions, and children in orphanages'.[115] Conscientious objectors provided another pool of bodies.

Double standards also marked the American response to Japanese human experimentation data. After capture, General Ishii Shiro offered documents and data of his experiments in trade for immunity from criminal prosecution. United States military intelligence was granted authority to make a deal in March 1947. Jeanne Guillemin suggests three reasons.[116] The first concern for American authorities in Japan was to secure order; in particular, a decision was made to keep the hierarchical structure of Japanese society largely intact. There was some suggestion that patronage of the biological warfare programme went right to the top, and so the 'immunity bargain was meant to protect the emperor'. Second, the data were judged to be of usable quality. Dr Norbert Fell, interrogator of Ishii, noted that 'data on human experiments, when we have correlated it with data we and our Allies have on animals, may prove invaluable, and the pathological studies and other information about human diseases may help materially in our attempts at developing really effective vaccines'; this attraction had the extra appeal of 'forbidden fruit'.[117] Third, the decision must be understood in terms not only of how American authorities read the past but also of how they viewed the future. Specifically, in an early intimation of the Cold War, the threats to Japanese stability were considered to be internal communist agitation and external Soviet influence.

Competition between the United States, Britain and the Soviet Union for the spoils of the scientific war also framed the post-war treatment of German experts.[118] While Operation Alsos had sought out and interviewed German nuclear scientists only to conclude that there was little the United States could learn from them, the Soviet Union ran its own equivalent raid, capturing several scientists who would be useful in the Soviet bomb programme, including Nikolaus Riehl and Gustav Hertz. More generally, the United States ran Operation Paperclip, which sought out scientists and engineers and moved them west. The policies of the United States and the Soviet Union towards German specialists had 'remarkable similarities', notes historian Andreas Heinemann-Grüder: both 'pursued their own economic interests', both evacuated scientists 'more or less under the threat of force', both ignored the Nazi past, or otherwise, of their prizes, and both acted largely to stop the other.[119]

German armament engineers, especially rocket scientists, were in special demand, which in turn enabled them to escape responsibility for wartime actions. The German rocket programme had many different components – top scientists and administrators, technicians, workers, blueprints, factories and machines – of which, crucially, some were more mobile than others. The elite scientists had passes to travel, and moved west. American troops who had occupied Thuringia took with them 1,300 missile workers, and from the Harz mountains they found and took blueprints; when the lands were handed over to Soviet troops, the less mobile or bulky components were seized: lower-status workers, technicians, raw materials, spare parts, unfinished V-1 flying bombs and unfinished V-2 missiles.[120] This unequal division of spoils would shape the post-war, Cold War world.

— 13 —

TRIALS OF SCIENCE IN THE ATOMIC AGE

A state of deep mistrust marked the Grand Alliance between the United States and the Soviet Union well before the end of the Second World War. In its defence of Russia, the Red Army lost more lives than any other armed force in the war. Stalin had wanted the United States and Britain to open an immediate second front to release the intense pressure and, when this did not happen, suspected a deliberate policy of delay.[1] Suspicion turned to non-cooperation as the Allied armies advanced in Europe. The Red Army was kept out of Italy, and British and American troops out of Romania, Bulgaria and Hungary. Already a post-war map of Europe was forming on the basis of occupation. Non-cooperation turned to competition in the race to seize German resources and expertise.

A second major source of distrust lay in how the Soviet leader was informed of the atomic bomb.[2] The Trinity test at Alamogordo had taken place on 16 July 1945. On that day the three leaders of the Grand Alliance, the new president, Harry S. Truman, Winston Churchill and Joseph Stalin, had begun to arrive at Potsdam, in occupied Germany, to plan the post-war world. Truman, who had only recently been briefed about the bomb, having been kept out of the loop by Roosevelt, informed Stalin of the possession of the weapon. Stalin, determined to display no sign of weakness, was poker-faced. He did know, however, that the need for Soviet military assistance had vanished in an atomic flash. The basis for the post-war settlement – the balance of blood lost on the battlefield – was shattered by this new power. What the American president did not know, then, was that Los Alamos had been successfully penetrated by Soviet spies – three times, independently. Stalin took the late announcement of the atomic bomb, therefore, not as an exchange of information between friends and allies but as an attempt to threaten the Soviet Union at the Potsdam negotiations, a strong-arm diplomatic gambit. His distrust increased.

The wartime tensions froze into the Cold War, a world organized around the competing ideologies and technological systems originating from two poles, East and West. The theory that guided the West's understanding of the East was supplied by a telegram sent on 22 February 1946 by George F. Kennan, a relatively junior diplomat serving in the United States embassy in Moscow. Kennan's 'Long Telegram' argued that Soviet external attitudes of hostility were determined by an internal need to justify the brutal conditions of its regime. Since there was nothing the West could do directly to resolve this hostility, a policy of 'long-term, patient but firm and vigilant containment of Russian expansionist tendencies' had to be followed.[3] This policy of containment defined the nature and endurance of the Cold War. Containment would be supplemented by lavish support for institutions embodying democracy and freedom outside of the East. Economically, the consequence was the Marshall Plan, which reconstructed Western Europe from 1947. Politically and militarily, the consequence was the establishment of NATO, an alignment of Western armed forces, in 1949. Each found a rival echo in the East. For science there would be equally profound reconstructions.

In this and the following chapters I will review the Cold War reconstruction of science under several headings: increased funding for science, especially for the sciences that contributed to military systems, the expansion of rival atomic programmes, the consequences for secrecy and openness in the sciences as national security became an overriding concern, reflection on scale prompted by the Manhattan Project as it was posited as a model for the organization of science, and the many ways that the Cold War as a global institution framed the global politics of science and technology. The Cold War 'military-industrial-academic complex' became the new working world for much important post-war science. This chapter and the next begin the task of understanding these working world sciences by focusing mainly on the atomic spaces and places. Chapter 15 reviews Big Science. Chapter 16 continues the exploration by examining the Cold War mutations of the Second World War radar sciences.

Permanent mobilization of science

Once nuclear weapons had been manufactured in a form capable of being dropped by long-range aircraft – doubly demonstrated at

Hiroshima and Nagasaki – the nature of preparing for war changed. The possibility of imminent devastation meant that all preparation had to be conducted before war between nuclear-armed combatants broke out. In the Second World War, new weapons had been improvised and old weapons steadily improved in wartime. After 1945, the necessity for a 'permanent state of readiness' implied that research and development must be carried out in advance, notes historian Alex Roland, in order that the 'arsenal for the next war . . . be invented, developed, and deployed today';[4] the consequence was the permanent mobilization of science and technology.

The more than tenfold increase in expenditure on science started in the Second World War was therefore sustained, and indeed increased further, during the Cold War. The mid-twentieth century therefore marks a step change in the funding of science. The funds for new military research translated into new, expensive technologies that both embodied and created the Cold War geopolitical order. This trend was reinforced by other considerations. In particular, there was no political will for a large standing army in the West. The alternative path followed was investment in the technologies of automation – electronics, computing, and so on – which enabled a smaller human component to operate a more devastating armed force.

The consequences of this increased funding are numerous and will be unpicked across the following chapters. Permanent and munificently funded mobilization of science raised questions about its organization, its control, the channels of funding, the particular sciences to be targeted, the effects of such funding on those sciences, and so on. There were also opportunity costs: the money not spent on other goods that went instead on military systems. In the 1960s, indeed, rival studies by the United States defence and civil science establishments – Project Hindsight and TRACES – came to conflicting conclusions about the contribution of pure research, the curiosity-driven open research supposedly starved by military investment. (We will see that this category of science, and its funding, was not so simply delineated in the Cold War.) Roland concludes that opportunity costs are not discernible: had investment in science for national security not been made, 'it is not clear that the government would have chosen to spend comparable amounts on the scientists' agenda'.[5] Not everyone has agreed. A lively discussion on whether defence spending in general, and military research and development spending in particular, has been detrimental to economic performance has been conducted since the 1950s.[6]

Frontier disputes

Despite being among the prime instigators of the new world, Vannevar Bush, and his allies among the American academic elite, attempted to preserve old pre-war values in a rearguard action to define the new organization by which the patronage would be channelled.[7] In November 1944, President Roosevelt had written to Bush, as director of OSRD, noting the 'unique experiment of teamwork and cooperation in coordinating scientific research and in applying existing scientific knowledge to the solution of the technical problems paramount in war', and requesting recommendations so that 'the lessons to be found in this experiment . . . be profitably employed in times of peace'. Specifically he asked for advice on four points:

> First: What can be done, consistent with military security, and with the prior approval of the military authorities, to make known to the world as soon as possible the contributions which have been made during our war effort to scientific knowledge?
> . . .
> Second: With particular reference to the war of science against disease, what can be done now to organize a program for continuing in the future the work which has been done in medicine and related sciences?
> . . .
> Third: What can the Government do now and in the future to aid research activities by public and private organizations? The proper roles of public and of private research, and their interrelation, should be carefully considered.
> Fourth: Can an effective program be proposed for discovering and developing scientific talent in American youth so that the continuing future of scientific research in this country may be assured on a level comparable to what has been done during the war?

In ending the letter, Roosevelt invoked a resonant metaphor from American history: 'New frontiers of the mind are before us, and if they are pioneered with the same vision, boldness, and drive with which we have waged this war we can create a fuller and more fruitful employment and a fuller and more fruitful life.' Bush lifted this quotation to serve as an epigram to his report, *Science: The Endless Frontier*, submitted to President Truman on 25 July 1945.[8] Bush's report was several things at once: a call for a new agency, a reflection on the impact of war on science, an invocation of resonant American imagery, and an implicit theory on the relationship between working worlds and something Bush called 'basic research'. The working worlds addressed, identified by Roosevelt, were medicine, defence and industry. In each of these Bush traced advances back to basic research – for example, stating that the 'striking advances in medicine during the war have been possible only because we had a large backlog of scientific data accumulated through basic research', while 'new scientific knowledge' was the ultimate source of 'new products'. 'There must be', wrote Bush, 'a stream of new scientific knowledge to turn the wheels of private and public enterprise.' Changing from a riverine to an economic metaphor, 'basic research' was 'scientific capital'; it must be accumulated and invested to reap profits:

> Basic research leads to new knowledge. It provides scientific capital. It creates the fund from which the practical applications of knowledge must be drawn. New products and new processes do not appear full-grown. They are founded on new principles and new conceptions, which in turn are painstakingly developed by research in the purest realms of science.

But, said Bush, there were several problems with the state of contemporary science. The war had turned scientists away from developing new scientific capital; rather the capital had been spent, the cupboards were bare: 'We have been living on our fat.' So where was new capital – basic research – to come from? 'We can no longer count on ravaged Europe as a source of fundamental knowledge.' Europe was out. Nor could the other traditional internal sources be relied upon: the endowment income, philanthropic foundation grants, private donations were all on a downward trend. Industry was 'generally inhibited by preconceived goals [i.e., profit], by its own clearly defined standards, and by the constant pressure of commercial necessity', which meant that 'satisfactory progress in basic science seldom occurs under conditions prevailing in the normal industrial laboratory'. Industry, then, was no reliable source. Therefore, urged Bush, it must be government that must recognize basic research as the source of progress and the rightful target of generous funding. 'Science' was a 'proper concern of government'. After all, it was

> basic United States policy that Government should foster the opening of new frontiers. It opened the seas to clipper ships and furnished land for pioneers. Although these frontiers have more or less disappeared, the frontier of science remains. It is in keeping with the American tradition – one which has made the United States great – that new frontiers shall be made accessible for development by all American citizens.

'Science has been in the wings', wrote Bush. 'It should be brought to the center of the stage – for in it lies much of our hope for the

305

future.' But, and here he was most insistent, it was essential that the basic research funded must be free of strings. 'The Government is peculiarly fitted to perform certain functions, such as the coordination and support of broad programs in problems of great national importance', he wrote, surely with the Manhattan Project in mind.

> But we must proceed with caution in carrying over the methods which work in wartime to the very different conditions of peace. We must remove the rigid controls which we have had to impose, and recover freedom of inquiry and that healthy competitive scientific spirit so necessary for expansion of the frontiers of scientific knowledge.
>
> Scientific progress on a broad front results from the free play of free intellects, working on subjects of their own choice, in the manner dictated by their curiosity for exploration of the unknown.

Thus Bush asked for the establishment of a single agency – for science is 'fundamentally a unitary thing' – administered by citizens ('selected only on the basis of their interest in and capacity to promote the work of the agency'), which would channel government funds to support basic research in colleges, universities and research institutes. What these institutions spent the funds on, as well as the overall 'policy, personnel, and the methods and scope of the research', was their choice entirely. 'This' freedom, insisted Bush, was 'of the utmost importance'. Funding must be lavish, free, and guaranteed over long time periods, for 'stability'.

With the spring of research freshly flowing – and certain other provisos met, such as the open dissemination of wartime data, the renewal and expansion of science education (especially for the generation in uniform) – the downstream wheels would move again, ensuring advances in health, national security and the promise of full employment. So the proposed agency – Bush's National Research Foundation – would fund, in effect, all the basic research needed to further medicine, new weapons and new products. The basic research would be conducted by free inquirers. This 'civilian group' would extend to include 'civilian-initiated and civilian-controlled military research', conducting 'long-range scientific research on military problems'. The civilian agency would then 'liaise' with military.

Science: The Endless Frontier was Bush's attempt to transfer the model of the remarkable successes of government-directed projects, exemplified by radar and the Manhattan Project, while protecting science's autonomy. To do this he invented 'basic research'. The category would not have needed to be invented if senior scientists such as Vannevar Bush had not seen government direction of scientific programmes as being a threat. To do so he subtly but crucially misrepresented the relationship between general science and working worlds. 'Basic research' may sometimes be, as Bush defined it, 'performed without thought of practical ends', but it is not ever disconnected from the practical world. The wartime case of cybernetics, discussed in chapter 16, will provide a clear and stunning example of this connection, but the point is in fact general.

Science: The Endless Frontier also failed as an intervention into policy-making. Bush's conservative vision was being challenged by an even more radical one. Harley Kilgore, senator for West Virginia, pushed not only for government funding of basic research, but for the planning by a National Science Foundation of all government research for the social good.[9] Government, in Kilgore's scheme, would have a free licence to exploit patents generated by government-funded science. Socialism!, cried industry. Bush's report was written, writes Daniel Kevles, 'to seize the initiative from Kilgore'.[10] Bush vs. Kilgore was joined on the same bill by a spat between Warren Weaver and *New York Times* science correspondent Waldemar Kaempffert. Weaver insisted, in a letter to the *New York Times*, that the atomic bomb was not a product of government science, but the result of the OSRD coordinating the 'practical application of basic scientific knowledge'; such knowledge was only produced 'by free scientists, following the free play of their imaginations, their curiosities, their hunches, their special prejudices, their unfounded likes and dislikes'.[11]

Yet in practice both Bush's and Kilgore's visions had only very limited influence. A National Science Foundation was indeed set up, and channelled federal funds to largely academic research – but it was late (1950), 'irreparably stunted', and outflanked by new, specialized patrons of much greater spending power.[12] Science policy in the United States was determined instead by bodies set up soon after the war's end: by the Atomic Energy Commission, the Department of Defense, the Office of Naval Research and the National Institutes of Health. The Department of Defense, justified by the Cold War, became the leading patron of science in general, taking in basic research, including scientific research at universities.[13] During the Second World War, defence spending on research and development had increased fiftyfold, but the Cold War sustained and surpassed that level of investment: the Department of Defense provided 80 per cent of government funding on research and development, a third of all industrial research, and three-quarters of research for fields such as electronics and aerospace.[14] In the 1930s, a 'National Institute of Health' had taken over previous government medical scientific work, such as hygiene laboratories, and had received philanthropic gifts of land in Bethesda, Maryland. Roosevelt had expanded the work of the institute. Now, post-war, the National Institutes of Health – institutes sprouted like mushrooms from the 1940s – were leading patrons and practitioners of medical research. 'Between the wars, most R&D had been done in industrial laboratories, supported by private and philanthropic sources', summarizes Larry Owens. By the early 1950s, 'the vastly increased amount of R&D, still performed predominantly in industrial and university laboratories, was paid for by the federal government – the bulk (around 83 percent) ostensibly for defense.'[15]

Did Cold War funding shape Cold War science?

The historian Paul Forman argued that this deluge of military funding transformed mid-twentieth-century science. (Readers will recall that Forman was the historian who argued that the Weimar political milieu directed and shaped physics. For Weimar Germany, read the post-war United States. It is a feature of Cold War literature that surface appearances can be deceptive, and an immediate subject of study might be a cover for deeper targets.) Forman's new argument has three levels, of increasing importance, and, it has to be said, decreasing evidential support. First, he demonstrates that the United States defence establishment was indeed a major patron of physical science. Private foundations, previously the pacesetters for American research, could not offer comparable grants to those available from the Atomic Energy Commission and the Department of Defense, and some withdrew from the natural sciences.[16] The Office of Naval Research, established in 1946 to plan, foster and encourage 'scientific research in recognition of its paramount importance as related to the maintenance of future naval power and the preservation of national security', in an act that symbolizes this shift in influence, took over 'a magnificent tract of 800 acres, and dozens of laboratory buildings', in Princeton abandoned by the Rockefeller Foundation.[17]

The patronage supported 'basic' research as lavishly as it did development and production of military systems. 'Through the 1950s, the only significant sources of funds for academic physical research in the U.S. were the Department of Defense and an Atomic Energy Commission whose mission was de facto predominantly military', writes Forman. The two bodies 'each provided roughly half of all funds for unclassified basic research in university physics departments'.[18] And basic science amounted to a mere 5 per cent of the military's total funding of research and development – although that 5 per cent was ring-fenced.[19] Academic science, while much smaller than industrial research, was therefore dependent. It was also of strategic importance: 'only the universities could both create and replicate knowledge, and in the process train the next generation of scientists and engineers', writes Stuart Leslie. After all, the 'universities provided most of the basic research and all the manpower for the defense industry.'[20]

In the second level of his argument, Forman suggests that the funding source made a difference to the direction in which science moved. There was a mutual orientation of interests between patron and physical scientists. The problems chosen were those of military interest; secrecy, the military necessity of classified information, was encouraged and enforced. Forman offers quantum electronics, specifically Charles Townes's and Gordon Gould's development of the maser and the laser, as his prime case studies. The maser and laser were examples of invention taking place in the large corporations of the post-war United States, in AT&T's Bell Laboratories and TRG Inc., and complemented by development in other labs. Laser research depended on new constellations of interdisciplinary research, involving physicists, electrical engineers, chemists, and materials scientists.[21] This diversity of firms and experts would ultimately encourage the application of the laser to many different purposes by the end of the twentieth century, from compact disc players and supermarket checkout technologies, to military weapons and navigation instruments, to basic scientific research. But, for Forman, 'quantum electronics', the science of masers and lasers, was a military creation. 'Altogether', he summarizes, 'program officers of the military funding agencies [were] ... powerful agents directing the advance of knowledge and forwarding the convergence of science and engineering toward military applications and hardware.'[22]

Forman's final, third, level is to argue that this redirection of physics shaped the nature of physical knowledge itself. His case is made by drawing an analogy between the social organization of physics and the characteristic object of physical study. Under military patronage, the former, the social organization, was distinctively large-scale, hierarchical teamwork. In the words of an academic administrator, speaking in 1953:

> There was a time when scientific investigation was largely a matter of individual enterprise but the war taught scientists to work together in groups; they learned to think in terms of a common project, they were impressed by the progress to be made through unified action. A notable

degree of this spirit has been transfused into the life of our larger universities . . . The scale of research and the complexity of its techniques have grown beyond anything imagined a few decades ago . . . It was the war that contributed principally to a major revolution in the method and spirit of laboratory investigations.[23]

The Cold War confirmed this pattern of physical science as directed teamwork – managed populations of physicists. The style of American Cold War physical science was utilitarian, pragmatic and instrumentalist. Forman then directs our attention to the remarkable similarity in shape between social organization and scientific object. Drawing together previous studies by scholars such as Sam Schweber and Andy Pickering, he notes that, through key specialties within physics – particle physics, nuclear physics, solid-state physics, and so on – ran 'a pervasive peculiarity . . . a new and widespread preoccupation with population redistribution' – the reorientation of nuclear spins is offered as a clear example – 'a kind of instrumentalist physics of virtuoso manipulations . . . in which refined or gargantuan technique bears away the palm'.[24] The techniques of population manipulation and management: 'It is just such a physics as the military funding agencies would have wished.'

So Forman is arguing that the great achievements of post-war physics must be understood not only as enabled by Cold War funding – not just as discoveries lying in the path of military application – but as being knowledge stamped by military interests. Other historians have agreed that Cold War funding was the decisive fact about the quantity and even the direction[25] of post-war physics but have disputed Forman's radical account of the deep shaping of physical knowledge.[26]

Controlling the atom

Concretely, the Cold War projects that mattered were electronic intelligence gathering, missiles and aircraft, early warning systems and, first and foremost, the production and maintenance of atomic bombs. A debate had waged between 1944 and 1946 on the extent to which a post-war atomic bomb should be under international control. Niels Bohr had visited the Manhattan Project in the winter of 1943 into 1944 and argued that the only safe world was an 'open world' in which international inspection would reassure nations about each other's nuclear ambitions.[27] Bohr's 'open world' was modelled on an ideal scientific community, which was multinational and in which

310

progress depended on shared information. Bohr's internationalist vision had an immediate, if varied, impact on the scientists, including the young physicist Ted Hall, who concluded privately that espionage for the Soviet Union was justified, and Oppenheimer, who concluded that war could 'not end without the world knowing' about the bomb: the 'worst outcome would be if the gadget remained a military secret'.[28] A subsidiary justification for Hiroshima, therefore, was scientific openness. Many scientists thought that the new United Nations should be the authority governing matters of atomic weapons and atomic power.

The Interim Committee that had met in May 1945, when it was clear that the atomic bomb was a fact, brought Bohr's open world arguments to the table. The committee was an encounter between the highest officials and politicians (excepting the president, who was represented by James Byrnes), led by Henry L. Stimson, and the elite scientists, including Vannevar Bush, Karl and Arthur Compton, Enrico Fermi, James Conant, Ernest Lawrence and Oppenheimer. The Interim Committee was split between those who favoured openness – Oppenheimer argued for a 'free interchange of information', Stimson floated the idea of an 'international control body' – and those who maintained that the Soviet Union could not be trusted; Lawrence argued that a 'sizable stockpile of bombs and material should be built up' with sufficient funds to 'stay out in front', while James Byrnes agreed, summing up: '[the United States must] push ahead as fast as possible in production and research to make sure that we stay ahead and at the same time make every effort to better our political relations with Russia.'[29] The Interim Committee agreed with Byrnes.

As soon as the war ended the scientists' movement tried to go public.[30] An Association of Los Alamos Scientists drafted a call for international control of atomic energy. Oppenheimer, to the surprise of the movement, agreed to the suppression of the paper.[31] Robert Wilson rewrote the document and saw that it was published in the *New York Times*. Meanwhile, the closed world of expanded atomic production and nuclear secrecy was growing in power and extent. The majority conclusions of the Interim Committee formed the basis of proposed legislation, the May–Johnson bill of October 1945, which would have granted military authority over atomic weapon development and punished breaches of secrecy with ten-year prison sentences and crippling fines. The military and political leaders were in favour, as were Ernest Lawrence and, wavering, Oppenheimer.

However, the scientists' movement effectively lobbied against the restrictions of the May–Johnson bill. When the bill was defeated, the movement threw its weight behind an alternative bill proposed by Senator Brien McMahon. The McMahon bill, which formed the basis of the Atomic Energy Act of 1946, placed atomic energy in the United States under civilian control. The Atomic Energy Commission was the body granted the nuclear monopoly and established by this act to administer programmes, taking over work at Hanford, Oak Ridge, the Argonne laboratory (the old Metallurgical Laboratory) in Chicago and Los Alamos, and the contracts for work at the universities of Berkeley, Chicago and Columbia. While it was civilian-controlled, the Atomic Energy Commission would become, as Forman noted above, a de facto military body because of the strategic centrality of nuclear weapons in the Cold War. A Military Liaison Committee ensured this influence. A civilian, David Lilienthal, former director of the Tennessee Valley Authority ('which had set the precedent for government involvement in funding and operating a large technological system'), was nevertheless placed in charge of the Atomic Energy Commission.[32] The McMahon Act also had the subsidiary effect of cutting Britain out of the nuclear loop: despite the Manhattan Project's inheritance from the tube alloys project, there was to be no debt paid, and no sharing of atomic knowledge even with the closest of allies.[33]

The question of the international control of atomic energy was settled in parallel. A plan for a United Nations atomic energy body had been agreed in principle. Truman had asked under-secretary of state Dean Acheson to work out a detailed proposal. Acheson called in turn on the advice of Lilienthal and Oppenheimer. The Acheson–Lilienthal Report, ready in March 1946, argued that all nation-states should renounce their sovereignty over atomic matters, investing it instead in an international Atomic Development Authority. The authority would rule over uranium mines, atomic power plants and nuclear laboratories, permitting peaceful uses of atomic energy and preventing the manufacture of atomic weapons.[34] With Oppenheimer as the translator, the Acheson–Lilienthal Report was Bohr's open world as a model for world government.

Truman and his advisors recognized the danger they had unleashed. In a deliberately destructive tactical manoeuvre, Truman charged a friend of James Byrnes, Bernard Baruch, with the task of taking the proposals to the United Nations. Baruch made sure that the details of the proposal for international control of nuclear power were unacceptable to the Soviet Union. (For example, penalties for violation would not be subject to Security Council veto.) Alternative Soviet proposals, to ban nuclear weapons, were equally unacceptable to the United States. The Baruch plan therefore led, as it was intended, to stalemate. The opportunity was missed. Containment and closure were preferred over openness. No strong international control of nuclear programmes was brokered. The result was that, even in a polarized, closed Cold War world in which atomic secrecy was taken seriously, there was no effective United Nations agency to guard against nuclear proliferation.

In the United States a Cold War policy of rapid nuclear armament was now followed. Nuclear tests resumed, at Bikini Atoll in the Marshall Islands, in July 1946. The responsibility for new designs, tests and the development needed to mass-produce nuclear weapons fell to the national laboratories: Los Alamos and the Argonne laboratory near Chicago. The support for these specialized laboratories was matched by funding for more general physics laboratories. High-energy physics trained generations of physicists who either later moved into direct defence research or formed a strategically available pool of expertise, or who, in their fundamental studies of the nature of matter and energy, developed the theory and techniques that underpinned nuclear research and development. Lawrence's laboratory at Berkeley was an immediate beneficiary. The trend of increasing size and energy of particle accelerators therefore continued into the post-war world. East coast academic physicists lobbied hard for matching facilities. Interested universities – Columbia, Cornell, Harvard, Johns Hopkins, MIT, Pennsylvania, Princeton, Rochester and Yale – clubbed together. This lobby group turned administrating body, the Associated Universities Incorporated, was given responsibility for running a new high-energy physics laboratory – the Brookhaven laboratory – built on Long Island and opened in 1947. Its facilities, such as the Cosmotron, which accelerated protons to energies greater than a billion electron volts (peaking in 1953 at 3.3 GeV), funded by the Atomic Energy Commission, made Brookhaven a powerhouse of post-war physics. Cold War funding confirmed American leadership in strategically central sciences.

The Soviet bomb

The Soviet Union had more nuclear scientists in the 1930s than any other country, with institutes such as Abram Ioffe's in Leningrad producing many of them. In common with physicists in Germany,

Britain, France and the United States, they had explored the exciting new phenomenon of fission. But investigation of military applications stalled, partly because of wartime mobilization for other projects and partly because the Nazi–Soviet pact discouraged the idea of developing atomic weapons for use against Germany.[35] Indeed, the bomb project would be thoroughly shaped by the character of science in the Soviet context. Science was held in high political regard – how could it not be when Marxism–Leninism was a science, indeed the foundational science? But there was a fault-line between official ideology and science when the latter acted as a rival form of authority. But the brute reality of political organization in the Soviet Union, in particular the immense power wielded by the leaders, especially Stalin, meant that favoured projects could progress with startling speed, fed by state resources and backed up by state terror.

Stalin had read the intelligence reports from Soviet agents in the British tube alloys and the American Manhattan projects: John Cairncross supplied the MAUD report, while Klaus Fuchs supplied other details. Deductions were also drawn from public sources, such as an article in the *New York Times* – passed on by Vladimir Vernadsky's son, who was at Yale[36] – and the telling evaporation of fission articles in the physics journals. But Stalin mistrusted both the intelligence and the rival authority of scientists. As the Red Army turned the war's tide at Stalingrad in 1943, Stalin approved the beginning of the Soviet atomic bomb programme. Igor Kurchatov, a graduate of Ioffe's institute, was placed in charge. Nevertheless, the Soviet leadership did not push the project until immediately after the devastations of Hiroshima and Nagasaki. Ideological harassment was suspended. Resources were devoted. While Christmas Day 1946 was not celebrated for religious reasons by the Soviet leadership, it certainly was for reasons of political and scientific power: the first reactor went critical in Moscow, merely four years after Fermi's success. Such progress probably doomed projects for international control even before the Baruch plan: the Soviet Union was well on the way to possessing nuclear weapons even as Soviet diplomats proposed banning them.[37]

In parallel with the Manhattan Project, nuclear cities sprouted to house the atomic bomb programme: a plutonium separation plant on the scale of Hanford at Cheliabinsk-40, Kyshtym, a Los Alamos equivalent in the atomic weapon laboratory of Aryzamas-16. The atom spies provided information that saved some time: a copy of the American uranium separation plants and a design for a plutonium bomb. But historian David Holloway argues that the greatest aid to speed was the availability, after 1945, of the uranium ore of East German and Czechoslovakian mines.[38] All of this activity was expertly orchestrated by Kurchatov, a systems builder of some genius.

On 29 August 1949, at Semipalatinsk-21 in Kazakhstan, the first Soviet atom bomb, a near copy of the American plutonium bomb, was detonated. No public announcement was made. However, the detection, in the following days, of radioactive debris in the atmosphere provided a tell-tale signal. The response in the West was shock and dismay: the West's monopoly, which had been expected to last until 1952, was over. President Truman had never believed that 'those Asiatics' could build a bomb at all.[39]

Many, but not all, historians agree that the irreplaceable expertise of physics in the design of weapons protected physicists, at least when compared to the fate of biologists. Certainly the ideological denunciations of physicists were interrupted during the Second World War – the 'survival of Soviet physics was the first example of nuclear deterrence', according to the great Lev Landau.[40] Arkady Timiarazev's campaign against Jewish theory was, temporarily, in retreat. Yet the brute power of the party was a different matter. In particular, the Soviet physicists had to deal directly with Lavrenty Beria, Stalin's chief of police, who headed the Politburo commission overseeing the Soviet bomb. 'I am the only person here who says what he thinks', the elderly Pavlov had told the physicist Peter Kapitsa in the early 1930s. 'I will die soon, and you must take my place.'[41] Kapitsa bravely – perhaps foolhardily – took up this invitation, and in October 1945, as the Soviet atom bomb project surged forwards, he wrote to Stalin:

Is the position of a citizen in the country to be determined only by his political weight? There was a time when alongside the emperor stood the patriarch: the church was then the bearer of culture. The church is becoming obsolete, the patriarchs have had their day, but the country cannot manage without leaders in the sphere of ideas . . .

Only science and scientists can move our technology, economy and state order forward . . . Sooner or later we will have to raise scientists to the rank of patriarch. That will be necessary because without that you will not make the scientists always serve the country with enthusiasm . . . Without that patriarchal position for scientists the country cannot grow culturally on its own, just as Bacon noted in his New Atlantis. It is time, therefore, for comrades of Comrade Beria's type to begin to learn respect for scientists.[42]

As an official, Beria had supervised the purging of enemies, the massacre of prisoners (such as the Polish officers at Katyn) and run the gulags; personally, he had tortured and raped. Now he was making sure the physicists worked hard towards their goal. 'It is true, he has the conductor's baton in his hands', Kapitsa complained to Stalin. 'That's fine, but all the same a scientist should play first violin. For the violin sets the tone for the whole orchestra. Comrade Beria's basic weakness is that the conductor ought not only to wave the baton, but also understand the score.'[43] Many of the stories told of Beria and the physicists are apocryphal – such as Beria being told that the atomic bomb could not be made without a theoretician's understanding of relativity,[44] or the time when terrified physicists mocked up a fake physical demonstration of 'neutrons' to appease the police chief – but they speak of a true tension between overwhelming arbitrary political might and a rival authority.

Contrasting fates of biology and physics in the Soviet Union

After the Second World War, ideological regulation resumed: Russian priority was celebrated (and where necessary invented), 'cosmopolitanism' (an anti-Semitic codeword) was attacked, and appeals to internationalism were suspect. Timiarazev renewed his campaign against idealism in physics. All the key sciences were planned to be subjected to ideological examination of high-level conferences, the guidance of which Stalin was minded to take a direct hand.[45] Trofim Lysenko, so adept at blending his science with manoeuvres for political influence, used the agricultural science conference decisively to consolidate his power. In July and August 1948, having privately confirmed Stalin's support, Lysenko addressed the meeting of the Lenin All-Union Academy of Agricultural Sciences (VASKhNIL), attacking Mendelian genetics as 'reactionary', 'idealist' and incompatible with Marxism–Leninism, while praising Michurinist biology as for its practicality, its responsiveness to people's demands and its materialism. On the last day, Lysenko opened with a short announcement:

> The question is asked in one of the notes handed to me. What is the attitude of the Central Committee of the Party to my report? I answer: The Central Committee of the Party has examined my report and approved it. (Stormy applause. Ovation. All Rise.)[46]

Lysenko had 'won' the staged debate. Rivals were blitzed. Historian David Joravsky argues that Lysenko's continued promise to solve the crisis in food production maintained his support.[47] The more recent study of primary sources by Nicolai Krementsov argues that Lysenko's 1948 success was comprehensive because of its being staged at a critical moment in the gathering Cold War. 'The essence of Lysenko's address was a juxtaposition of two opposing trends in biology: unscientific, idealist, scholastic, sterile, reactionary, anti-Darwinist Weissmannism–Mendelism–Morganism versus scientific, materialist, creative, productive, progressive, Darwinist Michurinist biology', notes Krementsov of the Cold War inflected speech. 'These two sets of antonymic labels obviously reflected the current socio-political situation: the escalating confrontation between the USSR and the West, or, as Lysenko phrased it, "two worlds – two ideologies in biology".'[48]

Lyensko could not have won so decisively without support from the top. And 'by far the most important factor in Stalin's decision to intervene on Lysenko's behalf . . . was the escalating Cold War', concludes Krementsov. 'Stalin used the competition between geneticists and Lysenkoists as a convenient pretext to announce a new party line in domestic and foreign policy: the final establishment of two opposing camps, Soviet and Western.'[49] Pollock is sceptical of this interpretation, noting that a 'speech given by Lysenko seems a particularly clumsy way to announce' the new policy.[50] But actually Pollock's own later studies, which show Stalin, far more than previously, recognized as 'concerned about ideas', a man who 'consistently spent time on the details of scholarly debates', lend support to Krementsov's argument that Stalin's intervention into biology was simultaneously a movement of the pieces on the Cold War chess board.[51]

The biological/agricultural conference of 1948 is infamous, but similar conferences were held to confirm ideologically correct interpretations of linguistics (Stalin directly intervened again), physiology (in praise of Pavlov), philosophy and political economy.[52] In 1949 it was the turn of physics. The conference was postponed. Interestingly, historians working from different sources disagree about why. Working with memoirs of protagonists, one view is that Igor Kurchatov told Beria or Stalin that the conference would impede progress on the Soviet atomic bomb. 'Leave them in peace', Stalin is supposed to have told Beria, 'we can always shoot them later.'[53] Another view, pieced together by historians working through the mass of documentation made available after the end of the Cold War, is that the conference was cancelled because bureaucratic infighting meant there was no consensus as to its purpose. The two views are not irreconcilable: Kurchatov's wish that his team were not to be distracted from working towards a nuclear weapon, and Stalin or

Beria's subsequent cancellation, merely provided the *coup de grâce* to an already troubled conference.[54]

Biology and physics would continue to have contrasting fates in the Soviet Union. Challenges to Lysenko, on the grounds, ultimately fatal to him, of practical failure to improve food and timber production, were made after Stalin's death in 1953. While Stalin's ultimate successor, Nikita Khrushchev, supported Lysenko too, his grasp on Soviet biology was gradually weakened, failed project by failed project. Genetics found a home in the Siberian science city of Akademgorodok, along with particle physics, computer science, environmental engineering and other specialties. Akademgorodok, a project of mathematician Mikhail Lavrentev, begun in the 1950s and housing eventually over 65,000 scientists in over thirty separate institutes, was a science city that was prefigured by – and dwarfed – Los Alamos.[55] Physics, having delivered the atomic bomb, received lavish patronage and other advantages. (One advantage was the preservation of a degree of autonomy: it is no coincidence that one of the most prominent Soviet dissidents, Andrei Sakharov, was also a nuclear weapons scientist.) This support was maintained by the Cold War competition to devise further weapons of devastating power.

The British bomb

By 1949, the year of the Soviet nuclear test, one other country was also making progress towards possession of its own atomic bomb. Most British nuclear weapons research had been folded into the American project. A cabinet committee, GEN 75, was set up in 1945 to continue the tube alloys work and to review the prospect of a British bomb. Its deliberations were given urgency in 1946 when the McMahon Act ended the sharing of nuclear secrets, including cooperation between the United States and the United Kingdom. Two prominent members of Clement Attlee's post-war Labour cabinet, Stafford Cripps and Hugh Dalton, opposed the expense and implications of a British bomb. They were trumped by the intervention of the foreign secretary, Ernest Bevin, who, delayed by a late and liquid lunch, burst into a meeting of GEN 75: 'That won't do at all' (speaking of Cripps and Dalton's arguments):

> We've got to have this. I don't mind for myself, but I don't want any other Foreign Secretary of this country to be talked at, or to, by the Secretary of State in the United States as I just have in my discussions

with Mr Byrnes. We've got to have this thing over here, whatever it costs. We've got to have the bloody Union Jack on top of it.[56]

An even more secret subcommittee of GEN 75 was set up, minus Cripps and Dalton, which approved the independent British bomb project in January 1947; Parliament was not told until May 1948.[57] The proximate cause of the British bomb was therefore the (attempted) preservation of Great Power status, and the cost was a blow to the democratic constitution.

William Penney, a British physicist who had been deeply involved with the Manhattan Project, was placed in charge of the weapons project, with the civil side led by John Cockcroft. Atomic laboratories and factories were built: a weapons laboratory at Aldermaston and an atomic energy research establishment on an old RAF airfield at nearby Harwell (both within reach of Oxford and London), complemented by another pair of facilities in distant Cumbria: a nuclear factory at Windscale and a prototype atomic power station, Calder Hall. At the time, the civil project – atomic energy – was assumed to be separate from the military project, but it is now clear that the two were thoroughly interlinked. Calder Hall, presented in 1956 as the first civil nuclear power station in the world, was engineered to produce military-grade plutonium.[58]

Britain was able to detonate its first nuclear test in October 1952 in the Monte Bello Islands, off the north-western coast of Australia. However, before then, nuclear relations between the United States and the United Kingdom were strained by the revelation in 1950 that Klaus Fuchs, who had worked at Los Alamos as part of the Manhattan Project and was now head of the Theoretical Physics Division at Harwell, was a Soviet spy. The exposure of Fuchs was soon followed by accusations against other, more low-level atomic spies of the Manhattan Project, including Julius and Ethel Rosenberg. While the husband and wife were certainly active communists, the strength of the espionage case against the Rosenbergs has been disputed ever since their trial and execution by electric chair in 1953.

Trials of science

The shock of the events of 1949 to 1950 – the Soviet bomb, the revelation of the atomic spies, and the communist victory in the civil war in China – created a viciously charged political atmosphere in the United States. Within this context three decisions were taken there that would shape the history of science in the Cold War: to investigate aggressively those accused of communist sympathies, to proceed with the hydrogen bomb, and to fight a war in Korea.

Membership of the Communist Party had always been meagre in the United States, peaking at perhaps 50,000 in the early 1940s. In the Cold War, those suspected of communist sympathies were subject to investigation before government and private panels, often charged on the basis of questionable evidence, sometimes by anonymous accusers, with, at least in private panels, no right to an attorney. Lives were ruined: an accusation alone could have the same meaning for suspicious minds as a conviction. Non-comformity was suspect. Guilt could be proved by mere association, or by opinions held rather than deeds done. President Truman had introduced loyalty reviews for federal employees. States and private industry would soon follow this lead. One in five employees would be checked by 1958. J. Edgar Hoover's Federal Bureau of Investigation lay at the heart of this internal intelligence-gathering operation. The other main agents were Senator Joseph McCarthy and the House Un-American Activities Committee. This body dated back to before the war, when it had first been set up to root out American fascists. In the early Cold War, the committee turned on communists, charging the spy Alger Hiss and going after Hollywood socialists. McCarthy, a demagogue, had announced in February 1950, to a folksy audience of a West Virginian Republican women's club, that he had a list of 205 communists known to be influencing the government, sparking rounds of suspicion and accusation – comparisons were made at the time to the witch-hunts of Salem.

McCarthyism affected teachers, film writers and army personnel, but also scientists. McCarthy himself stated that 500 of the 50,000 American men of science were communist. Just as in the Soviet Union, the internationalism of science seemed suspect. Restrictions were placed on free travel, at least for the targets of the witch-hunt. The chemistry professor Ralph Spitzer was seized in Europe in 1950 because he had read a paper on academic freedom in Prague.[59] The co-discoverer of carbon-14, Martin Kamen, had only dined with known communists, but this association was enough to get him fired, and he had to go to court to get a passport.[60] Linus Pauling, only marginally less famous than Albert Einstein and Robert Oppenheimer, was denied a passport in January 1952, because his 'proposed travel would not be in the best interests of the United States'.[61] Visa restrictions were also introduced, constricting the flow of scientists into the United States. Irène Joliot-Curie, a Nobel Prize winner, was denied

320

membership of the American Chemical Society in 1953 for being a communist.[62] Sensitive sites had special regulations. The Lawrence Radiation Laboratory was restricted, for example, to just three international visitors per year. Some scientists, such as David Bohm, a protégé, as we shall see below, of Oppenheimer (and Einstein), was forced to emigrate.

The two American scientists who were the most high-profile targets of anti-communist activity were Edward Condon and Robert Oppenheimer. Condon was an accomplished quantum physicist who had worked on magnetrons for radar as well as experiencing a brief stint on the Manhattan Project during the Second World War. After 1945 he took up posts as director of the National Bureau of Standards and research director of Corning Glass Company. A socialist, Condon was harassed by anti-communists, who were convinced he was a security risk. He sometimes managed to turn away the attacks with humour. A student of his recalled one line of questioning in front of a loyalty board:

> 'Dr Condon, it says here that you have been at the forefront of a revolutionary movement in physics called' – and here the inquisitor read the words slowly and carefully – 'quantum mechanics. It strikes this hearing that if you could be at the forefront of one revolutionary movement . . . you could be at the forefront of another.'[63]

But the slurs on Condon's character made by the House Un-American Committee had effect. His directorship of the National Bureau of Standards was 'undermined' while, at the instigation of Vice-President Richard Nixon, the removal of his security clearance cost him his post with Corning Glass.[64] Nevertheless, scientists' organizations fought back, protesting against Condon's treatment, and the American Association for the Advancement of Science symbolically elected him its president for 1953.[65]

The greatest target of anti-communist activity against scientists, however, was Oppenheimer.[66] The so-called Oppenheimer trial was as much about settling issues around the hydrogen bomb as it was about security risks. Robert Oppenheimer seems to have been largely apolitical until the years of the Great Depression, when he had dated Jean Tatlock, a communist and a daughter of a Berkeley professor of English. He had donated to party causes – not unusual among American anti-fascists – but had never been a party member (unlike his wife, Kitty, brother, Frank, and sister-in-law). The Federal Bureau of Investigation had opened a file on Robert Oppenheimer in autumn 1941, before his involvement with the Manhattan Project.

FBI surveillance continued until 1943, when the army took over this responsibility. General Groves had appointed Oppenheimer to lead Los Alamos in full knowledge of the FBI reports.

In 1943, Haakon Chevalier, a friend and assistant professor of French at Berkeley, had informed Oppenheimer that a Soviet sympathizer named George Eltenton was asking after information that might be passed on to America's ally, the Soviet Union. Oppenheimer had done nothing, immediately. Then in August, he had walked into the army security office in Los Alamos and reported the Chevalier incident, not naming the French professor but speaking of an 'intermediary' who was the link between Eltenton and 'three scientists'. Army surveillance of Oppenheimer had stepped up. In questioning by Groves, in late 1943 Oppenheimer had named Chevalier as the intermediary and spoke of just one scientist contact – here the sources disagree whether it was Robert himself or his brother Frank. Oppenheimer also had ratted on three students of his – Rossi Lomanitz, David Bohm and Bernard Peters – for communist activity.

After the war, the FBI had resumed investigation of Oppenheimer, compiling reports on his family ties to communism and wire-tapping the scientist's phone. In an interview in 1946, Oppenheimer had changed his line about the Chevalier affair, calling the description of three scientists a 'complicated cock and bull story'. At this time he had also been appointed to the Atomic Energy Commission's General Advisory Committee, and, despite FBI misgivings, his friends and allies on the committee – Ernest Lawrence, Vannevar Bush, Enrico Fermi, Bernard Baruch, Groves and James Conant – were persuaded to grant him full security clearance.

In June 1949, Oppenheimer testified before a closed hearing of the House Un-American Activities Committee. 'Dr Chevalier was clearly embarrassed and confused', recalled Oppenheimer, brushing off the incident, 'and I, in violent terms, told him not to be confused and to have no connection with it. He did not ask me for information.'[67] He gave evidence against his former student Bernard Peters, now a professor at Rochester. When the testimony against Peters was leaked, senior American physicists rounded on Oppenheimer: Victor Weisskopf wrote that Oppenheimer's behaviour was particularly awful, 'since so many of us regard [you] as our representative in the best sense of the word', while Condon, whom Oppenheimer had declined to support in 1948, bitterly commented: 'One is tempted to feel that you are so foolish as to think you can buy immunity for yourself by turning informer.'[68] Oppenheimer publicly retracted his comments on Peters.

But Oppenheimer made his deadliest enemies in the debate on thermonuclear weapons, which use the energy released by fusion of light elements such as hydrogen to produce a devastating power that stood in relation to the atomic bomb as the atomic bomb stood to conventional explosives. Edward Teller had promoted hydrogen bombs as early as the meetings of July 1942 that had sketched the first designs of nuclear devices. Then Oppenheimer had overruled the Hungarian and proceeded with fission. Teller had not forgotten.[69] Oppenheimer confirmed his opposition to the hydrogen bomb in 1945 and agreed with Arthur Compton when he wrote, on behalf of Oppenheimer, Lawrence and Fermi to Henry Wallace: 'We feel that this development should not be undertaken, primarily because we should prefer defeat in war to victory obtained at the expense of the enormous human disaster that would be caused by its determined use.'[70]

Again, in 1946, under Oppenheimer's chairmanship, the General Advisory Committee of the Atomic Energy Commission had favoured the production and improvement of uranium and plutonium weapons, while relegating the hydrogen bomb to a low priority. A sharp divide opened: in favour of making the hydrogen bomb were Teller, Lawrence (who had changed sides), William L. Borden, the joint chiefs of staff, air force leaders and Atomic Energy Commissioner Admiral Lewis Strauss; against were James Conant, Vannevar Bush, Hans Bethe, Robert Bacher and Oppenheimer.

Only following the Soviet bomb was President Truman told of the possibility of the 'super', Teller's hydrogen bomb. The General Advisory Committee of the Atomic Energy Commission argued the matter in October 1949. Oppenheimer and the committee rejected the weapon again, arguing that its 'use [carried] ... much further than the atomic bomb itself the policy of exterminating civilian populations', adding that its development would only spur the Cold War arms race.[71] Oppenheimer preferred a focus on tactical nuclear weapons, which led him into opposition with the air force (whose post-war influence depended on operating the strategic nuclear force). However the advice, channelled by the director of the Atomic Energy Commission, David Lilienthal, to the president, was dismissed. 'Can the Russians do it?', Truman broke in moments after Lilienthal had started, adding, after receiving nods, 'In that case, we have no choice. We'll go ahead.'[72] Teller had won.

Oppenheimer was the only one of the anti-hydrogen bomb crowd with a suspect background, and it was Oppenheimer's hydrogen bomb views which 'primarily' fuelled suspicions.[73] In 1950, following the Fuchs revelations, Paul Crouch, communist and paid informant, accused Oppenheimer of holding a Communist Party meeting in his Berkeley home in 1941. Lewis Strauss, who had taken an active dislike to Oppenheimer for complex moral, religious and political reasons, equated Oppenheimer's opposition to the H-bomb with something more sinister. 'If you disagree with Lewis about anything, he assumes you are a fool at first', said a fellow Atomic Energy Commissioner. 'But if you go on disagreeing with him, he concludes you must be a traitor.'[74] The United States Air Force and Lewis Strauss agreed that Oppenheimer must be removed; Borden, who by 1953 was convinced that Oppenheimer was a Soviet agent, did the leg-work.

In May 1953, Joseph McCarthy was sniffing round the case. In November 1953, Borden wrote a letter to the FBI charging that, between '1929 and mid-1942, more probably than not, J. Robert Oppenheimer was a sufficiently hardened Communist that he either volunteered espionage information to the Soviets or complied with a request for such information', and 'More probably than not, he has since been functioning as an espionage agent.' President Eisenhower was told of the accusations in December 1953. J. Edgar Hoover hinted that McCarthy would publicize the case but 'could be controlled', 'provided the appropriate action was taken in the Executive Branch'.[75] Eisenhower stripped Oppenheimer of his security clearance.

Oppenheimer's clearance was considered at two panels in the Atomic Energy Commission in 1954. First the Personal Security Board – members picked by Strauss – heard evidence, including the following statement from Teller:

> In a great number of cases I have seen Dr. Oppenheimer act . . . in a way which for me was exceedingly hard to understand. I thoroughly disagreed with him in numerous issues and his actions frankly appeared to me confused and complicated. To this extent I felt that I would like to see the vital interests of this country in hands which I understand better, and therefore trust more.[76]

Teller, in his memoirs, wrote of a wish to 'defrock' Oppenheimer 'in his own church'.[77] But many scientists did testify in favour of Oppenheimer. Too many, indeed, in the Personal Security Board's view:

> The Board has been impressed, and in many ways heartened by the manner in which many scientists have sprung to the defense of one whom many felt was under unfair attack. This is important and encouraging when one is concerned with the vitality of our society. However, the Board feels constrained to express its concern that in this solidarity

there have been attitudes so uncompromising in support of science in general, and Dr. Oppenheimer in particular, that some witnesses have, in our judgment, allowed their convictions to supersede what might reasonably have been their recollections.[78]

Oppenheimer lost that round 2–1, with the sole dissenter a scientist – who felt he could not return to Chicago having condemned Oppenheimer.[79] The judgement then passed up to the five-man Atomic Energy Commission. Two votes were taken. By a margin of four to one, the commission decided that Oppenheimer was 'loyal', but also by four to one it considered Oppenheimer a security risk.

The Oppenheimer trial is worth retelling at length because it demonstrates several features of the position of American science and scientists in the Cold War. First, it shows that senior scientists could be brought down by the hysterical forces of anti-communism, in this case a 'triumph of McCarthyism – really without McCarthy himself'.[80] Indeed, Bernstein notes that there was a certain equality before the law being demonstrated: Oppenheimer 'was being treated formally by the same loyalty-security standards that were applied to ordinary people'.[81] Second, as Bird and Sherwin conclude, the 'defrocking' of Oppenheimer marked the limits of the post-war movement in which 'scientists had been regarded as a new class of intellectuals, members of a public-policy priesthood who might legitimately offer expertise not only as scientists but as public philosophers'. Afterwards 'scientists knew that in future they could serve the state only as experts on narrow scientific issues . . . The narrowest view of how American scientists should serve their country had triumphed.'[82]

As was highlighted by the verdict of the Personal Security Board, scientists were criticized for supporting Oppenheimer as a moral person, whereas in the view of the board the proper attitude must be 'a subordination of personal judgment . . . as against professional judgment in the light of standards and procedures which have been clearly established by the appropriate process. It must entail a wholehearted commitment to the preservation of the security system.' Scientists in the Cold War must learn to be governed by security procedure. The board's 'case rested on Oppenheimer as a bureaucratic official within the state charged with efficient execution of the public will, with no authority to make decisions about what ends the government ought to pursue.'[83] The 'interrogation and judgment of Oppenheimer', therefore, argues Charles Thorpe, 'was simultaneously an appraisal of the role of scientist, and the relationship between science and the state, as these had developed' from the Second World War.[84]

Finally, the brutal basic politics that underlay the Oppenheimer trial was the question of developing the hydrogen bomb. This one, narrow interest – strategic nuclear weapons – trumped all others. Edward Teller had struggled with designing a practical fusion bomb, but in 1950, in a new round of examination of designs to which the mathematician Stanisław Ulam also contributed, a breakthrough was made. The detonation of a primary fission weapon would be used to compress and initiate the detonation of the secondary fusion device. By November 1952, a Teller–Ulam device – weighing over 60 tons – was tested on Eniwetok, in the Marshall Islands in the Pacific, producing a blast in the megaton range. Within nine months the Soviet Union responded by detonating a device which could be presented as a thermonuclear weapon. It was designed by Andrei Sakharov.

Atoms for peace or war

There was a sense by 1953, with the Soviet tests, that the atomic secret was now well and truly out. President Eisenhower responded with a different diplomatic tack. In December 1953 he announced to the General Assembly of the United Nations that it was time for the 'atomic colossi' to work together, to follow a 'constructive, not destructive path' and to use 'atoms for peace'. The Atoms for Peace movement had many purposes: it was designed to distract attention from the massive expansion in nuclear weapons production, it was a blow in the Cold War as a 'struggle for the minds and wills of men', it aimed to deflect charges of militarism aimed at the United States from Europe after the hydrogen bomb tests, and it fostered the development of a civilian nuclear industry.[85] In 1952 nuclear reactors, as John Krige notes, were 'still a military secret (embodied most notably in the navy's nuclear submarine) and a government monopoly'.[86] As the Atomic Energy Commission was discovering, secrecy and monopoly discouraged firms such as General Electric and Westinghouse from developing civil nuclear power; Atoms for Peace promised the lifting of some of these restrictions. Finally, Atoms for Peace was a pitch for the favour of international science. Radioisotopes had been distributed as a symbol of American good will: 'even the bottle-washings we throw away can be used literally for months of research over there', one scientist noted.[87] A great Atoms for Peace conference in Geneva in 1955 was a triumph, with an American reactor, flown in specially and afterwards sold to Switzerland, as its centrepiece. 'The presentation of the . . . reactor in Geneva was a masterpiece of marketing', writes John Krige. 'It was intended to demystify nuclear

326

power and to show that anyone and any nation could exploit it safely and to social advantage. The Soviets also had a chance to display their wares. The Cold War seemed to be settling into rituals of technological display.'[88]

The Atoms for Peace conference of 1958 provided the forum to announce the existence of a controlled nuclear fusion programme in the United States.[89] The idea of tapping the power of fusion for civil energy needs had been investigated in Britain, the United States and the Soviet Union, along with more extreme suggestions such as using hydrogen bombs to create artificial harbours in Alaska or new inland seas in Siberia. In 1951 a bogus report of 'nuclear fusion in a secret island laboratory' for the Argentinean dictator Juan Perón appeared in newspapers; this was the 'proximate cause' for the American fusion power programme, in that it spurred the development by Lyman Spitzer, Jr., of stellarators at Princeton, funded under the Atomic Energy Commission's Project Sherwood.[90] The Soviet tokamak designs, begun in 1956, would prove more influential.

The forced development of the hydrogen bomb had taken place against the background of a real, hot, shooting war in Korea. Ironically this conflict demonstrated that there were limits on the use of fission, let alone fusion, weapons. The Korean War also further boosted science within a military system. Korea had been part of the Japanese empire since 1910. With the latter's sudden collapse in 1945, the peninsula was occupied by the Red and United States armies, leading to a post-war division along the 38th parallel. The leader of the North, Kim il-Sung, repeatedly requested approval from the Soviet leadership for an invasion to unite the country, but did not receive it until Stalin, misinterpreting a statement by Dean Acheson, decided in January 1950 that the United States was willing to stand by.[91] North Korea made rapid gains in the first months of the war. However, General Douglas MacArthur ordered a bold amphibious landing at Inchon that cut the North Korean army from its supply lines. Soon it was the North Koreans who were encircled and near collapse. But communist China, with battle-hardened troops from its victory in its civil war, keen to impress Stalin, intervened.

At this point the question of the use of atomic weapons was asked. In one dangerous, early moment, President Truman seemed to suggest that the decision about what weapons to deploy rested with the commander in the field. This decision was hastily reversed, and MacArthur was removed for insubordination when he publicly opposed the change. While there were good tactical reasons not to use atomic weapons – the Chinese forces were dispersed, and European allies would be appalled[92] – the decision meant that, in practice, atomic weapons were confirmed as a class apart. Historian John Gaddis calls this recognition – 'we have to treat this differently from rifles and cannon and ordinary things like that' – 'revolutionary'.[93] The Korean War thus became a limited war and settled into a drawn-out conflict, featuring trench warfare reminiscent of the First World War.

The Korean War confirmed and deepened the relationship between American science and the state.[94] In 1950, the initial militarization of peacetime science had seemed to be reaching a plateau; indeed, the federal research and development budget was scheduled to fall slightly, from $530 million to $510 million.[95] In April 1950, two months before the North Korean invasion, a document drafted by Paul Nitze, NSC-68: United States Objectives and Programs for National Security, landed on President Truman's desk. It is another of the defining documents of the Cold War. It wrote of 'two centers': the 'free society' versus the 'slave state'.[96] International communism had a 'design': 'the complete subversion or forcible destruction of the machinery of government and structure of society in the countries of the non-Soviet world and their replacement by an apparatus and structure subservient to and controlled from the Kremlin'. The authors of NSC-68 feared a 'surprise attack' with nuclear weapons – after all, a 'police state living behind an iron curtain has an enormous advantage in maintaining the necessary security and centralization of decision required to capitalize on this advantage'. NSC-68 called for actions, both 'overt' and 'covert', that went beyond mere containment. 'Greatly increased general air, ground, and sea strength, and increased air defense and civilian defense programs', wrote its authors, 'would . . . be necessary to provide reasonable assurance that the free world could survive an initial surprise atomic attack.'

The Korean War seemed to confirm the analysis presented in NSC-68; the Soviet Union, by waging war by proxy, must have a wider design in mind. The increased defence spending – a threefold increase[97] – meant that the 'defense research and development budget followed the overall defense budget into the stratosphere, doubling to slightly more than $1.3 billion' in 1951, 'and rising higher'; that year, Department of Defense and Atomic Energy Commission contracts made up two-fifths of the total spend on industrial and academic research.[98] A President's Science Advisory Committee was introduced to assist in the governance of this expansion of science. 'Unlike that during World War II, the scientific mobilization during the Korean War had produced no miraculous new weapons', summarizes historian Daniel Kevles, but, reinforced by the Soviet bomb, 'it had generated a pervasive NSC-68 mentality, a devotion to "the ability to convert swiftly from partial to all-out mobilization" . . ., a commitment to an expansive technological readiness and to whatever big science that achieving that end might require.'[99]

— 14 —

COLD WAR SPACES

Big Science and 'Big Science'

'Big Science' has been a favourite category for historians of twentieth-century science. However, a useful and informative distinction should be made between the phenomenon of Big Science as an emerging style of organization and the labelling of 'Big Science' as something of concern. Although precedents can be found in earlier decades, even centuries, Big Science, qua style of organization, captures aspects of large scientific and technological projects of the Second World War and, especially, Cold War periods. 'Big Science', qua label, was a product of the period of transition of the long 1960s (a period that is the subject of chapter 17). The relationship between 'Big Science' and Big Science is one between text and context.

A basic model of the phenomenon of Big Science as a style of organization is captured by the five 'M's: money, manpower, machines, media and military.[1] The 'M's were often interlinked: the machines were expensive, as were the staffing costs to run them; in the Cold War, military patronage was a major, even dominant, source of this money, and once the scale of expenditure on science became comparable to other projects in the public eye then public justification and image management became essential. So, taking the Manhattan Project as an example, it was a case of Big Science according to this simple model because it cost a lot of money ($2.2 billion), involved tens of thousands of staff, centred around large machines (the calutrons, the reactors, the bomb assemblies), required management of its public representations (both through restricting public knowledge and through coverage of its results, such as John Hershey's *Hiroshima* of 1945), and was a thoroughly military project.

330

The five 'M' model works particularly well as a description of post-war physical science projects in the United States: the space science sites, such as the Jet Propulsion Laboratory at Caltech, or the particle accelerator laboratories, such as Brookhaven or the Stanford Linear Accelerator (SLAC). It is a less satisfactory checklist as soon as the focus is widened. So, for example, the post-war European nuclear science project of CERN is certainly an entity one would want to label as Big Science: the huge teams of researchers working with large and expensive particle accelerators have received plenty of media coverage, but there is no explicit military interest (apart from the 'standing reserve' argument that it keeps employed physicists with skills of potential military application). Likewise, as we shall see in later chapters, many life science projects of the second half of the twentieth century were big enough to warrant the title Big Science but did not revolve around single, expensive installations of machinery. Historians James Capshew and Karen Rader therefore make the useful distinction between sciences that are big in scale (such as those that are organized around centralized large single instruments) and those that are big in scope (such as scientific exploration, which need not be).[2]

However, the five 'M' model of Big Science is merely a useful checklist and does not tell us much about its distinctive organizational features. There are four. First, Big Science is goal-oriented science. No huge sum of money could be dedicated without a mission, without outcomes that could be articulated and measured. The very success of the Manhattan Project as an organized, mission-oriented project encouraged the view that it was a model for other scientific and technological projects. Second, the expense of Big Science leads to a concentration of resources, resulting in turn in a decreased number of leading research centres based around special facilities. In the nineteenth century, many nations had astronomical observatories that were broadly competitive. By the first half of the twentieth century, the trend towards expensive, wide-aperture reflecting telescopes meant that Mount Wilson (and later Mount Palomar) undoubtedly had the edge. Likewise, in post-war particle physics, the discoveries (and prizes) have generally gone to the teams working with the highest energies. Europe and Japan have struggled to compete with the facilities available in the United States.

As numbers of staff have increased and as the instrumentation has become more complex, so there has been an increased need for specialized expertise and for the work of such experts to be coordinated in an efficient manner. Therefore, third, Big Science was marked by a fine division of labour, hierarchically organized into groups, with group (and higher) managers. There has been a division, within the scientific and technical staff, into specialisms. A particle accelerator laboratory such as Brookhaven or CERN had distinct teams of theoreticians, instrument builders, experimenters, and so on. Getting these groups to work together – even to understand each other's subculture – was one of the greatest problems confronting the leaders and administrators of Big Science projects. Historian Peter Galison argues, for example, that, in order to work together, each subculture had to find ways of translating each other's specialized language at the points (often specific sites) where they met, like traders devising creole languages.[3] The trend towards fine division of labour could also be reinforced by the trend towards militarization, since both favoured the compartmentalization of work and knowledge, albeit for different reasons.

Finally, a high political significance was attached to mission-oriented expensive projects – to national health, military power, industrial potential or national prestige – although the proposed contribution varied. Along with political significance came media management. In the Cold War the political significance was, unsurprisingly, the extent to which the scientific or technological achievement could be presented as a triumph of a form of society. The very fact that the Cold War was not (except by proxy in Korea, Vietnam, Afghanistan, Nicaragua – the list, of course, is a long one) a fighting war but a war of symbols means that state resources were channelled into Big Science above and beyond the level of investment justified by the direct contribution of science to the building of weapons and defences. Space science largely falls under this analysis.

Science education

The political importance of Big Science in the Cold War ensured that there existed funds and the political will to expand scientific and technological education in the post-war years. In countries such as Britain, universities were encouraged to broaden their application pools. New universities were established. In other countries the pattern was similar. In the United States, the GI bill brought many into higher education, often from social and ethnic backgrounds that would have hindered entry before the war. There was, writes historian David Hollinger, an 'ethnoreligious transformation of the academy by Jews', although he considers the Cold War mobilization

332

of talent only partly responsible.[4] The effects reached down to science education at secondary-school level. In the United States, the philosophy of science education inherited from John Dewey was transformed in the 1950s, in the hands of leaders such as Joseph Schwab, into a call for inculcation into the methods and knowledge content specific to disciplines. 'To speak of using scientific methods outside of a disciplinary context was', for Schwab, writes historian John Rudolph, 'to no longer speak of science at all.'[5] The system mass-produced scientists, conversant within a discipline, reluctant to speak outside a discipline; knowledge was compartmentalized at an early age.

Security

The military character of post-war Big Science meant that administrators felt they had to enforce secrecy in the name of national security. Compartmentalization was largely accepted as a wartime measure among the Manhattan Project scientists, although characters such as Richard Feynman delighted in playing jokes that revealed some the absurdities of restricting knowledge.[6] Secrecy in arcane knowledge was certainly not new in the twentieth century.[7] Indeed, in the last century, restricting knowledge, such as that found in double-blind trials or in the anonymization of peer review, was made central to what was regarded as good scientific procedure. Scientists in practice can practise secrecy to protect ongoing research from competitors, hide bad results or trade secrets, or, in common with other creative pursuits, keep a project under wraps until complete.[8] But the negative consequences of secrecy also became a matter of concern and discussion from the middle of the twentieth century. Even someone of the stature and position of Lee DuBridge, president of Caltech, presidential advisor to Eisenhower and Nixon, might warn in 1949 of the consequences of military patronage of Big Science in the Cold War:

> When science is allowed to exist merely from the crumbs that fall from the table of a weapon development program then science is headed into the stifling atmosphere of 'mobilized secrecy' and it is surely doomed – even though the crumbs themselves should provide more than adequate nourishment.[9]

Complicated systems of classification were established to label and manage the secret knowledge generated by Cold War science. The quantities of information are staggering: the closed world of secret documents dwarfs the open world of our great libraries, a picture that remains accurate despite a temporary lifting of the veil at the end of the Cold War.[10] Peter Galison argues that this security blanket ends up stifling science, limiting innovation and wasting money and even leads to weakened rather than strengthened security.

Science and Western values

In this Cold War context it became essential to make explicit the nature and virtues of good science, and to demonstrate that it thrived within certain societies – 'free' or socialist – and not others. We have seen, in chapter 9, the argument that science was congruent with Soviet state philosophy. Symmetrically, we can now examine a liberal democratic variant. The start of this phase can be found in the middle of the Second World War. For example, in 1941 at the Royal Institution in London, between air raids, Richard Gregory, editor of *Nature* and president of the British Association for the Advancement of Science, announced seven Principles of 'scientific fellowship', which made clear that science thrived under conditions of intellectual freedom and independence.[11] Science's values were being made explicit because it was part of the fight against fascism. Before there was a threat, there had not been the need to state explicitly what needed protecting. But the view from 1941 was a transitional view, partly pointing towards the autonomy of a science community and partly looking back at pre-war (and indeed wartime) debates about the necessity to plan science for the 'progressive needs of humanity'.

In the Cold War, the need to make science's values explicit remained, but the form the expression took changed subtly, but crucially, with the times. The best exemplar is the 'norms' of science written out by sociologist of science Robert K. Merton. They are:

> Universalism . . . truth-claims whatever their source, are to be subjected to pre-established impersonal criteria: consonant with observation and with previously confirmed knowledge. The acceptance of rejection of claims . . . is not to depend on the personal or social attributes of their protagonist; his race, nationality, religion, class, and personal qualities are as such irrelevant . . .

> Communism . . . The substantive findings of science are a product of social collaboration and are assigned to the community . . .

> Disinterestedness . . . [and]

> Organized Skepticism . . . Science which asks questions of fact, including potentialities, concerning every aspect of nature and society may

come into conflict with other attitudes toward these data which may have been crystallized and ritualized by other institutions. The scientific investigator does not preserve the cleavage between the sacred and the profane, between that which requires uncritical respect and that which can be objectively analyzed.[12]

The first version of Merton's norm paper, published in 1942 in a journal article with the title 'Science and technology in a democratic order', was written as a defence of democracy against fascism.[13] In the Cold War, a similar but even more pressing demand was made to present science as an institution that could thrive only in a free Western society (despite the major patron being the state in the form of the defence funding bodies). Merton's Cold War versions of the paper included lengthy footnotes that detailed Soviet breaches of the norms. ('Communism' was sometimes rendered as 'Communalism'.) Merton's 'norms' thus showed how and why science would thrive in free, democratic societies and suffer in totalitarian societies, previously Nazi Germany and now Soviet Russia.

Nevertheless, some ethical expectations of science had changed. A comparison between Richard Gregory's principles of scientific fellowship and Merton's norms reveals a crucial shift: Merton's norms are descriptions of internal values, not external obligations. There was no talk of 'progressive needs'; rather, according to Merton, science would thrive in a liberal state by being allowed its freedom and autonomy and by policing itself. 'As the autonomy of science from external influences and demands was increasingly urged and defended ... it became all the more important that the moral qualities for which science was ostensibly a vehicle be seen as intrinsic to science', argues David Hollinger, taking the long view of Merton's norms; 'if certain approved imperatives were understood to be endemic to the very enterprise of science, society could rest more comfortably with the expansion of science.'[14]

Many Cold War commentators took up the task of showing that science either fostered democratic societies or thrived within them.[15] The Hungarian émigré chemist Michael Polanyi argued in his paper 'The republic of science' that 'in the free cooperation of independent scientists we shall find a highly simplified model of a free society'.[16] He compared the effectiveness of puzzle-solving of scientists acting independently with that of those being centrally directed and found that 'the pursuit of science by independent self-co-ordinated initiatives assures the most efficient possible organization of scientific progress'. Just as the invisible hand guided markets of independent traders, so science, according to Polanyi, would progress so long as planners (of Soviet or Bernal-style socialist stripe) were kept away: 'Any attempt at guiding scientific research towards a purpose other than its own is an attempt to deflect it from the advancement of science.' The journal that Polanyi's paper appeared in, *Minerva*, was funded by the CIA as part of the cultural Cold War.[17] Other texts, such as Don Price's *The Scientific Estate* (1965) and Harvey Brooks's *The Government of Science* (1968), served up other arguments on the congruence between science and democratic Western society. The decision to make science's values explicit, as well as the form such explicit statements took, were Cold War artefacts. They might be seen as a response to George Kennan's call in his 'Long Telegram': 'We must formulate and put forward for other nations a much more positive and constructive picture of the sort of world we would like to see than we have put forward in past.'

Cold War culture was polarized, not only East and West but outwards and inwards. Presenting science as compatible only with democracy was a projection outwards. But the celebration of the intrinsic virtues of the scientific community also had an inward audience. Indeed, the emphasis on a scientific 'community' itself was new. David Hollinger notes that the term 'scientific community' became widespread only in the 1960s, replacing the more individualist 'the scientist' or 'the scientists'; the switch from representations of science as an individualized pursuit to a communitarian one happened because of the 'revolutionizing of the political economy of physical science' in the wake of the Second World War and sustained by the Cold War.[18] In other words, Western science's dependence on Big Science military patronage raised questions about the autonomy of science which were resolved in part by explicit celebration of the intrinsic virtues of the scientific community. After all, public funds required public audit. And greater funds meant greater scrutiny. But science could resist some of this pressure if it could be accepted as self-policing. Hollinger calls this strategy 'laissez-faire communitarianism': 'Let the community of science alone.'[19]

Others were less sure that science and society could remain unscathed by the Cold War. The most extraordinary denunciation came in fact from the commander in chief, the president of the United States Dwight Eisenhower, in his farewell address to the nation of 17 January 1961. The mobilization necessary to meet 'a hostile ideology global in scope, atheistic in character, ruthless in purpose, and insidious in method' had led to a permanent rearmament that was becoming a deeply rooted, institutionalized power. He warned: 'we must guard against the acquisition of unwarranted influence, whether sought or unsought' by this entity, which he famously named 'the military-industrial complex'. 'Akin to, and largely responsible' for this danger, was 'the technological revolution during recent decades', from which emerged a second threat: the corruption of democracy by technocratic influence. 'In this revolution, research has become central, it also becomes more formalized, complex, and costly', said Eisenhower. 'A steadily increasing share is conducted for, by, or at the direction of, the Federal government.' He continued:

> Today, the solitary inventor, tinkering in his shop, has been overshadowed by task forces of scientists in laboratories and testing fields. In the same fashion, the free university, historically the fountainhead of free ideas and scientific discovery, has experienced a revolution in the conduct of research. Partly because of the huge costs involved, a government contract becomes virtually a substitute for intellectual curiosity. For every old blackboard there are now hundreds of new electronic computers. The prospect of domination of the nation's scholars by Federal employment, project allocations, and the power of money is ever present – and is gravely to be regarded.
>
> Yet, in holding scientific research and discovery in respect, as we should, we must also be alert to the equal and opposite danger that public policy could itself become the captive of a scientific-technological elite.
>
> It is the task of statesmanship to mold, to balance, and to integrate these and other forces, new and old, within the principles of our democratic system – ever aiming toward the supreme goals of our free society.

Eisenhower may, if his science advisor James Killian's evidence is accepted, have later privately retracted his warning.[20] However, the entity he named was real: the close interpenetration of defence industry and government in permanent mobilization, supported closely by a penumbra of dependent and interested parties, including universities and advice consultancies.

Systems thinking

The 'military–industrial–academic complex' (Senator William Fulbright's extended, and more accurate, label of 1967) encouraged a certain style of inquiry, in particular an emphasis on systems. Systems thinking has a long history, with origins in Enlightenment analyses and nineteenth-century technological projects such as railroads, telegraphy, electrical light and power.[21] But a distinctively managerial science of systems emerged during the Second World War.[22] Historian David Mindell has located its origin in fire-control radar projects of MIT's Rad Lab and Bell Labs in the United States;[23] I have made a case for finding systems, especially 'information systems', in British radar of the same period.[24] Agatha and Thomas Hughes distinguish between four forms of mid-twentieth-century systems approaches: operational research, systems engineering, systems analysis, and system dynamics.

Systems analysis, used in comparing, say, the deployment of long-range bombers with intercontinental ballistic missiles, was the specialty managerial science of consultancy think-tanks that worked closely with military patrons. The prime example and prototype was RAND.[25] In 1946, General Hap Arnold, leader of the United States Army Air Force, set up Project RAND in collaboration with Douglas Aircraft and in consultation with Edward Bowles of MIT. Douglas would leave in 1948, but the RAND Corporation, as it became, would continue and deepen its relationship with the new United States Air Force (an independent armed service from 1947). The United States Air Force provided RAND with funds and influence on decision-making. RAND provided the United States Air Force with 'independent objective analysis', helping it build its power within the new Department of Defense on the back of providing the strategic capacity to strike against the Soviet Union with airpower. The United States Air Force was the working world for RAND's science of warfare.

RAND's experts drew on diverse mathematical techniques – linear programming, game theory – and computer power to analyse the air force's work as a system with an aim of optimization. RAND became mathematical innovators too. George Dantzig offered the 'simplex method', while Richard Bellman, who left Stanford for RAND, developed linear programming into dynamic programming, ideal for optimizing the configuration of changing Cold War entities such as the air force and providing generalizable tools such as the minimax method. Game theory was another sharpened tool. Its core text, John von Neumann and Oskar Morgenstern's *Theory of Games and Economic Behavior* (1944), had been published during the war. Now, not least through von Neumann's continued military consultancy, games theory was adopted as a means of thinking about the Cold War. 'The emergence of the Soviet Union as "the enemy" and notions of a possible, single, intense exchange of nuclear weapons between the Soviet Union and the United States', notes historian David Hounshell, 'offered a nearly perfect parallel to the simple building block of game theory – zero-sum, non-iterative, two-person games.'[26]

Early on, RAND's aim was too general and ran into practical opposition. Its first large-scale systems analysis was a project to optimize the United States Air Force's strategic bombing of the Soviet Union. Pilots disliked its conclusions.[27] The next large-scale study, of air defence, also had little effect. However, more restricted systems analysis studies of the siting of bases, including in particular one by Albert Wohlstetter, were immediately persuasive. Likewise, Herman Kahn brought into RAND expertise of modelling scenarios using Monte Carlo sampling methods (which usually required electronic computers to run). Kahn would become a powerful voice on nuclear strategy in the 1950s, contributing to notions of a 'winnable' nuclear war, and, in particular, his 'ladder of escalation', discussed later.[28] RAND provided a channel through which civilian experts would shape Cold War policy and make the atomic devastation of a third world war rational.

RAND was the 'paramount thinktank of the Cold War', which, 'through its development of systems analysis ... helped to foster the pervasive quantification of the social sciences in the post-war era'.[29] RAND valued its independence of research just as highly as it valued the patronage it received and influence it gained. Indeed, two of the most influential economic analyses arguing that 'basic science' could not be left to the private sector, but must be state-funded yet kept independent of state influence, by Richard Nelson (1959) and Kenneth Arrow (1962), were written from within RAND.[30] Yet, as Paul Forman's arguments might lead us to expect, there was considerable mutual orientation between air force interests and RAND analyses.

Transistors: missiles, markets and miniaturization

Three Second World War technologies were central to the Cold War. One, the electronic computer, will be examined in detail in chapter 16. The other two were the atomic bomb and the rocket. Rockets had several potential uses in the Cold War. First, as small guided weapons, rockets could be used as anti-aircraft weapons or as remote-controlled flying bombs. In common with the demands of military aircraft (and, indeed, naval vehicles), missiles required small, light electronics to make up the guidance systems, essential to turn rockets into guided weapons. This drive for miniaturization created one of the most important trends of the Cold War, resulting in not merely military aircraft packed with sophisticated electronics but also a host of other military and civil spin-offs, from transistors, to integrated circuits, to computer chips.

The crucial scientific innovation – the invention of the transistor – has been presented as a triumph of basic research carried out within industry, at Bell Labs, but its development into the cheap, core component of the post-war semiconductor electronics boom, transforming the second half of the twentieth century, was driven as much by the Cold War as by consumer demand.[31] In 1938, Mervin Kelly, Bell's research director, had set up a team of scientists to conduct 'fundamental research work on the solid state' with a view to replacing the electromechanical switches found in every telephone exchange with something smaller, faster and more reliable.[32] The team of physicists included the Caltech- and MIT-trained William Shockley and Walter Brattain, who had been with Bell since 1929. They were joined by the talented John Bardeen, who many years later would join Marie Curie and Linus Pauling as the only scientists to win two Nobel prizes. The transistor would only be possible through 'multidisciplinary teamwork'.[33]

Shockley proposed an amplifier idea in 1939–40, but it did not work. Investigating the silicon photocell effect of fellow Bell scientist Russell Ohl in 1940, Brattain diagnosed a 'barrier' found between 'commercial' and 'purified' silicon, the first indications of P type and N type semiconductors.[34] Brattain and Bardeen, working with an element similar to silicon, germanium, figured that the effect could be exploited to produce an amplifying device, the point-conduct semiconductor amplifier. Shockley wanted co-credit. Spurned, he remarkably improvised a second way of making a transistor, a sandwich of N–P–N semiconducting material. Therefore in 1947, Bell Labs had a startling new electronic component to show off to the press and the military and to patent. A science fiction writer and Bell employee, John Pierce, named it the 'transistor'.[35] Bell presented the transistor as a triumph of teamwork. An iconic photograph showed the three men working together, even though Bardeen and Brattain were barely speaking to Shockley at the time.

Nevertheless, if it could both be made robust and be mass-produced, the transistor had the potential to replace triode valves as the core component in amplifying and switching circuits. By end of 1940s the 'military supplied about 15% of the Bell Telephone Laboratory's budget'.[36] In the middle of the Korean War, in 1951, the junction transistor nearly became a classified secret.[37] But Bell, concerned about anti-trust accusations, was able to convince the authorities that it was in the national interest to license the manufacturing rights.

The military, however, was the market that drove development. Shockley devised electronic proximity fuses (using transistors) for mortar shells. One of the first uses of transistors was as part of radar control systems, AN/TSQ units, for Mike-Ajax guided missiles.[38] The first fully transistorized electronic computer, TRADIC, was made for the United States Air Force. In the Cold War the military was willing to pay. AT&T, Bell Laboratories' parent company, also offered a captive market for switch applications.

The first transistor radios, using a component whose price had been driven down during military development, appeared at the end of the Korean War. Remarkably, the $25,000 licence to manufacture transistors had also been bought by a small Japanese company owned by Masaru Ibuka and Akio Morita, called Tokyo Tsushin Kogyo. Through experiment, they improved the semiconductor manufacturing process, and in 1954 launched a transistor radio firmly aimed at the commercial market. Ibuka and Morita chose a simple name for the American market: Sony. A Sony employee, Reona 'Leo' Esaki, would discover quantum tunnelling in 1957 and later share the 1973 Nobel Prize, which shows that the working world of Japanese semiconductor manufacture generated its own science.

A burgeoning commercial market began to sustain further investment for military purposes. Shockley decided to set up his own company: 'After all', he told his future wife (while his first wife was dying of cancer), 'it is obvious I am smarter, more energetic and understand people better than most of these other folks.'[39] The Shockley Semiconductor Laboratory was set up near Stanford University, recruiting electronics talent such as Robert Noyce and Gordon Moore. But Shockley was a poor manager of people, and morale plummeted. A group of eight left in September 1957 and founded Fairchild Semiconductor. This was the beginning of Silicon Valley, the proliferation of electronics and computing companies as part of the military-industrial-academic complex of the west coast. Fairchild's innovation of the integrated circuit – tiny components etched on a lightweight sliver of semiconductor – was perfect for missiles, aircraft electronics and spy satellites, as well as small computers.

From upper atmosphere to outer space

In 1945, it was obvious that the rocket would be an important military device of the near future, especially in combination with nuclear warheads. A race had developed, already discussed, between

341

the United States and Red armies to capture the Nazi rocket programme. Any concerns about Nazi backgrounds had been waved aside as Wernher von Braun's team had been brought to work under army contracts at White Sands Missile Range in New Mexico, the guided weapon unit at Fort Bliss, Texas, and the Redstone Arsenal in Huntsville, Alabama. The Redstone was a continuation of the V-2. Guided by RAND, the United States Air Force, which saw responsibility for strategic missiles as an extension of its strategic bomber force, set up a competing programme of rocket research.

Larger rockets that travelled into the upper atmosphere at great speeds before falling to earth were strategic weapons against which there was almost no defence. This had been the terror of the German V-2, which had only to travel a few hundred miles. Once rocket technology had improved to the extent that thousands of miles could be flown, the weapons could strike anywhere on the planet – they were intercontinental ballistic missiles (ICBMs). Both the United States (Atlas) and the Soviet Union (R-7 or SS-6) began ICBM projects soon after 1945. The Soviet ICBM programme was boosted in May 1954. The man placed in charge of design was Sergei Pavlovich Korolev, a veteran of amateur rocketry societies of the 1930s who had continued work in the gulag in the 1940s. His rehabilitation was another indicator of the indispensability of technical expertise to Soviet plans. The threat and capability of the ICBM meant that there was a strong military interest in sciences of the upper atmosphere in the early Cold War.

Finally, a missile capable of travelling around the earth was capable of orbiting the earth. Cold War imagineers – organizations such as RAND and individuals such as the science fiction author Arthur C. Clarke – drafted plans for different kinds of artificial satellite after 1945. Among uses imagined were commercial applications (such as communication satellites that would offer a service similar to undersea cables), scientific research (looking up to the heavens, down to earth, or simply exploring the space of low orbits), and military reconnaissance and surveillance. A combination of uses was not uncommon. Each of the three armed services in the United States had plans for satellites and jostled for favour and resources. In the Soviet Union, drawing on the tacit and explicit knowledge of captured German technicians, as well as careful study of facilities now in East Germany, missile development had continued apace. In May 1954, in parallel with the go-ahead for the ICBM, Korolev passed on to the Soviet leaders a *Report on an Artificial Satellite of the Earth*, written by a close colleague, Mikhail Klavdiyevich Tikhonravov.

The most urgent strategic problem facing the United States in the Cold War was knowledge of the state and distribution of Soviet military capabilities. There was an asymmetry here: the United States was more open than the Soviet Union, so the need for surveillance was probably felt less in Moscow than in Washington. Soviet spy networks seem to have been more effective than Western ones, although a sound comparison is almost impossible to make. Nevertheless, good information was the basis of good diplomacy, and for crucial questions – such as, where were Soviet strategic weapons? how many were there? – there was ignorance. From 1952, modified B-47 bombers began long-range surveillance flights over Siberia. From 1956, the specialist U-2 spy planes, complemented by high-altitude balloon flights (Project GENETRIX), provided information. But there were severe problems with these technologies: they violated Soviet airspace, they encouraged the Soviet Union to develop air defence systems as countermeasures, and they risked being shot down. (Indeed, an embarrassing diplomatic incident would occur in 1960: a U-2 was downed over Russia, and its pilot, Gary Powers, was captured.) A diplomatic proposal, 'Open Skies', of transparency over missile sites and freedom for aerial reconnaissance to check, was rejected.[40]

To resolve this issue, Eisenhower's advisors in 1955 came up with a clever plan.[41] The illegality of flying aircraft over sovereign airspace without permission was clear, but the legal status of orbiting satellites was not: after all, a satellite followed a path determined by gravity not, it could be argued, by geopolitics. By launching a scientific satellite, the principle of the 'freedom of space' could be established. Spy satellites, the solution to the problem of Cold War transparency, could follow in this scientific path. Science's image as international and above geopolitical squabbles was essential to the success of this scheme.

A suitable occasion to announce an American scientific satellite was immediately to hand. The International Geophysical Year (IGY), a coordinated set of geophysical research programmes involving 5,000 scientists, focusing mostly on Antarctica and the upper atmosphere, was being organized for the years 1957 and 1958.[42] The IGY provided the ideal backdrop for the presentation of scientific satellites as part of a public show of international scientific cooperation. In July 1955, James Hagerty, Eisenhower's press secretary, announced that the United States would launch a science satellite during the IGY. The announcement generated excitement and plenty of press coverage. In immediate response, a delegation of Soviet scientists, who happened to be at the Sixth International Astronautical Congress in

Copenhagen, announced that the Soviet Union would also launch a satellite during IGY. (Photographs of the event show the scientists and apparatchiks squeezed into what looks like a suburban front room, but was in fact in the Soviet Embassy in the Danish capital.) Their announcement generated almost no coverage at all. Korolev was charged with fulfilling the Soviet plan. The firm expectation, however, was that the United States would be the first nation in space.

The United States government was faced with a decision between the satellite proposals of the three armed services. It set up a committee, chaired by Homer Stewart of Caltech's Jet Propulsion Laboratory, to resolve the issue. The choice was between the army's Project Orbiter, which promised to launch a cheap basic satellite using the Redstone missiles developed by Wernher von Braun, a heavy satellite launched by the air force's Atlas ballistic missile, and the Naval Research Laboratory's scientifically sophisticated satellite, Vanguard, launched from an as-yet-undesigned modified sounding rocket. The Atlas ICBM was politically too important to tamper with, so the air force project was rejected. A combination of Vanguard satellite and the modified Redstone rocket (Jupiter C), which would have represented scientific sophistication with proven launch technology, was rejected because of the complications of inter-service rivalry. The Stewart Committee plumped for the navy's project. American success in space therefore depended on a rocket that existed only on paper. This diffusion of energy and focus was not paralleled in the Soviet Union, where Korolev assured Nikita Khrushchev that satellite launches would not interfere with ICBM development.

On 4 October 1957, in the middle of the International Geophysical Year, the Soviet satellite Sputnik was sent into orbit from a launch site in the Soviet Socialist Republic of Kazakhstan. It was a metallic sphere – 'the first Sputnik must have a simple and expressive form', recalled Korolev, 'close to the shape of natural celestial bodies'.[43] Even more expressive was a radio signal – a regular 'beep' – that was soon picked up by radio hams (as Korolev had designed it to be). The West was deeply shocked. Edward Teller's thought, that the United States had 'lost a battle more important than Pearl Harbor', was a commonplace.[44] The launch into space, only a month later, of a much larger satellite, Sputnik II, carrying the first living creature, Laika the dog, raised the level of alarm further. Sputnik II was the weight and dimensions of a nuclear weapon. Sputnik had announced in the most dramatic way that not only did the Soviet Union possess the ICBM, but it made any Western city defenceless against air-attack with nuclear bombs.

344

Sputnik suggested that the Soviet Union had overtaken the West in technological progress. Washington worried that Third World leaders would draw conclusions about the relative merits of capitalism and communism. The anxieties were stoked further by the fate of the West's own first attempted satellite launches: the navy's Vanguard exploded on launch-pad in December 1957, and would not be successfully put in orbit until March 1958. By then, a rush job, combining a Jet Propulsion Laboratory satellite, 'Explorer', on top of a Jupiter rocket, had launched from Cape Canaveral in January 1958.

There was a search for credit and scapegoats. Korolev's identity was kept a tightly held secret until after his death, supposedly to keep him safe from Western assassins, but really so that credit would be given to the system rather than the man. In the absence of a story of individual talent, some Western journalists speculated that Peter Kapitsa, the Cambridge-trained physicist held in Russia, must have been behind the triumph. In the United States, derision was heaped on Eisenhower, who was portrayed as out of touch. In fact, historian Rip Bulkeley suggests the previous Truman administration was more to blame, and the consensus now is that Eisenhower was in private content that the important strategic principle of the 'freedom of space' had been secured.[45] In the United Kingdom, the response to Sputnik was a national debate on the supposed failures of secondary education.[46]

Nevertheless, the surprise of Sputnik was a testament to the underestimation of Soviet capability in technical areas of high political value, rather than a result of Cold War secrecy. The Soviet Union had announced the Sputnik programme in public, and had given technical presentations on the satellite at conferences in the months before launch. It was not unknown. Indeed, in July 1957 the Soviets had approached the Naval Research Laboratory with an invitation to install an American 'radio frequency mass spectrometer' on the Soviet satellite; this proposal, made in the cooperative spirit of International Geophysical Year, had been rejected for 'technical' reasons.[47]

Sputnik prompted the establishment in 1958 of two institutions that would drive technological change in the 1960s. The first, in April, was a new government body in the Pentagon, the Advanced Research Projects Agency (ARPA), charged with rationalizing existing research and, more importantly, breaking armed service rivalries by supporting early-stage high-risk research and development that might lead to new multi-service military technologies.[48] ARPA supported a range of projects, including ballistic missile defence, nuclear test detection

345

and materials science, but especially information technologies. The agency had no laboratories of its own, and its funds strengthened the ties between universities such as MIT, Michigan and Stanford and surrounding defence industries. ARPA's Information Processing Techniques Office (IPTO) was particularly important in funding basic computer science, network ideas, command and control 'testbeds' and research into the faster components necessary for new generations of computers.[49] The most famous creation, the ARPANET, will be examined in a later chapter.

The second institution was a great civilian-led agency to compete and contrast with Soviet space successes in the space race. The National Aeronautics and Space Administration (NASA), established in July 1958, may have echoed the name of the National Advisory Committee on Aeronautics (NACA), which it replaced, but it was a different beast. Its first ten-year plan, articulated in 1960, encompassed 260 launches, among them orbital satellites, probes to the moon and the planets, and manned space flights, including some around – but not landing on – the moon.[50] The post-Sputnik boom in space-based planetary astronomy was sometimes viewed with 'hostility by some ground-based astronomers'.[51] Solar system astronomy was already a thriving, interdisciplinary field by the 1950s, an 'interwoven tapestry' of planetary astronomy, meteorology, geochemistry, geology and geophysics, with some large-scale university-based projects.[52] After Sputnik, it had to resolve the tensions between the extraordinary opportunities for investigation represented by NASA's patronage and the competing and expensive demands of human space flight.

The presidential election of 1960 was dominated by space. The Democrat candidate, John F. Kennedy, not initially an enthusiastic supporter of space programmes, chose space as a subject on which his Republican rivals were perceived as weak. The other big issue, civil rights, was muddied by Southern Democrats' intransigence. A decisive phrase in the election was the supposed 'missile gap' that Eisenhower had allowed to open. In fact, as U-2 flights had revealed, the Soviet Union by the end of 1959 had only six vulnerable missile sites, each requiring twenty hours to fuel.[53] Despite the resources pumped into NASA's first major programme, Project Mercury, the first human in space was a Russian, Yuri Gagarin, who orbited the earth on 12 April 1961. Kennedy responded in May 1961 with the announcement of NASA's Apollo mission: 'We choose to go to the Moon in this decade and do the other things, not because they are easy, but because they are hard.'

346

Antarctic space

Outer space was becoming defined as a neutral zone of international competition, frozen in place by Cold War rivalry, with activities justified by appeal to the values of international science. Interestingly, a close parallel can be drawn between this definition of outer space and Antarctic space. From the end of the nineteenth century to the 1950s, scientific research in Antarctica had been performed as part of explorations that aimed to ensure national territorial claims. National claims on slices of the Antarctic continent had been made by Britain (1908 and 1917), by France (1924), by Britain on behalf of New Zealand (1925) and Australia (1933), by Germany (1924), by Norway (1939, specifically to counter a German claim), by Argentina (1940) and by Chile (1942), both of which claimed territory in anticipation of a German victory. After the war, a British agency, the Falkland Islands Dependency Survey, a supposedly scientific body, had established Antarctic bases. 'The scientific fieldwork which is carried out in the Dependencies is not research for the benefit of colonial peoples', a civil servant had explained in 1952:

> It is done to maintain a UK interest . . . It is UK, not FIDS policy, that the activity should be maintained at a sufficient level to enable us to compete with our South American rivals, and it is inescapable that the receipt, coordination, working up, and publication of results of fieldwork at the bases must be regarded as integral to these activities.[54]

In this coldest theatre of the Cold War, the United States strategy was to make sure that any redivision of the Antarctic map excluded any sovereign rights granted to the Soviet Union. Freedom of Antarctic space might be best served by international neutrality, justified by science.

The International Geophysical Year provided the opportunity. Antarctica, along with space and the upper atmosphere, was the major location for IGY activity. The twelve IGY nations included seven with claims on Antarctic territory, the others being the United States, the Soviet Union, Belgium, Japan and South Africa. Fifty-five scientific stations were established. The Royal Society of London set up at Halley Bay on the Antarctic coast. The United States placed a station at the South Pole, a symbolic negation of the sector claims that converged at the pole. Not to be outdone, the Soviet Union set up a scientific base at the bleak 'pole of inaccessibility', the furthest spot from Antarctic coasts. Scientists photographed the aurora australis with new all-sky cameras, recorded cosmic ray showers (from which deductions about the upper atmosphere could also be made), drilled snow cores and undertook seismic studies of the depth of the Antarctic ice.

The cooperation of the International Geophysical Year laid the foundations for the Antarctic Treaty, which was signed in 1959 and came into force in 1961.[55] The important provisions of the treaty were Article I (demilitarization of the continent and a ban on dumping nuclear waste), Article II ('Freedom of scientific investigation in Antarctica and cooperation towards that end', requiring that stations would be open to inspection) and Article IV (freezing all sovereignty claims, with implications for biological and mineral resource exploitation), while Article IX.2 stated that the only countries that would be voting members under the treaty were those 'demonstrating [their] interest in Antarctica by conducting scientific research activity there'. In other words, nations could only have a say in Antarctic politics if they pursued science on the continent.

Science and Antarctic politics had become intimately linked for three reasons: science played a key role in defining the regime governing Antarctic decision-making (the Antarctic Treaty), science was used rhetorically to justify actions and cover other, often narrowly interested motives, and science would be essential to the design and implementation of regulations that followed the treaty for resource management and environmental protection. In conclusion, outer space and Antarctic space were both defined in a Cold War context as abstract international spaces, defined by legal treaties (an Outer Space Treaty would be signed in 1967), and the use of which would privilege science.

Global projects and the Cold War

Internationalism, sometimes sustained, sometimes fractured by Cold War polarities, was a feature of the post-war world. The World Health Organization (WHO) of the United Nations undertook a 'war against disease'. A campaign against malaria starting in 1955 was initially quite successful, before the problems of drug and pesticide resistance and a failure to build robust, sustainable, local organizations led to malaria sweeping back into areas whence it had formerly been eradicated.[56] Malaria remained a major agent of death in the twentieth century. However, another WHO campaign, vaccination against smallpox, which began on 1 January 1967, was spectacularly successful. Under the Intensified Smallpox Eradication Campaign, mass vaccination, which had been successful in Japan, Western Europe and North America, was extended to other parts of the globe. Swift responses were needed to combat outbreaks in Nigeria and India. The campaign was a slow process of learning how local populations could be surveyed and recorded and also depended on innovations in freeze-drying and mass-producing vaccine. After 1977, smallpox existed only in high-security laboratories. The resolution made at the World Health Assembly of 1965 to eradicate smallpox on a global scale had been driven not only by the technical opportunity to rid the world of a dread disease, but also by the choice of the United States to divert energies in this direction to project a particularly compelling Cold War message; the expansion of air travel was also a factor, since increased travel had heightened the risk of epidemics.[57] Cold War politics would continue to shape smallpox even after its eradication, determining which handful of laboratories might continue to cultivate the potential bioweapon.

The Cold War encouraged global perspectives. Fallout, the radioactive detritus of atmospheric atomic tests, was perhaps the first recognized global environmental hazard.[58] Data from the atmosphere, oceans and ice cores taken during the Cold War were integral to the recognition of anthropogenic climate change on a global scale.[59] Both will be considered in more detail later. But the Cold War also framed how another global issue was framed and addressed: so-called overpopulation and related threats to food supply. This framing can be seen in the history of developing high-yield crops, especially in the science-based, globe-spanning project known as the Green Revolution.[60]

The Rockefeller Foundation had been active in China before its access was ended by the 1949 victory of the communists over Chiang Kai-shek's nationalist forces in the civil war. The foundation's attention regarding foreign aid turned instead to Mexico. In 1941, a survey team of agricultural science experts reported that significant work on wheat rust and maize was justified. A research team, including the plant pathologist Norman E. Borlaug, was dispatched in 1943 to form the Mexican Agricultural Program, a semi-autonomous unit within the Mexican government funded by the Rockefeller Foundation.[61] It received political support from within Mexico by politicians and scientists who championed industrialization of agriculture over the radical land reform favoured under earlier Mexican regimes. The new moderate president, Ávila Camacho, also had particular hopes of supporting the development of the newly opened Yaqui valley area. This contingency, along with a certain middle-class prejudice, helps explain why wheat became a focus of research rather

349

than the traditional food of working Mexicans, maize and beans: wheat rust was a problem in the Yaqui valley; wheat also offered an export crop.[62] The Mexican Agricultural Program brought in Japanese short-stem wheat varieties, both from Japan and from cultivation in agricultural research stations in the United States. Borlaug and others worked hard to improve varieties, experimenting with crosses, guided by Mendelian genetics. The Rockefeller Foundation in turn learned that it could support large-scale research for agricultural industrialization in foreign countries.

The Rockefeller Foundation was therefore primed to respond to President Truman's call for action in his inaugural address of January 1949. Truman presented four points. Point 3 addressed the need for Western military organization to counter the 'false philosophy' of communism. The result was NATO. Point 4, also framed by the Cold War, addressed food and science:

> More than half the people of the world are living in conditions approaching misery. Their food is inadequate ... Their poverty is a handicap and threat both to them and to more prosperous areas ... The United States is pre-eminent among the nations in the development of industrial and scientific techniques ... Our imponderable resources in technical knowledge are constantly growing and are inexhaustible. I believe that we should make available to peace-loving peoples the benefits of our store of technical knowledge in order to help them realize their aspirations for a better life.[63]

The result was the Rockefeller Foundation redoubling its efforts in Mexico. Warren Weaver, collaborating with Mexican Agricultural Program scientists, summarized the position in a paper, *The World Food Problem*, submitted to the trustees in 1951. In it he echoed Truman's sentiments:

> The problem of food has become one of the world's most acute and pressing problems; and directly or indirectly it is the cause of much of the world's present tension and unrest ... Agitators from Communist countries are making the most of the situation. The time is now ripe, in places possibly over-ripe, for sharing some of our technical knowledge with these people. Appropriate action now may help them attain by evolution the improvements, including those in agriculture, which otherwise may have to come by revolution.[64]

Historian John Perkins calls this chain of reasoning 'population-national security theory', the Cold War logic that said overpopulation led to the exhaustion of resources, to hunger, to political instability, to communist insurrection, and to dire consequences for the interests of the West.[65] The trustees agreed with Weaver's argument. One of them, Karl T. Compton, president of MIT, wrote: 'I suspect that India may be fertile ground for activity in this field.'[66]

Shortly after India had won independence from Britain in 1947, it was devastated by partition as the Muslim lands broke away to form Pakistan. The first prime minister, Jawaharlal Nehru, at least in his first decade as leader of India, promoted industrialization, but not at the expense of aims for social equality. Intensification of agriculture nevertheless began, supported in part by the Ford Foundation. However, a combination of interconnected factors in the late 1950s – poor harvests, communist success in elections – tipped the balance in favour of the intensification and industrialization of agriculture. When Nehru died in 1964, his successor, Lal Bahadur Shastri, committed the country fully to the Green Revolution: the import of Mexican short-stem wheat varieties, further developed by Indian agricultural scientists such as Benjamin Peary Pal and Monkombu Samasivan Swaminathan, planted widely and fed with fertilizers manufactured by chemical industries. The startling increases in yields fed a burgeoning population and possibly staved off communist insurrection, but certainly promoted the power of elites at the expense of poor farmers.[67]

The limits of containment

So, in a Cold War context, the success of diverse projects across the globe became a freighted question. We will see in the next chapters how the techniques of comprehensive surveillance scrutiny – from exploring the bottom of the ocean to distinguishing between phenomena in the highest parts of the atmosphere – led to distinctive Cold War sciences. Historian Paul Edwards calls this state of global scrutiny, containment and control the 'closed world', a concept to which I will return.[68] Finally, however, the limits of this scrutiny and control need to be noted. The Cuban Missile Crisis of 1962 and the process of nuclear proliferation will illustrate.

In 1962 Nikita Khrushchev had ordered intermediate-range missiles to be deployed on the island of Cuba. It was once thought that this deployment was a bid to regain nuclear leverage after, contradicting the 'missile gap' rhetoric, the relatively meagre reality of Soviet missile forces had been revealed. Khrushchev's bluff had been called (he too had exaggerated the numbers and accuracy of Soviet missiles) and he was raising the stakes. A more recent interpretation, chiming in fact with Khrushchev's contemporary justifications, sees the deployment as an attempt to protect and preserve the revolution in Cuba, the success of which had surprised Moscow as well as Washington.[69] The previous year an attempted invasion of Cuba by the United States had foundered in the Bay of Pigs. The nuclear missiles, some of which were short-range and tactical, were a response. Unknown to the West, some nuclear missiles had been placed under local control. There were, in addition, Soviet submarines in the area, also with missiles under local command.

The missile bases had been revealed by scrutinizing surveillance photographs taken by U-2 aircraft. For fourteen days the significance of the missiles was debated, and responses were discussed. It is now well known that the world came very close to war between East and West. Only recently has it emerged that a fierce argument between the captain and his two senior officers on a Soviet submarine was swung in favour of not launching nuclear weapons only by one vote by one officer (Vasili Arkhipov, to whom the world should be very grateful). The Cuban Missile Crisis was only defused when Kennedy publicly promised that the United States would not invade Cuba, while privately also promising that Turkish intermediate-range missiles would soon be dismantled.[70]

The Cuban Missile Crisis marked a dramatic reversal of nuclear strategy. Eisenhower's strategy had been one dimensional: a promise of massive retaliation if provoked. The Single Integrated Operational Plan, the plan of attack, of 1960 called for 3,200 nuclear weapons used on 1,060 targets in the Soviet Union, China and allied lands, aimed at nuclear bases, military command centres, ports and cities, delivered by the United States Army, Navy and Air Force in a single combined attack.[71] Kennedy, listening to civilian strategic advisors, had moved away from massive retaliation towards favouring the 'rational' escalation ladder approaches exemplified by Herman Kahn's writings. Nuclear confrontation could be rationally managed through 'flexible response'. But rational management requires rational game-players and clear symbols sent, received and understood. The dreadful events of the Cuban Missile Crisis fatally undermined this model. 'What kept war from breaking out', concludes John Gaddis, 'was the irrationality, on both sides, of sheer terror.'[72] Henceforth, in practice there were sharp limits on the technocratic rationality promoted by strategic advisors. Robert McNamara, who perhaps more than anyone else embodied this rationality, was also the architect of the institutionalized irrational strategy: a return to Eisenhower's policy of massive retaliation, now known as Mutually Assured Destruction (MAD).

If the geopolitics of the Cold War were simply two poles, East and West, then containing nuclear proliferation might have been possible. But the reality of geopolitics was the manoeuvres of many interested nation-states within a polarized framework. The United States could not prevent the Soviet Union from developing the atomic bomb and the thermonuclear bomb. Possession of nuclear weapons conveyed status and influence. Neither superpower could prevent other nations eventually joining the atomic club: Britain (first test 1952), France (1960), Israel (probably ready by 1967), India (tested 1974), South Africa (possessed in the 1980s), Pakistan (1998) and North Korea (2006).

Each new nuclear nation has benefited from the transfer of nuclear knowledge, techniques, materials or personnel, despite the containment of the Cold War. Britain had experience of participation in the Manhattan Project, and then in 1954–5 helped the French project by providing crucial information (and samples of plutonium) while keeping the transaction secret from the Americans. Like the British, French politicians wanted the bomb as a signifier of status, following post-colonial humiliation in the battle of Dien Bien Phu (1954) and the effective symbolic demolition of Great Power status (of Britain and France) in the debacle of the Suez Crisis of 1956. The French, facing a rebellion in Algeria, were motivated to help Israel develop a bomb as a deterrent against Soviet-backed Arab nations.[73] This French decision was kept secret from Britain and the United States. Independently, a British civil servant, without informing his minister, sold Israel enough scarce heavy water – for £1 million via a Norwegian front company – to moderate the Israeli secret reactor at Dimona. It suited all nations' interests not to acknowledge the reality of an Israeli nuclear bomb, creating what historian Avner Cohen calls a dangerous state of 'nuclear opacity'.[74] Sometimes merely the resources and knowledge close to but not from active nuclear projects prove sufficient: West German technology exported to South Africa allowed the land of apartheid to build its nuclear weapons.[75]

— 15 —

COLD WAR SCIENCES (1): SCIENCES FROM THE WORKING WORLD OF ATOMIC PROJECTS

Science worked with the resources, networks of contacts and ways of thinking forged in the Second World War. The Cold War formed working worlds in which not only were these sciences sustained, they also, first, contributed to how the Cold War was conducted and, in turn, were inflected by Cold War values. This chapter and the next examine these Cold War sciences in detail.

Radiology

Let us start by reviewing the sciences of the working world of atomic projects. Scientists were among the first Westerners to visit the devastated cities of Hiroshima and Nagasaki. While others worked to rebuild Japanese society as an ally of the West in the Cold War, the scientists took the population of the cities as an object of study. The first teams confirmed that the radiation of the blasts had produced symptoms in the survivors, but also reported that residual radioactivity was not immediately harmful. In 1947, the National Academy of Sciences set up an Atomic Bomb Casualty Commission, funded by the Atomic Energy Commission, to continue the investigations, with a special emphasis on the longer-term genetic damage; 'if they could foresee the results . . . 1,000 years from now', Hermann Muller, discoverer of radiation-induced mutation in 1927, had warned in 1946, the Japanese victims 'might consider themselves more fortunate if the bomb had killed them'.[1] There would later be much controversy over the storage and study of biopsies and body parts, which the historian Susan Lindee has described as being akin to war trophies, material instantiations of victory.[2] In the immediate context of building a strong, democratic Japan as a bulwark against communism, historian John Beatty argues that the cooperation of Japanese scientists with the Americans in the Atomic Bomb Casualty Commission was encouraged for these reasons of Cold War diplomacy:

> A long-term study of atomic bomb casualties, in collaboration with the Japanese, affords a most remarkable opportunity for cultivating international relations of the highest type . . . Japan at this moment is extremely plastic and has great respect for the Occupation. If we continue to handle Japan intelligently during the next few years while the new policies are being established, she will be our friend and ally for many years to come; if we handle her unwisely, she will drift to other ideologies. The ABCC . . . may be able to play a role in this.[3]

The working world of the Cold War thus sustained investigations of the long-term damage of radiation on human genetics. The results were also shaped in this context. The principal investigator, a *Drosophila* geneticist named James Neel, drew only tentative preliminary, inconclusive inferences. But in the hands of science journalists, in the wake of the *Lucky Dragon* tragedy, in which Japanese fishermen were fatally contaminated by fallout from the 'Bravo' hydrogen bomb test of 1954, these results became concrete and falsely reassuring: 'Children of Japanese survivors of atomic bombs are normal, healthy and happy.'[4]

New methods, therapies and scientific objects came from radiobiology. Historian Alison Kraft's account of bone marrow transplantation provides excellent examples.[5] The acute sensitivity of blood to radiation was well known to pre-war researchers but became an especially urgent topic of research with the atomic bomb. During the Manhattan Project, physician Leon Orris Jakobson had noticed that mice injected with strontium-89 did not develop anaemia, while mice without spleens did. Radioactive strontium settled in the bones (due to the chemical similarity with calcium), so something from the mice spleens was protecting the blood. After 1945, radiobiological work of this kind expanded. Jakobson undertook spleen-shielding experiments in mice, suggesting the presence of a 'recovery factor' that restarted blood formation in the bone marrow after the mice were subjected to otherwise lethal whole-body irradiation. One outcome was new experimental methods for investigating irradiation: bone marrow placed back into animal models had measurable effects. Another outcome was a therapy: humans suffering from leukaemia could be treated, albeit through an immensely painful procedure in which whole-body irradiation destroyed the cancerous blood, and injection of bone marrow restored the capacity of the patient to make new cancer-free blood. The third outcome was the specification of a scientific object. There was intense debate over whether the 'recovery factor' was a hormone or a cell. Orthodoxy (and Jakobson) favoured a hormone. Nevertheless, new techniques, including the cytogenetic technique of chromosome marking developed at the Atomic Energy Research Establishment at Harwell, identified the 'recovery factor' as a cell. Indeed, this cell was a theoretical entity of the late nineteenth century, made manifest for the first time: the 'stem cell' materialized.

Tracer sciences: radiocarbon dating, metabolic pathways and systems ecology

The expanded nuclear programmes, which included reactors to make the enriched uranium and plutonium needed for nuclear weapons, generated, as a by-product, copious supplies of radioisotopes. Since the early twentieth century such radioisotopes had been used as tools in sciences such as chemistry, biochemistry and ecology, as well as in the clinic. In the 1930s, George de Hevesy, for example, had used radioisotopes to investigate the take up of potassium by plants. Now, however, the availability of cheaper materials, as well as the cachet of the atom, led to greater and more diverse uses. Two quick examples – radiocarbon dating and the tracing of metabolic pathways – will illustrate.

The identification in 1940 of a new carbon isotope, carbon-14, by Martin Kamen and Samuel Ruben was discussed in chapter 11 in the context of Lawrence's laboratory. The theory of radiocarbon dating was as follows. Nitrogen in the air absorbed neutrons produced in cosmic ray showers and decayed into carbon-14. The Chicago chemist Willard F. Libby, who had worked on uranium enrichment in the Manhattan Project, proposed that, since the atmospheric carbon-14, in the form of carbon dioxide, would be taken up by living plant tissues at a predictable rate, ceasing on the plant's death, then a measurement of the ratio of carbon-14 to carbon-12 would produce an estimate of age. The half-life of carbon-14, nearly 6,000 years, meant that this technique of radiocarbon dating was ideal for archaeology. While considerable work has been needed to calibrate the method, radiocarbon dating has become a core technique, transforming pre-historical disciplines and challenging historical disciplines.

The notion of a metabolic pathway or cycle is one of the central concepts of twentieth-century biochemical and biomedical science.

In the 1920s, the German scientist Hans Krebs had learned in the biochemical laboratory of Otto Warburg how to use manometers to measure rates of respiration in thin slices of tissue.[6] Krebs had used these techniques in the 1930s to identify the biochemical pathways that led to urea, and, after fleeing Germany for Britain in the face of Nazi anti-Semitic persecution, he had found the citric acid cycle (or 'Krebs cycle') that is at the centre of aerobic respiration, turning fats and sugars into energy and carbon dioxide.

With radioisotopes, the details of metabolic cycles and pathways could be traced by following the radioactivity with instruments such as Geiger counters. Combined with another wartime innovation in method, Archer Martin and Richard Synge's partition chromatography, developed at the Wool Industry Research Association laboratories in Leeds, radioisotope tracers were the crucial tools in the elucidation of the metabolic pathways of photosynthesis. It was known from the first decade of the twentieth century that photosynthesis proceeded in two phases, dark and light, as was the identity of some of the chemicals involved: types of chlorophyll and other pigments. Otto Warburg's team had applied quantum theory, too, calculating that perhaps four quanta of light were needed to reduce a molecule of carbon dioxide in *Chlorella*. In 1941, using radioisotopes from Lawrence's lab, Ruben, Kamen and others had followed oxygen-18 as it was taken up by the plant, in the form of H_2O, before being released again as oxygen. Then, in post-war Berkeley, between 1945 and 1950, a large interdisciplinary team of physicists, chemists, botanists and biologists, under Melvin Calvin, systematically developed and exploited carbon-14 as a tracer to find the many intermediates in the photosynthetic 'Calvin' cycle.

Radioisotopes therefore were tools useful in understanding biological processes at a cellular level. There were also applications on an ecological scale. As the *Lucky Dragon* incident demonstrated, atomic projects, especially atmospheric tests, raised global concerns about damage to human health and the environment. Such concerns will be examined in chapter 17. Here I want to focus on what historian Joel Hagen has described as the ways that the working world of atomic projects sharpened for ecologists 'an exciting new set of tools, techniques, and research opportunities'.[7] His principal case study is of the brothers Eugene and Howard Odum. Eugene was the older brother, a physiologist by training, capable of synthesizing and popularizing broad areas of research. Howard was the physical scientist who saw the world as an engineer might, but also perhaps the more willing to be unorthodox.

The primary achievement of the Odum brothers was to draw on Cold War patronage to establish system studies as the standard tool in ecology. The Office of Naval Research supported Howard Odum's 'monumental' study of the warm mineral Silver Springs in Florida to the tune of $20,000.[8] Silver Springs would become an ecology textbook classic. Furthermore, it provided the methodological tools for the brothers' remarkable atomic reef studies. Ecological study of Bikini had been underway since 1946. The Atomic Energy Commission funded Eugene and Howard Odum's 1954 research into the coral reefs of Eniwetok Atoll as a 'unique opportunity 'for critical assays of the effect of radiations due to fission products on whole populations and entire ecological systems in the field'.[9] The irradiation resulting from atomic tests flooded the reef ecosystem with artificial radioisotope tracers, whose movement could be followed. The reef as a whole was conceived as a 'system'; each component of the system would have its energy inputs and outputs measured.

The result was the drawing up of the 'energy budget' of an entire ecosystem, the reef. Along the way specific natural historical questions were answered: the algae in coral was determined to be a symbiont, not a parasite, and the reef as a whole was shown to be self-sustaining rather than heavily dependent on ocean nutrients.[10] But the project's significance did not end there. Both Silver Springs and Eniwetok Atoll were written up and widely circulated, not least through the Odums' textbook *Fundamentals of Ecology*. Tansley's 'ecosystems', interpreted by limnologist Raymond Lindeman in terms of trophic cycles in the 1940s, and in turn written up by the Odums in terms of energy circulation, finally triumphed over rival conceptions of nature's interconnectedness.[11] Energy budgets provided a standard language for analysing ecosystems. It is no coincidence that energy budgets were also the concern of the sponsor, the Atomic Energy Commission, and that systems thinking was encouraged by Cold War patronage.

Finally, geochemical cycles were also being articulated during the same early post-war period. Interestingly, in both ecosystem and geochemical systems there were links with pre-war Soviet ecology. Vladimir Vernadsky's biosphere ideas, published in Russian in the mid-1920s, influenced the British mid-century ecologist George Evelyn Hutchinson. (Vernadsky's son had moved to Yale, where Hutchinson worked – recall from chapter 13 that this connection had provided a pathway for nuclear news to pass from the United States to the Soviet Union.) Lindeman and the Odums were Hutchinson's students at one time, and historian Pascal Acot confirms the assessment of Hutchinson as 'the missing link' between Vernadsky's work and ecosystem ecology.[12] Geochemical cycles were mapped out in detail from the 1950s to the 1970s by researchers such as Robert Garrels, in whose hands 'movements in the lithosphere-hydrosphere-atmosphere' ended up 'resembling those in a factory, with "pipes" carrying different elements from one repository to another, the whole being driven by the earth's radiogenic heat and solar energy'.[13] This was a picture of the earth as interlocking systems. The movement of materials in some of these earth systems took place over millions of years – far too long for radioisotope tracer methods to be applied directly. However, the Cold War concern for knowing how radioactive releases might move through the earth's land, sea and air encouraged this more general, systems-style understanding.

Quantum electrodynamics

Moving now from attention to the global scale back to the microphysical scale, the continuing nuclear projects provided the justification for training further generations of nuclear researchers. Provided with resources – cyclotrons, linear accelerators, money for buildings, salaries, meetings, and so on – for investigating the physics of high energies, this pool of strategically skilled scientists articulated one of the most successful physical theories ever devised: quantum electrodynamics. They also described a veritable zoo of exotic new fundamental particles, which in turn called for theoretical interpretation and explanation.

By the 1930s, Fermi's theory of beta decay and Yukawa's theory of nuclear forces provided descriptions of fundamental processes, but attempts to describe a relativistic quantum field theory ran into mathematical problems, specifically divergence.[14] During the Second World War, the theoreticians had been mobilized for atomic and radar projects. After 1945, they thrived in the Cold War hot-housing of physics. Princeton physicist John Wheeler, in 1945, in the gap between working on the Manhattan Project and Teller's hydrogen bomb, gave an overview of four fundamental forces – electromagnetic, weak, strong and gravity – and identified 'interesting and exciting areas of research'. Research into cosmic ray showers, confirmed in accelerator experiments, suggested the existence of a new type of middle-sized particle, pi (π) mesons, or pion, to be distinguished from Yukawa's meson, now labelled a mu (μ) meson, or muon. A classification that now made sense split the 'hadrons' (neutron, proton, pion) from the 'leptons' (electron, muon, neutrino).

Advances in theory were prompted by the focused attention given to the fine structure of electron levels in the hydrogen atom. Willis Lamb and Isidor Rabi, making use of the microwave techniques developed during wartime radar research, carefully measured the electron levels when the hydrogen atom was placed in a magnetic field and found a discrepancy between their data and predictions made using Dirac's formulae.[15] Hans Bethe interpreted the 'Lamb shift' – the difference revealed, demonstrated and measured by Lamb between two electron levels – by using a mathematical tool called 'mass normalization'. Bethe's approach was generalized by Julian Schwinger and Richard Feynman, while Freeman Dyson argued that the capacity for a theory to undergo this 'renormalization' was a way to pick out good theories.[16] Dyson presented his arguments in the familiar form of a mathematical theory, while Feynman in addition offered a series of graphical sketches. As these 'Feynman diagrams' became taken up by physicists, adapted to local interests as they were, so a generation slowly began to picture theory in a new way.[17] However, the more immediate result was that, by the early 1950s, there existed a theory that drew together quantum mechanics and special relativity – quantum electrodynamics (QED) – which has been extraordinarily successful in prediction and has served as a model for later theories.[18]

'Symmetry, gauge theories and spontaneous symmetry breaking', writes historian Sam Schweber, were 'the three pegs upon which modern particle physics rests'.[19] Gauge theories were those theories in which a mathematical description of energy would not change under certain transformations. (An analogy might help here: the height of a mountain above the sea floor does not change when the sea level changes: the 'theory' that a mountain's total height can be found by subtracting the depth of base below sea level from the height of the peak above sea level is gauge-invariant with respect to where sea level is drawn – that is to say, what gauge of height is chosen.) QED was a gauge theory under transformations known as symmetry group U(1). Symmetries are kinds of transformations – when you look in the mirror, up stays up and down stays down, but left and right swap: that's what mirror symmetries do. The development of gauge theories encouraged theoretical physicists to look closely at how their equations, and the physical world, stayed the same or changed under different transformations. If it was the same, they spoke of symmetry conservation. If it was different – in other words, if symmetry was not conserved – they spoke of a breaking of symmetry.

Tsung-Dao Lee and Chen Ning Yang were two Chinese-born physicists working in American east coast academia in 1955. The sons of a Shanghai chemical industrialist and a mathematician, respectively, Lee and Yang had similar educations, eventually being supervised by physicist Ta-You Wu during the Japanese war years. Both secured study places in the United States before communist victory in the civil war. Lee and Yang asked whether parity – the symmetry that said the physics would not change if viewed in a mirror – was conserved. Together they demonstrated that, while parity conservation was not to be doubted in strong and electromagnetic interactions, it was unproven in weak. With the assistance of a third Chinese-born physicist at the National Bureau of Standards, Chien-Shiung Wu, Lee and Yang's speculation was shown to be on the mark: parity was not conserved in weak interactions.

A theoretical generalization of the requirement for symmetry by Yang and a colleague at Brookhaven, Robert Mills, predicted massless particles of integer spin (particles with integer spin are known as 'bosons'). However, when the symmetry was broken, the particles that were predicted possessed mass. This line of argument was developed by many theoretical physicists from the late 1950s over several decades.[20] Examples include the proposals by the American Steven Weinberg in 1967 and the Pakistani Abdus Salam in 1968 that electromagnetic and weak forces could be understood in the same theoretical framework: that a 'unified' theory could be given of these two fundamental forces.

Quarks and the particle zoo

Another host of theoretical models came from using symmetries to propose ways of classifying the zoo of particles emerging from cosmic ray and particle accelerator experiments. Two theoretical physicists independently proposed a classification of hadrons in 1961.[21] They were two of the strongest personalities in post-war physics. The first was New Yorker Murray Gell-Mann, prodigy and son of Jewish emigrants. The second was Yuval Ne'eman, an ex-Israeli soldier who as a military attaché had been supervised for a doctorate by Abdus Salam at Imperial College, London, before returning to Israel to help devise the Israeli nuclear bomb. The classification, based on the SU(3) symmetry, what Gell-Mann called, fancifully, the 'eightfold way', demanded constituents, sub-particles of hadrons. Gell-Mann called these theoretical entities 'quarks'. The first three quarks would be labelled up, down, and strange. Quarks came in different colours, a value that cancelled out when put together in observable hadrons. Different combinations gave different particles in different theories.

The proliferation of theoretical models was kept in check by communication with experimentalists, who in turn had to communicate with a third community, the instrument builders.[22] In 1947 George D. Rochester and Clifford C. Butler, physicists in Patrick Blackett's team at Manchester University, had captured and published some unusual cosmic ray tracks in a cloud chamber. More of these 'V particles' (the tracks looked like a 'V'; the particles are now called K-mesons) were found in cloud chamber photographs in California. Many more particles followed. Gell-Mann's and Ne'eman's classification was an attempt to put the particles into order. In the pursuit of higher and higher energies achieved, and of higher numbers of interactions recorded, the research tipped quite quickly away from the natural sources (cosmic rays) to artificial sources (particle accelerators). Furthermore, American facilities such as Brookhaven and at Stanford gained a competitor as Europeans, recovering from war, learned to work together at CERN. A new American rival was added in 1967 when work started outside Chicago on a national accelerator laboratory, funded by the Atomic Energy Commission. From 1974 it was known as 'Fermilab'. The national and international competition drove energies reached by particle accelerators even higher.

In 1968, the Stanford Linear Accelerator (SLAC) ran experiments at 20 GeV, energies high enough to begin to get into protons and neutrons and probe into what constituents might be within; the comparison has been made with experiments in Rutherford's Cavendish of the early twentieth century that delved into the atom to find the hard nucleus.[23] Again evidence accumulated for hard constituents, this time within the proton. Feynman suggested 'partons', but soon equivalence was being sustained between these experimental phenomena (a statistical regularity within SLAC measurements) and Gell-Mann's 'quarks' (a constituent of a mathematical symmetry group). Reversing the process, quarks were now declared real, even though they could not be directly observed.

On 11 November 1974, researchers at SLAC and at Brookhaven simultaneously announced the discovery of a new particle. The SLAC team called it a psi (ψ) particle, and the Brookhaven team called it the J, so now it goes by the unusual joint name of J/ψ. Again the observation prompted selection among the many theoretical models, in this case elaborations of electroweak theories to embrace quarks. The theory that made most sense proposed a fourth quark, carrying 'charm'. The combination was then reversed again to announce the 'discovery' of the charm quark. Tidying up the theory led to the prediction of two further quarks, which were 'discovered' in 1977 (the bottom quark) and 1994 (the top quark), both at Fermilab.

Nevertheless, not all the prizes in particle physics went to the big accelerator laboratories of the United States. In Europe the pre-war leaders in physics had been Germany, the Soviet Union, Britain and France, probably in that order. The Soviet Union was behind the Iron Curtain, Germany devastated, and France recovering. British physicists enjoyed relatively good conditions, prestige, contacts and escalated expectations.[24] Despite massive government debts, a programme of provision of nuclear physics facilities had provided large capital grants to universities to build accelerators: Oliphant's Birmingham, for example, had constructed a 1.3 GeV proton accelerator, Chadwick's Liverpool a 400 MeV proton-synchrotron, and so on at Cambridge, Glasgow, Oxford and Manchester.[25] Further facilities had been provided to the government laboratories at Harwell. British physicists, therefore, had been loath to take the lead in rebuilding European physics, even though the joint facilities in the United States were already making the single-university academic site uncompetitive. Britain had therefore responded to continental European proposals for a joint project in 1949–50 with tepid 'cooperation without commitment'.[26]

In 1951, a split had been readily apparent between French, Italian and Belgian physicists, who wanted a big new accelerator, and others, including senior British physicists, who preferred a traditional organization: leadership of Niels Bohr from a new institute at Copenhagen, perhaps with the Liverpool cyclotron as a central facility. By mid-1952 British involvement had looked in doubt, as the Treasury opposed open-ended commitment to a European project. By the time lobbying by younger physicists, who disagreed with their senior's preference for Copenhagen, had changed minds and Britain joined the European Organization for Nuclear Research (CERN), it was too late to steer policy. In a nutshell, then, British scientific relations, in its hesitancy and insistence on opt-out clauses, had anticipated British political relations in later European projects.

A series of impressively large accelerators were built at CERN's central laboratory, eventually straddling the French–Swiss border near Lausanne. A 600 MeV synchrocyclotron built in 1957 was comfortably more powerful than the Liverpool machine. A 28 GeV proton synchrotron followed two years later. The scale of subsequent colliders was astonishing: particles travelled around a 7 kilometre tunnel in the 300 GeV Super Proton Synchrotron commissioned in 1976, while electrons and positrons ran in opposite directions in a 27-kilo-metre ring before smashing in the 100 GeV Large Electron-Positron Collider (1989). The 14 TeV Large Hadron Collider, which reaches even higher energies by accelerating and smashing the larger protons, came into service in 2008; although it suffered a severe breakdown, it was up and running again in 2009. These were immensely expensive investments, made possible by the post-war economic boom but ultimately justified by, first, the strategic importance of fundamental physics in putative future technologies and, second, a commitment to build a Europe as a strong cooperative supranational entity that would break from the war-torn past.

Back to the development of particle physics theory: the success of quantum electrodynamics suggested that QED should be the model to follow in any successful account of strong nuclear forces, now understood as forces between quarks. Quarks were understood to be fermions, particles with half-integer spins. As such they obeyed the Pauli Exclusion Principle: just as two electrons (also fermions) could not occupy the same place, so two or more quarks of the same kind could not sit together. Yet the post-war accelerators were turning out particles (such as the $\Omega-$, which made sense as three strange quarks, or the $\Delta++$, three up quarks) which were precisely that. The invention of a new quantum value – colour – meant that the Pauli Exclusion Principle could be evaded by these particles. Quantum chromodynamics (QCD) was built as a theory of the nuclear force between quarks transmitted by gluons, with several features elucidated by theoreticians. One odd feature is confinement: single quarks would not be seen. David Gross, Frank Wilczek and David Politzer published in 1973 on another feature, asymptotic freedom. Put together, the strong nuclear force performed like almost unstretchable string between quarks: so loose at small scales as almost to disappear, so tight at extremity as to bind quarks together. Evidence for QCD was sifted from collision data from CERN, SLAC and the Deutsches Elektronen-Synchrotron (DESY) at Hamburg, Germany.

Accelerators are complex, ever-shifting assemblages of equipment and people. Getting the best out of a large accelerator is a matter of teamwork, management and incremental, occasionally dramatic, improvement. Carlo Rubbia, for example, persuaded others that CERN's Super Proton Synchrotron could be converted into a proton–antiproton collider. Other innovations involved new techniques that made the operation more efficient. We will examine Georges Charpak's revolution in detection in the context of the impact of computerization later. But Simon van der Meer's invention of stochastic cooling, when applied to Rubbia's refashioned collider, contributed to CERN's greatest triumph: the detection of two massive particles, the charged W and the neutral Z in 1983.

Standard models

The electroweak theory combined with QCD made up what physicists began to call the 'standard model', an account of three fundamental forces (weak, strong, electromagnetism) but not the fourth (gravity), for which Einstein's general relativity, a different sort of theory, provided the best account. The standard model made predictions that were testable. In particular, the Higgs boson, a very heavy fundamental particle, was the principal target of, and scientific justification for, the planned Large Hadron Collider. 'The standard model is one of the great achievements of the human intellect', writes historian Sam Schweber. 'It will be remembered – together with general relativity, quantum mechanics, and the unravelling of the genetic code – as one of the outstanding intellectual achievements of the twentieth century.'[27] But the standard model is also, Schweber reminds us, not complete: there is no definitive explanation of why quarks have particular masses or why physical constants take the values they do. Going beyond the standard model requires investigation of energies out of reach of conceivable generations of earth-bound accelerators. These energies, however, were exceeded in phenomena in distant astronomical objects and in the earliest moments of the universe.

Benefiting from a desire among some academic physicists to apply their war-developed tools of analysis to distant – in both a material and a cultural sense – subjects, cosmology became a respectable, mainline, boisterous specialty after 1945. Physicists such as George Gamow, Ralph Alpher, Hermann Bondi and Thomas Gold and astronomers such as Fred Hoyle inherited cosmological models derived from general relativity and the Hubble's observations of receding galaxies. Born in Odessa, George Gamow had worked at the great centres of early twentieth-century physics – Göttingen, the Cavendish, Copenhagen, Lev Landau's institute – before defecting with his physicist wife to the West in 1933 at a Solvay conference. Gamow was refused security clearance to work on the Manhattan Project, probably because of his Soviet background, but was consulted for the United States Navy on high explosives instead.[28] Gamow's project, begun in the late 1930s, was to combine nuclear physics with Alexander Friedman's relativistic models.[29] After 1945 Gamow and colleagues, especially Ralph Alpher, described a hot big bang as a nuclear oven cooking the light elements – hydrogen, helium – that the first stars would burn. Gamow would briefly be drawn into the hydrogen bomb project.

The big bang theory had problems, however: Hubble's data pointed towards a younger universe than seemed possible (younger, for example, than the geologists reckoned the earth to be), and no mechanism seemed to account for the formation of the heavier elements. The big bang soon had a rival theory, hatched in Britain by Bondi, Gold and Hoyle: the steady state universe. Bondi was an Austrian who had been interned as an enemy alien on the Isle of Man during the Second World War. He therefore did not follow the paths taken by similarly talented young physicists, into nuclear projects or radar or their spin-offs. Tommy Gold, a fellow Austrian, had met Bondi in the British internment camps. Together Bondi and Gold published a paper in 1948 that argued for a universe that would look the same not only from any point in space but also from any point in time. The result was a universe with no beginning and no end, yet exhibiting continual expansion (as Hubble's observations demanded). Two months later the atheist Yorkshireman Fred Hoyle sketched a similar picture, with the added phenomenon of continuous creation, the necessary fine-scale appearance of matter to fill the spaces left by expansion.[30] Continuous creation, for Hoyle, was preferable to Abbé Lemaître's Christian science of a single, creative origin of all matter. In the Soviet Union even continuous creation was 'unscientific, as well as ideologically illegitimate', and both cosmologies were rejected there.[31]

It is a feature of cosmologies, broadly understood, that a culture's most prestigious values are naturalized by being written into cosmological origin stories. It is perhaps no surprise, therefore, that the cosmologies devised in the middle of the twentieth century had at their heart the scientific values of the working world of nuclear projects. Choosing between the scientific cosmologies on offer in the 1950s depended upon observations made from a science that grew directly from the working world of the other major wartime project, radar: radio astronomy.

— 16 —

COLD WAR SCIENCES (2): SCIENCES FROM INFORMATION SYSTEMS

Radar, recall, was built into an extensive 'information system'. Transmitters sent out pulsed radio beams and receivers picked up echoes, but that was merely the beginning of the process. The data collected were repeatedly sifted, as useful 'information' was separated and abstracted as it passed from receiver stations, through filter rooms, to operations rooms, where military decisions could be made after surveying the massively reduced representation of the world. Numerous skills were required. At Britain's Telecommunications Research Establishment, physicists and engineers might devise and design radio transmitting and receiving equipment and figure out how to reduce and represent data. Others, working closely with military officers, would work out how to fit equipment with existing military hardware: aircraft, ships, coastal reporting stations. As we have seen, this working world had already produced its own science, operational research. In this chapter I review other radar and information and global sciences.

Radio astronomy

Another new science was substantially started by extending the radar surveillance of the sky further into astronomical spaces. Before the Second World War, isolated small-scale radio astronomy projects had begun in the United States. Karl Jansky, an electrical engineer investigating sources of interference for Bell Laboratories, found cosmic sources of radio noise in the early 1930s, including one identified with the centre of our galaxy in Sagittarius.[1] An amateur, Grote Reber, with his hand-built dish telescope, had confirmed Jansky's source and identified others, among them ones in Cassiopeia and Cygnus as well as the sun. However, Jansky and Reber's pioneering work was matched and rapidly overtaken by the radio astronomy invented by scientists returning to academic study after wartime radar work. Post-war, radio astronomy became an autonomous specialty, differentiated from experimental physics, electrical engineering and optical astronomy, and pursued in highly organized Big Science projects in Britain, Australia, the Netherlands and the United States.

In the Netherlands, Jan Hendrik Oort began his astronomical investigations in 1945 with the excellent German war-surplus radar dishes. In Australia, a government establishment, the Radiophysics Laboratory of the Commonwealth Scientific and Industrial Research Organization (CSIRO), under ex-TRE and ex-MIT Rad Lab physicist Edward G. Bowen and Joseph L. Pawsey, developed radar astronomy into radio astronomy. In Britain, physicists leaving the Telecommunications Research Establishment brought with them equipment, contacts, and experience of scanning the skies with radar. At Cambridge University, an extramural site of the Cavendish Laboratory was set up by ex-TRE physicist Martin Ryle in the countryside. At Manchester University, ex-TRE physicist Bernard Lovell set up an outpost of Patrick Blackett's physics department in a former university botanical station south of the city, at Jodrell Bank. Both tapped contacts for the loan (and gifts) of war-surplus radio equipment. They set about investigating the phenomena noted during wartime when echoes from aircraft had had to be distinguished from sources of radio 'noise'. At Cambridge the early focus was on solar noise. At Manchester it was on radar studies of meteors – or, rather, of the trail of ionized gases left in the wake of a meteor that reflected radio waves for several minutes. (During the Second World War the Army Operational Research Group, based at TRE, had tried to distinguish reflections from meteor trails from incoming V-2 rockets.) The research soon uncovered previously unknown daytime meteor showers; radio astronomy was already contributing to the stock of knowledge held by optical astronomy.

Competition between the Sydney, Cambridge and Manchester groups encouraged each to pursue different technical strategies to explore the unknown radio universe, a process that David Edge and Michael Mulkay argue reveals the social character of scientific specialty formation and growth.[2] At Cambridge, Martin Ryle's team focused on building interferometers, instruments that matched the resolution of large radio telescopes by combining the measurements taken by several small antennae. Ryle's group built a series of

368

interferometer telescopes in the 1950s and 1960s. Interferometers were especially effective at mapping the radio sources and thereby revealing information about their distribution. At first these sources were described as 'radio stars', not unreasonably, since it was well known that the sun was a radio source. However, only some of the radio stars congregated along the galactic plane, suggesting that many were extra-galactic objects. The second and third Cambridge catalogues of radio sources, '2C' and '3C', were published in 1955 and 1959 respectively. Ryle used data from the surveys, in particular the relationship between number and brightness of sources, to argue that they did not fit the steady-state model of the universe. A fierce controversy followed, with the steady-staters both criticizing the data and arguing that their model could be tweaked to fit.

At the Jodrell Bank Research Station, Lovell's team concentrated on research that could be done effectively with large single 'multipurpose' dishes.[3] The first one they built in the late 1940s was a bowl of wire, 218 feet in diameter, which sat on the ground reflecting radio waves to a receiver mounted on a pole. It had been designed to search for cosmic ray echoes – a research interest of Blackett. By 'happy accident', recalled Robert Hanbury-Brown, 'they had built a powerful instrument with which . . . to study the radio emissions from space'. This transit instrument was a proof of the concept of large dish radio telescopes.

Lovell's team then imagined a fully steerable version, possessing a dish of great collecting power and capable of targeting any point in the sky. With a 250-foot steel mesh bowl and a mount powered by electric motors, this was a huge – and expensive – project. The government was asked for £154,800 for the 'construction of a large radio telescope for investigating galactic and solar radio emission, meteoric phenomena, aurorae, lunar and planetary echoes'.[4] Like other Big Science projects, the Jodrell Bank Radio Telescope won sufficient backing because it provided solutions for other people's problems as well as a fine scientific instrument. British optical astronomers, specifically the elite professional Royal Astronomical Society, were sold on the idea of the instrument sparking a resurgence of British astronomy unrestricted by inclement weather. The government decision-makers applauded a 'Great Public Spectacle', a visible flagship, a carrier of 'national prestige', a prestigious scientific project suitable for a nation otherwise struggling with post-war winds of change.

In 1954, with the radio telescope project underway, the radio astronomers requested design changes. There were several reasons. Astronomers at Harvard University had discovered that cosmic emissions on the 21-centimetre band from neutral hydrogen could be detected. Dutch astronomers, including Oort, swiftly demonstrated that, by mapping hydrogen, the most common element in the universe, extraordinary details of galactic structure could be revealed. Oort's maps of the mid-1950s showed the separate arms of the Milky Way for the first time. (The ability to investigate a wavelength as small as 21 centimetres was a direct result of wartime innovation: the techniques of generating and measuring near-microwave wavelengths such as 21 centimetres were precisely those developed for radar systems such as Lovell's Home-Sweet-Home.) For the Radio Telescope to operate at smaller wavelengths, a denser, therefore heavier, mesh was required. The design changes therefore meant more money was needed.

As costs spiralled, other, Cold War, justifications for the Radio Telescope were proposed.[5] Lovell lobbied the armed service departments with arguments that the telescope could be used as an early warning system, or that radio stars could be used by missiles in a form of astro-navigation. In October 1957, only two months after the telescope was completed, in debt, the instrument was deployed in a highly publicized tracking of the Soviet satellite Sputnik. Not only were there very few instruments capable of tracking the satellite and its ICBM-like rocket, also in orbit, but the use of a British instrument to locate a Soviet satellite, while the United States panicked, also sent a politically useful message. When the debt was finally cleared it was by philanthropist and rabid anti-communist Lord Nuffield, who had justified his earlier financial support for the telecope with the hope that it 'might one day be directed if need be on the steppes of Russia'.[6]

Radio astronomers, therefore, were able to exploit Cold War concerns to secure scientific facilities on a scale that would not have been possible otherwise. A project for the United States Navy that would have dwarfed even the Jodrell Bank telescope provides another example. The telescope designed for Sugar Grove, West Virginia, had a planned span of 600 feet and would have weighed 36,000 tons. Congress, spooked by Sputnik, awarded $79 million, and construction started. It was cancelled in 1962. However, the spiralling costs were only one cause. Another was that Sugar Grove, while publicly a radio telescope, was secretly also designed to be an eavesdropping instrument, recording radio transmissions from Russia as they echoed back from the moon. The success of spy satellites removed this secret justification for its existence.

Combinations of 'black' (secret) and 'white' (public) science were typical of Cold War projects. Historian David van Keuren, who uses this colour terminology, has described another fascinating example of this combination.[7] The GRAB – Galactic RAdiation Background – satellite was launched on 22 June 1960. GRAB was part the first astronomical satellite and part the first intelligence satellite, a forerunner of the famous Corona series. Van Keuren emphasizes that the need for cover gave scientists far more leverage than might have been expected (say, from a reading of Forman's 'mutual orientation' of interests). He quotes the GRAB astronomer Herbert Friedman: 'We weren't destitute for opportunities. But the fact that the intelligence people were happy to have a cover for what they were doing made it opportune for us to move in there.'

The astronomers were pleased with the GRAB results: a demonstration that solar flares produced X-rays, which ionized the upper atmosphere, blocking out radar transmissions, was both useful and basic research. The 'intelligence people' regarded GRAB as a 'revolution', providing information on the number and character of Soviet radars (including the revelation that the Soviets had an anti-ballistic missile radar) – information that the Strategic Air Command put to immediate use, plotting attack flight paths into the Soviet Union. Both sides were happy. The scientists in particular gained resources and data they would not otherwise have had. The whole story was declassified only in 1998, after the Cold War. In retrospect the name 'GRAB' should perhaps have raised suspicions!

Like light, radio waves pass through the earth's atmosphere. Optical and radio astronomy could therefore be land-based projects. The atmosphere is not transparent to other parts of the electromagnetic spectrum. The development of ultraviolet, X-ray and gamma ray astronomy therefore depended on space technologies. Again there was often a link to national security. For example, the first X-ray astronomical satellite, Uhuru, launched in 1970, was designed by Riccardo Giacconi's team at American Science and Engineering, Boston. Giacconi and his team had previously worked on using X-ray instrumentation in satellites to detect X-ray flashes from atmospheric nuclear tests in the early 1960s.[8] Indeed, the first cosmic X-ray sources were found in this search for clandestine nuclear tests.

Radio astronomy therefore grew out of the working world of radar and was sustained as a form of Big Science by the resources made available during the Cold War. The scientific return has been extraordinary. Bell Telephone Laboratories, still concerned with radio interference with their systems and also developing an interest in satellite communication, built a large unusual horn-shaped radio antenna for use in the Echo and Telstar satellite programmes. In 1964, two Bell employees, Arno Penzias and Robert W. Wilson, using this acutely sensitive radio antenna, measured a faint but pervasive residual radio hiss that they could not locate or eradicate. Word reached astrophysicists at Princeton, among them Robert Dicke, who in the 1940s had predicted that remnants of the big bang would be detectable as a pervasive microwave hiss. Observation and interpretation were connected in what is now recognized as the detection of the 3 K cosmic microwave background radiation, core evidence for the big bang and further empirical difficulty for the steady state theory.

Radio astronomy also produced novel astronomical objects. In the 1950s, it was known that many of the radio sources listed in the Cambridge surveys were small in diameter. (The Jodrell Bank Radio Telescope contributed to this finding.) Some radio sources were matched to optical objects. In 1963, using measurements made with the large Parkes telescope in Australia during an occultation by the moon of a radio source, Dutch-American astronomer Maarten Schmidt at Mount Palomar identified source 3C273 with an odd bluish star. Schmidt's investigations of 3C273's spectrum revealed an extraordinarily high red shift, suggesting that the source lay at a great distance. This 3C273 was a member of a new class of astronomical object: very distant and, to be detectable at all, intrinsically very bright. Indeed, these 'quasars', quasi-stellar objects, astronomers concluded, must be among the most violently energetic objects in the universe.

In July 1967, the young Cambridge radio astronomer Jocelyn Bell had noticed some 'scruff' in a pen-recording of a radio source in the small constellation of Vulpecula. Cleverly thinking to check the fine structure of this 'scruff', she found a very fast regular repeating pulse of radio waves. Bell's team leader, Anthony Hewish, led the investigation, determining that the source was of cosmic origin (but ruling out 'little green men'). After many such checks had been conducted in private, the discovery of 'pulsars' was announced in late 1967. The speed of the pulses gave clues about the size of the transmitting object – the faster, the smaller – and these pulsars were very fast. Thomas Gold, in 1968, argued that pulsars were spinning neutron stars, super-compressed hitherto hypothetical bodies in which the atomic structure had been crushed to a dense sea of neutrons. By the late 1960s, therefore, radio astronomy, barely two decades after it grew from wartime radar, had provided scientists with a universe populated by novel entities as well as strong evidence of a Big Bang.

Cybernetics

Another new science, cybernetics, was a result of direct reflection on the working world of anti-aircraft radar research. Bombers approaching at high speed gave human anti-aircraft gunners very little time to register the position of a target and direct fire. This acute problem was certainly an obvious focus for technical innovation, some of which was led by Warren Weaver for the NDRC;[9] it did not seem at first glance the likely foundation of a new science. Norbert Wiener, ex-child prodigy, PhD in mathematics from Harvard at the age of eighteen, was an accomplished middle-aged MIT mathematician when he was given the problem of radically improving anti-aircraft technology in 1940.[10] Assisted by electrical engineer Julian Bigelow, Wiener considered the problem of feeding back radar echo data on the position of an approaching aircraft as a way of automating anti-aircraft fire. He called his device an 'anti-aircraft predictor'. The task was not as simple as merely extrapolating the path of the aircraft, as one would of the smooth path of a cannon ball (Wiener knew his ballistics anyway). Rather, the pilot's actions, conscious and unconscious, continually shook the flight path. Wiener and Bigelow reported their breakthrough insight:

> We realized that the 'randomness' or irregularity of an airplane's path is introduced by the pilot; that in attempting to force his dynamic craft to execute a useful manoeuver, such as straight-line flight or 180 degree turn, the pilot behaves like a servo-mechanism.[11]

The pilot's consciousness could be represented in the action of the anti-aircraft machinery, and the whole could be generalized by focusing on negative feedback, corrective, loops. The anti-aircraft predictor's capture of the pilot's psyche would interest, reported Wiener already with eye to the future, the 'physiologist, the neuropathologist, and the expert in aptitude cases'. 'More to the point', Peter Galison reminds us of the direct task at hand, 'it suggested that a more refined AA predictor would use a pilot's own characteristic flight patterns to calculate his particular future moves and to kill him.'[12]

'It does not seem even remotely possible to eliminate the human element as far as it shows itself in enemy behavior', recalled Wiener. 'Therefore, in order to obtain as complete a mathematical treatment as possible of the over-all control problem, it is necessary to assimilate the different parts of the system to a single basis, either human or mechanical.'[13] Wiener's reflections on his anti-aircraft device thus led him to a foundational project: to dissolve disciplinary boundaries in the natural, human and physical sciences and recast a new science, 'cybernetics', in terms of feedback. With the Mexican neurophysiologist Arturo Rosenblueth, Wiener and Bigelow let the civilian world gain some sense of this scheme in a 1943 paper, 'Behavior, purpose, and teleology'. Wiener also began working with two computing pioneers, Howard Aiken and John von Neumann. The group established a Teleological Society. 'We were all convinced', recalled one member, 'that the subject embracing both the engineering and neurology aspects is essentially one.'[14] The Josiah Macy Jr. Foundation stumped up funds for a series of cybernetics conferences. In 1948 Wiener published *Cybernetics* (the neologism of the title was based on the Greek for 'steersman'). Its subtitle named the project: 'control and communication in the animal and the machine'.

Core exemplars, central texts and foundational meetings form the heart of a new discipline. Cybernetics was a project for a new 'universal discipline', one which subsumed and reinterpreted others – psychology and engineering, among many others.[15] Cybernetics also blurred the boundaries between living and non-living systems. 'We believe that men and other animals are like machines from the scientific standpoint because we believe that the only fruitful methods for the study of human and animal behavior are the methods applicable to the behavior of mechanical objects as well', Rosenblueth and Wiener had argued in 1950. 'Thus our main reason for selecting the term in question was to emphasise that as objects of scientific enquiry, humans do not differ from machines.'[16] Such remarkable ambition was significantly encouraged and sustained in a Cold War working world where the integration of humans and machine in military systems was a central problem.

Cybernetics was taken up in different ways in different societies. In Britain, W. Grey Walter, a neurophysiologist who had worked on radar during the war, built robot tortoises that could seek out 'food', interact with each other, and even provide a model of 'affection'.[17] Walter characterized the brain as an electrical scanning device, not unlike radar. Soviet scientists attracted by cybernetics faced the problem that their chosen subject was denounced by ideologues. 'Cybernetics: a reactionary pseudo-science arising in the USA after the Second World War and receiving wide dissemination in other capitalist countries', ran the definition in one mid-1950s Soviet dictionary: 'Cybernetics clearly reflects one of the basic features of the bourgeois world-view – its inhumanity, striving to transform workers into an extension of the machine, into a tool of production, and instrument of war.'[18] After the death of Stalin in 1953, Soviet cyberneticians were able to withstand ideological attack, partly by learning to present cybernetics in appropriate rhetorical form, but mostly because they could claim that cybernetics in general, and the computer specifically, were essential to Soviet Cold War defence projects.[19]

Computers and the Cold War

Historian Paul Edwards argues that the computer was the central tool and metaphor of the Cold War. His argument is best seen in the way the computer was placed at the centre of the post-war American early warning system. The story of how it happens also illustrates some of the contingencies of Cold War projects. In 1944, Jay Forrester, an electrical engineer at the Servomechanisms Laboratory of MIT, was developing a digital version of an analogue flight simulator. This became Project Whirlwind, a vision of digital computing as a component of centralized control systems. When rising costs caused the first sponsor, the Office of Naval Research, to get cold feet, Forrester hawked the project around other sponsors. By turns it became a solution for problems of logistical planning, air traffic control, life insurance calculations and missile testing, before Forrester reached the air force. The air force, pondering a proposal from another MIT professor, George Valley, saw Project Whirlwind as the centre of an air defence system.

Forrester's Whirlwind became the electronic computer at the centre of SAGE, the Semi-Automatic Ground Environment. SAGE was a computer-controlled air defence system of national coverage, in which incoming radar information would be passed to a combat centre, which in turn would control direction centres. 'The work of producing SAGE', notes Edwards, 'was simultaneously technical, strategic, and political.'[20] Technically, SAGE demanded innovations in online representation of electronic data and real-time computing. (Existing stored-program computers did not operate in real time; rather, computing tasks were completed in batches.) SAGE required new techniques of data transmission and reception (such as modems) as well as video and graphical displays. Magnetic core memories, a considerable advance over older Williams tubes and delay lines, were developed for the Whirlwind. The main beneficiary in business terms was IBM, which received the substantial contract for providing the machines and turned its access to government-funded cutting-edge military technology into post-war domination of the computer market.

Strategically, acceptance of SAGE demanded a transformation in the organization and values of the armed services. SAGE pointed towards automated, centralized, computerized control of warfare. SAGE went against traditional military values of human command and the delegation of responsibilities for interpreting how an order be fulfilled.[21] The SAGE project was also an important step in confirming the mutual orientation, in Forman's terminology, of the military patron and the academic laboratory: the air force changed in following Forrester's vision of computerized control, while, in this case, MIT changed in becoming, functionally, a research and development arm of the United States Air Force. Politically, SAGE illustrates Edwards's broader argument: the computer not only served as a tool at the centre of Cold War systems but also provided a metaphor of control in a contained, 'closed world'.[22]

Military agencies were the primary, generous patrons of computers. It was the Ministry of Supply, responsible for British atomic projects, that stumped up most of the cash for the first electronic-stored program computer built in 1948 at Manchester University by ex-radar scientists. The first Soviet stored-program computers, starting with the MESM built by Sergei Lebedev's team in 1951, were needed to meet military demands from ballistic missile, atomic weapon and missile defence projects.[23] However, it was the combination of military demand and industrial innovation in the United States that produced the world-changing developments in computing in the second half of the twentieth century. In general, sustained by the continuing Cold War and boosted further by burgeoning commercial applications, computers became smaller and cheaper to make and operate.

In the process, science has changed. In parallel with Edwards's argument, the changes can be said to have a dual nature: computers have become revolutionary tools for scientific research but also a guiding metaphor for new disciplinary programmes. Examples of computer-as-tool can be found in radio astronomy, X-ray crystallography, meteorology,[24] high-energy physics and, indeed, well before the end of the century, in every science. Examples of sciences that took the computer as a metaphor guiding research programmes were artificial intelligence, cognitive psychology, immunology and, in a subterranean but profound way, the reinterpretation of genetics as a science of codes.

Computers as tools of revolution in science

Particle accelerator laboratories and astronomical observatories were both places where complex expensive equipment was set up to generate masses of data that could be simplified and rearranged to produce representations of the natural world. These representations, stand-ins for nature, were the subjects of scientific work. A large optical telescope, such as the 200-inch instrument at Palomar, produced photographs which could be measured, compared and combined. By the 1980s, chemical photography was being replaced with the digital read-outs of charge-coupled devices (CCDs). By then the computer was already the tool guiding the control of large telescopes' move-ments as well as the means by which data were handled.

Optical astronomy was following the lead taken by radio astron-omy. Computers, such as the Ferranti Argus used with the Jodrell Bank Mark II telescope, were deployed to control the movement of telescopes from the early 1960s. But the greatest leap in performance came with the use of computers to process the immense amount of information generated. The techniques of aperture synthesis, devel-oped primarily by Martin Ryle's team at Cambridge, provide the most startling example. Data from small, widely spaced telescopes could be combined in a way that mimicked the capabilities of very large telescopes. Indeed, by waiting for the earth's rotation to sweep the telescopes through space while observing the same patch of sky, the capability of a telescope many thousands of miles in equivalent aperture could be matched, at least in terms of resolution. But the reduction of these data by hand, or even by mechanical calculator, would have been far too costly in terms of hours of human labour. The future of radio astronomy depended on computers.

Computers had been deployed even earlier in nuclear weapons science, X-ray crystallography and high-energy physics. Developing the atomic bomb had raised critical questions, such as how much energy would be released effectively. Direct experimental investiga-tion was ruled out: no measuring instrument could withstand the atomic blast long enough to report. Nor did theory provide a way forward: the equations generated were intractable. For the thermonu-clear bomb, John von Neumann and Stanislaw Ulam offered instead an artificial version: simulation by numerical methods, named 'Monte Carlo' by their collaborator, Nicholas Metropolis.[25] They tried the simulation out on the ENIAC, a fast and flexible wartime calculator, in the late 1940s, but the technique was only fully exploited on the early stored-program computers, such as the MANIAC built at Los Alamos.

In X-ray crystallography, the process of reducing data involved, first, making careful measurements of the position and dimensions of the black spots found on photographs of X-ray diffraction patterns and, second, a fourier transformation, a mathematical procedure which entailed calculating and summing large numbers of trigonometrical series.[26] This calculation, described by A. Lindo Patterson in the 1930s, was laborious even for simple crystalline molecules. As larger molecules drew interest, so the computing task became even more time-consuming. Two physicists working in Lawrence Bragg's X-ray crystallography laboratory in interwar Manchester developed a bespoke tool to assist the work: Beevers–Lipson strips, cards holding the relevant values of sines and cosines. Placing the appropriate cards together, the sum could be more easily calculated. X-ray crystallographers also knew that a heavy element within a molecule could be used as a peg around which structure could be more easily determined. X-ray crystallographers worked iteratively: taking a diffraction picture, deriving partial indications of the structure, drawing conclusions, trying a new angle, and repeating the process.

As larger and larger molecules were tackled, the Beevers–Lipson method was translated first to punched cards and then to electronic stored-program computers. Linus Pauling at Caltech and Bragg's teams had used punched cards. Oxford X-ray crystallographer Dorothy Hodgkin took up punched-card methods after her derivation of penicillin structure.[27] But for her new project, the three-dimensional structure of vitamin B_{12}, she turned in the early 1950s to electronic computers, both analogue instruments, such as Ray Pepinsky's X-RAC, which had been built with Office of Naval Research and private philanthropic funding, and digital stored-program computers, such as the Ferranti mark 1 at Manchester and the SWAC, built at the University of California, Los Angeles, for the National Bureau of Standards. Likewise, Max Perutz and John Kendrew at Cambridge would use the Cambridge EDSAC computer (as well as time begged on industrial and military computers) to derive structures of haemoglobin and myoglobin, respectively.[28] These were massive molecules of great interest to life scientists. Reduction of the data to resolve structure at the order of a tenth of a nanometer required many hundreds of hours of computer time – far longer than would have been feasible by hand. Perutz, Kendrew and Hodgkin would all receive Nobel prizes to mark the impact of the computer as a tool for molecular biology.

In high-energy physics, there were differing views about the proper use of computers. Galison argues that two distinct traditions can be found in microphysics.[29] In the 'image' tradition, what was sought were picturing technologies that generated mimetic representations, ideally 'golden events': the capture, for example, of an interesting particle decay on a single cloud chamber photograph. In the second, 'logic' tradition, the preference was for counting techniques: the clicks of a Geiger–Müller counter accumulating as data that could be statistically analysed. As the scale of physics increased after the 1940s, both traditions embraced the computer as a means of reorganizing work, but in different ways. Within the image tradition, a physicist such as Luis Alvarez at Berkeley might insist that human judgement (at least of the professional physicist) must remain, while at CERN, scientists in the logic tradition, such as Leo Kowarski, might aim for thoroughgoing automation. At stake was the role of the individual physicist in Big Science projects: a mere machine-minder or a creative partner. Nevertheless, from the mid-1960s, argues Galison, the traditions became merged; in instruments such as the SLAC-SBL Solenoidal Magnetic Detector, computerization was central, mediating between other parts of the complex system.[30] At CERN, Georges Charpak's multiwire proportional chamber, developed in 1968, made the detection of particles an electronic event, replacing photography, inviting further computerized control and analysis, and transforming the experimenters' art.

The computer, by creating a simplified accessible manipulable representation of nuclear reactions, had been an essential tool of the nuclear weapons scientist. The computer, by speeding up calculation, made aperture synthesis and three-dimensional modelling of large proteins feasible. The computer, as control of a particle detector, could count many millions of events. The computer was, in this sense, a revolutionary tool for astronomy, molecular biology and particle physics; a similar case can be made for its causing a qualitative shift, a 'phase change', in other data-rich subjects.[31] Indeed, in Douglas Robertson's view, the computer became, in the second half of the twentieth century, the most important tool in science: 'more important to astronomy than the telescope', 'more important to biology than the microscope', 'more important to high-energy physics than the particle accelerator', 'more important to mathematics than Newton's invention of the calculus', and 'more important to geophysics than the seismograph'.[32]

There is something right and something wrong with this view. Robertson is wrong in the sense that astronomy, say, has always been, largely, the science of inscriptions – whether photographs, pen-recordings or outputs from CCDs – made by telescopes. He is also wrong to argue that there was an unambiguous qualitative shift: computerization nearly always automated previous methods of mechanical and human labour processes of calculation; the transition was often relatively smooth.[33] Robertson, however, is right in his argument that what instruments are capable of doing, and what reduction and analysis is capable of achieving, is now defined by computing power. This has made the sciences of the late twentieth and early twenty-first centuries strikingly more similar to each other – sciences of a digital working world – than perhaps they were in earlier incarnations.

The computer, argues Robertson, vastly extended the scientists' ability to see. Certainly new ways of visualization in the sciences have been dramatically transformed. At MIT from the mid-1960s, for example, computer scientists developed ways to display molecules in three dimensions, allowing chemists to interact with the representations.[34] These methods at first complemented and later supplanted other ways of molecular model building, such as using sticks and balls, plasticine, cardboard or paper cut-outs. Medical imaging technologies provide even more startling examples. From the mid-1960s, Godfrey Hounsfield, an ex-radar engineer working for Electrical and Musical Industries (EMI), developed a method of focusing X-ray beams and recording the data digitally.[35] By combining the data from many thin slices and using a computer to piece the data together again, Hounsfield and his colleague A. J. Ambrose experimented with producing images of the inside of brains. An early experiment, with a cow's head from a local butcher, failed to produce a good image; but they were delighted when a second cow's head, this time from a kosher butcher, who had bled the animal to death rather than bludgeoning it, was revealed in intricate glory. The EMI company, flush with money from the success of the Beatles, invested in this technology of computerized tomography (CT) scanning, which found direct medical uses, locating a brain tumour on its first scan of a patient in October 1971. Hounsfield and an independent scientist, Albert Cormack, received a Nobel Prize in 1979.

Along a similar timescale, CT has been joined by other medical imaging techniques. Nuclear magnetic resonance imaging (MRI), which had the advantage of not requiring X-rays, instead had its roots in the precision measurements of magnetic moments of protons and neutrons in the early Cold War by physicists such as Isidor Rabi and Felix Bloch. Nuclear magnetic resonance then became a tool for inorganic chemists in the 1950s and organic chemists in the 1960s; only in the 1970s did it become a method of medical imaging, developed in competition between British and American groups.[36] Like CT, MRI is expensive, reliant on computer power (even the first CT scan required 28,000 readings), and indicative of a scaling up of hospital medicine in the West.

The 'computer began as a "tool", an object for the manipulation of machines, objects and equations', summarizes Peter Galison. 'But bit by bit (byte by byte), computer designers deconstructed the notion of a tool itself as the computer came to stand not for a tool, but for nature.'[37] The Monte Carlo simulations had been one, precociously early, indication of how computer models could 'stand in' for nature. In other sciences, the computer became a dominant organizing metaphor. Brain and mind sciences provide examples: while CT and MRI might deploy the computer as a tool for processing images, in artificial intelligence and cognitive psychology the computer stood for nature.

The computer as mind

There have been two, often competing, approaches to modelling human brains and minds since the mid-twentieth century. In the first, bottom-up approach, the structure of the brain was directly modelled, in much reduced form, in hardware. Warren McCulloch and the young unstable prodigy Walter Pitts had written a paper, published in 1943, which treated the nerve cells as switches, and argued that phenomena of the mind (for example, memory) could be found in feedback loops within networks of neurons and that networks could operate like logical circuits. McCulloch and Pitts established that their networks of neuron-switches were fully equivalent to Turing's universal machines. Von Neumann and Wiener had read this paper with great interest.[38] After the war, following in their footsteps, a young Harvard student, Marvin Minsky, assembled with electronics student Dean Edwards and psychologist George Miller a neural network machine (the 'Snarc') from war-surplus supplies and a small grant from the Office of Naval Research. The Snarc simulated a rat learning the route through a maze, and amounted to both a materialization of behaviourist theories of stimulus and reward and an application of cybernetic theories of the brain.

Frank Rosenblatt, working at Cornell University, developed the neural net-as-learning device even further, into the theory and practice of what he called 'perceptrons'. By 1958, Rosenblatt was making great claims about their achievements, telling the New Yorker that:

[The perceptron] can tell the difference between a dog and a cat, though so far, according to our calculations, it wouldn't be able to tell whether the dog was to the left or the right of the cat. We still have to teach it depth perception and refinements of judgment.[39]

Yet this bottom-up approach already had a critical and equally hubristic rival. The idea that a computer could not, at least under favourable conditions, be distinguished from a human mind had been discussed with subtlety and wit by Alan Turing in a paper for the philosophical journal *Mind* in 1950. In it he had raised and dismissed many objections to the claim that machines could think. In the hand of American scientists, many of whom were riding a tide of Cold War and philanthropic patronage in the early 1950s, the step was made from Turing's question 'can machines think?' to the questionable claim 'machines can think'. Rather than bottom-up constructions of simplified brains, the second approach labelled symbol-manipulating software as minds in action. The main protagonists were John McCarthy, Herbert Simon, Allen Newell and Minsky, who jumped ship from neural net research in 1955.

John McCarthy came from a family of Marxists, an encouraging upbringing, perhaps, for regarding ideas as material products. He trained as a mathematician and worked at Bell Labs, with Minsky under Claude Shannon, in 1953. He gained direct experience of stored-program computers while working for IBM in the summer of 1955, after which he taught at Dartmouth College, New Hampshire, a site of pioneering computing experimentation. Minsky and McCarthy persuaded the Rockefeller Foundation to cover the cost of a summer workshop at Dartmouth in 1956. 'Every aspect of learning or any feature of intelligence', ran the proposal for the workshop discussion, 'can in principle be so precisely described that a machine can be made to simulate it.' The workshop was a gathering of ambitious men: Minsky, McCarthy, Shannon, Nathaniel Rochester of IBM, the former assistant to Norbert Wiener, Oliver Selfridge, Trenchard More, Arthur Samuel, Herbert Simon and Allen Newell. They declared the existence of a new discipline, coining a new name: 'artificial intelligence'.

Although many attendees thought the 1956 conference a relative failure, the aspirations expressed were sky high. The only concrete result presented was Logic Theorist, a program written by Simon, Newell and J. C. Shaw. Herbert Simon was a Cold War political scientist who had practical experience of Marshall Plan administration and Ford Foundation projects.[40] In books such as *Administrative Behavior* (1948) he had presented his theory of 'bounded rationality', proposing that people in organizations did not consider all the possible options, making a choice that maximized their interests (as neo-classical economic theory assumed) rather than decisions being taken by starting with acceptance criteria, reviewing few possibilities, and often choosing the first that passed the criteria. Simon had speculated further that the human mind worked in a similar way: heuristic rule-bound problem-solving.

Logic Theorist ran on the JOHNNIAC computer at RAND, where Newell worked. It modelled mathematical deduction, starting with the axioms and deriving the theorems of Bertrand Russell and Alfred North Whitehead's *Principia Mathematica*. A 'clear descendant of Simon's principle of bounded rationality', it did so by following heuristics:[41] 'Our theory is a theory of the information processes involved in problem-solving', the programmers argued, as they explicitly rejected the bottom-up approach. It was 'not a theory of neural or electronic mechanisms for information processing'. Symbol manipulation was how the mind operated, they reasoned, so a machine manipulating symbols was mind-like. Simon later recalled his view of Logic Theorist's significance, a stroke that cut the Gordian knot of post-Descartes philosophy: 'We invented a computer program capable of thinking non-numerically, and thereby solved the venerable mind/body problem, explaining how a system composed of matter can have the properties of mind.'[42]

Turing, in his *Mind* paper of 1950, had speculated that a computer might pass his imitation test in fifty years. Newell and Simon, in 1958, went further:

1 ... within ten years a digital computer will be the world's chess champion, unless the rules bar it from competition.
2 ... within ten years a digital computer will discover and prove an important new mathematical theorem.
3 ... within ten years a digital computer will write music that will be accepted by critics as possessing considerable aesthetic value.
4 ... within ten years most theories in psychology will take the form of computer programs, or of qualitative statements about the characteristics of computer programs.[43]

Artificial intelligence's promises attracted scepticism from some quarters of academia. Hubert Dreyfus attacked the 'artificial intelligentsia' in a RAND pamphlet, *Alchemy and AI*, in 1964, later presenting the criticisms at length in his *What Computers Can't Do* (1972). Joseph Weizenbaum, author of a lovely program, ELIZA, which seemed to converse intelligently but was just a box of simple tricks, wrote about the limits in *Computer Power and Human Reason* (1976). Yet, through the Cold War, the optimism for achieving artificial intelligence in the short-term returned as regularly as waves to the shore: feats such as natural language use or intelligent pattern recognition were forever merely a decade away.

The sustenance of optimism for near-future success in artificial intelligence was a Cold War effect. 'Virtually all' the funds for artificial intelligence research in the United States came from the Information Processing Techniques Office (IPTO) of the Advanced Projects Research Agency of the Department of Defense.[44] The ideology of automated command and control favoured AI projects. IPTO's leader, the visionary computer scientist J. C. R. Licklider, had written in an influential essay, 'Man–computer symbiosis', in 1960:

> The military commander ... faces a greater probability of having to make critical decisions in short intervals of time. It is easy to over-dramatize the notion of the ten-minute war, but it would be dangerous to count on having more than 10 minutes in which to make a critical decision. As military system ground environments and control centers grow in capability and complexity, therefore, a real requirement for automatic speech production and recognition in computers seems likely to develop.[45]

Artificial intelligence promised a high-risk, high-gain solution to the problems of automating military decision support. The automation of natural language – production, translation, understanding – was needed by military commanders but also by the secret eavesdroppers on Soviet communications. Many Cold War patrons would be lured by AI. RAND considered Newell and Simon's work 'important for developing and testing theories about human intelligence and decision-making and building [its own] computer programming capabilities', and soon set up a Systems Research Laboratory to 'examine how human-machine systems perform under stress', simulating parts of SAGE.[46] In turn, the research agendas of some of the best technical universities in the United States – MIT, Stanford, Carnegie Mellon – were further aligned along Cold War interests. In contrast, bottom-up neural net approaches, fiercely attacked by the symbol-manipulators, struggled to find funds.

Nevertheless, artificial intelligence would mutate in the face of criticism. One approach was to simplify the representations of the working world even further until it became tractable with the techniques to hand. Terry Winograd's SHRDLU, developed at MIT in 1968–70, was the best example: a 'microworld' of simple shapes that the machine could reason with and move about. A second approach was to model the narrow but deep knowledge of scientific specialties. Dendral, built at Stanford University from 1965, was an early example. It was a collaboration between the computer scientist Edward Feigenbaum, the molecular biologist Joshua Lederberg and the chemist Carl Djerassi, among others, with the aim of automating an organic chemist's ability to draw conclusions about molecules from the output of a mass spectrometer. For Lederberg this was one response among many to an 'instrumentation crisis in biology': 'an immense amount of information is locked up in spectra ... which require the intensive development of a "man–computer symbiosis" for adequate resolution.'[47] Since diagnostic expertise was expensive to make and replicate, there followed considerable commercial interest in this approach. An expert system for mimicking doctors' judgements about infection treatments, Mycin, for example, followed Dendral in the early 1970s.

Mind as a computer

Artificial intelligence saw the computer as mind. Cognitive psychology saw the mind as a computer, and in doing so broke the rules of earlier psychology. 'You couldn't use a word like "mind" in a psychology journal', Herbert Simon wrote of the dominance of behaviourism, 'you'd get your mouth washed out with soap.'[48] But in books such as George Miller's *Language and Communication* (1951) and especially Miller, Eugene Galanter and Karl Pribam's *Plans and the Structure of Behavior* (1960), a manifesto for cognitivism was developed.[49] The latter was 'the first text to examine virtually every aspect of human psychology, including instincts, motor skills, memory, speech, values, personality, and problem solving', writes Paul Edwards, 'through the lens of the computer analogy'.[50] Cognitive psychology would thrive on Western campuses from MIT to Sussex. Closely connected with this research programme was the revolution in linguistics instigated by Noam Chomsky at MIT from the mid-1950s. Natural language competence was the achievement of the application of deep-lying heuristics, a universal grammar, common to all humans. Information processing had its modern birth barely a decade before in the sciences of the working world of radar; by 1955 the model had been naturalized. Universalism – whether it was Rostow's economics, Turing's machines or Chomsky's grammar – was a distinctive shared feature of projects of the Cold War.

The work of computing

However, artificial intelligence, with its grand statements about solving age-old philosophical problems and promising machines that matched the greatest human achievements, was a tiny component of the growth from the 1950s of computer science, a true science of the post-war working world. Aspray sees computer science as an 'amalgam' of four intellectual traditions: the mathematical or logical (which thrived largely on inherited scholarly status), hardware engineering (universities were a major source of computer prototypes until the 1960s), software engineering (which saw continuities with civil engineering) and the experimental 'science of the artificial'.[51] Computer science, as previous sciences had done, set up and experimented with artificial representations of working world problems. 'Computer science', wrote Simon, Alan Perlis and Newell in *Science* in 1967, was the 'study of computers'.[52] Computers, in turn, were materializations of the organization of human work.[53] In *The Sciences of the Artificial* (1968), Simon defended this account of the relationship between science and world, one which I have tried to generalize.

'Programmer' became a new occupation in the 1950s, and faculties in computer science sprang up from the late 1950s to teach generations of university students.[54] The first machines had had to be programmed by technicians at the most basic level, little above feeding in strings of zeroes and ones. To make programming easier, high-level computer languages were developed, important ones including FORTRAN (designed for scientists, the name a contraction of 'formula translation'; the language came with IBM computers from 1957) and COBOL ('Common Business Oriented Language', which owed its success to the insistence of the Department of Defense that the United States government would only buy COBOL compatible computers, 1959).[55] The informal anecdotal techniques of programming were made explicit in textbooks such as Donald Knuth's *The Art of Computer Programming* (1968), as was a defining focus on algorithms. Many programmers, in a rapidly expanding industry, were nevertheless trained on the job, a route that encouraged women to be programmers.

The stored-program computers of the 1950s were typically large mainframes. They would be programmed by punching out a batch of cards which would be taken to the mainframe building. Users would return later to see if their program had run well or not. By the late 1950s the cost of a computer had dropped somewhat, enough for many medium-sized businesses, large laboratories or smaller government departments to own their own machines, but still computers followed this same mainframe pattern. It would change in the 1960s, when the scale of computing tasks reached a crisis point

Code Science: DNA and the rewriting of molecular biology

The proliferation of information technologies, and the high status attached to them during the Cold War, encouraged talk of 'information' and 'codes'. The most remarkable example of this rhetorical turn was the transformation in the concepts and techniques of the sciences of molecular biology and inheritance. By 1940 there were three unresolved problems in genetics.[56] We have already seen one of them: the relationship between genetics and evolution was articulated from the 1930s to 1950s in the Evolutionary Synthesis. The second, broadly a relationship between genetics and development, would remain problematic for much of the century. Certainly in the mid-century most embryologists were sceptical of the capacity of the chromosomal theory of the gene to account for a whole life history of an organism.

The third outstanding problem was the chemical composition of the gene. In the words of Hermann Muller, the genes must be 'autocatalytic', capable of duplicating themselves, and 'heterocatalytic', able to control the formation of other substances found in living creatures. Chromosomes were made up mostly of protein with a little nucleic acid. Deoxyribonucleic acid (DNA) was a highly unlikely candidate, since it was considered to have a rather dull structure of uniformly repeating nucleotides. DNA, the reasoning went, probably served as scaffolding, supporting the true genetic substance. Proteins, however, were incredibly diverse in structure. We have seen how pre-war philanthropic programmes encouraged the 'protein paradigm', the working assumption that genes were, in some as yet unknown way, composed of proteins.[57]

Progress came when scientists focused on manipulating and experimenting with micro-organisms in sophisticated ways. Micro-organisms can multiply at astonishing rates, and some process simple foods, making them useful laboratory subjects. We have already seen how George Beadle and others worked with *Neurospora* to show that genes affecting nutrition controlled the formation of enzymes. Also in the 1940s, Oswald Avery at the Rockefeller Institute in New York was working with *Streptococcus pneumoniae*, or pneumococcus, a bacterium that came in two forms, rough and smooth.[58] The smooth form was virulent, the rough form not. They could be cultured in medium and could be injected into mice. From work that had been done in a London pathological laboratory by Frederick Griffith in the 1920s it was known that mice injected with live rough bacteria and dead smooth bacteria still perished. What was more, live smooth bacteria could now be found in the mouse blood. Somehow the virulence had been transferred and inherited. Avery asked: what substance was responsible for this transformation? Avery's team worked hard to extract samples of the 'transformative principle', growing pneumococci in vats of broth, spinning them in centrifuges, and treating them with brine and enzymes to remove sugars and most proteins to produce a fibrous substance. They ran test after test to give clues about the substance's chemical nature: everything pointed towards DNA.

In the 1944 paper that summarized the work Avery was cautious. In private he wrote to his brother:

> If we are right, & of course that's not yet proven, then it means that nucleic acids are not merely structurally important but functionally active substances in determining the biochemical activities and specific characteristics of cells – & that by means of a known chemical substance it is possible to induce predictable and hereditary changes in the cells. This is something that has long been the dream of geneticists . . . Sounds like a virus – may be a gene.[59]

Avery was not a geneticist, he was 'by training a physician', 'not exactly a biochemist, but an immunologist and microbiologist'.[60] The phage group of Salvador Luria and Max Delbrück was also interdisciplinary. From 1939, electron microscopes, built in the United States by RCA to imported designs, began to reveal the shape of phages ('Mein Gott!', exclaimed one professor, quoted by Judson. 'They've got tails').[61] The phage group agreed in 1944 to focus attention on seven particular bacteriophages of the human gut bacterium E. coli, which they labelled T1 to T7. Electron microscopy showed T2, T4 and T6 to look like little watch-towers, with box heads and spindly legs. In an elegant experiment of 1943, Luria and Delbrück showed that some E. coli cultures were resistant to phage while others were not, and concluded that bacterial mutation must have taken place. By the late 1940s and early 1950s the phage group had techniques to work with mutation (phages as tools, E. coli as subject) and speculations about the action of phages: 'I've been thinking', one of them wrote, 'that the virus may act like a little hypodermic needle full of transforming principles.'[62]

Yet the phage group remained convinced that protein was the genetic substance. But in 1952, group members Alfred Hershey and Martha Chase used radioisotope tracers – phage protein marked with radioactive sulphur, phage DNA marked with radioactive phosphorus – to show that the T2 phage's 'transforming principle' was the DNA. A young phage group member was reading a letter from Hershey describing this experiment in Oxford – his mentor, Luria, a pre-war Marxist, had had his visa for travel from the United States refused.[63] The young man was James D. Watson.

The great global upheavals of the 1930s and 1940s brought scientists from many disciplines together and into play in the DNA story, albeit in some unexpected ways. One effect, already discussed, had been to disperse the great quantum physicists away from Central Europe. Delbrück had arrived in the United States, where he had hitched up with Luria, another émigré. The journey of Erwin Schrödinger was even more extraordinary. He had left Germany because he despised Nazism and drifted to Ireland, where he lived with his wife and mistress in a *ménage à trois*. In Dublin he wrote a book, with the bold title *What is Life?* (1944), in which he gave a quantum physicist's analysis of how mere molecules might produce the inherited order of life. By analogy with the Morse code, he spoke of a 'code-script' whereby the rearrangement of a molecule might provide 'the miniature code [that] should correspond with a highly complicated and specified plan of development and should somehow contain the means to put it in operation'. Perhaps, he wrote, it might be an 'aperiodic crystal'.[64] The precise influence of Schrödinger's book on molecular biology is a subject of controversy among historians;[65] undoubtedly it was read by physicists and turned some towards biology.

Donald Fleming compares the caution of biologists such as Avery with the ambition and gusto of the physicists turned biologists and concludes that the latter's willingness to strike directly at the molecular centre of what life is came from the intellectual culture of émigré physics.[66] It was transferred, like a T2 injecting its host, from the phage group to Watson. Other key players in the DNA story, notably Francis Crick (who had worked on mines for the Admiralty) and Maurice Wilkins (who took part in the Manhattan Project), were also physicists who had turned to biology. Fleming suggests that two forces were at work: a desire to turn from the sciences of death to the sciences of life and, less grandiose, a movement of 'loners or small-team men who lamented the passing of do-it-yourself physics' in the age of cyclotrons.[67] The phage group also inherited some features of Bohr's Copenhagen, some social (an openness), some intellectual (perhaps specifically the notion of complementarity) and some cultural (not least the lived experience of scientific revolution).

With experiments by the early 1950s pointing towards DNA as the genetic substance, the precise structure of DNA became much more interesting. (However, what we now see as its obvious relevance is an artefact of retrospection; formal genetics was seen at the time as sufficient, not requiring knowledge of the material structure of the gene.) Three groups followed this interest. At King's College London, the skills and endurance of Rosalind Franklin – preparing crystalline samples of DNA and producing sharp photographs of X-ray diffraction patterns – led the way.[68] Franklin's boss was Maurice Wilkins, although she did not see the relationship as such. In California, Linus Pauling, fresh from his triumphant demonstration of the alpha helix form of some proteins, but hounded by McCarthyites, turned his attention to the structure of DNA in the summer of 1951.[69] Finally, the American James Watson, picking up European post-doctoral experience after being brought up in the phage group, shared his enthusiasm for DNA with Francis Crick at the Cavendish Laboratory, Cambridge.

Watson and Crick were convinced that DNA must have a helical structure. Franklin, in her laboratory notebooks, dismissed the idea. Nevertheless, Franklin's X-ray diffraction photographs were the best source of data, and the Cambridge duo, in an underhand manner, sneaked a look at the latest photographs via contact with Wilkins. If Watson's memoir (1968) is to be trusted, then he viewed science as a race in which the prizes went to winners, and sharp practice was fair game.[70] Maurice Wilkins, as he testified in his own memoirs (2003), had a completely opposite view of the nature of science. It had been 'great community spirit and co-operation' in the King's College Laboratory that produced the 'valuable result' of sharp diffraction photographs.[71] Cooperation was the mark, for Wilkins, of productive scientific communities. He recalled:

> Francis and Jim asked me whether I would mind if they started building models again. I found this question horrible. I did not like treating science as a race, and I especially did not like the idea of them racing against me. I was strongly attached to the idea of the scientific community.[72]

For Wilkins, cooperative science meant pooling knowledge, sharing photographs. Franklin's view was that these were her results, for her use. The controversy over credit in the DNA story was therefore at root a divergence over the proper social character of science.

Crick and Watson did indeed start building models again, using the King's College data to constrain their dimensions. They knew DNA contained nucleotide bases: guanine (G), cytosine (C), adenine (A) and thymine (T). By cutting out precise card shapes of the bases and arranging them, they matched G to C and A to T. (A clue to this pairing would soon be recognized in the biochemical finding of Erwin Chargaff of the one-to-one ratio of G to C and A to T in DNA. Crick seems to have known Chargaff's result without immediately dwelling on its significance. Biochemistry and molecular biology were antagonistic and often talked past rather than to each other.) Crick and Watson settled on a double helix structure, with paired bases on the inside. The result was published in *Nature* on 25 April 1953. Articles describing the X-ray diffraction photographs were published alongside. Journalist and seasoned Cavendish publicist Ritchie Calder wrote up the story for the newspapers.

'We wish to suggest a structure for the salt of deoxyribose nucleic acid (D.N.A.)', is how Watson and Crick had opened their short paper. 'This structure has novel features which are of considerable biological interest.'[73] 'It has not escaped our notice', they remarked near the end, 'that the specific pairing we have postulated immediately suggests a possible copying mechanism for the genetic material.' It was a structural explanation of autocatalysis: the double helix could be unzipped and then two copies rebuilt from each half.

The elucidation of the structure of DNA is – now – regarded as one of the stellar achievements in science in the twentieth century. However, despite initial newspaper coverage, its fame has grown in intensity rather than shining brightly since 1953. There are three concentrically packed factors fuelling this. First, the relationship between DNA, as the chemical of genes, and the cells and bodies it affected needed to be worked out, which took time. How, for example, did DNA, the stuff of genetics, relate to proteins, the stuff of bodies? Second, the wider world needed to be interested in that relationship, and indeed in the science of molecular biology. The promise and then delivery of the promise of biotechnology from the 1970s into the twenty-first century would retrospectively brighten the light of Watson and Crick's discovery. Finally, in the process, the relationship between germ plasm and soma, to use tactically useful archaic terms, was redescribed within a powerful new discourse, a way of speaking about a subject that encourages certain ways of thinking while closing others. You may be wondering why the famous story of DNA is in a chapter about Cold War science. The direct politics of the Cold War did shape the story slightly: Linus Pauling, for example, might have seen the King's College photographs had he not been denied permission to travel by anti-communists. But this is relatively uninteresting historical contingency. Rather it is in the redescription of life in terms of 'information discourse',[74] a Cold War language, that interests us here.

Among the readers of Watson and Crick's paper was George Gamow. 'I am a physicist, not a biologist', he wrote to the pair. 'But I am very much excited by your article . . . and think that [it] brings Biology into the group of the "exact" sciences.'[75] In particular, Gamow saw the task of determining how four nucleotides related to the twenty amino acids that in different combinations composed proteins as a cryptanalytical problem; and Gamow knew the right people to have first crack:

> at this time I was consultant in the Navy and I knew some people in this top secret business in the Navy basement who were deciphering and broke the Japanese code and so on. So I talked to the admiral, the head of the Bureau of Ordnance . . . So I told them the problem, gave them the protein things [the twenty amino acids], and they put it in a machine and after two weeks they informed me that there is no solution. Ha![76]

If there was to be no two-week solution, Gamow would have to work harder. He proposed that triplets of nucleotides – such as AGC or TGA – were numerous enough $(4 \times 4 \times 4 = 64)$ to cover the twenty amino acids. He wrote papers, controversial among biologists, which presented the problem of DNA-protein specificity in terms of 'information transfer, cryptanalysis, and linguistics'.[77] He gathered like-minded scientists together to tackle what was now seen as the 'coding' problem. His co-worker, Martynas Yčas, a Russian émigré working for the United States Army, defined the coding problem as understanding the 'storage, transfer, and the replication of information'.[78] (Gamow's network, which included Francis Crick, even had a special tie, and they called themselves the 'RNA Tie Club'). With the help of Los Alamos colleague Nicholas Metropolis, Gamow ran Monte Carlo simulations of different coding approaches on the MANIAC computer. This work saw the coding problem as a cryptanalytical one to be tackled with the resources of the Cold War security state; the process was black-boxed, with the input (four letters for nucleotides) and output (twenty amino acids) the only considerations. This theoretical attack on the coding problem was largely unsuccessful, but it did have the effect of rephrasing genetics in terms of information. Most famously, this rephrasing was expressed in 1958 in Francis Crick's 'sequence hypothesis' and 'Central Dogma'.

The former stated that 'The specificity of a piece of nucleic acid is expressed solely by the sequences of its bases, and that this sequence is a (simple) code for the amino acid sequence of a particular protein.' While the Central Dogma stated:

> Once 'information' has passed into protein it cannot get out again. In more detail, the transfer of information from nucleic to nucleic acid, or from nucleic acid to protein may be possible, but transfer from protein to protein, or from protein to nucleic acid is impossible. Information means here the precise determination of sequence, either of bases in the nucleic acid or of the amino acid residues in the protein.[79]

Others – even biochemists – would learn to speak in the same informational language.

In 1960, biochemists at Wendell Stanley's Virus Laboratory at Berkeley, including Heinz Fraenkel-Conrat, succeeded first in reconstituting Tobacco Mosaic Virus (TMV) from its constituent RNA core and protein coat – greeted in the press as the creation of artificial life – but also in establishing the sequence of 158 amino acid residues that made up the protein. (RNA is another nucleic acid; one difference between RNA and DNA is that the thymine (T) in the latter is replaced by uracil (U) in the former.) Stanley described this achievement as 'a Rosetta Stone for the language of life'.[80] With Frederick Sanger's first sequencing of the amino acid residues of a protein, insulin, in the early 1950s, the project had output information to work with.[81] Sanger had emplyed a range of techniques: first, conventional chemical techniques were enlisted to break off amino acids of the ends of the insulin molecule and identify them; second, enzymes were used to break up the protein molecule at specific places; third, chromatography was deployed to spread the pieces out, and then radioactive tracers were used to label and identify the fragments.[82] The result, announced in 1949, was not only knowledge of the sequence of amino acids in the insulin molecule, but also an argument that the sequence of amino acids in proteins was specific.

Scientists in the 1950s began to trace the biochemistry of protein synthesis, opening Gamow's black box. They also spoke the language of information, cybernetics, command and control. The Pasteur Institute in Paris provides a clear example. The institute was home to two laboratories crucial to the growth of post-war molecular biology: Jacques Monod's laboratory of cellular biochemistry and François Jacob's microbiological laboratory within André Lwoff's Department of Bacterial Physiology.[83] Monod and Jacob began to understand genetic control of protein synthesis in cybernetic terms, rewriting the life sciences in terms drawn, as we have seen, originally from the analysis of anti-aircraft operations. Leo Szilard seems to have been the critical link in this chain of influence: he had talked at length about the ideas of Norbert Wiener and Claude Shannon with Monod.

Specifically, Monod and Jacob proposed the idea of 'messenger RNA', short-lived RNA molecules that, in their language, carried information from nucleic acid to protein. Monod had a stock of *E. coli* mutants, some of which could ('lac+') and others could not ('lac–') develop on lactose. Jacob had developed a 'zygotic technology', techniques which shook bacteria apart at different moments of bacterial sexual conjugation. This technique had shown that the genes were transferred from 'male' to 'female' bacterium in sequence, and amounted to a method for mapping the gene. Putting their resources together, and working with visiting Berkeley scientist Arthur B. Pardee, Jacob and Monod argued in the so-called PaJaMa article in the new *Journal of Molecular Biology* that their *E. coli* experiments demonstrated, first, that there must be a chemical messenger from gene to cytoplasm and, second, that a 'repressor' molecule blocked the synthesis of a lactose-eating enzyme. The cybernetic debt is clear in this reminiscence of Jacob's:

> We saw this circuit as made up of two genes: transmitter and receiver of cytoplasmic signal, the repressor. In the absence of the inducer, this circuit blocked the synthesis of galactosidase. Every mutation inactivating one of the genes thus had to result in a constitutive synthesis, much as a transmitter on the ground sends signals to a bomber: 'Do not drop the bombs. Do not drop the bombs'. If the transmitter or the receiver is broken, the plane drops its bombs.[84]

The Pasteur Institute scientists cooperated and competed with Sydney Brenner, Francis Crick and others to identify the chemical messenger. The messenger in *E. coli* was found to be short-lived strands of the nucleic acid RNA. And, as Jacques Monod said, what is true for *E. coli* is true for an elephant.[85] Joint papers announced the discovery in *Nature* in May 1961.

From the late 1950s, the National Institutes of Health had grown and attracted a galaxy of biochemical talent to well-funded laboratories. In 1961, Marshall W. Nirenberg and his post-doc Heinrich Matthaei, working at Bethesda, and inspired by the Paris techniques and the cell-free approach of Paul Zamecnik, used an *E. coli* cell-free system – ultracentrifuged fractions of *E. coli* – to synthesize a protein from a known sequence of nucleotides. The sequence of the synthetic RNA was UUU and the protein was polyphenylalanine. There followed a rush of ingenious biochemical experimentation to discover the rest of the genetic code. The experiments depended on an extraordinary array of techniques to build up and break down nucleic acids, including enzymes to assemble nucleotides into sequences, such as DNA and RNA polymerases or DNA ligase, and enzymes to destroy them again, such as DNAase and RNAase. The scientist who, more than any other, made these chemical tools was Har Gobind Khorana, an Indian émigré biochemist working at the University of Wisconsin.

The period of the completion of the code, from Nirenberg and Matthaei's announcement of 1961 to 1967, witnessed a flourishing of code talk, now packaged for popular consumption. Carl Woese wrote in *The Genetic Code* (1967) of 'informational molecules' and DNA and RNA as being like 'tapes' and 'tape readers'; Robert Sinsheimer, in his *The Book of Life* (1967), described human chromosomes as being 'the book of life' containing 'instructions, in a curious and wonderful code, for making a human being'; this was the informational discourse, devised in the early Cold War in Los Alamos, CalTech, Cambridge and Paris, now becoming a world language.[86] And if life could be read, written or copied, it could be re-edited. In this way molecular biology's rephrasing of biological specificity as information invited the imagination of the biotechnological industries of the late twentieth century.

The cracking of the genetic code was an achievement of international networks of scientists, with centres in California, Cambridge and Paris. It was the crowning achievement of this second wave of molecular biology. On a theoretical axis, molecular biologists now held that genes were made of DNA, and that DNA encoded information that determined replication and protein synthesis; the cellular machinery was described in detail, along with the biochemical pathways for replication, expression, and so on; enzymes, the protein catalysts, were isolated and related to sequences.[87] On a technical axis, new methods of manipulating DNA were devised. By 1970 Har Gobind Khorana was able to achieve the first complete chemical synthesis of a gene.

Global science (1): Plate tectonics

Another revolution of the 1950s and 1960s, the acceptance of the theory of plate tectonics in geophysics, was also the achievement of a dispersed network of scientists. It, too, was framed by the Cold War.[88] Furthermore, the data gathering justified by the Cold War contributed to a global science that would be essential to the later recognition of global climate change. I finish this chapter with a brief examination of these last Cold War sciences.

I have discussed in an earlier chapter the chilly reception of Alfred Wegener's theory of continental drift from the 1910s. After the Second World War, the context for theorizing about large-scale movements of the earth's crust was transformed, first, by new sources of evidence, paid for largely by military patrons, and, second, a disciplinary realignment in which geophysics gained in credit compared to field-based geology, largely on the back of physics' prestige after 1945.[89] An early source of new geophysical evidence was studies of land-based palaeomagnetism: the mapping of the orientation of magnetic fields frozen in rocks of different ages.[90] Patrick Blackett's group, which moved from Manchester to Imperial College, London, was active here, as was ex-radar scientist Keith Runcorn's group, which moved from Cambridge to Newcastle. Palaeomagnetic research suggested that the magnetic poles of the earth had wandered, and even reversed. Untangling the meaning of this information was challenging, but within it would be clues to past positions of the earth's rocks.

But it was from oceanography that the crucial data would be generated. Detailed knowledge of the contours and consistency of the ocean floor was essential to submarines carrying the nuclear deterrent. But the military–geophysics relationship, write historians Naomi Oreskes and Ronald Doel, 'was not simply one of increased practical application, but of vastly increased funding for geophysical and geochemical work, which spawned a greatly expanded institutional base and largely determined the priorities of the discipline.'[91] Bodies such as the Office of Naval Research therefore poured resources into oceanography, supporting institutions such as the Lamont–Doherty Geological Observatory of Columbia University, as well as research at Woods Hole on the east coast and the Scripps Institution on the west. These oceanographical investigations traced a worldwide network of oceanic ridges, for which explanations based on the presumption of both fixed and mobile crust were offered.[92] Mobile crust theories made two testable predictions. The first, independently suggested by Fred Vine and Drummond Matthews, was of the existence of alternating magnetic polarizations of rock in bands parallel to the ridge, caused when the crust spread out under conditions of reversing north and south poles. The second, by J. Tuzo Wilson, was of transform faults detectable by seismology.[93] There had been a massive increase in global seismological recording in the 1960s, including the establishment of the World-Wide Standardized

Seismograph Network, as part of the Cold War requirement to detect atomic tests.[94] These data could locate tremors theorized as being caused by moving crust.

When these predictions were borne out by oceanographic sampling of the magnetic polarity of the sea floor near ridges and by earthquake measurements, the idea of mobile components of the earth's crust gained considerable ground. It was worked up into a fully fledged theory of 'plate tectonics' – which asked how moving crust would behave on a spherical earth – by Jason Morgan at Princeton and Dan MacKenzie at Cambridge, both geophysicists. Plate tectonics provided a model of the earth of giant slow convection currents in the mantle pushing the plates into, under and over each other, driving mountain-building, causing earthquakes and, over the millennia, changing the face of our planet.

The success of plate tectonics in making such diverse knowledge 'cohere'[95] further boosted the prestige of geophysics over traditional geology. Subjects such as mineralogy, historical geology and paleontology were now squeezed out of geology curricula to make room for more physics-inspired methods teaching.[96] But also since the 1960s, there has been a shift from geology to 'earth science', as comparative data came in from missions to the moon and bodies in the rest of the solar system. 'Earth science', writes historian Mott Greene, is 'best seen as a subdivision of planetology'.[97]

Global science (2): Global warming

One of the most important effects of the Cold War for twentieth-century science (and beyond) was to encourage the globalization of data recording. We have seen examples of this in oceanography and seismology. Another can be found in the atmospheric and oceanographic recordings of past and current climate change.

The idea of a greenhouse effect, the trapping of warmth by the blanket of the earth's atmosphere, goes back at least to John Tyndall in the nineteenth century. The Swedish chemist Svante Arrhenius had made calculations of the expected warming of the earth due to added carbon dioxide, from industry, in 1896. He had been sanguine about the results:

> We often hear lamentations that the coal stored up in the earth is wasted by the present generation without any thought of the future . . . [However, by] the influence of the increasing percentage of carbonic

acid in the atmosphere, we may hope to enjoy ages with more equable and better climates, especially as regards the colder regions of the earth, ages when the earth will bring forth much more abundant crops than at present, for the benefit of a rapidly propagating mankind.[98]

In 1938, a professional steam engineer and an amateur meteorologist, Guy Stewart Callendar, had read a paper to the Royal Meteorological Society arguing that added carbon dioxide had caused a 'modest but measurable increase in the earth's temperature'. The following year he warned of humanity conducting a 'grand experiment', and he continued to publish on the linkage until the early 1960s. Historians Mark Handel and James Risbey suggest Callendar was ignored because of the distraction of the Second World War, Spencer Weart suggests he was ignored because he was an amateur, while James Fleming, citing references to Callendar's work by professional meteorologists, argues that he was not ignored at all.[99]

Certainly what would count as evidence for global warming required a plausible claim to global coverage. This was where the Cold War was important. Consider four dimensions. First, the Cold War encouraged sophisticated, mathematical meteorology such as that pursued by Carl-Gustav Rossby's group at Chicago or the use of computers to calculate future weather patterns at the Institute of Advanced Study at Princeton. Partly this was a continuation of the historical patronage of meteorology by the military. The outcome was a concentration of computer resources, funded through bodies such as the Office of Naval Research, on weather modelling and prediction. General Circulation Models (GCMs, usually rendered now as Global Climate Models), computer calculations of changing climate, were first attempted in the mid-1950s by Norman Phillips at Princeton.[100]

Second, the question of absorption of infrared radiation in the atmosphere had direct relevance to defence research. Heat-seeking missiles or other projects to track missiles by their exhaust temperatures required an understanding of how infrared travelled and was absorbed in the air. Gilbert Plass, at the Lockheed Corporation, working with Johns Hopkins University on Office of Naval Research funds, did such missile research by day, and by night, using the same data, made calculations of the global warming caused by human activity: $1.1°$ per century was the estimate Plass gave in 1956.[101]

Third, the Office of Naval Research and the Atomic Energy Commission supported Roger Revelle's oceanographic research at the Scripps Institution into how fast the earth's oceans turned over (this had implications for the spread of nuclear fallout). Revelle calculated that the slowly circulating oceans could not absorb the carbon dioxide pumped into the atmosphere; the gas would linger in the air. 'Humans are carrying out a large scale geophysical experiment', he wrote, echoing Callendar, but speaking from a much more powerful position, 'of a kind that could not have happened in the past nor be reproduced in the future'.[102]

Fourth, the International Geophysical Year may be associated with the Russian coup of Sputnik, but it was primarily a programme of interdisciplinary Antarctic and atmospheric research. This included work started in 1956 at the Little America base, Antarctica, to record baseline levels of atmospheric carbon dioxide far away from specific sources of the gas. Charles David Keeling, at the end of the International Geophysical Year, moved his measuring equipment to another almost pristine location: the summit of Mauna Loa on the Hawaiian Big Island. Keeling's work was first funded through the IGY, then by the Atomic Energy Commission, and then, after a short hiatus, by the National Science Foundation. Drawn as a (now famous) graph, Keeling's data showed the clear and steady rise of levels of carbon dioxide in the atmosphere.

'Without the Cold War there would have been little funding for the research that turned out to illuminate the CO_2 greenhouse effect, a subject nobody had connected with practical affairs', writes historian Spencer Weart. 'The U.S. Navy had bought an answer to a question it had never thought to ask.'[103] I would go further: it took a global conflict to create the particular conditions to identify global warming as a global phenomenon.

Part IV

Sciences of our World

Part IV

Science in our World

TRANSITION: SEA CHANGE IN THE LONG 1960s

The long 1960s, the period roughly from the mid-1950s to the mid-1970s, was one of transition. On one hand the massive post-Second World War growth in science was maintained. Funding was high, and there was a rapid expansion of higher education, not least in the technical and physical sciences, in response to the demand for expertise relevant to burgeoning civil and military working worlds, from building chemical plants and finding and testing new medicines and foods, to Cold War projects for new missiles, radar, nuclear reactors and warheads, eavesdropping networks and jet aircraft.[1] Across the Western world, new universities were established and old ones reformed and expanded to produce experts of all kinds. On the other hand, voices of discontent became louder. The Cold War chill passed into a period of relative thaw. Meanwhile, experts clashed in public, campuses became theatres of demonstration, and science was caught up in broader movements of social and cultural change. One historian has characterized the period as an 'age of contradiction', which has some mileage.[2] Plenty of contrasts and ironies can be found.

In the following I explore some of these transitions and contrasts. First I look at how social movements related to science in different ways, particularly in the areas of civil rights, nuclear projects and environmentalism, and also how violent clashes could result in the form of campus protest. Then I examine how the tensions were expressed within science, for example in contrasting ideals of how science should be pursued and organized. I review two cases of new biomedical technologies – the contraceptive pill and psychiatric drugs – that have received contradictory assessments in the years since their introduction: are they tools of liberation or control? In all these disputes authority was questioned, disputed or made to justify itself anew. Further contrasts are displayed by the case of cybernetics, a Cold War tool that could be applied to produce both environmental warnings and, in the case of China's one child per family policy, justifications of extraordinary social engineering. Finally, as the mood of the long 1960s soured, I draw attention to a strange neo-catastrophism that could be found in sciences that were disparate but shared the possession of a public dimension.

Science and social movements

Social movements were highly visible features of long 1960s politics and culture, on which there is an immense literature. This secondary literature and first-hand primary accounts allow a number of features of social movements to be made out. First, social movements had a distinctive fluid, network form. While there were prominent spokespeople, heroes and revered ancestors, each social movement was a patchwork of sometimes short-lived organizations and campaigns. What gave a social movement cohesion was a rough consensus on ultimate targets, such as the removal of racist prejudice or nuclear disarmament. The presence of targets strongly promoted a polarized culture that structured much of the literature, speeches, actions and identities of the movements. Social movements thus shared an oppositional tone. Second, social movements learned from each other, exchanging and transmitting members, ideas and techniques.[3] The relevant social movements here include, but are not restricted to, the civil rights movement, anti-nuclear movement, anti-Vietnam movement, political activism typified by umbrella groups (such as Students for a Democratic Society), new environmentalism and feminism. Each movement, but in particular the civil rights movement of the 1950s, became a model for later movements, just as they in turn drew inspiration, techniques and other lessons from even earlier tides of activism, including the anti-slavery campaigns of the eighteenth and nineteenth centuries.

Equipped with a sense of these terms, we can now see how science figures in the second wave. Science and scientists featured in social movements in three relationships. First, certain scientists and sciences were objects of criticism because they were seen within social movements as tools of their opponents. Second, places where science was done became theatres for social movement demonstration. Third, scientists as activists were contributors to social movements. This third relationship took two forms: their science could be incidental to their involvement in a movement or, most significantly, it could be the cause, the tool, the object and subject of activism. The cases considered below involve all three of these relationships.

The civil rights and anti-nuclear movements furnish candidates for science as an object of criticism. In the civil rights case, experts contributed to both sides, in support of segregation and in questioning claims of racial difference.[4] The environmental-hereditarian controversy continued after civil rights legislation was enacted. Arthur Jensen and the transistor pioneer William Shockley appealed directly to reactionary public sentiments in the 1960s and the 1970s, an appeal rebutted, also in public, by critics. The end effect in these kinds of confrontations was 'socially visible' disagreement. The 'principal audience of the debate was not so much the scientific community itself but the lay public', comments philosopher of science and politics Yaron Ezrahi. 'The contestants were naturally led to invest much effort in building indirect evidence through which science is made more socially visible in order to persuade the public that their opinion is more representative of the true scientific consensus than that of their rivals.'[5]

At first glance, anti-nuclear campaigns also provide a case study of scientific activity as the target of a social movement's criticism. Britain's Campaign for Nuclear Disarmament (CND), for example, chose the UK's nuclear weapons laboratory, Aldermaston, as the terminus of its Easter marches. From the days of the Manhattan Project, scientists offered the 'most serious resistance to the use of the Bomb', but their critique was the abuse of science rather than of science per se.[6] Nevertheless, expressed in anti-nuclear debates were the seeds of a critique of such use/abuse instrumentalism. While science was seen by some as a neutral tool that was being abused rather than well used, for CND nuclear science was a tool it would rather did not exist in the world. A case could thence be made that, from Leo Szilard onwards, nuclear control or disarmament campaigns were one source of questions about the neutrality of science. Furthermore, scientists' arguments could be appropriated and reinterpreted as more generalized critiques. For example, when the Greater St Louis Committee for Nuclear Information argued that Edward Teller had 'a vested interest in arguing that atomic fallout was not harmful', Teller responded by arguing that CNI member Edward U. Condon's claim that fallout was dangerous was itself politically motivated;[7] controversy generated public accusations that science followed interests.

But the moral crusade rhetoric of CND and similar bodies supplanted rather than complemented scientist-led critiques of nuclear weapons policy. Before 1958, many prominent interventions had been led by scientists. Examples include the Chicago scientists' opposition to the use of the bomb before Hiroshima, the foundation of the *Bulletin of Atomic Scientists*, the Russell–Einstein manifesto of 1955, and the subsequent first Pugwash conference of 1957. The Russell–Einstein manifesto had railed against the hydrogen bomb, and had invited its immediate audience, 'and through it the scientists of the world and the general public, to subscribe to the following resolution:

> In view of the fact that in any future world war nuclear weapons will certainly be employed, and that such weapons threaten the continued existence of mankind, we urge the governments of the world to realize, and to acknowledge publicly, that their purpose cannot be furthered by a world war, and we urge them, consequently, to find peaceful means for the settlement of all matters of dispute between them.

It carried the signatures of eminent scientists: Max Born, Percy W. Bridgman, Albert Einstein, Leopold Infeld (a Polish physicist, pre-war colleague of Einstein's, who had been hounded by anti-comminists in Canada), Frederic Joliot-Curie, Hermann J. Muller, Linus Pauling, Cecil F. Powell, Joseph Rotblat and Hideki Yukawa, as well as Bertrand Russell. The petition organized by Linus Pauling in 1957–8 was signed by 11,038 scientists from forty-nine countries, including thirty-seven Nobel laureates.[8] Scientists were not so prominent after 1958.

Social movements learned from each other. In many ways, environmental activists appropriated the roles and arguments of activist nuclear scientists. For example, Rachel Carson repeatedly drew parallels between radiation and pesticides in arguments in *Silent Spring* (1962). She could be confident that her reference to *Lucky Dragon*, the Japanese fishing vessel contaminated by fallout, would be familiar to her audience – an audience that had sat, terrified, through the Cuban Missile Crisis and could translate from the effects of one known insidious invisible contaminant (fallout) to make another unknown (pesticide pollution) meaningful and alarming. Carson is an exemplary figure, showing how key individuals could orchestrate publicly visible disagreement between experts. It is worth digging into her biography to find out why she could take up such a powerful position and role.

Rachel Carson was born in 1907. Her father was a travelling salesman and her mother was keen on natural history.[9] She attended the Pennsylvania College for Women, where her mentor was Mary Scott Skinker, who encouraged a change in major to biology. She then went to Johns Hopkins University in 1929, and was able to study too at the Marine Biology Laboratory at Woods Hole on Cape Cod, where, as she wrote, 'one can't walk far in any direction without running into water'. She wrote a masters thesis on catfish kidney embryology. By now the principal breadwinner of an impoverished family, Carson wrote books, starting with *Under the Sea-Wind* of 1941, which received good reviews but made poor sales in a year of the more deadly maritime events of Pearl Harbor. Her next book, *The Sea around Us* (1951), was an immediate success, a high-point in popular science writing; it won prizes, was serialized in the *New Yorker*, and secured Carson's reputation.

In the early 1950s, at the Fish and Wildlife Service (FWS), Carson had been able to sit at the centre of three networks. First, the reports of different experts such as oceanographers, marine biologists, ornithologists and ecologists crossed her desk, an obligatory passage point in the FWS's review process. Second, through her contacts with bodies such as the Audubon and Wilderness societies, Carson was in touch with naturalists and nature writers. Finally, through her skilful agent she could tap the resources of the publishing world. This position, combined with an enviable gift of expression, provided the basis for the publishing successes of *The Sea around Us* and *Silent Spring*. More importantly, Carson was ideally placed to orchestrate the public display of expert difference: 'Perhaps even more important than the particular conflict she wrote about – between chemical and biological insect control', write sociologists Andrew Jamison and Ron Eyerman, 'was the presentation of conflict itself.'[10] This staging and demonstration of expert disagreement is at the heart of *Silent Spring*.

With evidence gathered from her networks, Carson was able to describe the devastating and indiscriminate effects of insecticides such as DDT (which had been sprayed from the air across swathes of Long Island in an attempt to control the gypsy moth) and chlordane (used against the fire ant). There had existed expert knowledge about the deleterious effects of DDT as early as 1946, but assessments and arguments were internal and not readily visible from outside.[11] Carson drew public attention to disputes between experts. It is a testament to her considerable rhetorical skill that she made the death of destructive insects into a matter of public concern. (For example, there is very little in the book about birds, but by opening the text with a chilling fantasy in which a town in spring lies silent because the song birds have starved, a powerful sympathetic connection had been made.) There was enough chemistry present to make her case.

She called for moderation and the use of biological controls as an alternative to chemical pesticides.

Carson was subject to a sustained and organized attack by the chemical industry, including an attempt to prevent publication through the courts brought by the manufacturer Veliscol.[12] Reviewers from the industry side were extremely hostile (one review was titled 'Silence, Miss Carson!'). Monsanto penned a rather ingenious parody of the opening chapter of *Silent Spring*, 'The desolate year', which asked its readers to 'Imagine . . . that by some incomprehensible turn of circumstances, the United States were to go through a single year completely without pesticides.' The results were straight from a B-movie horror, as the insects ate America:

> Genus by genus, species by species, sub-species by innumerable sub-species, the insects emerged. Creeping and flying and crawling into the open, beginning in the southern tier of states and progressing north-ward. They were chewers, and piercer-suckers, spongers, siphoners, and chewer-lappers, and all their vast progeny were chewers – rasping, sawing, biting maggots and worms and caterpillars. Some could sting, some could poison, many could kill.[13]

Opinion-shapers across the Western world were bombarded with cut-tings. (My copy of these reviews came from the National Archives in London; they had originally been sent by the chemical industry lobby to the Duke of Edinburgh, who forwarded them, with expressions of concern, to the relevant government department.) Carson was, never-theless, able to wield political influence of her own, testifying, before her early death by cancer, to congressional hearings on pesticides that led to changes in regulation. The impact of *Silent Spring* also marked a 'turning point' in the debate about chemical carcinogenesis, point-ing to later formulations of industrial products as pervasive threats and risks.[14]

Carson was an insider: trained as a scientist, respectful of science, a defender of the authority of science. But she was also critical of elite, disengaged science, as the following excerpt of a speech made in 1952 suggests:

> Many people have commented with surprise on the fact that a work of science should have a large popular sale. But this notion that 'science' is something that belongs in a separate compartment of its own, apart from everyday life, is one that I should like to challenge.
>
> We live in a scientific age; yet we assume that knowledge of the sci-ences is the prerogative of only a small number of human beings, isolated and priest-like in their laboratories. That is not true. It cannot be true.

The materials of science are the materials of life itself. Science is part of the reality of living; it is the what, the how, and the why of everything in our experience. It is impossible to understand man without understanding his environment and the forces that have molded him physically and mentally.[15]

Implicitly, and increasingly explicitly, there was a new politics of science here, in which expertise should be made democratically responsive: pesticide use must be made a 'public issue, not a technical issue decided in expert arenas often subject to industry influence' and hidden from public view.[16] In *Silent Spring*, Carson argued that the public that 'endures' pesticide effects had the right to know and an obligation to act. Some readers violently disagreed, as evidenced by the following letter published in the *New Yorker* in June 1962:

Miss Rachel Carson's reference to the selfishness of insecticide manufacturers probably reflects her Communist sympathies, like a lot of our writers these days. We can live without birds and animals, but, as the current market slump shows, we cannot live without business. As for insects, isn't it just like a woman to be scared to death of a few little bugs! As long as we have the H-bomb everything will be OK. PS. She's probably a peace-nut too.[17]

But many other readers were carried along by Carson's writing. The members of social movements, in this case the new environmentalism of the 1960s and 1970s, provided the core readership and audience for Carson's public orchestration. New readers, in turn, became potential new members, a growing audience that could be upset, concerned and eventually curious about rival expert claims.

Social movements were institutions that provided the resources necessary to sustain a more general scrutiny of expertise. Scientist turned environmentalist Barry Commoner, in *Science and Survival*, addressed this moment explicitly: 'If two protagonists claim to know as scientists, through the merits of the methods of science, the one that nuclear testing is essential to the national interest, the other that it is destructive of the national interest, where lies the truth?'[18] The fact that the 'thoughtful citizen' has to ask 'How do I know which scientist is telling the truth?' 'tells us that the public is no longer certain that scientists – all of them – 'tell the truth'. The situation in which a concerned witness is confronted by a spectacle of expert disagreement was replicated many times. If asked to choose, the witness was faced with a difficult choice between two experts, each claiming to 'know as scientists'. The slippage identified by Commoner, the slide from challenges to some scientists to doubt in science ('all of them') was invited by this situation. It is a situation that called yet again for the production of more experts, even experts in the sociology of science.[19]

Carson's and Commoner's campaigns were projects orchestrated so that scientists were organizing critically to observe science. Jerry Ravetz nailed this moment: industrialized science was provoking its 'opposite, "critical science"', a 'self-conscious and coherent force'.[20] The new breed of science journalists, notably Daniel S. Greenberg, helped others to analyse science, treating science like any other politically interested branch of the economy.

Campus protest

A second slippage was a common feature of social movements. Competition for activists' attention, time and resources, in combination with the loose organizational structure of movements, encouraged movement between movements. For example, there is some evidence that campaigns against the war in Vietnam weakened disarmament activism.[21] Alternatively, social movements could run together if the targets proved flexible enough to coordinate action among very different groups. Philosopher Andrew Feenberg's account of the May '68 events can be translated into these terms. Surveying more broadly the 'dramatic shift in attitudes toward technology that occurred in the 1960s', it was 'not so much technology', he notes, 'as rising technocracy that provoked public hostility'.[22] In Paris, in particular, students interpreted the university as a technocratic society in miniature that could form the basis of a common cause with workers' movements and with French middle strata.

In May 1968 student demonstrations closed universities, 10 million strikers joined them, and opposition to the establishment 'exploded among teachers, journalists, employees in the "culture industry", social service workers and civil servants, and even among some middle lower level business executives'.[23] Here positive notions of 'autonomy' acted as a common coordinating thread. Calls for autonomy were not calls for severance from society but identification with the 'people' against technocratic masters. 'While the May Events did not succeed in overthrowing the state', summarizes Feenberg, the ferment starting on the French campuses 'accomplished something else of importance, an anti-technocratic redefinition of the idea of progress that continues to live in a variety of forms to this day'.[24] When revolutionary ideals were scaled back to 'modest realizable goals', a successful new micropolitics of technology emerged.

410

Like the Paris streets, American campuses became theatres for anti-technocracy protest. At Berkeley, the free speech movement, which began in 1964 and grew from civil rights campaigns, launched a tide of student activism. By the following year Berkeley campus was a centre of anti-Vietnam protest and organization. The Vietnam War, notes Feenberg, 'was conceived by the US government and sold to the public as a technical problem American ingenuity could quickly solve'.[25] Edgar Friedenberg's response to the call by the president of Berkeley for universities to be 'multiversities', putting knowledge at the disposal of society's powers (not least the military), was to call instead for the university to be 'society's specialized organ of self-scrutiny'.[26] Mario Savio advocated direct action: 'You've got to put your bodies upon the gears and upon the wheels, upon the levers, upon all the apparatus – and you've got to make it stop.'

In among this general, amorphous complaint about technocracy were sharp, specific arguments against war research at universities. At Princeton, military-sponsored research was fiercely debated from 1967, pitting activist engineers such as Steve Slaby against Cold War scientist Eugene Wigner, who had likened the actions of the Students for a Democratic Society (SDS) to those of Nazi youths; however, protest fizzled after one final summer of strikes in 1970 and a committee (chaired by Thomas Kuhn) reported that Princeton had relatively little military-sponsored research.[27] At Stanford, student and faculty protests against secret contracts and classified research began in 1966. 'The extent of Stanford's classified research program', centred at the Applied Electronics Laboratory and the Stanford Research Institute (home to counter-insurgency projects), writes historian Stuart Leslie, 'although common knowledge among the engineers, shocked an academic community still coming to terms with the Vietnam War.'[28] In 1967, a Stanford 'student-run alternative college', The Experiment, called for the indictment of university officials and trustees for 'war crimes', while The Experiment and the local SDS chapter organized anti-war marches and campaigns.[29] April 1969 saw the occupation of the Applied Engineering Laboratory by protesters.

SDS had also organized a small sit-in against the Dow Chemical company at MIT in November 1967, but it was federal defence contracts at the university that triggered vehement opposition.[30] MIT received more defence research and development grants than any other university. Its Lincoln and Instrumentation laboratories, specializing in electronics and missile guidance technologies respectively, as well as the independent but adjacent MITRE labs, were very much part of the Cold War 'first line of defence'.[31] Nevertheless,

411

at MIT, the conversion in 1967 of the Fluid Mechanics Laboratory, from missiles, jet engines and re-entry physics to environmental and medical research, provided an exemplar for the protestors, although it appealed to the laboratory workers on more pragmatic grounds.[32] In January 1969 MIT faculty members called a strike intended to 'provoke "a public discussion of problems and dangers related to the present role of science and technology in the life of our nation"'.[33] The protesters' manifesto of 4 March called for 'turning research applications away from the present emphasis on military technology toward the solution of pressing environmental and social problems'.[34] A tense standoff between protesters and Instrumentation Laboratory boss Charles Stark Draper was broken by a riot, featuring police dogs and tear gas, in November 1969. In the cases of both MIT and Stanford the moderate protesters won. The Instrumentation Laboratory was divested from MIT to become the independent Charles Stark Draper Laboratory in 1970, while the Stanford Research Institute was also divested, and its campus annexe, the theatre of protest, closed. The radicals had wanted conversion. All the divested Cold War laboratories prospered under continued defence patronage and with continuing ties with the adjacent, if now formally independent, universities.[35]

The campuses and laboratory plazas were indeed theatres of demonstration. Furthermore, the establishment of new bodies indicates that much else was at stake. The British Society for Social Responsibility in Science, set up in 1969, brought together remnants of the older radical generation of the 1930s, such as Joseph Needham, with the new generation of left-of-centre and radical academic scientists of the 1960s.[36] MIT students and faculty initiated the Union of Concerned Scientists (UCS) in 1969. In a link back from Vietnam to disarmament, what started as a concern about campus contributions to the war in South-East Asia shifted in the 1970s to a critique of nuclear safety issues.[37] UCS was particularly active in the second Cold War period of the 1980s, organizing a report that in many ways was an echo of Pauling's 1957–8 petition.[38] The UCS, alongside the Scientists' Institute for Public Information (SIPI, formed in 1963) and Science for the People (SftP, founded in 1969 'as a group dedicated to finding ways to take political and social action against the war in Vietnam'), has been studied by the sociologist Kelly Moore, who has offered an interesting general argument. Moore argues that 'public interest science organizations', such as UCS, SIPI and SftP, were an institutional response to a severe quandary posed by the mixture of political activism and the sciences:

412

Activist scientists had to be politically critical of science without sug-gesting that the content of scientific knowledge might be tainted by non-scientific values ... More specifically, they faced two related problems. First, their activities and claims threatened to fragment professional organizations that represented 'pure' science and unity among scien-tists. Second, once the discussion became public, it threatened to reveal the subjective nature of problem choices, methods, and interpretations because it focused attention on the relationship between sponsors of science and scientific knowledge.[39]

Such tensions were reconciled, argues Moore, through the forma-tion of public interest science organizations. They 'made serving the public interest relatively permanent and durable, obfuscated how political interests affect scientific knowledge, and helped preserve the organizational representations of scientific unity: professional science organizations.' In other words, UCS, SIPI and SftP functioned to preserve the purity of bodies such as MIT, the American Association for the Advancement of Science, and the American Physical Society, respectively. Also clearly revealed in Moore's study is the extent to which this institutional response was provoked by the problem that 'attention was increasingly being drawn to multiple interpretations of scientific evidence – certainly not a desirable state of affairs for a profession that relies more so than others on the presentation of una-nimity on rules, methods, and interpretations.'[40]

Self-criticism

It is a commonplace that the post-war baby boom generation held attitudes in opposition to their parents' generation. More significant is the observation that the baby boom generation identified and ana-lysed themselves as different.[41] Self-consciousness emerged as a theme in elite intellectual thought partly as a reaction against overbearing systematization. The dominant social science was quantitative and scientistic in method and, in the words of David Hollinger, 'triump-phalist' in spirit, 'marked by the buzzwords modernization theory and the end of ideology'.[42] But the books of Daniel Bell and Walt Rostow, while governing policy, were not the texts deemed influ-ential among the members of social movements in the long 1960s. The texts and authors that were influential had a common theme of overbearing structural determination and the limits on responses of the individual. Examples are Herbert Marcuse's *One-Dimensional Man* (1964), Marshall Berman's historical examination of radical

413

individualism and the emergence of modern society, *The Politics of Authenticity* (1971), and Jacques Ellul's 1950s book *La Technique*, published as *The Technological Society* (1964). Ellul argued that 'science has become an instrument of technique': 'Science is becoming more and more subordinate to the search for technical application'; like all other human affairs, science had been assimilated, an 'enslavement' that became entrenched only in the twentieth century.[43]

Yet alongside jeremiads against overbearing control were published celebrations of individual triumph in science. The point to grasp is that these two interpretations of science provoked each other.

Tensions between individualistic entrepreneurship and communal ideals

In James Watson's bestselling *The Double Helix* (1968), an account of his and Crick's discovery of the structure of DNA, readers found protagonists who behaved like ordinary human beings rather than following some higher moral code. *The Double Helix* cannot be read simply as an account of what 'really' happened; as Jacob Bronowski was quick to observe, the structure of the narrative resembles a fairy tale, complete with heroes, quests and wicked witches. While there is an authenticity in Watson's account of scientists 'in real life', we should also scratch beneath the surface and ask what else was being championed.

Watson's protagonists were individualistic, entrepreneurial, willing to bend rules and to slight colleagues to get ahead. Such agents could be found elsewhere in the long 1960s: historian Arthur Marwick has noted that individual cultural entrepreneurship was a prominent feature of the long 1960s.[44] Indeed, such entrepreneurialism explains why new social movements, especially the counterculture, took the forms it did. Watson's protagonists are from the same mould. And they had lasting influence. The exponent of privately funded biotechnology Craig Venter was in California in his early twenties when he read *The Double Helix*. A recent hagiographical sketch significantly informs us: 'Years later Venter would complain that he had no mentors . . . If there was one, he said, it was the Watson of *The Double Helix*.'[45] The point is not that the Watson persona stands in contrast to that of the cultural movers in the long 1960s. Rather, in 'doing his own thing', he is similar. Watson's protagonists provide just the models of behaviour, the 'heroes', necessary for 1970s commercialization in the biosciences, discussed in the next chapters.

More accurately, the individualistic entrepreneurship and more communal ideals of the organization of science coexisted, uncomfortably, in the long 1960s, with Maurice Wilkins, in this instance, representing the latter.[46] In the early Cold War, the communal model had been captured in the form of Merton's CUDOS norms. It was expressed in its managerial form in the emergence of Big Science as a style of organization. But 'Big Science' as a term was coined in the 1960s, as a term of abuse. As a self-critical term, 'Big Science' originated in the 1960s reflections of Alvin Weinberg, an administrator of a central laboratory of the Cold War, Oak Ridge, as well as in Derek de Solla Price's 1960s science of science.[47] It could have been coined as a label of description in any earlier decade of the century. Watson's account of individuals making great discoveries proved popular precisely because it stood in contrast to science as a large-scale organized, largely depersonalized activity. Watson's protagonists were therefore also useful fictions, useful in providing a semblance of individualism in the far less individualistic pursuit characteristic of Big Science.[48] Science was fully a part of what Brick labels the 'socialization of intellect in the new mass universities', as well as part of the system of Big Science;[49] Watson's individualistic self-portrait emerges as a means of resolving tensions. I have already discussed, in chapter 14, how the rise of Big Science encouraged talk of 'scientific community' not just because science was more communal but also because its precarious autonomy could better be defended.[50] Likewise, Brick argues that the condition of the socialization of the intellect, the organization of intellectual life in formal institutions relying on public funds and engaged with public policy formation, provoked questions about the consequences of socialization for knowledge.[51] But, if science was a community, then did it not therefore have communal responsibilities?

Talk of crises and revolutions

Furthermore, talk of a scientific community came with talk of 'crisis' within the community. The philosopher Thomas Kuhn provided a specific sense of crisis in *The Structure of Scientific Revolutions* (1962). Kuhn's book would become, by far, the most cited and discussed twentieth-century book about science. Kuhn had taught 'case study' history of science, through the close examination of primary sources, at Harvard in the years following the end of the Second World War at the invitation of James Conant.[52] In *The Copernican Revolution* (1957) he had written a monograph following this

415

approach. But in *The Structure of Scientific Revolutions* he generalized. In his model, scientists worked under a 'paradigm', which provided exemplars of how scientific puzzles might be solved as well as of approved methodologies. However, some anomalous phenomena could not be accounted for under a paradigm. If anomalies withstood attack, then they triggered the development of full-blown 'crises'. The deep, implicit assumptions of paradigms would be drawn to the surface, made explicit and criticized. The outcome might be a new paradigm. Since a new paradigm held new rules about what counted as acceptable science, old and new paradigms were incompatible; specifically, they were 'incommensurable'. 'Normal science', science done under a paradigm, was the achievement of a social community. Crisis was therefore an experience of the scientific community. Kuhn famously came to loathe how his terminology – paradigms, revolutions – was appropriated, or in his eyes abused, by his many readers outside of the philosophy of science.

Nevertheless, it is striking how often diagnoses of a more general crisis in science and technology were made in the long 1960s. There was a 'crisis' in computer programming in the same year (1968) as there was revolution on the campuses.[53] There was even a self-diagnosed 'crisis in sociology'.[54] At the meeting on Social Impact of Modern Biology in 1970, Maurice Wilkins had reminded the audience of the 'crisis in science today [which] has not only direct bearing on the question of our survival but is of deep significance in relation to our fundamental beliefs and our value-judgments'.

> The main cause is probably the Bomb: scientists no longer have their almost arrogant confidence in the value of science. At the same time non-scientists increasingly and openly question the value of science. There are extremists who go further and object to rational thought as a whole.[55]

Wilkins portrayed the scientific community as deeply split over its response to this growing 'breakdown in confidence in reason' among the many. Following the use–abuse model, Wilkins concluded that it was vital not to 'over-react', lest this lead to the 'overall condemnation of science'; rather he urged his audience to be socially responsible and choose to pursue science that, to borrow Peter Medawar's phrase, provided 'imaginative uplift' and, more significantly, 'self-knowledge [, which] should greatly influence our values. Science is valuable . . . in terms of the self-knowledge that it gives.'

Other elite 'leftish' scientists made similar arguments at the same meeting. Fresh from his widely publicized involvement in the Paris

'68 events, Jacques Monod proposed that science, as a 'strictly objective approach to the analysis and interpretation of the universe ... [which] must ignore value judgements', was destroying any and all 'traditional systems of value'.[56]

> Hence modern societies, living both economically and psychologically upon the technological fruits of science, have been robbed, by science itself, of any firm, coherent, acceptable 'belief' upon which to base their value systems. This, probably, is the greatest revolution that ever occurred in human culture. I mean, again, the utter destruction, by science, by the systematic pursuit of objective knowledge, of all belief systems, whether primitive or highly sophisticated, which had, for thousands of years, served the essential function of justifying the moral code and the social structure.

This 'revolution', argued Monod, was 'at the very root of the modern mal du siècle', especially among 'the young'. When Monod looked out from the Pasteur Institute to see the 'revolution' on the streets in '68, one has to imagine him thinking the events were caused by science.

Wilkins was not the only contemporary commentator to identify 'crises'. Barry Commoner, for example, had written in *Science and Survival* (1966) of the crises of modern biology, by which he meant a science that was being torn apart by the conflict between traditional organismic science, derived from natural history, and an aggressive new molecular biology. In history of science, Paul Forman's discovery and account (1971) of 'crises' in science in Weimar Germany, discussed earlier, date from this moment, and may be considered as part reflections on his contemporary scene. Forman had finished his PhD at the University of California, Berkeley, in 1967, and had witnessed directly the response of scientists in a 'hostile intellectual milieu'.

But Wilkins's 'crisis' is interesting because he portrayed it as a momentous condition afflicting the sciences more broadly. The point is not whether his diagnosis was correct. The quantitative evidence, for example, is problematic. Amitai Etzioni and Clyde Z. Nunn reported the results of the Louis Harris poll that the proportion of public expressing 'great confidence' in the people 'running science' (not of course a direct measure of confidence in science) had dropped from 56 per cent (1966) to 37 per cent (1972).[57] Etzioni and Nunn held that the 'falling away from science' was 'part of a general lessening of faith in American institutions and authorities rather than a major anti-science groundswell ... from religion to the military, from the press to major US companies [a]ppreciation for all of them, without exception, has fallen'.

Pills and the biomedicalization of everyday life

Crises could be invented by technical experts. Crises also offered working world problems that in turn generated new science and technology. The 'population crisis' was a term invented by demographers (and only later publicized in works such as Paul Ehrlich's bestselling *Population Bomb* of 1968), and the oral contraceptive, 'the pill', offered a technocratic solution by family planners. In fact, the story of the pill richly illustrates the transitional themes of this chapter. In the 1910s birth control clinics had been illegal in the United States, while the posting of contraceptives to physicians was still classed as 'obscene' until 1936, and thirty states prohibited the sale of contraceptives as late as 1960.[58] Yet in 1955, 70 per cent of white married American women under forty had practiced some form of contraception, such as the diaphragm, condoms or the rhythm method.[59] The pill would be entering a crowded marketplace. The pill nevertheless differed from existing contraceptive methods in crucial respects: it would be the first chemical medication marketed at healthy individuals, offering, like the diaphragm, women's control over their own fertility but without the indelicate attentions involved. Pharmaceutical companies were unwilling to provoke the sizable Roman Catholic market by developing an oral contraceptive alone. The project therefore required the management skills and finances of other interested parties.

The two most important parties were the independently wealthy birth control champion Margaret Sanger and the Population Council, an elite group concerned with population control. Sanger, who had worked as a nurse in poor districts of New York City in the 1910s and seen at first hand the pressure brought by a combination of motherhood and poverty, wanted a reliable form of contraceptive. She 'envisioned a "birth control pill"', writes historian Elizabeth Siegel Watkins, 'because a pill that could be taken at a time and place independent of the sex act would place the control of contraception solely in the hands of women, where it belonged.'[60] A male pill was not considered: it would be technically difficult (but not impossible), suffered from a dearth of male research subjects, and went against the 'social convention' that classed 'contraception, like pregnancy and childrearing . . . [as] a feminine responsibility'.[61] Significantly, Sanger trusted science and organized medicine to fulfil her vision and provide a pill.

Sanger was also concerned about apparent overcrowding in the cities in the Far East, which she had witnessed on travels to Japan, China and Korea in the 1920s.[62] She was therefore broadly sympa-

418

thetic to the core concern of post-war demographers who regarded 'overpopulation' in the Third World as a severe problem. John D. Rockefeller III funded the Population Council from 1952 to examine ways of reducing human fertility. Members were drawn from the elite tiers of science, including Detlev Bronk (president of the Rockefeller Institute and the National Academy of Sciences), Karl T. Compton (head of MIT), Frederick Osborn (president of the American Eugenics Society) and Lewis Strauss (Oppenheimer's nemesis on the Atomic Energy Commission).

Sanger sought out a scientist who was an expert in steroid hormones and in reproductive physiology. She found Gregory Goodwin Pincus, who worked at the Worcester Foundation for Experimental Biology in Shrewsbury, Massachusetts. Pincus had been researching production methods of the steroid cortisone, used for seemingly miraculous arthritis treatments from the late 1940s, for G. D. Searle and Company, but he had also done past work on mammalian sexual physiology.[63] Sanger and Pincus discussed exploiting the effect of the hormone progesterone, the presence of which prevented an ovary from releasing an egg. Progesterone could be extracted from animal organs, but at the rate of 1 gram from 1 ton of organs it was prohibitively expensive. Chemists at two companies worked to find cheaper synthetic alternatives to natural progesterone, starting with an extract of Mexican yams. Carl Djerassi's team at Syntex and Frank Colton at G. D. Searle succeeded in 1951 and 1952 respectively. The companies foresaw a limited market in gynaecological treatments – they had little interest in a controversial contraceptive.

Sanger persuaded her colleague Katherine Dexter McCormick, who had married into the harvester company fortune, to provide the substantial funding for Pincus to turn the synthetic hormones into 'the Pill'. McCormick, too, a graduate of MIT, trusted science to deliver.[64] For medical trials, Pincus sought out a Catholic obstetrician-gynaecologist at Harvard Medical School, John Rock. Trials took place in the United States (including psychotic patients at Worcester State Hospital) and Puerto Rico, where the main field tests for contraceptive effect took place in 1956. By 1958, 830 women in Humacao, Puerto Rico, and Port-au-Prince, Haiti, had taken the Enovid pill. The Food and Drug Administration (FDA) approved the prescription of Enovid for contraception in May 1960, the first drug intended for regular use by healthy women.[65] This approval was therefore a step towards the biomedicalization of everyday life.[66]

This first phase of the history of the Pill was marked by high levels of trust in the experts involved. Patients, guided by positive coverage in the press, requested it from physicians, who followed the positive recommendations of advertising and the urging of the industry's marketers, the 'detailsmen'. Contraceptive use increased at the margins. The popular press connected the Pill, a revolution in contraception, with changing permissiveness in society more generally and the so-called sexual revolution, a moral panic about the behaviour of unmarried youth specifically. Pearl S. Buck, the author, wrote in *Reader's Digest* (in the article 'The Pill and the teen-age girl', 1968):[67] 'Everyone knows what The Pill is. It is a small object – yet its potential effect on our society may be even more devastating than the nuclear bomb.' In fact, as the sexological studies of Alfred Kinsey of the 1940s and 1950s had shown, there was plenty of unmarried sex taking place before the pill, and there is no evidence from the 1960s that the pill was a cause of change.[68]

Where there was change was in the relations of medical-scientific authority to its publics. Anecdotal and then research claims about links between the use of the Pill and incidence of blood clot embolism and cancer circulated from the mid-1960s. But these were debates within the relatively closed walls of academic publication and the medical press. As with Carson and *Silent Spring*, an author and book, in this case Barbara Seaman's *The Doctor's Case against the Pill* (1969), orchestrated the debate as a public controversy. Senator Gaylord Nelson led congressional hearings on the matter, at which male experts were heard but female users were denied a voice. The tone of the debate was now intensely critical. What previously had operated as trusted links between patron, scientist, pharmaceutical company and physician now looked seamy and interest-driven. Crucial in this process was the orchestration of public controversy, which supplied an expanding media with copy, who in turn were responding to its readership. Among this readership were social movements, in this case, specifically, the D. C. Women's Liberation, who found a cause, as one prominent member recalled:

> we had heard that there were going to be these hearings on the birth control pill on the Hill, so about seven of us went . . . When we got there, we were both frightened, really frightened, by the content and appalled by the fact that all these senators were men [and] all of the people testifying were men. They did not have a single woman who had taken the pill and no women scientists. We were hearing the most cut-and-dried scientific evidence about the dangers of the pill . . . Remember all of us had taken the pill, so we were there as activists but also as concerned women . . . So while we were hearing this, we suddenly said, 'my

God', and we – because in those days you did things like that – raised our hands and asked questions.[69]

This increased willingness to question authority by the women's movement led to a specific change: the statutory provision of more information, 'an important turning point in the doctor–patient relation',[70] and was indicative of a broader shift in the expert–public relation more generally. 'A generation earlier, Margaret Sanger and Katherine McCormick ... [had achieved their goal when] they enlisted the help of scientists and physicians and encouraged these experts to develop a solution to the problem of birth control ... [and had] hailed the oral contraceptive as a scientific triumph for women and gladly entrusted contraception to the hands of physicians', summarizes Elizabeth Siegel Watkins. 'In contrast, the women's health movement rejected the hegemony of the medical-pharmaceutical complex and instead advocated lay control over the delivery of health services ... [while interpreting] the pill as representative of patriarchal control over women's lives.'[71]

A second development which illustrates the complex interaction between chemical innovation, challenges to authority and social movements was the deinstitutionalization of the severely mentally ill and the transformation of psychiatric expertise in the post-war years. The severely mentally ill, in Europe and the United States, were confined to asylums in great numbers. In mid-century asylums, psychiatrists attempted to treat the patients with insulin coma therapy (in which the schizophrenic patient went into a state of shock on injection of insulin, to emerge a more manageable person) or electro-convulsive therapy or lobotomy (which achieved similar results with electrical or surgical means). In the 1950s, new psychopharmacological substances contributed to a transformation of psychiatry.[72] In the 1940s, the French pharmaceutical firm Rhône–Poulenc, a world leader in tropical infections, took the stain methylene blue and tried it first as an anti-helminthic (anti-worm) candidate drug and then its phenothiazine nucleus as an anti-histamine. Some anti-histamines were known to have sedative or stabilizing effects. Paul Charpentier produced a new series of phenothiazine compounds in December 1950, including a chlorinated one, chlorpromazine. Remarkably, when rats were given the new substance they did not climb ropes (a sedation test), not because they were sedated but because they were indifferent.[73]

Henri Laborit, a Parisian doctor who had experimented with anti-histamines as stabilizers during operations, asked whether human

421

patients could also be made indifferent. When in early 1952 news of chlorpromazine reached Hôpital Sainte-Anne, the largest psychiatric hospital in Paris, with the University of Paris's department of psychiatry in its grounds, Jean Delay and Pierre Deniker tried it on their psychotic patients. The cures, reported in a very quickly written paper, were almost incredible.[74] While an older generation of psychiatrists were sceptical (some considered the complete treatment of psychosis impossible), a younger generation began to prescribe the anti-psychotic drug. This use of chlorpromazine was not especially in the wishes of the companies involved: Rhône–Poulenc viewed psychiatrists as a poor market, and Smith Kline and French, which had taken it up in the United States, had sought a licence for the drug as an anti-emetic.[75] Nevertheless, Smith Kline and French's 'Thorazine' (chlorpromazine) made $75 million in the first year after its introduction in 1955, in competition with tranquilizers such as Wallace Laboratories' Miltown and Ciba's reserpine.

With chlorpromazine and reserpine, a new science of psychopharmacology seemed possible, with responses to doses measured and evaluated (the chemical psychiatrists felt a particular need to demonstrate effectiveness through controlled and publicized evaluative tests because of the dominance of their field by psychoanalysts); news that other substances, such as Sandoz Pharmaceuticals' LSD (Albert Hofmann's discovery of 1938, set aside until revisited in 1943), could bring on psychosis, while the first two drugs removed it, seemed to confirm this new sense of control and measurement.[76] The availability of drugs that controlled psychosis was a powerful complementary force to social therapies that had started the unlocking of asylums. Large hospital asylums, seen as authoritarian structures that engendered institutional neuroses, were criticized from the late 1940s; this, says historian David Healy, created 'the perfect setting for the emergence of pharmacotherapy', and the use of chlorpromazine tackled problems that social milieu approaches did not: 'drugs made it possible to do therapy.'[77]

A much more radical criticism of old psychiatric therapies came from the anti-psychiatrists, a movement led by the very different figures of the Scot R. D. Laing and the Hungarian-American Thomas Szasz. In *The Myth of Mental Illness* (1960), Szasz denounced the scientific pretensions of psychiatrists and was particularly suspicious of state-led psychiatry. Hospitalization of the insane was, for Szasz, the collective trampling of the rights of the individual. In books such as *The Divided Self: An Existential Study in Sanity and Madness* (1960), *The Self and Others* (1961) and *The Politics of Experience* (1967),

Ronald Laing set out his arguments for madness as a socially induced condition, and not necessarily a negative one. From 1965 Laing guided a therapeutic community at Kingsley Hall, in east London, in which therapists would live with patients, who would be neither restrained nor made indifferent with drugs such as chlorpromazine. The authoritative hierarchy between doctor and patient was being radically challenged.

Anti-psychiatry had a resonant message in the long 1960s. 'In 1969, following visits to the country by Szasz, Laing and other anti-psychiatrists, Japanese students revolted and occupied the department of psychiatry at Tokyo University', Healy records.

> This was a revolt against biological psychiatry, against the dominance of Tokyo in the Japanese hierarchy, and against authority in general. Hiroshi Utena, the professor of psychiatry at the university, was forced from office by student disapproval of biological research he had done in the 1950s. The university department was to remain occupied by students for ten years, bringing all research to a halt.[78]

While in Paris in 1968, it was Jean Delay, regarded by the students as a symbolic figure of hierarchical society, who attracted critics of chemical lobotomies.[79] Writes historian David Healy:

> Chlorpromazine for him was a symbol of the capacity of psychiatry to restore order to the world. But an older world order based on hierarchy and deference was giving way to a new world, which had no sympathy for him, a world that chlorpromazine had done much to create, in ways he could never have understood. The students, guided by Laborit's view that chlorpromazine caused indifference, proclaimed it a means of buttressing a madness-inducing social order. Where chlorpromazine had been hailed as liberating the mad from their chains sixteen years earlier, it was now seen as the dreaded camisole chimique (or chemical straitjacket) and its self-professed creator was a target for the revolutionaries.[80]

While in this instance the protestors got their prize (Jean Delay resigned), in general, psychopharmacology, the management of mental illness through prescribed drugs, is the world we live in. 'The clash of a rationalist psychiatry with a romantic anti-psychiatry did not lead to a triumph of either', concludes Healy, 'but rather led to the takeover of both by a psycho-pharmaceutical complex.'[81]

If one response to medical-scientific controversy was to wrest control away from experts, and a second was to transform expertise, then a third was to build new types of expertise. Bioethics and evidence-based medicine are sciences of the working world of medical-scientific controversies. In 1972, the conditions under which the Tuskegee Syphilis Study took place came to light.[82] From 1932, for forty years, black patients in Tuskegee, Alabama, had been left untreated so that the later stages of syphilis could be studied. This practice had continued even after penicillin, available in the 1940s, provided an effective cure. These shocking revelations came soon after other published accounts of dubious human experimentation, such as Maurice Pappworth's *Human Guinea Pigs* (1967) on cases in Britain and Henry K. Beecher's 1966 exposure of deliberate hepatitis infection of patients at Willowbrook State School for the Mentally Ill and the injection of elderly patients at the Jewish Chronic Disease Hospital with cancer cells.[83] In the wake of Beecher's article, organizations in the United States such as the National Institutes of Health and the Food and Drug Administration felt 'forced to design consent forms and institute ethical review boards'.[84] In the National Research Act of 1974, which responded to Tuskegee, research institutions receiving federal money had to enforce the principal of informed consent and set up review boards to monitor human experimentation. These administrative bodies created a demand for persons trained in bioethics. Bioethics as a discipline was also encouraged by the increasing expense of Western medicine, either through innovations in instrumentation, such as dialysis machines (from 1960), or procedure, such as kidney (1967) and heart (1967) transplants. Bioethicists provided professional answers to questions of life and death.

Evidence-based medicine, a more recent term, was the proposal of a Scottish physician, Archie Cochrane, an expert in the epidemiology of tuberculosis at the University of Cardiff. In *Effectiveness and Efficiency: Random Reflections on Health Services*, published in 1972, he set out the case for systematic use of randomized clinical trials as the essential guide to choosing between medical treatments. (Cochrane had been taught statistical methods by Bradford Hill, organizer of the first randomized clinical trial in 1948.) Cochrane's proposals were picked up and developed further by David Sackett and others at McMaster University, Canada.[85] This self-analysis was a variant of the long 1960s phenomenon of self-scrutiny.

Cybernetics and the management of populations

The Cold War shaped the long 1960s by encouraging the development of techniques that, when turned inwards, became instruments of critique. Sometimes this critique was played out within the sciences, as when Cold War oceanographic data provided ammunition for the plate tectonics revolution. Sometimes the critique was played out on a more traditionally political stage. Sometimes the techniques were tools of criticism, sometimes the same techniques were tools of control. Cybernetics, for example, contributed its core techniques, the analysis of feedback loops, to the models used by the Club of Rome to identify the limits to growth.[86] The Club of Rome was an international group of businessmen, scientists and politicians put together by Italian industrialist Aurelio Peccei. The MIT computer scientist Jay Forrester, architect of Project Whirlwind, provided a computer simulation of five global systems (natural resources, population, pollution, capital and agriculture) called 'World 1'. Forrester's protégé, Dennis Meadows, developed further models, Worlds 2 and 3. In each the world ran out of resources, leading to system collapse. Published as *The Limits to Growth* in 1972, the Club's conclusions made global environmental collapse a political issue. The models were heavily criticized at the time and since: relying on input data of dubious quality and making many assumptions about the relationships between factors, the quantitative computer models offered a false sense of precision.

Other environmentalist interventions were also shaped by systems science. Howard Odum, co-leader with his brother of an ecology directly imprinted with the systems approach of atomic science, widened the scope of his arguments still further with *Environment, Power and Society* (1971), which addressed not only the environment but matters of war and religion too in the course of arguing for the recognition of limits to industrial growth.[87] In Britain, E. F. Schumacher, author of *Small is Beautiful* (1973), was a trained economist who had participated in J. K. Galbraith's American Bombing Survey of Germany before working for the National Coal Board as an analyst of systems.[88] Staying in Britain, Edward Goldsmith, inheritor of a banking and hotel fortune, sought in systems theory a reformation of science. 'In Goldsmith's view', writes historian Meredith Veldman, 'modern science could not address the impending systemic breakdown of industrial society, because it was itself a product of that society.'[89] Goldsmith bankrolled the publication of the *Blueprint for Survival* (1972), a text in the spirit of the Club of Rome, and was canny enough to secure the signatures, prior to publication, of thirty-six scientists and academics, including Julian Huxley and C. H. Waddington.[90]

Goldsmith's view about the nature of science is especially ironic given what we now know, through the work of anthropologist

Susan Greenhalgh, about the influence it had in China.[91] Western science in pre-revolution China had a great deal of intellectual prestige. Under Mao Zedong, the People's Republic of China remained scientistic – how could it not be when the guiding philosophy of Marxism–Leninism–Mao Zedong Thought was a science? Indeed, for a while, there was considerable continuity in attitudes towards science. If there has been 'one constant in China since the middle of the nineteenth century', concludes historian Benjamin Elman, 'it is that imperial reformers, early Republicans, and Chinese communists have all prioritized modern science and technology.'[92] Neverthless, the social sciences and most of the natural sciences became subservient to the Communist Party in the 1950s.[93] Scientific objects, such as the famous 'Peking Man' skullcap and other bones discovered in the 1920s near Beijing and lost during the Second World War, were reinterpreted to demonstrate the truth of Friedrich Engels's socialist explanation of the evolution of humans from apes.[94] In the period of the Cultural Revolution (1966–9), anti-scientific, anti-rational and anti-Western attacks were at their height, as the youthful Red Guards led a vicious campaign of criticism of authority. (The wave of anti-authority action was, notice, a global phenomenon.) Many 'scientists and university professors were harassed, humiliated, paraded in the streets, and physically abused, sometimes to the point of death', observes Greenhalgh. 'Teaching and research were severely curtailed, publication of professional journals ceased, scholarly manuscripts, files, and libraries were destroyed, and ties with the outside world were completely severed.'[95] Experts, from geologists to geneticists, were sent to learn from the peasants – an extraordinary phenomenon with surface similarities to the 'citizen science' advocated in the West later in the century. Einstein's theories of space-time, as in the Soviet Union in the 1930s, were attacked in China in the early 1970s as idealist and insufficiently in line with the thinking of Lenin and Mao.[96]

The exception was military science. Some defence scientists, by providing the People's Republic of China with rockets, nuclear weapons, submarines and other military systems, were not only protected but prospered. Qian Sanqiang had studied under the Joliot-Curies in Paris in the 1930s and 1940s before returning to China in 1948, where he joined the Communist Party and oversaw the Chinese nuclear bomb programmes. Qian Xuesen went to MIT and led a highly successful American career: in the late 1940s and early 1950s he was Goddard Professor of Jet Propulsion at Caltech and director of its Jet Propulsion Laboratory. He then became the target of McCarthyite agitation and in 1955 was allowed to leave the United States for

China, where 'as the top military advisor to Mao and Zhou [Enlai], Qian quickly became the most powerful scientist in the country'.[97]

Qian Xuesen protected and promoted the career of his protégé, the missile systems scientist Song Jian. Song Jian had joined the party in 1948, before revolutionary victory in 1949, providing him with excellent political credentials. In 1953 he was sent to the Soviet Union, where he learned from the master of control theory A. A. Feldbaum and gained a doctorate.[98] On the Sino-Soviet split in 1960, Song returned to China and worked in the protected and lavishly funded Seventh Machine Building Ministry, home of Chinese missile control science.

In 1976 Mao died. There followed a period of infighting, in which the moderate Deng Xiaoping emerged victorious over the hard-line Gang of Four, led by Mao's widow. Deng introduced a new policy of Four Modernizations, in agriculture, industry, science and technology, designed to restore China's prosperity. Crucially, Deng faced a problem of legitimacy. Modern science was uniquely able to offer a potent new source of legitimacy, one which provided distance from Mao's regime while benefiting from the strong continuing scientistic values of Chinese culture.[99] In a symbolic act, the rehabilitation of modern physics was signalled by a 'grand ceremony' to mark the 1979 centenary of Einstein's birth.[100] Furthermore, population issues – a fraught topic in the Mao era, during which worries about population growth were equated with the class enemy, Malthus – became articulable.

The working world of Chinese governance, in 1978–80, found population to be a problem for which science could be asked to provide a solution. At this crucial juncture, the only science left standing was defence science. In perhaps one of the most consequential contingencies in twentieth-century science, Song Jian attended the Seventh Triennial World Congress of the International Federation of Automatic Control in Helsinki in mid-1978. There he met two Dutch control theorists, G. J. Olsder and Huibert Kwakernaak, who had developed Club of Rome-style cybernetics-based population science.[101] He spent a week or two in Holland carrying on the conversation, imbibing the core idea of *Limits for Growth* (probably not even reading the text at this time), before returning to China. Song Jian, on the basis of this brief encounter, was now convinced that cybernetic control science provided the scientific tool for modelling the Chinese population. He also probably picked up a copy of Goldsmith's *Blueprint for Survival*, since he would later borrow some of the imagery.

427

Song Jian put together a team at the missile ministry, including the control theorist and systems engineer Yu Jingyuan and the programming expert Li Guangyuan.[102] Together they produced models and a quantitative picture of Chinese population growth that presented the issue as a 'crisis' only manageable by allowing Chinese parents to have just one child. At the decisive policy meetings of 1978 to 1980, Song Jian was able to present his sinified cybernetics of population as 'scientific' compared to two rival claimants, a Marxian statistics of population and a Marxian humanism which opposed the one-child policy.[103]

Therefore, springing from Club of Rome-style arguments as interpreted by contingently powerful missile scientists, a one child per couple policy was introduced in China, leading to traumatic effects in Chinese society, especially in rural China, where family farming required two children for survival; other consequences include coercion, sterilization programmes, violence against infant girls, an aging population, and a gender gap of 120 boys to 100 girls.[104] The one-child policy marks the resurgence of technocratic influence in China from the 1980s to the present, a veneration of science at the expense of humanistic influences.

There were other strange linkages between the Cold War establishment and environmentalist ideas. James Lovelock built his technical authority on the development of ionization detectors for gas chromatography, which attracted patronage from NASA, before he spent this intellectual capital on promulgating Gaia, the theory of the earth as organism embraced by many environmentalists. Lovelock's Electron Capture Detector, which could identify DDT at parts per trillion, was crucial to the provision of evidence in support of Carson's claims.[105] The Cold War provided the new environmentalism with critical tools and evidence. Or, shifting fields again, RAND techniques for assessing the management of research and development contributed to a critique of top-down hierarchical centralized authority.[106] Radical variants of technocratic tools, such as operational research, were proposed.

Critical voices were therefore generated partly from within the Cold War establishment, a feature noticed in passing by several commentators. In his review of the relations of science, social movements and the long 1960s, Everett Mendelsohn has also emphasized the stranger linkages and sympathies that made and crossed the oppositional culture.[107] Thus Lewis Mumford's despairing The Pentagon of Power (1970) found critics in conservative historians of science such as Gerald Holton, but Mumford's jeremiad was also in sympathy with the concerns articulated in Eisenhower's military-industrial complex speech, the president's farewell address to the nation of January 1961, which in turn echoed C. Wright Mills's arguments in the *Power Elite* (1956) (Mills was hardly a political bedfellow of Eisenhower). Historian Ed Russell has noted the striking similarities between Eisenhower's and Carson's analyses.[108]

For the generation growing up in the 1960s, the images of science and technology were 'contradictory'. The generation were free to enjoy benefits (domestic technologies, 'high-tech music', synthetic drugs) while consuming critical texts (Kuhn, Feyerabend, Carson, Ehrlich, Commoner, Illich, Schumacher) and recognizing the 'loss of innocence' of science made vivid by anti-nuclear and anti-Vietnam movements. Howard Brick has emphasized how contradiction was a feature of much of what was distinctive about the long 1960s. The period is marked by short-term turmoil, a feeling of profound movement, a sense of failure, and a few modest successes.[109]

Neo-catastrophism

The souring mood of the 1960s was matched by a curious vogue for scientific theories featuring violent change, especially in sciences in which debates were often played out in public. Examples can be found in evolutionary biology, palaeontology, astronomy and climate science. Just as Lyell's nineteenth-century uniformitarianism resonated with wider political values (in his case, reform as a middle way between radical change and conservative stasis), so, we might wonder, what public values did gathering neo-catastrophism in 1960s and 1970s public sciences chime with? Indeed, Lyell's uniformitarianism was a direct target of the young evolutionary theorist Stephen Jay Gould in 1965.[110] Gould's intervention was part of the continuing project of the Evolutionary Synthesis, making sense of the fossil record in neo-Darwinian terms, while at the same time being critical. Niles Eldredge and Gould (students of Norman Newell, who had in turn worked with George Gaylord Simpson) argued in 1971 that the fossil record revealed not gradual evolution but episodes of extraordinarily rapid change. They called this theory 'punctuated equilibrium'. The 'Burgess shale' fauna, uniquely well-preserved fossils from Cambrian rocks in Canada, first described by Charles Walcott in the first decades of the twentieth century, became a battleground of reinterpretation.[111]

Gould would become one of the foremost popular writers on

science in the twentieth century. Likewise the violent controversy over the extension of the population genetics to draw conclusions about human societies – sociobiology – was largely argued out in public. Edward O. Wilson's attractively illustrated *Sociobiology: The New Synthesis* (1975) included just one chapter on humans, but the effect was to provoke a backlash from those critics, mostly on the political left, such as Gould and Wilson's Harvard colleague Richard Lewontin, who saw it as a project to naturalize social inequalities.[112] One effect of this public battle has been to give extraordinary public prominence, via popular texts such as Richard Dawkins's *The Selfish Gene* (1976), to otherwise obscure arguments over the relationship between gene populations and behavioural patterns, for example 'altruism', articulated by mathematical geneticists such as William Donald Hamilton. It is interesting to note that this further extension of materialist explanation was energized by long 1960s social movements – in this case the criticisms of sociobiology by the left and by feminists – and has continued to thrive in the individualistic decades since.

Meanwhile in palaeontology, the ever popular subject of the demise of the dinosaurs received a neo-catastrophist reworking.[113] In 1980, the father and son team of Berkeley nuclear physicist Luis Alvarez and geologist Walter Alvarez, with others, argued that a chemical analysis of rocks lying between the Cretaceous and Tertiary periods provided evidence of a meteorite collision of devastating, global consequences. Specifically, the measurement was of iridium, a rare metal found in meteorites, at this so-called K–T boundary. The scenario depicted was of dinosaurs being wiped out by the cataclysmic effects of the impact, including a global chilling of climate brought on by dust in the atmosphere. This imagery came direct from descriptions of the 'nuclear winter', an environmental catastrophe that mid-1970s commentators depicted as following a worldwide nuclear war.

In 1969, Nigel Calder, the son of fellow science journalist Ritchie Calder, scripted a BBC television series called *Violent Universe*. It drew particularly on discoveries of radio astronomy – pulsars, quasars, and the cosmic microwave background radiation – to present 'exploding stars', 'exploding galaxies' and the 'exploding universe'.[114] 'What is at stake is nothing less than the imminent portrayal of a new and more vivid picture of the universe we inhabit, as astronomers routinely investigate cataclysms in the sky', Calder wrote. 'It turns out that we live in a relatively peaceful suburb of a quiet galaxy of stars, while all around us, far away in space, events of unimaginable violence occur.'[115] Modern astronomy, it seems, was much like watching the Vietnam War on television from the safety of the suburban living room.

Nigel Calder, editor of the popular magazine *New Scientist* in the mid-1960s, also had a hand in the transition to neo-catastrophism in climate science. The transition was guided by several strands of evidence. First, investigations by Chicago scientist Cesare Emiliani, a student of Harold Urey, of the ratio of oxygen isotopes in foraminifera, tiny ocean-dwelling shelled creatures, generated indicators of past ocean temperature over 300,000 years.[116] Emiliani disputed geologists' claims that the ice ages came and went only gradually. Second, computer models also pointed towards the possibility of sudden climatic change. In 1961, the MIT modeller Edward Lorenz had shown not only that models were intensely sensitive to initial conditions, but also that some simple climate equations permitted the sudden flipping from one state to another.[117] (Much later, in 1979, Lorenz would famously capture this sensitivity by asking: 'Does the flap of a butterfly's wings in Brazil set off a tornado in Texas?')

Third, measurements in the Antarctic, published in 1970, showed that the West Antarctic ice sheet was held back only by a relatively flimsy ice shelf at the rim. Again the possibility of sudden change was raised. Planetary science, from the revelation of a runaway greenhouse effect on Venus, pictured in the Soviet Venera missions of the 1960s and 1970s, to the recognition that the Martian atmosphere flipped between two stable states (skies clear and skies filled with dust, as seen by Mariner 9 in 1971), provided comparisons with earth. In the mid-1970s, the idea of sudden climate change found popular expression. Nigel Calder, for example, warned in 1974 of an imminent 'snowblitz', the triggering of a new ice age.[118] Climatologists followed, with books such as Stephen Schneider's *The Genesis Strategy: Climate and Global Survival* (1976), addressing the public.[119] As previously discussed, by the 1960s scientists had not only inherited the nineteenth-century theories of greenhouse effect and ice ages but had also gathered evidence of the steady accumulation in carbon dioxide. By the early 1970s, scientists had models and examples of sudden catastrophic climate change. Putting them together, they asked whether incremental human environmental influences could trigger catastrophic climate change.

Oddly, the possibility of catastrophe was being raised in these public sciences during a period when the relationship between East and West was stabilizing, self-perpetuating, and even thawing to a degree. After the Cuban Missile Crisis there was tacit understanding that reconnaissance flights by spy satellites would be tolerated.[120]

In 1963, the United States and the Soviet Union agreed the Limited Test Ban Treaty, which outlawed atmospheric, underwater and outer space nuclear tests. This built confidence, leading towards the Nuclear Non-Proliferation Treaty of 1968. Under the treaty, states deemed to be in possession of nuclear weapons – the Soviet Union, the United States, the United Kingdom, with France and the China signing later – undertook not to transfer the technology and to move, on some unspecified timetable, towards nuclear disarmament. Other signatory states would receive help with civil nuclear ambitions. India, Israel, Pakistan and North Korea have not signed the treaty. Further nuclear arms agreements followed between the United States and the Soviet Union: the Strategic Arms Limitation Talks (1972), the Anti-Ballistic Missile Treaty (1972) and SALT II (1979).

The Anti-Ballistic Missile Treaty was, notes Gaddis,[121] 'the first formal acknowledgement, by both sides . . . that the vulnerability that came with the prospect of instant annihilation could become the basis for a stable, long-term, Soviet–American relationship'. However, the consequences of the Biological Weapons Convention, signed in 1972, suggest a less sanguine interpretation. Under the convention, signatory countries were committed 'never in any circumstance to develop, produce, stockpile, or otherwise acquire or retain' biological agents or toxins as weapons. There was no provision for mandatory verification. The following year the Soviet Union launched Biopreparat, a civil cover for a huge biological warfare programme, employing around 9,000 scientists and technicians, developing science cities such as Sverdlovsk. Factors behind this duplicitous rearmament include lingering Soviet suspicions of American motives, a power shift within the Kremlin that favoured the military, a decision deliberately to exploit American disarmament, and the extraordinary influence of a particular scientist, the biochemist Yuriy Ovchinnikov, on Leonid Brezhnev.[122] The long 1960s marked continuities as well as transitions.

— 18 —

NETWORKS

While the quantity of scholarship is increasing, there is no doubt that the historical writing on science post-1970 is patchy, with some subjects attracting attention and other, important, topics awaiting research. The topic has nevertheless been taken up by commentators on the contemporary world – journalists, sociologists, anthropologists. Several of the most influential commentaries have emphasized change – even radical breaks – over continuity. A veteran of German Green activism, Ulrich Beck, revitalized the sociology of modern technological societies with his identification of 'risk society', a break with the past.[1] Whereas formerly, argued Beck, threats were external to society and were in principle calculable (think of how an insurance company calculates the risk of loss of ships at sea), now risks were generated from within society and were often incalculable.

Science and technology, once part of the solution, were now, according to Beck, part of the problem, still essential to managing risks but creating further risks in the process. Nuclear projects to solve energy crises, geo-engineering to solve global warming, genetic engineering to solve food crises, and computerized buying and selling to resolve market uncertainties have all been held as illustrations of the risk society at work. Industrial society, once best characterized by how it produced 'goods', was now best described by how it produced 'bads', manufactured uncertainties. In the language of this book, risk society has become a working world generating new sciences that reflect on and seek to manage risks (perhaps ultimately unsuccessfully). Epidemiology, the statistical study of disease, for example, has flourished from the mid-twentieth century, a landmark achievement being the identification of the link between smoking and the risk of lung cancer through Austin Bradford Hill and Richard Doll's studies

433

of the 1950s and 1960s. A counter-epidemiology, lavishly funded by the tobacco industry, which pursued a strategy of undermining the epidemiological certainties by funding science that created uncertainties,[2] provides a macabre twist on this process of sciences springing from working worlds of risk. With the proliferation of reflexive sciences, 'society becomes a laboratory with nobody responsible for the outcomes of experiments.'[3]

Another analysis, by sociologists Michael Gibbons, Helga Nowotny and colleagues, posits that a change in how science was produced occurred in the last decades of the twentieth century.[4] In the first, earlier phase, which they label 'mode 1', scientific problems had been set by academics, solved within disciplines (by work sited in hierarchical institutions), and managed by the norms and processes (such as peer review) of autonomous science. Now, they say, science is increasingly produced in a second way, 'mode 2', in which problems are set by contexts of application (not controlled by academics), solved by often temporary transdisciplinary combinations of experts, typically outside academia in a diverse array of kinds of sites, and forego autonomy for appeal to social, political and commercial interests.

Both Beck and the mode 2 sociologists have been criticized for presenting ahistorical pictures of abrupt breaks, unsustainable when compared with a detailed, nuanced history of modern science, technology and society.[5] So risks that were internally produced and difficult to quantify existed before the risk society. Likewise, the notion that science was once academically controlled, autonomous and confined largely within disciplines was only ever a 'partial reality'; 'more importantly', notes historian Benoît Godin, 'it has been a rhetorical creation on the part of scientists anxious to justify their social position,'[6] Science may never have been 'pure', but parts of it were certainly made to look that way.

It is understandable that sociologists, positioned and funded as consultants to decision-makers, turn the contrast up in order to produce analyses that have clarity at the expense of historical nuance. Science, located in working worlds, has always, typically, been more mode 2 than mode 1.[7] Nevertheless, the question remains: even if we dismiss stories of abrupt breaks as exaggerations of a more complex and continuous history, what changed about science in the final third of the twentieth century, going into the twenty-first?

Five trends

I think there are five big trends. First, the old dream of engineering life took concrete and extraordinary form as the post-war projects of molecular biology were privatized in the new biotechnology boom starting in the early 1970s. Some of this was new: tools to edit genes, patents on living organisms, a self-proclaimed language of risk and danger, and the ability to tap into the immense quantities of investment capital flooding the world in the wake of the freeing up of exchange markets and the jump in petro-wealth in the 1970s. New sciences, it seemed, chimed with new economies. But some was old; indeed much of the new biosciences looked like the old chemistry – consultancies, entrepreneurial professors, dense links with industry, the raising of capital for research, and so on.

The second trend was very closely connected to the first: biomedical funding of research, through corporations, charities and national institutes, expanded vastly, meaning that the working worlds of diseased (and, crucially, also healthy) bodies were as generative of science in the last third of the twentieth century as the working world of military systems. For example, in the United States, growth in funding through the National Institutes of Health, as well as through NASA and the National Science Foundation, pushed the relative proportion of federal funds for military projects in the 1970s down to a Cold War low of 20 per cent.[8] To take just one disease, the funding of the national cancer institutes grew from a mere few hundred thousand dollars to over a billion during Nixon's war on cancer of the 1970s.[9] I say healthy in addition to diseased bodies because part of this trend was the further biomedicalization of everyday life. The results of the new bioscience were everywhere: gene discoveries in the newspapers, bio-prospecting across the world, a new picture of the tree of life, the publication of whole genomes, and on and on.

The third trend concerned the ways that small and networked computing power became not only a ubiquitous tool, and method of organization, of science but also contributed to, first, networks becoming a repeated motif of scientific theory and, second, networks becoming a subject for study for new sciences. Automation of scientific work was a feature of both physical and life sciences. This leads us to the fourth trend: the scaling up of science continued, but in a manner that generated immense demand for data processing. Examples could be found at CERN or the Hubble Space Telescope orbiting the earth, or the Human Genome Project and its commercial rivals.

Punctuating these trends were events. The end of the Cold War – and the beginning of a more diffuse 'War on Terror' – meant that the working worlds for scientists, especially physical scientists, changed. (However, such was the institutional investment in these working worlds that a kind of institutional momentum carried projects forward, and the stresses were often catastrophically revealed in controversial moments, such as the cancellation of the Superconducting Supercollider.) Another kind of punctuating event was the emergence of new diseases, in particular AIDS in the 1980s.

This brings us to the fifth trend, which relates to public involvement in the decision-making processes around science and science-based enterprises. There have always been alternative forms of lay expertise and lay knowledge, relevant to working worlds, even during mid-twentieth-century heights of formal, accredited expertise.[10] What has happened since the 1970s, in a fraught, contested and incomplete process, is the growing expectation that these lay expertises might be tapped or appropriated in official decision-making. The disputes, compromises and collaborations between professional and lay expertise in AIDS science are exemplary. In the 1990s and 2000s, at least in Britain, this expectation went under the title of 'public engagement'. Driving this trend were apparently contradictory wishes: one for democratic control and the other for a mode of legitimation when traditional props failed. Neither wish has been fulfilled. However, the legacy has been a framing of new developments in science and technology – such as nanotechnology – in the language of public threat.

The biotech boom

The central metaphors of post-war molecular biology – information processing, codes, programming – encouraged the idea that living organisms could be reprogrammed or re-engineered. There were predictions of genetic engineering as soon as the genetic code had been completed in 1967.[11] In the early 1970s, several technical breakthroughs that enabled DNA to be manipulated – cut, spliced, replicated, edited – with much greater ease than before were necessary but not sufficient conditions for the development of genetic engineering. This research took place almost entirely in academic settings in the United States. However, it was funded largely through the National Institutes of Health, which expected a practical justification and eventual application.[12] The early genetic engineering techniques were a science of the working world of medicine.

The first set of crucial techniques came from Stanford University Medical School, where the biochemist Paul Berg, with two postdoctoral researchers, aimed to use bacterial viruses to carry a piece of foreign DNA into a cell. (A third researcher at the same institution, Peter Lobban, a graduate student, seems to have been trying a very similar project at the same time.)[13] In 1971, Berg's team succeeded in cutting up the loops of DNA of the lambda virus and a monkey tumour virus called SV40 using an enzyme, Eco RI, which came from Hebert Boyer's laboratory at the University of California, San Francisco. Berg then added new complementary ends to the snipped pieces. When mixed together the ends joined up, and the result was a new spliced loop of hybrid DNA. Berg intended to use the lambda phage as a vector to insert foreign DNA into *E. coli*, the workhorse model organism for molecular biology but also a common human gut inhabitant. Berg suspended his work as news of the experiment raised concerns and alarm.[14]

In 1972, an even more effective way of splicing DNA was figured out when the precise character of the ends of DNA cut by 'restriction enzymes' (the proteins, such as Eco RI, used by bacteria to cut up and destroy unwelcome genetic material) was understood. The ends were not clean cut, but ragged, leaving a small sequence of nucleotides sticking out. These sequences would be attracted to their counterparts. The ends were therefore 'sticky' in an understood way. They could be 'spliced'. The next important development exploited this new splicing method. In 1972, Herbert Boyer (with Robert Helling) and Stanley Cohen (at Stanford University Medical School, with his technician Annie Chang) took a plasmid (a loop of DNA) called pSC101, spliced in a gene for resistance to an antibiotic, and then saw the hybrid plasmid taken up by *E. coli*. The new genetically engineered *E. coli* – the first genetically modified organism – was resistant to the antibiotic. The following year, the team introduced DNA from *Xenopus* toads into *E. coli* by the same method.

The experiment was significant on three broad fronts: biological, commercial and ethical. Biologically, Boyer and Cohen had combined genetic material from utterly different species. 'For the first time', reported the Cambridge molecular biologist Sydney Brenner, 'there is now available a method which allows us to cross very large evolutionary barriers and to move genes between organisms which have never been in genetic contact.'[15] The genomes of new organisms could be pieced together, life re-engineered. Commercially, the experiment raised golden opportunities. In the right environments bacteria will replicate at astonishing speed. If genes from higher organisms – such as humans – could be introduced into bacteria, then the bacteria could be used to synthesize useful proteins, perhaps on a lucrative scale: bacteria as factories. François Jacob, in *Logic of Life*, had written:

> The cell is entirely a cybernetic feedback system . . . And it works virtu-ally without any expenditure of energy. For example, a relay system that operates a modern industrial chemical factory is something that consumes almost no energy at all as compared to the flux of energy that goes through the main chemical transformations that the factory carries out. You have an exact logical equivalence between these two – the factory and the cell.[16]

Stanford University took out patents, on behalf of Cohen and Boyer, on the methods and the specific plasmid involved.

Ethically, the response to the recombinant DNA techniques con-sisted of rounds of discussion and, after a brief hiatus, the introduc-tion of some restrictions on experiment. The National Academy of Sciences recommended that scientists meet to discuss biohazards. The Asilomar Conference on Recombinant DNA, organized by Paul Berg, took place near Monterey, California, in 1975. The participants at Asilomar decided on self-imposed restrictions – different levels of containment for different types of experiment, and some experi-ments ruled out entirely. A summary statement, authored by Paul Berg, David Baltimore, Sydney Brenner, Richard O. Roblin III, and Maxine F. Singer, was published in June 1975. The alternative to self-regulation would probably have been far more restrictive direct government-imposed regulations. Erwin Chargaff noted that it was 'the first time in history that the incendiaries had summoned their own fire brigade', adding that 'the bishops of Asilomar were first in denouncing the heresies of which they themselves were guilty'.[17]

Nevertheless, the commercial response to genetic engineering was initially cautious.[18] A small, specialist firm, called Cetus, had been established in 1971, advised by Joshua Lederberg and joined later by Stanley Cohen. In 1975, in response to the leaps forward in tech-nique, Cetus was bullish about the future:

> We propose to do no less than stitch, into the DNA of industrial micro-organisms, the genes to render them capable of producing vast quanti-ties of vitally-needed human proteins . . . Cetus proposes to make these proteins in virtually unlimited amounts, in industrial fermenters . . . We are proposing to create an entire new industry, with the ambitious aim of manufacturing a vast and important spectrum of wholly new micro-bial products using industrial micro-organisms.[19]

To exploit the Boyer–Cohen patents, Boyer joined with the venture capitalist Robert Swanson to set up another new company, Genentech, in 1976. Other small start-ups followed, such as Genex Corporation in Maryland and Biogen. These small companies sought to survive and grow.

Investment in biotechnology, however, came primarily from multinational companies, especially pharmaceutical, chemical and oil corporations. These companies made products which might be threatened by genetically engineered alternatives. They therefore invested defensively in order to preserve domination of existing markets and to provide, through access to scientists and seats on boards of directors, 'a window' on fast-emerging technology.[20] Eli Lilly contracted with Genentech to produce insulin; Standard Oil of Indiana and Standard Oil of California were the first major investors in Cetus stock; Hoffmann-La Roche, the Swiss pharmaceutical giant, and ICI, the British chemical combine, were both early investors in intramural and extramural recombinant DNA science. A supply of insulin was essential to the lives of sufferers of diabetes, and had been extracted by processing tons of animal organs since Banting's day; insulin made by genetically engineered micro-organisms promised a second, perhaps highly profitable source of this marketable substance. The multinationals were therefore pushed as well as pulled into investment. Traditional manufacturing was beginning to decline in the United States relative to developing countries, and the collapsing trade balance in the products of traditional manufacturing could be offset if new more research-intensive products could be found. The oil crisis of 1973–4, as well as inflation in Western economies, exacerbated these problems. Pharmaceutical, chemical and oil companies therefore poured investment into recombinant DNA companies as a new strategy. (It is interesting that the oil crisis also prompted Japan to greatly increase its investment in research and development.)

The confidence and resources of investors were enough to carry the commercialization of recombinant DNA even as the science was revealing that the expression of genes and the nature of the genome were more complex than hitherto imagined. In 1977, for example, through work by Phil Sharp (MIT) and Richard Roberts (Cold Spring Harbor), it became clear that eukaryote genes were often not continuous sequences of bases but discontinuous: spaced between the 'exons', the nucleotides of the gene that coded the amino acids were seemingly gibberish 'introns', non-coding regions. Introns made snipping out eukaryote genes, splicing them into a prokaryote bacterium, and simply expecting results in expressed proteins foolish. Furthermore,

439

'transposons', genes that could copy themselves multiple times in a genome, were found to be widespread, making the genome a much more dynamic entity than many had thought. 'Jumping genes' had first been found by Barbara McClintock, working with maize, in the late 1940s; their 1970s rediscovery vindicated McClintock, who, as a woman espousing a radical scientific theory in a male-dominated context, had been marginalized.[21] Such complex intellectual challenges, alongside the public safety concerns, 'might have acted as brakes on development', concludes historian Susan Wright, had it not been for the pull of private investment and the push of government deregulation.[22]

The result was that, first in recombinant DNA science and then throughout molecular biology, a picture of public-funded academic research was replaced by one of privately funded science. The shift was facilitated by government policies: cuts in capital gains tax that encouraged venture capital and deregulation more generally. The Bayh–Dole Act of 1980 (formally the Patent and Trademark Amendment Act) unambiguously handed the patents arising out of federally funded research to the universities, small businesses and non-profit organizations that undertook it.[23] 'Biotechnology' became the preferred term among investors around 1979.[24] In 1982, a peak year, $111 million was invested in biotech equity. Martin Kenney estimates the cumulative investment in biotech at $3 billion by 1984.[25] Genentech produced somatostatin, a small peptide hormone, in 1979, the same year as the company, with Eli Lilly, began animal tests of genetically engineered human insulin. In October 1982, human insulin would become the first American product of genetic engineering.[26] Particular commercial excitement surrounded the production of the small protein interferon, which promised to have a role in therapies for diseases from flu to cancer.

The consequences of the privatization of knowledge

Susan Wright argues that the ethos of scientists working in the field of recombinant DNA changed as it went from public to private: 'as soon as genetic engineering was seen as a promising investment prospect, a fundamental deviation from traditional scientific practice occurred and a movement toward a corporate standard took place.'[27] The changing ethos can be seen in three areas: how results were communicated, how intellectual property was held, and in the commercialization of the academic campus.

440

Results began to be announced outside the normal internal scientific channels of peer review and scholarly journal. The successful production of somatostatin was announced at a congressional hearing, part of an argument to speed up deregulation; the announcement of the successful expression of human insulin was made at a press conference in 1978; Biogen's trumpeting of its commercial production of interferon in 1980 covered up the fact that quantities produced were almost insignificant, that the interferon was neither pure nor proven to be biologically active, and that the methods were unoriginal.[28]

In terms of property, according to Merton's norms, knowledge and ideas had supposedly been held communally as the possession of the scientific community. Research objects and tools, which were as much a form of intellectual property as a theory, were also passed around the scientific community as research dictated. Indeed, these networks of friendship and exchange were the organizing sinews of biological research communities, as Kohler discusses for the case of *Drosophila*.[29] For recombinant DNA research the picture of communally supportive and community-defining networks of supply were also present, at least in the early years. Richard Roberts, in his Nobel autobiography, recalled, for example:

> A key factor in our restriction enzyme success was a highly talented technician, Phyllis Myers, who joined me in 1973. She became the keeper of our enzyme collection and a valuable resource to scientists around the world. We constantly sent samples to other researchers and were inundated with visitors. Every meeting at Cold Spring Harbor brought a few people carrying tubes of DNA to see if we had an enzyme that would cut it. Three quarters of the world's first restriction enzymes were discovered or characterized in my laboratory. I made a lot of friends in those days!

Yet even in the early days the profit motive was not far away. 'In 1974 I had tried unsuccessfully to convince Jim Watson that Cold Spring Harbor should start a company to manufacture and sell restriction enzymes', Roberts also recalled. 'He declined, thinking there was no money to be made.' Such commercial entities would have been marginal in the early 1970s. Within a few years, commercial values were central. An illustrative example was the controversy over the KG-1 cell-line, which had passed, according to traditional norms of sharing, out of the University of California, to the National Cancer Institute, to the Roche Institute of Molecular Biology, Hoffmann-La Roche and Genentech. Hoffmann-La Roche, which saw KG-1 as a commodity, sought through patents to own the cell-line. A row blew up – a battle between old and new norms.[30] It was eventually settled out of court in 1983.

Intellectual property was not something to be held communally but something to be patented and held by profit-seeking individuals or companies. The Boyer–Cohen patent of 1974 marked this turn. In 1980, the case of *Diamond* vs. *Chakrabarty* was decided by the Supreme Court. It established the precedent that patents could be held on novel organisms. Ananda Chakrabarty, a biochemist at the General Electric Company, had not used recombinant techniques when he bioengineered a bacterium to consume oil – an organism with a role, he hoped, in cleaning up oil slicks – but the ruling had general application.[31] The first patent on a living animal was awarded to Harvard University in 1988, for a transgenic mouse that was genetically engineered by Philip Leder and Timothy Stewart to be especially susceptible to cancer.[32] Patenting became widespread, perhaps even the norm.

With private ownership came an emphasis on secrecy, to protect trade secrets. Merton's norm of openness held that scientists should allow knowledge to flow freely. You 'no longer . . . have this free flow of ideas', complained Paul Berg in 1979. 'You go to scientific meetings and people whisper to each other about their company's products. It's like a secret society.'[33]

Finally, as Martin Kenney has mapped out, many university campuses were transformed: professors of biology became entrepreneurs, or at least deeply involved in commercial affairs, graduate students were channelled towards the companies, undergraduate science further routinized, and commercial values prospered.[34] A huge and diverse array of contracts brought academic biology people and bodies into relationships with corporations. Examples include Harvard Medical School and Monsanto and Du Pont, Massachussetts General Hospital and Hoechst AG, and MIT and the Whitehead Foundation (Edwin Whitehead, founder of Technicon Corporation, was an immensely wealthy supplier of precision instruments to clinical laboratories). Of course, in chemistry or electrical engineering such commercial contacts were routine. Indeed, the consultancy form of relationship can be considered quite traditional. However, there is a difference: the biotech entrepreneur-professors typically remained in post, taking on dual jobs, and therefore were able continually to tap a stream of results and young scientists.[35] Research agendas were mutually realigned, not, as in the era that Forman describes, towards military or public interests, but increasingly towards private interests. The biotech campus example has been copied by other disciplines and

442

in other countries. The result has been a worldwide transformation in the ethos of research in higher education.

Big pharma

While biotechnology hit the headlines, the traditional chemistry and biochemistry-based pharmaceutical industry remained vastly more significant in terms of producing the medicines, new and old, used throughout the world. Indeed none of the famous new drugs of the 1980s and 1990s – Prozac, Losec, Viagra, Retrovir, AZT – were produced by biotech companies.[36] Previous interwar waves of consolidation had lead to combines such as IG Farben in Germany and Basler IG, out of Sandoz, Ciba and Geigy. From the 1970s a dizzying new wave of consolidation rearranged the world's pharmaceutical sector into new giant multinational combinations. Ciba and Geigy merged in 1970. In 1996, Ciba–Geigy merged with Sandoz to form Novartis. A repercussion was the formation of Syngenta, an agricultural chemical and biotechnology multinational made from the agribusiness parts of Novartis and AstraZeneca, itself an entity formed in 1999 out of parts of old combines, demerged units of ICI of Britain and Astra of Sweden. The peak year for mergers was 1989, when SmithKline Beecham, Bristol-Myers Squibb and Marion Merrell Dow all formed.[37] The last-named would merge between 1994 and 1999 with Hoechst, Roussel and Rhône–Poulenc to form Aventis Pharma. In 2000, GlaxoWellcome (itself a product of a 1995 merger) combined with SmithKline Beecham to form GlaxoSmithKline. The path is complicated, but the end point was a small number of very large multinationals which spent immense sums of money moving drugs from research, through the 'pipeline' of development, preclinical testing, the phases of clinical trials (for safety, effectiveness and dosage) and regulatory approval, to become prescribed products.

The consolidation of the pharmaceutical and wider chemical industry was driven by the increased expense of research and development. Only large multinational firms, so the argument went, had the scope to catch inventions as they appeared, the resources to push drugs through the pipeline, and the scale to profit from diverse global markets. One of the reasons the pipeline was so expensive was changing regulatory attitudes to drugs. In the United States, only certain drugs (the addictive ones) needed prescription before 1951, when the Durham–Humphrey amendments to the 1938 Food, Drug and Cosmetic Act separated prescription drugs (including nearly all new medicines) from over-the-counter drugs. Yet there were plenty of loopholes. However, in 1962 the 1938 Act was amended again in response to the thalidomide crisis. Thalidomide was a tranquillizer developed in Germany in the 1950s and given to pregnant women in Germany, Britain and other countries as a means of countering morning sickness. (The United States, driven rather by individual discretion than by following the regulations, refused to approve thalidomide.) Children born from these mothers often had birth defects. Directly in response to the thalidomide issue, 'drug therapies suddenly became "risky" in a way they hadn't before', and randomized clinical trials were fully endorsed as the way to check the safety as well as the efficacy of new medicines.[38] Since the trials are very expensive, the pipeline was a longer process that only large companies could afford to navigate. 'The 1962 amendments were passed as part of an effort to guard people from the unfettered forces of capitalism', notes David Healy of the ironies of reformers often producing the opposite of what they set out to achieve. 'It is a moot point whether the reforms have fostered instead the growth of a psychopharmaceutical complex whose power to penetrate markets is now all but comprehensive.'[39]

Let's quickly review the history of some of the blockbuster drugs of the late twentieth century. The ideal medicine, from a pharmaceutical manufacturer's point of view, would be one that needed to be taken regularly and often to manage a medical condition that was widespread in wealthy Western societies. In the second half of the twentieth century this meant pills for high blood pressure and cholesterol, heart disease, some cancers, some infections, depression, diabetes and stomach ulcers. Rarely, a drug would find favour through powerful individual advocacy, such as the unpatentable lithium as a treatment for depression promoted by the Danish psychiatrist Mogens Schou.[40]

More frequently, a drug treatment was the product of sophisticated biochemical understanding and industrial investment. James W. Black would win the Nobel Prize for the development of drugs at ICI in the late 1950s (the beta-receptor antagonist, or 'beta blocker', propranolol, for use against heart disease and hypertension) and at Smith Kline & French (searching for substances to block histamine-induced acid secretion in the stomach) from the mid-1960s. The Swedish firm Astra considered refocusing research to improve Black's anti-ulcer substances, before deciding on a different course. (Glaxo did pursue this strategy, leading to another immensely profitable drug, Zantec.) Ulcers were considered to be the effect of hydrochloric acid attacking the stomach lining. In the mid-1960s Swedish researchers decided to search for a substance that stopped acid production rather than rely on existing drugs that neutralized acid once produced. Many variants of compounds were generated and tested on rats and fistula dogs, eventually settling on the most promising candidate. The development and marketing by the Swedish firm Astra AB of omeprazole, under the name Losec, produced the biggest-selling drug of the 1990s.

Black's research, and the blockbuster drugs that flowed from it, depended on an analytical understanding of the chemical pathways of the body. In a similar fashion, a more detailed and specific picture of neurotransmitters, such as that of the function and actions of serotonin, led to other profitable drugs. In particular, the best-known and most widely prescribed selective serotonin reuptake inhibitor (SSRI), fluoxetine, was developed by Eli Lilly from 1970 and marketed as Prozac in the United States from 1987.

Receptor-binding theory also prompted a realignment of the relationship between the pharmaceutical industry and psychiatry. In the mid- to late twentieth century, the dominant model of schizophrenia, for example, shifted in the 1950s and 1960s from a contested psychoanalytical explanation to the strictly biochemical 'transmethylation hypothesis' (the suggestion that neurotransmitters chemically changed within the body had the effect of bringing on psychoses), before this theory was swept aside as rival dopamine gained ground. Receptor theories benefited from the war on drugs, both by making LSD research unfashionable and by channelling funds into radio-labelling techniques, such as those developed by Solomon Snyder at Johns Hopkins University for the D-2 (dopamine) receptor. 'When the dopamine hypothesis entered the fray, the latest technologies appeared to stunningly confirm the reality of mental illness, and they did so in a hypothesis that broke free from the entanglements of the transmethylation hypotheses with the counter-culture', writes David Healy. 'Moreover without access to radio-labels and vacuum manifolds, critics could not even enter the debate.'[41] Healy notes also two other advantages. The theory was aligned to commercial interests, first, because 'receptor binding techniques also produced a common language for psychiatry and the pharmaceutical industry, allowing the advertising power of the industry to support mainstream psychiatry', and, second, because receptor approaches allowed a shift in testing drugs from animal models (which were slow, as well as wasteful of animals) to receptor binding assays (which were fast, capable of simultaneous testing of thousands of candidate drugs). The psychiatric profession, on the defensive after the criticisms of the anti-psychiatrists, were likewise receptive to receptors.

A similar reaction to criticism within psychiatry towards the specific and the measurable, towards the biochemical and away from social explanations and therapies, can be seen in the adoption of DSM-III, a 1980 handbook 'widely regarded as marking a revolution in American and world psychiatry'.[42] The first *Diagnostic and Statistical Manual of Mental Disorders* (DSM) was published by the American Psychiatric Association in 1952, and the second in 1968. Apart from the classification of homosexuality as a disease (retracted in a revised DSM-II in 1973), these early editions were relatively uncontested. DSM-III was much more controversial, an indicator of what was at stake in classification, especially when the boundaries and realities of phenomena of normal and abnormal mental life were up for grabs.[43] DSM-III was the product of a fierce argument between different interest groups, in particular analysts (who wanted the neuroses included), sceptics of psychoanalytical approaches (who doubted the analysts' categories), the insurance and pharmaceutical companies (which desired recognized classifications of mental disorders for different reasons) and regulators (which needed accountable objects to regulate). The outcome, a compromise, was that the neuroses familiar to working analysts were included, by name, but were defined according to operational criteria. This in turn encouraged biochemical theories and solutions for mental disorders, as well as the use of randomized controlled trials to test them. Both pointed towards the increased influence of private pharmaceutical industries in defining and managing our mental lives. 'Biobabble', says Healy of the 1990s, 'replaced psychobabble.'[44]

Cancer science

The techniques of genetic manipulation were applied in a dazzling array of new biomedical developments. I will sample two of these: cancer science and systematic biology. The development of recombinant DNA techniques was implicated in a change in the theory of the onset of cancer, or oncogenesis.[45] Cancer had been seen as a diverse set of diseases that had in common the uncontrolled growth of cells. This explanation had grown from Monod and Jacob's theory of gene regulation: cells had structural genes that coded for factors that could take off the cell's brakes to growth; normally these genes would be suppressed by regulatory genes that coded for a suppressor; if this regulation failed, or was turned off, then uncontrolled growth of cells followed.[46] What triggered failure was unclear. Nixon's war on cancer, in which the target-oriented generous funding of Big Science, so successful in the working world of defence, would be applied to the working world of medicine, exhausted the idea that virus infection might be main cause. In the early 1980s, the regulatory vision was replaced by what historian Michel Morange calls the 'oncogene paradigm', a new set of theories, working methods and assumptions which said that cancers were caused by a small number of genes, oncogenes, mutating. Proto-oncogenes – human genes that looked very similar to the genes in cancer-causing viruses – were discovered. The first human oncogene was found at the University of California, San Francisco, by Michael Bishop and Harold Varmus. The key tool was the transfection array – extracts of tumour DNA transferred into an array of normal cells.

The other key tools for cancer research were the mouse and the cell-line. The bodies of mice, bred for various susceptibilities to cancerous tumours, stood in for the bodies of humans. While not much less complicated than a human body, a mouse body was manipulable and controllable in ways far beyond those reasonably possible for a human body in the name of science. The investigation of the rejection of transplanted tumours in mice had shaped immunology in the early twentieth century, while experimentation with inbred strains of mice had forced 'the acceptance of the notion that the fate of tumours (and grafts more generally) depended only on the genetic constitutions of donor and host'.[47] Cancer research by the 1930s had focused on strains of mice and tumour cell-lines, which could be maintained indefinitely in culture, and swapped between researchers. In the 1940s, genes for histocompatibility – the degree of antigenic similarity between tissues of different individuals, and therefore an indicator of the likelihood of acceptance of a graft – were found in highly inbred mice. These genes, clustered together, were called the Major Histocompatibility Complex (MHC). In the 1970s, a human homologue of the MHC was found on chromosome 7. The genes of the human MHC coded for antigen proteins, crucial for immune response. These were called human leukocyte antigen (HLA) genes, claimed as the genetic basis for the body's discrimination of self from non-self.[48] Understanding HLAs vastly improved the success rate of human organ transplantation and prompted the discovery of soluble protein factors (such as interferon in 1957 and interleukin-2 in the mid-1960s) that were early target products for the new biotech companies.

Systematic biology

The increasing range and power of genetic techniques effectively ended a dispute in systematics that had been rolling through the 1950s, 1960s and 1970s. At the centre of the dispute had been divergent views about how the classification of living organisms, from the largest taxa of kingdoms, through phyla, classes and families, to species and subspecies, should be made. Broadly speaking, one approach, phenetics, made many measurements and justified the placing of organisms in the same taxon on the basis of the numerical closeness of these measurements.[49] Another approach, cladistics, forcefully advocated by the German systematicist Willi Hennig, argued that only common descent mattered. All the creatures on the same branch of the evolutionary tree, descending from a common ancestor, formed a monophyletic group or 'clade'. Good taxa were clades. Both sides had to work with an inherited system of classification, essentially the revised Linnaean system, which generated the names essential to coordinating work in the life sciences. The dispute generated more heat than light. Philosopher David Hull uses the debate to argue that science itself can be said to evolve, while remaining a partly social process.[50]

The development of 'rigorous, quantitative methods for erecting classifications' was, considers biologist Douglas Futuyma, the 'first and foremost' cause of the 'resurgence of phylogenetic research' in the last third of the twentieth century.[51] (Although he also agrees with Hull that 'the integration of phylogenetic systematics with other fields of evolutionary study would have happened faster if nonsystematicists had not recoiled from the "warfare" among adherents to different systematic doctrines'). Three other causes were the availability of new kinds of data, especially molecular data (which did not demand time-consuming immersion into traditional taxon-specific systematics), the 'resurgence of interest in macroevolution' (marked by theories such as that of punctuated equilibrium), and individual advocacy of a new synthesis.

Cladistics held that phylogeny should be the guide to classification of the living world (Hennig called it Phylogenetic Systematics, the title of his influential English language book of 1966). New molecular techniques, in particular developments of Frederick Sanger's sequencing techniques of the 1950s, allied with increased computing power, allowed phylogeny to be traced in new and accurate ways. In particular, assuming that mutations changed DNA at a known and steady rate, differences between sequences of nucleotide bases of DNA of two organisms could be used as a measure of how far back in time their last common ancestor lived. This idea of a molecular evolutionary clock had been proposed in the mid-1960s by Linus Pauling and his French postdoctoral colleague Emile Zuckerkandl.[52]

One of the most spectacular discoveries of twentieth-century science – the discovery of Archaea – came from the exploitation of this idea. Since all living cells make protein, the comparison of molecules common to the protein-making process would reveal clues about the so-called universal tree of life, the map of ancestral relationships of all living creatures. Specifically, proteins are made in cells by ribosomes, which have two characteristic structures, one large (the large sub-unit, LSU) and one small (the small sub-unit, SSU). In each sub-unit are RNA molecules (ribosomal RNA, or 'rRNA'). LSU and SSU rRNA are found in all cells and are large molecules, which change slowly over immense evolutionary timescales. The sequences of SSU rRNA were therefore judged to be excellent sources of evidence for constructing the universal tree of life.

In the mid-1970s Carl Woese, with George Fox, of the Department of Genetics and Development at the University of Illinois, painstakingly assembled – by hand – sequence data of ribosomal RNA extracted from cells from organisms as diverse as mouse sarcoma cell-lines, yeast, duckweed, E. coli, and odd methane-processing microorganisms, and also from cell bodies such as chloroplasts. They were funded, although never sufficiently according to Woese, by NASA and the National Science Foundation.[53] They noted the history of the great divisions in systematics:

> Classically, what was not plant was animal. The discovery that bacteria, which initially had been considered plants, resembled both plants and animals less than plants and animals resembled one another led to a reformulation of the issue in terms of a yet more basic dichotomy, that of eukaryote versus prokaryote. The striking differences between eukaryotic and prokaryotic cells have now been documented in endless molecular detail. As a result, it is generally taken for granted that all extant life must be of these two basic types.[54]

Herbert Copeland in 1937 had placed bacteria in their own kingdom,[55] and subsequent research had widened the distinction further. But, on the basis of differences between the sequences, Woese rejected the distinction between prokaryotes (no nucleus) and eukaryotes (nucleus present) as the primary, most fundamental phylogenetic grouping. Instead, the molecular sequences pointed towards not two, but three 'basic types' of life. The eukaryotes – animals, plants, fungi

– were one type; many of the familiar bacteria formed a second; but the odd methanogens formed another, as different from bacteria as bacteria were to us. Woese proposed three 'urkingdoms', later called the 'domains' – the eukaryotes, bacteria and the 'Archaea'. The three-domain big picture of the universal tree of life was as controversial as it was revolutionary. Many resented the fact that the 1977 article was accompanied by a press conference, a press release and press coverage that made hay out of Woese's phrase, the discovery of a 'third form of life'. In the opinion of Ralph Wolfe, the colleague who had supplied cultures of the crucial methanogens to Woese, 'in hindsight, the press release polarized the scientific community and the majority refused to read the literature, delaying acceptance [of Archaea] . . . for perhaps a decade.'[56] While attacked by the established grand experts in systematics, such as Ernst Mayr, further molecular evidence has supported the existence of Archaea and therefore the three domains of life. However, Woese's methods had stretched the limits of such a manual approach to the comparison of sequence data. Future methods would make much more use of computers. Data would also be made available via electronic digital networks.

Networks, computers and science

Science has always followed the networks that have been built to make our world. Some of these networks have taken centralized, hierarchical forms. Examples include the postal service, the administrative arms of the state, the electrical supply system, and telegraph and telephone communications. Others have been more decentralized. Examples are the road system and networks of trade by monetary exchange. While it is clearly false to say that there has been a transition from centralized, hierarchical models to decentralized networks, there has been a shift in emphasis from the former to the latter, particularly from the 1960s onwards. The shift, its causes and consequences, has generated an immense commentary.[57] A particularly important agent of change has been the construction and use of packet-switching computer networks. Such networks, from the ARPANET – funded by the military blue skies research agency ARPA from 1966 – to the internet, have been implicated in social change, among them changes in how scientists organize and conceive their work. The nature of the implication, however, is complex, with influences going in both directions.

The first four ARPANET nodes were at academic sites: the

University of California at Los Angeles, the University of California at Santa Barbara, the University of Utah and Stanford Research Institute, with the software contract awarded to Bolt, Beranek and Newman. Between 1968 and 1970 the network became operational. Further links followed, especially with universities receiving lots of Department of Defense funding and with crucial Cold War research centres such as RAND, RAND's spinoff computer programming outfit SDC (System Development Corporation), the Ames Research Center, and MITRE in Massachusetts. Academic scientists were early users of the ARPANET, especially following a demonstration in 1972 in Washington which showed over a thousand delegates how computer power and digital data as far afield as Paris could be accessed.[58] In late 1971, in a development unanticipated by the designers, a capability to send packet-switched messages – email – over the ARPANET was invented. By 1982, the military authorities, nervous about the mixed civil–military use of the ARPANET, split off a separate Milnet.

While computer scientists dominated early use of the ARPANET,[59] other scientists, initially attracted by the intended purpose of the network, the shared access to powerful computers, became users in the early 1970s. Historian Janet Abbate notes academic physicists using UCLA's IBM 360/91, chemists accessing the molecular mechanics system at the University of California, San Diego, and a partnership in computational chemistry between the University of Chicago and the Wright-Patterson Air Force Base, while ARPA ran its own seismology and climatology projects.[60] Nevertheless, ARPANET through the 1970s remained relatively unexploited. One factor behind the shift of the model from resource-sharing to communication lay in how scientists exploited this underused infrastructure.

Restrictions on access to ARPANET also encouraged scientists to exploit other networks. An important one was BITNET, developed from 1981 by IBM system administrators at the City University of New York and Yale University.[61] In the last third of the twentieth century, collaboration between scientists increased. The average number of authors per paper increased from fewer than two to more than four, while collaboration between institutions increased too.[62] For example, among scientists and engineers receiving National Science Foundation funding, cross-institutional collaboration went from being a rarity in 1975 to two-fifths of projects in 2005.[63] Partly this trend was due to the multidisciplinary and teamworking character of Big Science. But it also correlated to the rise of email, which made communication between scientists cheaper.[64] Even in mathematics, a stereotypically solitary pursuit, new patterns of collaboration co-evolved with the internet, perhaps undermining traditional research schools.[65]

However, patterns of collaboration were also shaped by historical cultures of working together and sharing material that were specific to research fields. Some specialties, such as seismology and oceanography, were geographically distributed by nature. Under certain conditions, such as during the Cold War, when detection of nuclear detonations encouraged the World-Wide Standardized Seismograph Network in the early 1960s, this natural tendency to distribution could flourish into global networks of data collection.

Some of the most significant examples of inherited cultures of sharing come from the genetics research communities that clustered around particular model organisms. We have seen how the exchange of material, such as *Drosophila* mutants, was the glue holding together twentieth-century research communities. Among geneticists investigating maize, *Zea mays*, materials and unpublished data had been shared on the model established by R. A. Emerson and his Cornell students, including George Beadle, from 1929. A *Maize Genetics Cooperation Newsletter* had followed, which in turn was the model for other publications such as the pre-war *Drosophila Information Service* and the post-war *Worm Breeders Gazette*, for the *Caenorhabditis elegans* nematode community. By the mid-1980s, for example, the nematode researchers shared sequence data, discussed below, via BITNET.[66] The appearance of cheap computerized communication networks provided the opportunity to transfer cultures of sharing from a paper-based medium to an electronic one. If the pull came from cheap available communication, the push that moved scientists to exploit networked computing came from the necessity to manage and make accessible immense quantities of data generated by sciences in the last third of the twentieth century.

A revealing document of visions of increased use of computing power within the life sciences is Joshua Lederberg's 'An instrumentation crisis in biology', written to persuade Stanford University administrators in 1963. Involved with NASA's early attempts to develop biochemical ways of detecting signs of life, Lederberg wrote that the molecular biologist had to 'face some of the most profound issues of biology' with tools that were 'astonishingly primitive by the standards of instrumentation' (such as the precision measurement instruments typical of physics or even physiology), made little use of mass data processing, and should learn the approaches of 'man–computer symbiosis'.[67] This invocation of J. C. R. Licklider's Cold War manifesto for automation was not only a reminder of the 'informational' turn in molecular biology, but also a pointer towards an important trend: towards large-scale projects that looked to large-scale data processing for techniques to manage the process of scaling up.

(Not all significant projects to scale up genetics would depend on the adoption of the latest techniques of automation. Christiane Nüsslein-Volhard and Eric Wieschaus's team, which traced how thousands of genes were expressed in the development of *Drosophila*, from embryos to adult flies, achieved their results through heroic scaling up of relatively conventional techniques, reliant on the manual and visual skills of individuals to prepare specimens and identify mutant strains. Productivity may have increased 'a hundredfold',[68] but, as Nüsslein-Volhard recalled in her Nobel autobiography, 'on the whole the EMBL [the European Molecular Biology Laboratory, at Heidelberg] of that time, with its strong emphasis on expensive high tech experimental set ups, was not the best place for us, and sometimes it struck us how strange it was to discover very exciting things and know at the same time that there was not a single person in the entire institute outside of our lab who would appreciate it.')

Sequencing and multiplying: the scaling up of biomedical science

The trend towards automation, and its implications, can best be seen in the scaling up of efforts to write out the specific sequences of nucleic acids. Frederick Sanger, following his success in identifying the amino acid sequence of the protein insulin, had turned to sequencing nucleic acids.[69] His first attempts had been with the single-stranded RNA of viruses. He then turned, when new techniques came to hand, to the DNA of bacteriophage, in particular one known as ΦX174. Rather than merely breaking down the DNA into parts, by 1975 Sanger had developed techniques using DNA polymerase to string together new DNA strands each missing in turn one of the four bases, A, C, T, G. Each strand would end at the length of DNA at which the missing base was due. The next trick was to order the strands by length. Sanger did this by a variant of chromatography called gel electrophoresis. School coloured-ink chromatography sorts molecules by size because smaller molecules are dragged along faster by the liquid spreading through blotting paper than larger ones. Gel electrophoresis worked on the same principle, with the difference that the blotting paper was replaced by gelatine and the drag was by electric current. Sanger had labelled the DNA strands with radioactive markers. By reading along the gelatine with an instrument to detect radioactivity, the DNA sequence was 'read off'.

Sanger had sequenced the 5,375 nucleotide bases of ΦX174 by 1977. The human mitochondrion (17,000 bases) and the bacteriophage lambda (50,000 bases) were also sequenced in Cambridge with manual methods of matching and combining individual sequence strand information, in 1981 and 1982, respectively. Soon after Sanger's achievement of dideoxy sequencing (named after a clever way of labelling the ends of strands), two Harvard scientists, Walter Gilbert and Allan M. Maxam, had developed a similar method, using chemicals rather than enzymes, also announced in 1977; 'Since then', summarizes Horace Judson, 'the two methods have been standardized, speeded up, in large part automated.'[70]

The other, connected, type of scaling up was to move from the genetics of simpler to those of more complex organisms. Viruses were the simplest. Prokaryotes, such as the model organism of much molecular biology, the bacterium E. coli, were more complex. But eukaryotes, the organisms with large, nucleated cells, were more complex still. The phenomenon of linkage had provided the means to generate genetic maps of eukaryotes, such as those of the fruit fly Drosophila, in the first half of the twentieth century. Linkage in humans, despite J. B. S Haldane and Julia Bell's discovery in 1936 of correlation of haemophilia and colour-blindness in pedigrees,[71] was far harder to trace for the simple reason that breeding in humans was far harder to manipulate and control than in fruit flies. Only nine human linkages had been established by 1967.[72] Investigation of genetics, while grounded in the working world of human disease, focused on more tractable model organisms. In the mid-1960s the fly was joined by the worm. Sydney Brenner, joint head with Francis Crick of the cell biology division of the Medical Research Council's Laboratory of Molecular Biology (LMB) at the University of Cambridge, had chosen the nematode worm Caenorhabditis elegans as a fast-breeding organism with a simple visible anatomy (every cell could be seen under the microscope), easy to keep and breed in the laboratory. Brenner wanted to investigate the genetics of development, so he needed a multi-cellular organism (ruling out yeast as a model) and one simpler than the fly. Brenner's group started looking for worm mutants.

In 1969, John Sulston, a British chemist trained in nucleic acid synthesis who had spent time as a postdoc at the Salk Institute in California, joined Brenner's group, initially to work on the chemistry of the worm's nervous system.[73] Soon the work was more developmental, following the growth of the worm, including its rudimentary nervous system, cell by cell. This work went hand in hand with the employment of new visualization techniques, either imported, such as the Nomarski differential interference contrast microscope that revealed individual nuclei of cells in living tissue, with no need to stain, or exported, such as the confocal microscopes later developed at the LMB. The worm's development was also traced through laser ablation (knocking out single cells early in its life and watching the consequences) and genetics (looking for mutants with abnormal numbers of cells).

By the late 1970s, people passing through Brenner's laboratory had set up their own worm labs: Bob Horvitz at MIT, Bob Waterston, later at Washington University in St Louis, Judith Kimble at the University of Wisconsin-Madison. To communicate and coordinate research, annual gatherings took place at Cold Spring Harbor. Material was shared. As we saw earlier, the distributed cooperative network of worm researchers made use of new distributed means of communication, such as BITNET, to keep in touch. In 1983, inspired by accounts of investigating genetic maps in *Drosophila*, Sulston persuaded the LMB to pursue a physical map of the entire *C. elegans* genome. This relied on making in vivo clones, cut-up extracts of DNA inserted into bacteria, multiplied up, compared and stored. Sulston recalls:

> I saw that by digitizing the [clone sequence] information it would be possible to hunt for overlaps automatically in a computer. Instead of crawling along the genome clone by clone, you could map the whole thing in one go. That was the point at which I discovered for myself the power of genomics. I had just finished the embryonic lineage, and was looking for something else to do. And at that point I was driven by an obsession that to map the worm genome was the right thing to choose. Just as the lineage had been a resource for developmental biologists, a genome map would be a godsend to worm biologists looking for genes.[74]

With Sanger's sequencing assistant, Alan Coulson, and computing expert Richard Durbin, Sulston developed automated techniques for reading and matching clone sequences.

But even then there was a difference in scale between the project to generate libraries of clones of worm genes, the mapping of the worm genome, and the project to sequence the entire 100 million bases. This project, the worm genome sequence, required riding the wave of ambition in the mid-1980s to write out the human genome.

Again the push came from the availability of new techniques – a necessary but not sufficient condition for accounting for the human genome projects of the last decades of the twentieth century. First, methods were developed to work with smaller parts of the human genome. In 1967, Mary Weiss, who had worked with Boris Ephrussi, announced that, with Howard Green at New York University School of Medicine, she had used a tumour virus to carry chunks of human chromosome into mouse cells. 'Mouse/human hybrids, by effectively transplanting a few human chromosomes into a new cell type', conclude Doris Zallen and Richard Burian, 'have permitted detailed study of the organization of genes on human chromosomes and provided a substantial stimulus to research in human genetics – research that previously had been stymied by the difficulty of conducting research on humans.'[75] Likewise, X-rays were used to break up the chromosomes in such hybrid cells for further study.[76] When combined with fluorescent staining of mammalian chromosomes, achieved by Torbjörn Caspersson and colleagues at the Karolinska Institute near Stockholm in 1970, the mapping of human chromosomes via somatic-cell hybridization became a tractable and growing project. The achievements of the interwar fly school were now being matched for the human subject.

Another set of methods grew from the observation that restriction enzymes cut up the DNA of different individuals at different points. Variation in these lengths – called restriction fragment length polymorphisms (RFLPs) – provided insight into genetic variation of human individuals and families. Yet other sets of methods tackled the problems of dealing with tiny amounts of genetic material. Mary Harper's 1980 method of in situ hybridization, capable of allowing radioactive tracer methods to pick out genes present in the genome in only a singe copy, was one such set.[77]

The remarkable polymerase chain reaction (PCR) was another invaluable new method. In 1969, Thomas Brock and Hudson Freeze of Indiana University had described a new species of bacterium, *Thermus aquaticus*, sourced from the hot springs of Yellowstone National Park. Evolved to live in the extreme heat of the springs, *T. aquaticus* possessed the cellular machinery, including a DNA polymerase ('Taq'), capable of operating at temperatures at which most cells would fail. Kary Mullis, a researcher at the San Francisco biotech firm Cetus Corporation, had the idea of exploiting this capacity to make copies of pieces of DNA, a process that could be multiplied exponentially. Mullis, a rabble-rouser, was given the space to experiment in lines of future profit in the spaces created by the new biotech industry. Yet the transition from idea to working 'experimental system', let alone a tool that could work outside the laboratory, was complex and required many more players than the lone Mullis.[78] Cetus patented PCR (a move challenged in the courts by Du Pont, whose lawyers argued that all the constituent steps were well known), and Hoffmann-La Roche in 1991 bought the intellectual property rights for $300 million. PCR was a revolutionary tool, making working with DNA affordable: it '"democratised the DNA sequence" and allowed (to paraphrase Erwin Chargaff) "the practice of molecular biology without a permit"'.[79]

A third set of methods was the further automation of sequencing. Leroy Hood, working with a 'regiment of scientists and technicians' at Caltech and the nearby Applied Biosystems, Inc., a spin-off from Hewlett-Packard, found ways to automate fluorescent dye sequencing. When combined with an improved way of conducting gel electrophoresis – using pulsed electric fields, which meant that longer segments of DNA could be sorted effectively – a powerful array of sequencing methods was available by the mid-1980s. Automated sequencing techniques were also being developed rapidly in Japan, drawing on its own expertise. Led by University of Tokyo biophysicist Akiyoshi Wada, the effort brought in 'corporations of manifest ability to forge technologies of low unit cost and high in quality', such as Fuji Film, Hitachi, Mitsui and Seiko; Wada in 1986 reflected that such automation 'could well turn out to be the equivalent of the Industrial Revolution in biological and biochemical laboratories'.[80]

Public vs. private? Human genome projects

New and refined techniques only made whole genome sequencing possible; the changing working worlds of science in the 1980s provided other compelling justifications. The first was a jostling for advantage between increasingly entrepreneurial universities. The University of California at Santa Cruz was a relatively new campus, ambitious to find a prestigious scientific project of its own. Robert Sinsheimer, the chancellor of UCSC and a molecular biologist of some repute (he had mapped the phage ΦX174 that Sanger had sequenced), had been frustrated when the University of California had lost out to Caltech in the chase for philanthropic funds for a huge new telescope.[81] Caltech's triumph (the Keck Foundation's massive grant of $70 million to build a new generation telescope in Hawaii) had put the Hoffmann Foundation off a similar astronomical venture with UCSC.[82] In 1984 Sinsheimer calculated that an ambitious project, this time in the life sciences, might recapture the philanthropic interest. He organized a

457

workshop on the technical aspects of a human genome project held in May 1985. Sinsheimer's vision, a human genome project, was not picked up by the University of California, being 'thwarted by internal politics', but his report was circulated and the idea was spread.

Second, in parallel, the United States Department of Energy, successor to the Atomic Energy Commission and therefore the paymasters of the network of large, expensive national nuclear laboratories, also sought to organize the field. Not only did scientists at the nuclear laboratories have deep experience in researching the genetic effects of radiation, but also the Department of Energy wanted to diversify its research programmes in order to be in a strategically stronger position as the Cold War waxed and waned.[83] Charles DeLisi, physicist and administrator of the Department of Energy's Office of Health and Environment, convened a workshop in March 1986 at Los Alamos.

At the Los Alamos workshop, Harvard scientist Walter Gilbert likened the project of sequencing the human genome to the search for the Holy Grail.[84] Medical interests, capable of generating enthusiasm in the United States Congress, now became prominent. Renato Dulbecco wrote an editorial in *Science* in March 1986 declaring that cancer research would be expedited by a complete sequencing of the human genome. The National Research Council of the National Academy of Sciences, chaired by Bruce Alberts, sponsored a report, *Mapping and Sequencing the Human Genome*, which appeared in February 1988; the report, the considered view of the American science establishment, wrote up the project in glowing terms. The National Institutes of Health stepped in, persuading James Watson to leave the agreeable lifestyle of director of Cold Spring Harbor to lead the far more politically driven National Center for Human Genome Research. 'I would only once have the opportunity to let my scientific life encompass a path from the double helix to the three billion steps of the human genome', Watson told colleagues.[85]

Likewise, in the United Kingdom, the Wellcome Trust was beginning to flex its financial muscles. Molecular biology had been supported by the Medical Research Council, a government body. The Thatcher government had been persuaded to support a UK Human Genome Mapping Project, but showed little enthusiasm for full sequencing. The Wellcome Trust, however, was a philanthropic organization, directed by the decisions of its independent trustees, and, following the sell-off (in 1986 and 1992) of Wellcome shares at near the peak of the market, sat on immense resources to invest in biomedical research. By 1990 the Human Genome Project of the United States was in place, with funds to develop methods by starting with simpler organisms and the aim of completing the human sequence by 2005. In 1992 the Wellcome Trust increased its funding, establishing a new flagship laboratory, the Sanger Centre, outside Cambridge. The Medical Research Council took a back seat. The human genome project was emerging as a network of laboratories, with different roles and projects. The estimated final price tag was some $3 billion.

Full sequencing of the human genome was controversial. Many scientists worried that money spent on sequencing was at the expense of other, cheaper, more effective research. Biomedical researchers, including those funded by the National Institutes of Health, feared that the Department of Energy's involvement would shift genomic research into 'mindless Big Science sequencing', marked by loss of control by scientists, bureaucratization, and goal-oriented research that stifled rather than inspired. Others saw the attack on the whole sequence as wasteful because it meant tracing out the repetitive, apparently meaningless 'introns' that filled the genome between the interesting gene-coding regions.

Among the first beneficiaries of the human genome project were the nematode worm sequencers, primarily a division of labour between John Sulston's team at the Sanger Centre and Bob Waterston's group at Washington University, St Louis. They equipped themselves with Applied Biosystems Inc.'s automated sequencers and began racking up sequence data. 'The best way to drive the science', recalled Sulston, 'was to get the sequencing machines going, cheaper and faster, and get the data out so that all the theorists in the world could work on the interpretation.'[86] This meant resisting attempts (through, for example, hacking around software locks) by Applied Biosystems to gain proprietary control of the data. The complete nematode worm sequence was published in 1998.

Commercial interests in the cornucopia of information of the human genome, in particular exploiting genetic markers of disease in marketable tests, now began to express themselves in a powerful fashion. The routes of expression were diverse. Some came from government. The new head of the National Institutes of Health, Bernadine Healy, an appointee of George H. W. Bush, wanted the aggressive patenting of small fragments of genes, expressed sequence tags (ESTs), that had been generated by a then-NIH researcher, Craig Venter. James Watson opposed the patenting, as did the scientific advisory committee of the human genome project, chaired by Paul Berg.[87] In the fallout, Watson had to resign in April 1992. A second route was via some of the philanthropic organizations involved: the administrators of the Wellcome Trust, in particular its head Bridget Ogilvie, were far keener to patent than the researchers.[88] But a third route came from the researchers themselves. Sometimes the human genome project is portrayed as a race between public and private interests. This is vastly oversimplistic. Even John Sulston, usually presented as an archetypal defender of open, academic science, for example, considered a proposal from Frederick Bourke, an investor, to take the worm sequencing work private.[89]

Nevertheless, the private companies set up to sequence for profit ended up driving much of the human genome project. Dismayed by the handling of the EST affair, Craig Venter left the NIH system to set up his own enterprise. It came in two parts. On one side was the Institute for Genomic Research (TIGR), a research centre in Rockville, Maryland, that in principle was not-for-profit. On the other, Wallace Steinberg, an investor in healthcare, paid $70 million to support TIGR in return for his new company, Human Genome Sciences, to have the right to any marketable opportunities coming out of TIGR. TIGR made an impressive start by using whole-genome shotgun methods to write out the first complete sequence of a free-living organism, the bacterium *Haemophilus influenzae*, in 1995. Venter regarded the ESTs, snippets of DNA information, as providing a quicker way to mapping the human genome. In 1997, Venter fell out with Human Genome Sciences and in 1998 found a new backer in the automated sequencer company Applied Biosystems (or, rather, in its parent company, Perkin-Elmer), to found a new company, Celera Genomics.

An example of the implications of the patenting of genes comes from breast cancer research conducted in Britain and the United States. A gene, known as *BRCA1*, that indicated a higher risk of developing breast cancer had been found by a Berkeley geneticist in 1990. In 1994, Mike Stratton, working at the Institute of Cancer Research, south of London, and collaborating with Mark Skolnick, a university-based researcher at Utah with access to the nearby library of worldwide genealogies maintained by the Mormon Church, identified a second gene, *BRCA2*. The relationship between Stratton and Skolnick had broken down months before the announcement, the issue being the plans of Skolnick and Walter Gilbert's start-up company, Myriad Genetics. Stratton persuaded the Sanger Centre to devote resources to sequencing the area around the gene, looking for its precise location and for any mutations. They were successful, and published in *Nature*, but not before Skolnick put in a patent application.[90] The result was that Myriad could market tests for the *BRCA2* mutations at the cost of many hundreds of dollars a go. Money collected for cancer research would need to be channelled to patent lawyers to contest Myriad's claims. Historian Jean-Paul Gaudillière judges that the human genome projects sparked a revival in genetic explanations of cancer more generally.[91]

Examples such as the *BRCA2* gene encouraged the non-privately funded human genome project scientists to maintain their commitment to placing sequence data fully in the public domain. In fact there was a spectrum of beliefs. In the United States, where the model of entrepreneurial science was championed most strongly, many publicly funded scientists favoured or even looked forward to private benefits. In the United Kingdom, John Sulston's view was traditionally progressive. 'So what is wrong with leaving the work to a company?', he asked. 'Simply that, to the extent that the data are fundamental and important, they should be available to all on equal terms.'[92] In 1996, the partners in the public human genome project agreed the so-called Bermuda Principles, named after the conference location. These principles promised fast release of data, not only the raw sequence but also the annotation of the genome (which added biological and especially medical meaning to the data), to the public domain. The Bermuda Principles confirmed and strengthened the model of deposition of genomic data in public databases, in particular in the big three (GenBank in the United States, the European Molecular Biology Laboratory Data Library in Heidelberg, and the DNA Data Bank in Japan). Open deposition is typical in some sciences (protein biochemistry, for example, has the Protein Data Bank, open since 1971) but not in others (epidemiology, for instance, which has masses of data but also patient confidentiality issues).[93]

Public domain science could also be promoted, counter-intuitively, by private interests. The big pharmaceutical company Merck, for example, bankrolled the generation of an EST database as a defensive measure against its potential future rivals, the smaller genomic companies. Such public release would make patenting much more difficult, if not impossible. On the other side, Venter's private human genome project aimed to patent between two and three hundred genes, possibly more in the future as shareholders exerted pressure to maximize profit on genomic information. In response, the Wellcome Trust, now fully convinced of the need for a public domain genome, upped its funding, which in turn strengthened the resolve of federal human genome funding agencies such as the National Institutes of Health.

The relations between the human genome projects were now

written up by journalists as being a race. Press management became crucial for both sides, but perhaps more so for Celera, which did not have public databases of successful sequences to point to as direct evidence of achievement, and were also seeking a stock-market launch. Venter's plan, on the surface, promised an early win in the race. However, his strategy – the whole genome-shotgun approach – had major drawbacks: while fast early on, it would inevitably be hard to complete, with major computing problems foreseeable in the final stages.

Nevertheless, the fanfare of Venter's press conferences provoked the public human genome project to change tack. Rather than systematically assembling the sequences piece by piece, aiming for complete coverage, a rapid dash to a 'working draft' was now authorized in order to match the private project's similar plan. This change in strategy had uncomfortable consequences: the public human genome project became concentrated in five big centres as other smaller centres, unable to keep up, fell away. (The five were the Department of Energy's Joint Genome Institute in Walnut Creek, California, the Baylor College of Medicine in Houston, Washington University in St Louis, the Whitehead Institute in Cambridge, Massachusetts, and the Sanger Centre outside Cambridge, England). International collaboration was also a victim. German and Japanese sequencing was sidelined. Tensions caused by national differences were in fact never absent from the human genome project.[94] Indeed, the overall effect was to make explicit the American pre-eminence. Sulston's impressions of a visit stateside are historically resonant:

> I felt uncharacteristically nervous in the big room with its central table and the staffers arrayed in ranks around it. I had been summoned and had obeyed, just as a century earlier colonial leaders had been summoned to London and had obeyed. There were no tapestries or banners, indeed no decoration at all in the utilitarian government-issue room, but this was a theatre of power in the greatest empire on earth today.[95]

The beneficiary of the acceleration in the human genome projects was the sequence machine manufacturers, Applied Biosystems. Both sides had to purchase more of the new high-speed, even more automated capillary gel machines, the 3700, which cost $300,000 each. While they may have had machines in common, the extremes of the two sides, in 2000, were aggressively hostile to each other, as this exchange recorded by a journalist illustrates:

> 'The people from the Wellcome Trust do not believe this shit that they are saying', says Venter, leaning back on the chair in his office at his

462

Rockville headquarters, near Washington. 'It's much easier to demonise us than justify the hundreds of millions of dollars they have wasted. I have been a thorn in their side for some time because I keep coming up with breakthroughs. Having a rival in any sense is unacceptable to them. They are trying to destroy someone they view as a serious competitor. We did not set out to sequence the human genome just to make John Sulston and the Wellcome Trust look bad. What is it about Celera that has to be stopped – because we are trying to treat cancer, find treatments for disease?'

On the other side of the Atlantic, in his bare office in the shiny new Sanger Centre near Cambridge, Sulston takes an implacably hostile view of his enemy. 'Global capitalism is raping the earth, it's raping us. If global capitalism gets hold of complete control of the human genome, that is very bad news indeed, that is something we should fight against. I believe that our basic information, our "software", should be free and open for everyone to play with, to compete with, to try and make products from. I do not believe it should be under the control of one person. But that is what Celera are trying to do as far as they can. They want to establish a monopoly position on the human sequence. Craig has gone morally wrong.'[96]

Yet, within the month of these exchanges, the apparent completion of the 'working draft' was announced at carefully choreographed press conferences featuring, on either side of the Atlantic, President Bill Clinton and British Prime Minister Tony Blair. The date, 26 June 2000, was chosen because it was a free date in the two leaders' diaries. The publication of the sequence data, also incomplete, was a more fraught affair. While the two sides had been forced to agree to publish simultaneously, the first choice, an appearance side by side in the journal *Science*, was scuppered by Celera's insistence on less than full release of data. Michael Ashburner, a fly geneticist who was joint head of the European Bioinformatics Institute and who was independent of the two human genome projects, circulated a letter around the scientific community denouncing *Science*'s chosen path.[97] The public human genome project decided to publish its paper in the rival *Nature*. Both appeared in February 2001.[98] An irony pointed out by the venerable science journalist David Dickson was that the public genome project's results appeared in the privately owned *Nature* (published by Macmillan), while the privately backed genome results appeared in the not-for-profit *Science*, the journal of the American Association for the Advancement of Science. Leaders of the public project, including Sulston, felt strongly that Celera had free-ridden on the back of publicly deposited sequence data.[99] Certainly, without the public sequence there would have been no private sequence in February 2001, but not vice versa.

A golden age of natural history?

We are interested in the human genome because we are human. And because we fall sick, and have to pay, somehow, for expensive medicines, there is potential profit in the human genomic information. But, viewed from the perspective of a broader history of science, genome sequencing is perhaps just as significant for transformations in our knowledge of the full scope of life on earth. As sequencing became cheaper, and the software for comparing and making sense of sequence data became more sophisticated, so it became possible to compare more organisms. From the 1980s, phylogeny went from a state of virtual invisibility in evolutionary biology journals to ubiquity two decades later.[100] Various specific arguments have been made on the basis of such phylogenetic evidence, regarding, for example, the size of the ancient human population in Africa, the causes of speciation and diversity, the co-evolution of viruses and hosts, and the underappreciated prevalence of horizontal gene transfer, but particularly on the relationships between branches of the tree of life. Phylogenies, often on the basis of ribosomal sequences or other specific well-conserved genes such as *rbcL* in the chloroplast, have been produced for all major clades, and for many smaller ones. A host of wonderful detail has been published; skunks, for example, should be placed in a separate family to weasels. The revelations have not been without conflicts, such as between the authors of new phylogenies of seed-bearing plants[101] and old traditional systematists, such as the Soviet-Armenian Armen Takhajan.

But perhaps the greatest surprises came from an application of PCR. PCR, with its ability to pick out and amplify minuscule amounts of DNA, has found many profitable uses and has become a standard, automated procedure. University biology students now treat PCR as mundane. Yet, as a tool, PCR has also found diverse tailored applications. In particular, it has been used to make in situ amplification of nucleic acids direct from the field, thereby sampling all the organisms present. Whole new kingdoms of life have been found using this method of culture-independent PCR (ciPCR).[102] Life has been found in almost all extreme environments, including deep in the earth's crust. Furthermore, seawater and soil were found to contain many more organisms, especially tiny micro-organisms, than had ever been suspected. The most common life forms on earth (possibly the minuscule ocean bacterium SAR11, also known as *Pelagibacter ubique*), in this way, were only discovered late in the twentieth century. There may be more Archaea, the fundamental clade only identified in the 1970s, than bacteria. The late twentieth (and early twenty-first) century was a golden age not only for micro-biology but for natural history more generally. The discoveries are also startling demonstrations of the case that science knows what it can represent. Before ciPCR, the organisms science studied were those it could cultivate with the laboratory methods of the time. Not all micro-organisms grow well in culture. After ciPCR, life on earth was known to be different because different techniques of amplification, manipulation and representation were in place.

— 19 —

CONNECTING ENDS

Many of the features of the human genome projects – the politics of very big science, the tensions between national and international projects, the issues involved in scaling up, the interweaving of science and information technologies – could be found in the largest late twentieth-century projects of the physical sciences, including the great particle accelerator laboratories and space projects such as the International Space Station and the Hubble Telescope. But there were differences too. The entrepreneurial spirit, long ingrained in chemistry and newly championed in molecular biology, was less evident in physics. Moreover, the relevant working worlds for the physical sciences remained the animation of military and civil technological systems, whereas for the life sciences it was the working world of sustaining human health.

The end of the Cold War reshaped the military working world. Four factors nevertheless continued to support investigations into physics on a large scale. First, the assumed hierarchy of the science placed physics as the most 'fundamental' of the sciences, followed by chemistry, biology and the human sciences. Frequently challenged in the name of anti-reductionism, disciplinary holism and humanism, and from a host of other positions, this assumed hierarchy remained an assumption in the judgements that guided resources to scientific projects. Second, an implicit (occasionally explicit) model of technology as applied science concluded from this first factor that fundamental discoveries in physics would continue to be a major source of innovation. The timeline of development may be long, but it was assumed that new economies would be built from the latest discoveries of physics, just as lasers and solid-state electronics were built from the discoveries in quantum physics in the first half of the twentieth century. Third, international competition and collaboration continued to drive choices made between physical science projects. The details of the competitors may have changed – instead of the early twentieth-century constellation of Germany, France and Britain, or indeed the Cold War axis of the United States and the Soviet Union, we might look in the twenty-first century to Europe, Japan and the United States, or the United States and China – but the competitive dynamic remained. Finally, a sense of the increased pace of change in modern societies provided a justification for permanent mobilization. Twentieth-century wars and economies moved too fast to equip later; training scientists and providing expensive research facilities for the physical sciences had to be done up front.

Big particle accelerators in a changing world: CERN and the SSC

These considerations had justified the construction of flagship laboratories such as the KEK (the National Laboratory for High Energy Physics), part of the Tsukuba Science City built in the 1960s in Japan; Brookhaven, Fermilab and the Stanford Linear Accelerator in the United States; and CERN on the border between France and Switzerland. The Large Electron-Positron (LEP) collider at CERN came into operation in 1989 and sent its particles along a tunnel 27 kilometres in circumference. Such a large instrument demanded the coordinated work of thousands of people, playing many different roles, in order to operate. Furthermore, CERN was an exercise in European collaboration, which added extra issues of coordination and communication.

The giant accelerators at CERN generated huge amounts of data. Computers were needed to handle this information, as well as to control the actions of the myriad physics instruments that surrounded the particle beams. The wide distribution of computers across the CERN site made the laboratory a promising place for networking techniques to develop. In the 1970s these techniques had only been needed in-house, but in the 1980s the question of how to connect to the outside world arose. By 1989, after some deliberation over which form of networking protocol to adopt, CERN chose to connect to the internet, using TCP/IP. By 1990, 'CERN became the biggest internet site in Europe in terms of traffic', facilitated by IBM funding the main United States–Europe internet link there, and 'the laboratory's expertise was widely called upon by other organisations wishing to jump on the Internet bandwagon'.[1] By the turn of the decade, then, CERN had become a good place from which to launch a new networked hypertext project.

Tim Berners-Lee was unusual even before he moved to CERN. His parents, Mary Lee and Conway Berners-Lee, had worked at Ferranti in the 1950s, thus making him a second-generation computer scientist. Tim must have seen the Ferranti mainframes, since, as he recalled, he had built cardboard models of them as a child: 'The main features were that you could push paper tape in one side and pull paper tape out of the other side, and there was a clock in the middle. And that's a pretty good model of a computer.'[2] He had been educated at Emanuel School, London, and Queen's College, Oxford, where he repeated his childhood achievement, building a working computer out of an old television and a keyboard from a discarded adding machine. In 1980, aged twenty-five, Berners-Lee was working for a small start-up outside Southampton while he applied for, and got, a temporary job at CERN. He was confronted with a massive laboratory, with a moving army of physicists, all dependent on paperwork. In the 1980s, the documents were archived in a traditional hierarchical way, through the system 'CERNDOC'. But Berners-Lee had developed for himself a different archival structure: with his program Enquire, a name he took from the Victorian compendium *Enquire Within upon Everything*, he linked documents in an arbitrary, non-hierarchical way.

The World Wide Web – initially just Berners-Lee's solution to CERN's document-handling difficulties – was Enquire writ large: an arbitrary structure of hypertext pages transportable over the internet. The idea of hypertext was familiar to savvy computer scientists; Berners-Lee credits Vannevar Bush's Memex machine idea as well as 1960s computer visionaries Doug Engelbart and Ted Nelson as influences.[3] In March 1989 he outlined his idea in a memorandum under the title 'Information management: a proposal'. In it, he wrote precociously that the 'problems of information loss may be particularly acute at CERN . . . but in this case CERN is a model in miniature for the rest of the world . . . CERN meets now the problems that the rest of the world will have to face soon.'[4]

Was Berners-Lee profoundly insightful or lucky in this claim? For in its accuracy lies an explanation for the spread of the Web. If true, we would have to accept the idea that laboratories such as CERN, in its extreme division of labour, multilingual fluidity and high technology, provide a model of society-to-be. Going further, we might want to add a fifth factor supporting investment in large advanced physics laboratories: they play a functional role, representing probable future societies in miniature and providing the reflective and inventive space to solve the problems arising in such models.

However, this role was certainly not widely expected in 1989. Berners-Lee's proposal first received a puzzled reception. 'Vague, but exciting', was his boss's response (scribbled by hand).[5] But the wider CERN management was just then considering the likely problems to be encountered in the next expansion of the laboratory, the construction of the Large Hadron Collider, and information management was near the top of the list of priorities. Good timing therefore secured Berners-Lee's proposal limited support. Various names were kicked around – 'Mine of Information' and 'Information Mesh' among them – before Berners-Lee and colleague Robert Cailliau settled on 'World Wide Web'.[6]

Ironically, the lukewarm support from the CERN management contributed to the Web's success, as Berners-Lee was forced to tap the enthusiasm and resourcefulness of the wider programming community: unable to develop the software in-house, he released toolkits and let enthusiasts do much of the work. The CERN server (http://info.cern.ch) went public in 1990, and the subsequent story is one of appropriation and diffusion, as users discovered the Web's potential and wrote software to view its contents. It was this process that decisively shaped the Web – indeed, its 'invention' should be located as much with the developers as with Berners-Lee and CERN. The first browser was written by Nicola Pellow, a maths student doing a sandwich course at Leicester Polytechnic. Within a decade, firms based on selling browser software, not so different from Pellow's, would be worth billions of dollars. In 1993 the Web grew by 350 per cent, mostly in the United States. A different mode of governance arose, the World Wide Web (W3C) consortium of interested parties. CERN, in the process, backed out: the laboratory was being distracted from its core concern, particle physics.

CERN had been funded as a demonstration of European cooperation, rebuilding the devastated communities of continental European physics both as a symbol and as a practical way of competing with post-war laboratories in the United States. In the 1950s, and intermittently in subsequent decades, scientists discussed an even more utopian Very Big Accelerator, which would also be both a symbol (of internationalism) and a practical tool, pooling all nations' contributions to build as powerful an accelerator as possible.[7] In the 1980s, the idea of a very big accelerator was revived, this time not as an internationalist but as an American nationalist project. There were two immediate prompts. First, the great discoveries in 1983 of the W+, W– and Z particles predicted by electroweak theory had been made in the collision data generated by the European accelerator at CERN. Second, in the same year, the next generation ISABELLE proton–proton collider in the 200 GeV range under construction at Brookhaven was cancelled. CERN's success was blamed. Physicists in the United States argued that their lead would only be regained with the construction of a new, even larger, accelerator, the Superconducting Supercollider (SSC).

In 1986 Maury Tigner, a physicist at the Lawrence Berkeley laboratory, drew up a design for the SSC.[8] Using superconducting magnets, the proposed accelerator was to reach energies of 40 TeV (in comparison, CERN's Large Hadron Collider, then in the earliest stages of planning, would aim to reach energies of about 14 TeV). The cost of Tigner's design was estimated at $3 billion, which soon rose to $4.4 billion. As Big Science physics facilities increased in scale and expense, so a recurrent question that had been asked was: who should run them? In the Manhattan Project, General Groves had been Robert Oppenheimer's superior. In the post-war laboratories, the extent of the authority of the Atomic Energy Commission had been challenged, and physicists had retained control over laboratories such as the Stanford Linear Accelerator (SLAC). With the projected Superconducting Supercollider, the same question of management arose.[9] The University Research Association's view was that academic scientists should ultimately be in charge, along the model of the joint-university committee managing Brookhaven. The Department of Energy (successor to the Atomic Energy Commission) argued that physicists did not have the management skills to oversee a project that was as expensive as a brand new aircraft carrier. Further, the Department of Energy knew that ceding control to physicists would upset congressional interests. Indeed, congressional interests determined where the SSC would be located: lobbying brought it to Waxahachie, Texas – hardly a hot spot of science.

The Department of Energy ran a competition for bids in 1988. The winner marked a compromise: the University Research Association would team up with two companies with a track record of contributing to large-scale defence projects: the Sverdrup Corporation (which specialized in bridges, highways, rocket and space shuttle test facilities) and EG&G Intertech (detection instruments, lots of defence contracts). Nevertheless, there swiftly developed two distinct organizational cultures, the defence administrators and the more academic physicists. Tigner did not join the project, so his outstanding design

470

was lost. The budget continued to overrun, reaching $8.25 billion by 1991. Nevertheless, the SSC retained political backing, for example from President George H. W. Bush.

Yet one of the primary factors supporting very large physics laboratories was changing. The first term of Ronald Reagan (1980–4), coinciding with the first prime ministerial years of his ally in Britain, Margaret Thatcher, brought in a much more aggressive stance to the Cold War. (Some scholars speak of a Second Cold War of the early 1980s, with the First Cold War lasting from the late 1940s to the early 1960s.) Reagan boosted defence spending, renewing the military-industrial complex through projects such as the Strategic Defense Initiative (SDI, or 'Star Wars'), an overly ambitious 1983 plan for space-based missile defence systems. SDI required the development of new lasers, and other equipment, that existed only as speculative designs. As a result funds were pumped into military research and development laboratories. In comparable terms, the 'defence build-up of the 1980s actually pushed military R&D spending (in constant dollars) past the record levels of the mid-1960s', notes historian of Cold War science Stuart Leslie.[10] (In response, in 1985 President Mitterrand of France promoted Eureka, a symbolically civil European programme.)

The early 1980s was also a peak of anxieties over the possibility, indeed likelihood, of nuclear war. In the West, it was a time of resurgent nuclear protest and the making of apocalyptic films such as *Threads* (1984) and *The Day After* (1983). In the East, too, there was fear. Indeed, it is now known that the world came very close to nuclear war in November 1983. In March 1983, Reagan had described the Soviet Union (in a speech to the National Association of Evangelicals) as an 'evil empire', and fifteen days later had outlined SDI, repudiating Mutual Assured Destruction and arguing that Soviet backwardness in computer technology opened the possibility of complete missile defence, rendering nuclear weapons obsolete. There was panic in Moscow: 'Reagan's speech persuaded the increasingly frightened Soviet leaders that [missile defence] was about happen', even though the technology was in fact speculative, writes historian John Gaddis. The leaders were convinced 'that the great technological potential of the United States had scored again and treated Reagan's statement as a real threat'.[11] In November of that year, the NATO allies conducted a defence exercise, Able Archer 83. The reports of the gathering forces, channelled by the Soviet intelligence agencies, caused Soviet leader Yuri Andropov and his aides to 'conclude – briefly – that a nuclear attack was imminent'.[12] A spy reported the scare back via London to Washington. Reagan, probably shocked, changed his tone: his speech of January 1984 suggested that 'Jim and Sally' and 'Ivan and Anya' could get along.

The Cold War was coming to an end. Mikhail Gorbachev, an insider convinced of the rottenness of the Soviet system, became the Soviet general secretary in 1985. In 1989 the Berlin Wall was dismantled. Gorbachev had ordered the soldiers to remain in their barracks. There would be no Soviet tanks on the streets, unlike Hungary (1956) or Czechoslovakia (1968). Germany reunified in 1990, and the Soviet Union dissolved in 1991. Triumphalist historians have argued that Reagan's tough stance, especially his refusal to compromise over SDI and the subsequent realization among Soviet leaders that the Soviet Union could not match the United States in the strategically critical area of defence research and development spending, caused the change in Soviet policy. Outspending in science and technology, it follows, ended the Cold War. But triumphalist historiography has been challenged. Alternative explanations see the end of the Cold War as a largely unexpected result of attempts at internal change, often recalling Alexis de Tocqueville's remark that 'the most dangerous moment for a bad government is generally that in which it sets out to reform'. Raymond Garthoff, for example, argues that it was the Soviet leaders' recognition of how badly their system had failed, a decay from within, that ended the Cold War.[13] Indeed, in this analysis, increased defence spending prolonged the Cold War.

In the heady days of the rapid dismantling of the Eastern bloc, there was speculation, even an expectation, of a peace dividend. Certainly Cold War justifications did not routinely convince. In June 1992, a surprise congressional vote to terminate the Superconducting Supercollider was carried by 232 votes to 181. While a second vote was in favour of continuing, the vulnerability of the SSC was exposed. As part of the cycle of elections that brought in President Clinton in 1992, a Congress intent on deficit cutting attacked the SSC again. The crucial vote in October 1993 was lost, and the Superconducting Supercollider was cancelled. Historian Daniel Kevles compares the fate of the SSC with that of the Human Genome Project – both vastly expensive, the first a flagship of the physical sciences, the other of the life sciences.[14] Their contrasting fates speak not only of the post-war swing from physical to life sciences, but rather more of changes in their respective working worlds: the burgeoning working world of human health versus the working world in transition at the end of the Cold War.

With the Superconducting Supercollider cancelled, the largest

472

particle accelerators in the world were the Tevatron at Fermilab at Batavia, Illinois, and the projected Large Hadron Collider at CERN. The Tevatron had become operational in 1983 and, with a series of upgrades, was capable of accelerating and colliding protons and antiprotons to energies of nearly 2 TeV. It was expected to be closed when CERN's Large Hadron Collider became operational. But delays to the Large Hadron Collider led to Tevatron being maintained.

At CERN the Large Electron Positron collider, with its 27 kilometre tunnel, had come online in 1989. In the following two decades, experiments using the LEP – four particle detectors, ALEPH, DELPHI, L3 and OPAL, each filling a large building – probed the accuracy of the standard model. In 2000, the LEP was switched off and the tunnel given over to the Large Hadron Collider. The scientific justification for the Large Hadron Collider was that it would reach energies at which the most important, but as yet unobserved, particle predicted by extensions to the standard model, the Higgs boson, would probably be detectable. There was also the chance of unexpected discoveries, perhaps clues to the mystery of dark matter, a missing mass deduced from astronomical observation and cosmological theory. The projected cost of the Large Hadron Collider, at least before its troubles, was €3 billion,[15] broadly comparable to the SSC.

Spectacular science: space telescopes and planetary probes

Space science projects, such as the Hubble Space Telescope, have also been 'big science of the very biggest kind'.[16] The scientific justifications for an optical telescope of a reflector design placed in space were that the observations would contribute to cosmology, its resolving power would exceed that of earth-based telescopes, and the absence of obscuring atmosphere would permit measurements across the spectrum, from ultraviolet to infrared and even sub-millimetre wavelengths. When it was first seriously discussed in the mid-1960s, the price tag of $1 billion was evidently more than philanthropy could provide. There would, it was assumed, be no post-war Carnegie or Rockefeller. Instead federal patronage was sought.

The process of funding and designing the Space Telescope was 'ad hoc', 'disordered', pulled around by all kinds of departmental and political interests. 'The Space Telescope should thus be viewed in many ways as the concrete realisation of the interests supporting the program', writes its historian Robert Smith, 'and in particular of the coalition building that its advocates engaged in win approval for the program'.[17] For example, optical astronomers of the 1960s did not like it: they did not need to go into space (unlike X-ray astronomers) and they did not wish to cede control of their facilities to adminis-trators, in this case NASA. In response, an 'all things to all people' approach was pursued by Lyman Spitzer, Jr., who led the efforts 'to bundle as many interests as possible into the proposed instrument' and 'evangelize among the astronomical community'.[18]

Nevertheless, the Space Telescope was relegated to the position of a second-tier priority project by 1972. In financially straitened times, it was as good as dead. The following battle for political approval, between 1974 and 1977, featured 'vigorous lobbying' by the indus-trial contractors involved, supported by lobbying efforts of Spitzer and John Bahcall, both from Princeton but representing a broader astronomical community. This marked the forging of what might be called an astronomy-industrial complex, an alliance of interests between traditional astronomy and large-scale industrial corporations in support of the Space Telescope. A compromise was reached, which illustrates the way that political negotiations could shape expensive scientific projects: the mirror would be smaller (lighter, cheaper), the launch later, and international partners found (indeed the European Space Agency became a junior partner). Building began in 1977.

NASA, in the meantime, had succeeded with an extraordinary series of scientific probes to planets in the solar system. Pioneer 10, launched in 1972, travelled through the asteroid belt, returning the first close-up pictures of Jupiter on a fly-by in December 1973. Its sister craft, Pioneer 11, launched in 1973, visited Jupiter and Saturn in 1974 and 1979, respectively, passing through the rings of the latter planet. Another pair of NASA craft, Viking 1 and 2, each consisting of an orbiting and landing section, visited Mars in 1976. The special interest here was whether or not signs of life could be detected in the frigid Martian 'soil': a gas chromatograph sought signs of organic molecules, and dilute nutrients were added to see if Martian organ-isms might metabolize them. In preparation for these experiments, NASA had essentially created a new specialty, exobiology, from scratch through targeted funds. (Woese's discovery of Archaea was one event shaped by this funding.) The results did not support the existence of Martian life. (NASA would later court controversy, when in 1996 it unveiled a meteorite, ALH 84001, in which it claimed Martian microbe fossils could be seen.)

Further projects followed where the Pioneers had led. Voyager 1 and 2 were both launched in 1977. Like Pioneer 2, Voyager 1 passed by Jupiter and Saturn. Signs of past volcanism were discovered on Jupiter's moon Io on a fly-by in 1979. Voyager 2 not only visited Jupiter and Saturn but also became the first human-made probe to Uranus and Neptune. Io's volcanism was observed to be still active, with plumes of material photographed reaching into space. Beautiful photographs of the moons of great planets were relayed back to earth. The Pioneer and Voyager missions had revealed that the largest moon in the solar system, Saturn's Titan, had a thick atmosphere. The Cassini–Huygens mission, which reached Saturn in 2004, dropped a probe (Huygens) through the atmosphere, photographing possible oceans on the way.

Another later mission was Galileo, an orbiter which arrived at Jupiter in 1995. Galileo had been up for approval at precisely the same time as the Space Telescope, in 1977. Astronomers and NASA officials worked very hard to present the two projects as complementary. However, the Space Telescope was approved, while Galileo was – temporarily – rejected. The Space Telescope – named the Hubble Space Telescope in 1983 – was finally launched in April 1990. At $2 billion, it was 'probably' (at least at the time) 'the most costly scientific instrument ever built'.[19] This assessment assumes that projects such as the Large Hadron Collider are not single instruments.

To the horror of all involved in the project, the first Hubble images showed that the 2.4-metre mirror had been ground inaccurately. The aberration had been missed by the firm – Perkin-Elmer (incidentally, recall from chapter 18, the parent company of Applied Biosystems Inc.) – polishing the mirror. At first an acutely embarrassing incident for NASA, the problem became a triumph of innovation, spurring a creative solution (a corrective lens, placed over the Space Telescope by visiting astronauts, an act that has been used to justify human spaceflight) and the development of new image processing techniques. These techniques were later transferred to improve mammogram imaging: 'the flaw in Hubble's original mirror turned out to be a window of opportunity for early detection of cancer.'[20] Indeed, Bettyann Kevles sees this transfer as symbolic of a broader change: 'The separation of disciplines that characterized the way the physical sciences worked during the middle decades of the twentieth century became obsolete in its closing decade.'

Images from the Hubble Space Telescope, many taken with the Wide Field and Planetary Camera 2 and processed with an aesthetic eye drawing from a multicoloured palette, were judged to be beautiful. Some of them – such as the picture released in November 1995 of the so-called pillars of the Eagle Nebula, an area of gas, dust and star formation – have become iconic images of science. Big Science projects in democracies had to connect with a broader public if politicians were going to commit support to them. NASA, post-Apollo, has walked a tightrope, offering scientists unheralded opportunities in astronomy, delivering news stories (including beautiful images) that the public like and consume, while keeping the tight budgets workable and industrial contractors coordinated, and avoiding promising too much. This context has shaped both NASA's successes and failures, including the Space Shuttle disasters of 1986 and 2003.

The reporting of science

In the United States, from the 1960s to the 1980s, the reporting of science was dominated by what Sharon Dunwoody describes as an 'inner club', a 'small group of newspaper, magazine and wire reporters', 'no more than 25 or 30 journalists . . . a closely knit, informal social network', that together 'have largely determined what the public has read about significant scientific events'.[21] Smaller analogues of the 'inner club' can be found in other democratic countries. The 'inner club' described by Dunwoody formed around the coverage of the early space race, meeting up at Cape Canaveral or Houston and, given they were often untrained in the sciences, pooling information. Working and socializing together: the 'major emphasis of the club as it evolved was on cooperation'. Collectively the inner club could evaluate scientific information more accurately. Also, since 'the quality of a reporter's work was judged by his editor in relation to what other reporters were producing' (editors, after all, also had no scientific training), the 'reporter . . . could satisfy his editor not by "scooping" his fellow science writers but by producing the same story each day'.[22] The disadvantages of this cooperative behaviour, rare among journalists and a social response to the unusual expert character of their subject matter, were twofold: the reporting tended to be homogeneous (a scientific subject was reported the same way across different outlets) and selective (only subjects collectively chosen were covered at all).[23] Since the print media were the primary means by which non-scientists found out about science, these effects on mediation shaped the relations between science and its public audience.

Science communication, especially since the 1960s, has been shaped by other factors in addition to the politics of Big Science and the social phenomenon of the 'inner club'. For various reasons, the number of media outlets has grown and diversified. Small local papers may have folded, but magazine titles have increased, and new media have appeared, not least via the internet and the web. Science in the late twentieth and early twenty-first century received more media coverage than ever before. While many of these outlets had specialist audiences, all media had to serve the interests of an audience if they were to be read or watched. Science reporters in major newspapers, nearly always dependent on advertising revenue which correlated with audience size, therefore pitched stories knowing that they were in competition with other non-science journalists for space and well aware that the editor's decision would be guided by readers' interests. This set-up produced science reporting dominated by a few subjects: health foremost, followed by other topics, including space, dinosaurs, and 'discoveries' and 'breakthroughs' more generally. Museums followed analogous pressures. When the Academy of Natural Sciences of Philadelphia put on a new dinosaur exhibit in the mid-1980s, public attendance soared; other museums soon copied the approach.[24]

The relationships between the scientific scholarly publications and the general press have developed in interesting ways in response to public interest. Just as 'health' was the best issue for an aspiring journalist to tag a story to get past an editor, so the leaders in developing new relationships were the general scholarly publications on medical science. In 1967, the gastroenterologist Franz Ingelfinger took over editorship of the *New England Journal of Medicine*. Two years later he announced a new editorial policy, the so-called Ingelfinger Rule: he would not accept for publication any findings previously announced via other media.[25] The announcement of the rule was a response to two changes in science communication: first, Ingelfinger was alarmed by the rising 'numbers of journals, prior publication in free trade publications, and the general [post-war] proliferation of science coverage in the media[, which] presented a dangerous threat to [his] journal's capacity to print original content'; second, the gathering critical voices of the long 1960s called for public knowledge of medical science, and Ingelfinger introduced the rule as an expression of the journal's public responsibility.[26] The 'rule created a formal dependence of the lay press on medical literature for information', concludes Jennifer Toy. 'It supported the notion that the only medical research information communicated to the public should be that published in a journal and thus approved by peer review.'

The move has been partially successful: many news stories since the 1970s have originated in embargoed publication in the general science and medicine scholarly journals – the *New England Journal of Medicine*, *The Lancet*, the *British Medical Journal*, *Nature*,

477

Science – but, of course, the control has by no means been complete. The scientific publishing sector, especially journals, has been under increasing pressure from two directions: the growing sophistication of commercial marketing, especially of drugs, and the apparently free communication offered by the internet. 'Ghost authoring', in which a high-profile professor puts his or her name to a report generated by unnamed writers contracted by companies, became an issue in the first decade of the twenty-first century, as have articles on new drugs that were written up like scientific papers but were in fact advertisements.[27]

In 2000, three scientists – Patrick O. Brown, Michael B. Eisen and former NIH director Harold E. Varmus – wrote an open letter calling for 'the establishment of an online public library that would provide the full contents of the published record of research and scholarly discourse in medicine and the life sciences in a freely accessible, fully searchable, interlinked form', the establishment of which, they argued, 'would vastly increase the accessibility and utility of the scientific literature, enhance scientific productivity, and catalyze integration of the disparate communities of knowledge and ideas in biomedical sciences'. In particular, they complained about the privatization of journal content: 'We believe', they wrote, 'that the permanent, archival record of scientific research and ideas should neither be owned nor controlled by publishers, but should belong to the public and should be freely available through an international online public library.' The trio therefore pledged to 'publish in, edit or review for, and personally subscribe to only those scholarly and scientific journals that have agreed to grant unrestricted free distribution rights to any and all original research reports that they have published, through PubMed Central and similar online public resources, within 6 months of their initial publication date.' Their lead was followed by 34,000 other scientists, who added their signatures to the letter. In 2003, with backing from the Gordon and Betty Moore Foundation (in other words, Intel wealth, recycled as philanthropy), a new journal, the *Public Library of Science* (PLoS), was launched. By the end of the first decade of the twenty-first century, *PLoS One*, and its penumbra of specialist titles, threatened to rival in status *Nature* and *Science*.

Cold fusion

Nevertheless, accusations of 'science by press release' have occasionally arisen, and mark attempts to maintain perceived norms of scientific behaviour. Several examples have been discussed above, in the development of biotechnology and the human genome projects, but the most spectacular came from the fuzzy border between chemistry and nuclear science. On 23 March 1989, two chemists from the University of Utah, Stanley Pons and Martin Fleischmann, held a press conference where they announced they had achieved 'cold fusion'.[28] Hot nuclear fusion – the release of energy from the merging of two lighter nuclei – had been pursued since the 1950s as a potential source of power. It was immensely expensive, required gigantic apparatuses to control the plasma, and repeatedly disappointed its backers. Now two chemists were claiming that they had produced 'excess heat' in a test-tube containing heavy water and a palladium electrode. In the run up to the press conference, Pons and Fleischmann had heard that another local scientist, Steven Jones from Brigham Young University, was producing similar results. The University of Utah's administrators and patent lawyers, with an eye on the prize, insisted that Pons and Fleischman rush their results out. The chemists agreed. Following the press conference and the headline news stories, scientists around the world dropped their work and attempted to replicate cold fusion. Some found odd phenomena. Many found nothing. While the resulting debacle was embarrassing for science, the events that led up to it, while extreme, were not different in kind from science as it was sometimes conducted in entrepreneurial universities: pressure from administrators, secrecy, the involvement of patent lawyers, and the announcement by press conference.[29] Disciplinary politics also played a part in the ferocity of the backlash; after all, here were two chemists at a small university apparently achieving something that among the largest Big Science physical laboratories had failed to match despite lavish funding.

Chaos

In the case of cold fusion, the phenomenon (unless it only worked in Utah) was a will-o-the-wisp, never substantiated, controlled or, to the satisfaction of the scientific community, replicated. A similar extraordinary level of hype could create a bubble of expectation around other phenomena. The case of 'chaos' was exactly contemporaneous with the cold fusion story. Mathematicians had been investigating the odd solutions of nonlinear equations for well over a century. Nonlinear equations can often not be solved directly, and therefore approximate answers, using numerical methods, provided

479

solutions. Numerical methods were repetitive and time-consuming to follow by hand. Therefore, as the speed of electronic computing increased, along with better memories and imaging software, so modelling by numerical methods became an increasingly attractive and powerful technique. Between the 1960s and the 1980s, computer scientists such as Mitchell Feigenbaum (Los Alamos) and Benoît Mandelbrot (at IBM) investigated iterative methods (such as numerical methods), adding a pictorial dimension to the mathematicians' findings.

When a parameter varied, some solutions to nonlinear equations moved from order to disorder, or 'chaos'. When the map of solutions was represented graphically the edges were crinkly at every level of resolution – what the mathematicians had called 'fractals'. In the late 1980s, computer graphics had reached a point in development where, with suitable colouring, these maps were extraordinarily vivid. (Actually, I prefer Mandelbrot's first computer-generated pictures: with the rough quality of old computer print-outs, fragments of the Mandelbrot set are glimpsed like a new continent through fog.)[30]

In 1987 a science writer, James Gleick, published *Chaos*, a best-seller that promoted the notion that a new, fundamental science was brewing. In its universal promise, foundational rhetoric, popular appeal and fringe enthusiasm, it was reminiscent of cybernetics in the immediate post-war years. Actually, it was bundle of many existing specialties – turbulence studies (from the working world of aeronautics), so-called catastrophe theory, the mathematics of nonlinear analysis, and so on. Attempts to build a discipline from this foundational moment have largely stumbled. Complexity theory, for example, while it has prophets – appropriately based in the desert, at the Santa Fe Institute – and converts, is an enthusiasm of a sect rather than the dogma of an establishment.

Nevertheless, the chaotic phenomena of interest, both to professional scientists and popular commentators, were being generated from the working worlds of twentieth-century science. A prime example would be Edward Lorenz's 1960s investigation, discussed in chapter 17, by computer-based numerical methods of the nonlinear equations of meteorology, his discovery of 'strange attractors' and the implications for weather prediction. Lorenz would later (in 1979) ask the famous question of whether a butterfly flapping its wings in Brazil could start a tornado in Texas, a cliché of both popular chaos writing and a parable of unpredicted consequences for environmental commentators.

The sciences and politics of global climate change

By the late 1960s and early 1970s, the evidence for anthropogenic climate change consisted of knowledge of the increase of greenhouse gases in the atmosphere, evidence of past, rapid climate change from a range of studies (ice core samples, deep-lake pollen samples, field geology, planetary science), and computer models of global climate circulation, including ones modelling the greenhouse effect. Investigations also raised questions about the potential instability of the West Antarctic ice sheet (J. H. Mercer in 1970) and the possible effects of increased dust and smoke, and even aircraft con-trails. 'It was not the weight of any single piece of evidence that was convincing', concludes Spencer Weart, 'but the accumulation of evidence from different, independent fields.'[31] A consensus was certainly forming within the professional science community. In 1977, the National Academy of Science of the United States was asked to survey the evidence and judge the reliability of models; its Committee on Climatic Variation reported back an estimated increase in global temperatures of 1.5° to 4.5° over the coming century. Interestingly, this range estimate has remained stable even as the underlying models have changed quite dramatically.[32]

Anthropogenic climate change raised thorny questions of translating scientific consensus into international political action. As argued earlier, the Cold War had encouraged a deepening of data gathering on a global scope, as well as accelerating the provision of fast computing tools needed to handle the data. The scientists, too, were already organized at a global level, through initiatives such as the World Meteorological Organization's World Weather Watch (from 1963) and the joint WMO/International Council of Scientific Unions' Global Atmospheric Research Program (from 1967). In 1985, an assembly of experts at Villach, Austria, announced an international consensus: that between 2000 and 2050 there would be 'a rise in global mean temperature . . . greater than any in man's history'; Weart calls the Villach conference a 'turning point'.[33]

By 1987, the advocates of global action to combat global warming had an exemplar of a way forward. In 1985, measurements by scientists working for the British Antarctic Survey of ozone in the upper atmosphere had revealed a depletion of the gas. Stratospheric ozone absorbs ultraviolet radiation, shielding life on earth. The developing 'ozone hole' was reasoned to be the consequence of the chemical reactions of chlorofluorocarbons (CFCs), a class of gases used as refrigerants and propellants in spray cans across the world. Despite industrial

481

inertia and widespread use, the political will was found in 1987 to sign the Montreal Protocol, which effectively led to the phasing out of CFCs.[34]

Effective replacements existed for CFCs. There was no simple solution to the production of carbon dioxide, the most important greenhouse gas, generated by power stations, internal combustion engines, and many industrial processes, as well as by the earth's vegetation. Nevertheless, a model of tackling the ozone hole was followed when in 1988 the WMO and other United Nations agencies set up the Intergovernmental Panel on Climate Change (the IPCC) as a means of generating consensus scientific statements as a prelude to international agreement of a suitable protocol. Bert Bolin, chair of the IPCC, argued that the IPCC represented a separation of the responsibilities of scientists and politicians: 'Only scientists can grasp the intricate interactions ... of the global environment', Bolin said. Their role therefore was to 'present available knowledge objectively [to policymakers, who were in turn responsible for making political decisions] based on a combination of factual scientific information as provided by the IPCC and [their own] value judgements'.[35]

The form of the IPCC was a compromise, the best worst organization for making consensus statements. Governments set the agenda and owned the process, ensuring saliency, while scientists were authors, which, plus peer review, ensured credibility. Its division of labour between scientists and political appointees was much more complicated than Bolin suggests. The statements took the form of assessment reports, each representing a mammoth coordination of scientific and political labour. The IPCC itself, a political body, approved an outline, governments nominated experts who were allocated to four working groups, an IPCC Bureau selected authors of each working group's report, the report was drafted and sent out to peer review, a second draft was made, followed by combined expert and governmental review, a final draft was made and followed by more governmental review, before final acceptance by the working groups and the IPCC and publication. The phrasing of key sentences – the use, for example, of terms such as 'likely' – would be wrangled over for months.

The first IPCC *Assessment Report* was published in 1990. Its negotiated conclusion was relatively mild: the world was warming, but it would take another decade to be sure that the cause of warming was artificial amplification of the greenhouse effect. In the run up to the Rio summit of 1992, a coordinated attack on the IPCC's claims was made by industrial interests. The Global Climate Coalition, which channelled funds for this attack, included as members the oil giants (Royal Dutch/Shell, Texaco, British Petroleum and Exxon) and the automobile manufacturers (Ford, DaimlerChrysler), as well as other interested parties such as Du Pont and the Aluminium Association. At the United Nations Conference on Environment and Development at Rio, a Framework Convention on Climate Change was agreed. It was not a strong document, as, for example, it did not include binding commitments to greenhouse gas emission reduction. But it did set out a forward path.

In 1995, the second *Assessment Report* of the IPCC was ready. Again its conclusions were marked by compromise: 'The balance of evidence suggests that there is discernible human influence on global climate.' Again the politicians met. This time the result was the Kyoto Protocol (1997), Article 3 of which read:

> The Parties included in Annex I shall, individually or jointly, ensure that their aggregate anthropogenic carbon dioxide equivalent emissions of the greenhouse gases listed in Annex A do not exceed their assigned amounts, calculated pursuant to their quantified emission limitation and reduction commitments inscribed in Annex B and in accordance with the provisions of this Article, with a view to reducing their overall emissions of such gases by at least 5 per cent below 1990 levels in the commitment period 2008 to 2012.

The Kyoto Protocol needed to be ratified by enough individual nations' governments to come into force. Yet there were sharply diverging attitudes. At Kyoto the United States had only been willing to concede to a gradual reduction to 1990 levels, European governments wanted faster action, while China sat on the sidelines. In the United States, the Senate rejected the Kyoto Protocol by a vote of 95–0, arguing that it could not sign up to an agreement that placed no limits on emissions by developing countries. George W. Bush renounced the Kyoto Protocol in 2001. The third (2001) and fourth (2007) assessment reports of the IPCC announced firmer consensus agreements on the existence and dangers of global warming.

The result, in the early twenty-first century, was a stalemate, a mismatch between the recognition of the implications of an international scientific consensus and the political will to act. However, the process of scientific fact-making found in the IPCC was not only shaped by the hybridity of the IPCC's constitution. Geographer David Demeritt argues that the reliance of the IPCC working groups on model simulations (rather than other ways of assessing climate change), as well as the parameters used in the models, reflected choices governed by

483

assumptions that were not acknowledged and made explicit. Having made the facts, he suggests, the governance of climate change followed. 'In this way', he concludes, 'a socially contingent form of scientific knowledge is being shaped by an emergent international policy regime that, in turn, is being constructed and legitimated by this same body of scientific knowledge.'[36] Spencer Weart, in contrast, is much more positive in his assessment. It was an 'extraordinary novelty' that global warming 'became a political question at all', given that it was invisible, a prediction of future change, 'based on complex reasoning and data that only a scientist could understand'. 'It was a remarkable advance for humanity that such a thing could be the subject of widespread and vehement debate', which he ascribes to the 'worldwide advance in democracy . . . The democratization of international politics was the scarcely noticed foundation upon which the IPCC . . . took its stand.'[37]

Global change emerged as a dominant working world in the late twentieth and early twenty-first century. Understanding the phenomenon of climate change, as well as working out policies for mitigation and adaptation, requires interdisciplinary science. Other, interrelated types of global change have proved similarly complex. As an illustration, here Rita Colwell describes her work:

> My scientific quest to understand cholera began . . . in the 1970s, when my colleagues and I realized that the ocean itself is a reservoir for the bacterium *Vibrio cholerae*, the cause of cholera, by identifying the organism from the Chesapeake Bay. Copepods, the minute relatives of shrimp that live in salt or brackish waters, are the hosts for the cholera bacterium, which they carry in their gut as they travel with currents and tides. We now know that environment, seasonal, and climate factors influence copepod populations, and indirectly cholera. In Bangladesh, we discovered that cholera outbreaks occur shortly after sea-surface temperature and height peak. This usually occurs twice a year, in spring and fall, when populations of copepods peak in abundance. Ultimately, we can connect outbreaks of cholera to major climate fluctuations. In the El Niño year of 1991, a major outbreak of cholera began in Peru and spread across South America. Linking cholera with El Niño/ Southern Oscillation events provides us with an early warning system to forecast when major outbreaks are likely to occur.[38]

In this case the clue to managing cholera came from combining the sciences of microbiology (*Vibrio cholerae*), marine biology (copepods), including far-flung field studies (Chesapeake Bay, Bangladesh, Peru), oceanography (sea temperatures, El Niño), astronomy (tides), global climatology (El Niño/Southern Oscillation) and medicine (cholera).

El Niño was the name given to a cyclical change in currents and temperatures in the east Pacific, with global climatic consequences. The correlation of droughts in India and Australia had been noticed in the 1890s, and a century of accumulated data, combined with modelling, has been used to trace out many linkages and effects. El Niño increased in frequency in the late twentieth century, which scientists connected to global warming. In the above example, then, cholera prediction, in addition, depends on the sciences of global warming.

Biodiversity and bio-collecting

Global climate change has also begun to escalate the great extinction, comparable, palaeontologists told the world, to the ones at the end of the Permian and Cretaceous eras, but this time caused by human action. In 1979, Norman Myers published startling estimates of the rate of disappearance of rainforest. These data prompted a cadre of senior biologists – Edward O. Wilson, Myers, Peter Raven of the Missouri Botanical Garden, Jared Diamond, Paul Ehrlich and Thomas Lovejoy – to organize against rainforest destruction.[39] The second half of the twentieth century saw the transition – involving many continuities – from an imperial model of resource conservation (discussed in chapter 7) to a global bureaucratic model of 'biodiversity' management. Edward O. Wilson had written a policy paper on 'The biological diversity crisis' in 1985, given a keynote speech at a 'National Forum on BioDiversity' in 1986, and edited the proceedings *BioDiversity* (1988), by when the term 'was one of the most frequently used terms in conservation literature'.[40]

Biodiversity, defined by Wilson as 'the totality of hereditary variation in life forms, across all levels of biological organization', was equally adaptable to all levels of human bureaucratic organization – the global (the Convention on Biological Diversity was adopted at the Earth Summit at Rio, 1992), the regional (for example, a European Community Biodiversity Strategy was adopted in 1998), the national (for example, the United Kingdom Biodiversity Action Plan dates from 1994), and the local (for example, the city where I live, London, has its own Biodiversity Action Plan, as does the borough, the 'Hackney Biodiversity Action Plan'). These bureaucracies provided the working worlds for most conservation scientists in the 1990s and 2000s.

In parallel to the transition from imperial resource management to global bureaucratic management of life on earth, there has been a transition in its economic corollary, from colonial and imperial resource exploitation to post-colonial 'bio-collecting'.[41] During the early post-colonial years, independent non-settler states had nationalized the colonial research facilities. While the states devoted relatively high proportions of national income to education, including the training of scientists and technicians, the typical economic policy – import substitution – had 'provided few opportunities for local technicians to innovate in the application of science'; furthermore, agriculture, which had been exception to this rule, was transformed during the Green Revolution towards a 'reinforced dependency on foreign technical advisers and [foreign] industrial inputs'.[42] Furthermore, the effect of World Bank and International Monetary Fund policies has been to keep these nations in their colonial roles as providers of raw resources.

In the 1970s, declining commodity prices, population growth and a rise in the cost of imported industrial goods all led to an increase in public debt, which led to cutbacks, for example in science, as well as short-term policies of replacing forest with plantations: notably for beef (Brazil) and palm oil (Indonesia).[43] It was in this context that the Convention on Biological Diversity had been signed at Rio in 1992. No longer would biological resources be regarded as a global commons. Instead, developing nations were assigned the property rights to biological resources located within their borders. With the forests a reserve of future commercial value, protected by property rights, the balance would be tipped, so the justification went, away from clearance and plantations towards preservation, controlled extraction and local benefit. Furthermore, closely following Rio, an agreement on Trade-Related Aspects of Intellectual Property Rights (TRIPS) was reached in 1993–4 as part of the General Agreement on Tariffs and Trade (GATT), talks designed to open up global markets. Powerfully lobbied for by corporations such as Pfizer, TRIPS obliged all countries to adopt the patent systems typical of industrialized countries.[44]

Bio-collecting has developed within this changing regulative world. Bristol-Myers Squibb, the big pharmaceutical firm, fought a legal battle, tooth and nail, trying to establish its claim over the name 'taxol' for the anti-cancer drug derived from the yew species *Taxus brevifolia*; the working worlds of human health and conservation in the case of taxol worked in opposite directions, presented in the media as 'environmentalists' versus 'cancer patients'.[45] Or, take the case of another tree with interesting chemical properties, neem.[46] Following *Silent Spring*, there had been a demand for new biodegradable insecticides. American timber importer Robert Larsen knew of

the traditional uses of the oil of the neem berry in India – an antibacterial agent, a contraceptive and a fungicide as well as an insecticide
– and took samples to his Wisconsin laboratory in 1971. He patented
the neem extract, azadirachtin, in 1985 and sold this property right
on to the corporation W. R. Grace & Co. in 1988. The company
turned neem oil extract into a product, 'Bio-Neem', also known
as 'Margosan-O', and in India organized an Indian firm to process
neem. With TRIPS, India was obliged to respect patents, such the one
for W. R. Grace's neem extract, while traditional knowledge went
largely unrecognized and unrewarded. Activist-scholar Vandana
Shiva called this relationship 'bio-piracy'.[47]

In response to obligations under the Convention on Biological
Diversity to respect the property rights of nations of origin of biological resources, agrichemical and pharmaceutical firms have responded
with a range of strategies. Some have signed partnership agreements
with countries. In 1991, Merck, for example, anticipating the international settlement, had teamed up with Costa Rica's Instituto Nacional
de Biodiversidad (INBio). Costa Rica gained percentages of sales,
access to Merck technology, and training, both of local scientists and
technicians and of knowledgeable villagers as local prospectors.[48]
Merck gained access to INBio's collection of 2.5 million specimens,
with more collected from the field all the time. Following the convention, other companies signed similar deals, made arrangements
directly with tribal communities, or became brokers dealing with the
discoveries of indigenous bio-prospectors. 'Unbenownst to most',
writes Bronwyn Parry of the 1990s, 'in the quiet confines of board-
rooms, laboratories and warehouses, away from the public gaze,
executives, scientists, and technicians have been involved in a worldwide collection project unrivalled in scope since colonial times.'[49]

Another response has been to seek to avoid dealing with countries
at all, thereby side-stepping the costs of property rights. Furthermore,
the task of extraction of biological resources and the ability to
manipulate them have been powerfully multiplied by the application of recombinant DNA techniques. 'Genetic resources may now
be rendered in a variety of progressively less corporeal and more
informational forms: as cryogenically stored tissue samples, as cell-
lines, extracted DNA, or even gene sequences stored in databases',
notes Parry, and, 'in these new artifactual forms, genetic resources
become infinitely more mobile and hence more transmissible.'[50] The
biotech company Phytera, for example, paid seven European botanical gardens for rights to mine herbaria for samples.[51] The final strategy was an extension of the approach that had first been developed

in the 1920s by Vavilov. In the last decades of the twentieth century, international collections of seeds of key crops, preserving local strains and variations, became 'gene banks'. A sharp tension existed between those who wanted access to gene banks to be free of charge (in line with science's norms) and those who wanted to charge for access (advocated notably by the NGOs representing developing countries' interests).[52]

We have seen how the working world of global governance generated much interdisciplinary research. Interdisciplinarity was a much remarked upon theme of science in the second half of the twentieth century. I have discussed various examples. (Recall, for example, Dendral, the artificial expert system that was a collaboration of computer science, molecular biology and chemistry, or the example of cholera prediction based on copepod surveys.) Interdisciplinarity was also encouraged because of a model of innovation that prospered as university administrators promoted entrepreneurial activity. This model held that radical novelty was most likely to appear at the boundaries and overlaps of disciplines. Interdisciplinarity at mid-century – 1930s molecular biology or 1940s radio astronomy, neuroscience in the 1960s – had not been promoted in the name of profit to the extent of interdisciplinarity at the century's end.

Participatory science and public engagement: from AIDS to BSE

The question of what range of relevant expertise should be considered when responding to an issue was raised in the last decades of the twentieth century. 'One of the most striking aspects of the conduct of AIDS research', notes Steven Epstein, was 'the diversity of the players who have participated in the construction of credible knowledge'.[53] In 1981, doctors in New York and California had reported unusual cases of a rare cancer, Kaposi's sarcoma, and a rare form of pneumonia among gay men. Within the year similar cases were reported among intravenous drug users, haemophiliacs and the Haitian community. A wasting disease, 'slim', was reported in Africa, in particular Uganda. Different names were attached to the disease, among them Gay-Related Immune Deficiency (GRID) and the homophobic popular term 'gay plague', and only from 1982 was it called Acquired Immune Deficiency Syndrome (AIDS). Reportage stoked anxiety and fear. Scientists in France and the United States raced to identify the cause of the disease. Within a day of each other, in April 1984, scientists at the Pasteur Institute and Robert Gallo of the National Cancer Institute announced their independent isolation of a virus: the French team called in LAV (Lymphadenopathy Associated Virus), the Americans called it HTLV-III (Human T-lymphotropic Virus III). Patents were taken out on Gallo's work. It took a year of argument to establish an identity between LAV and HTLV-III. In 1986 the International Committee on the Taxonomy of Viruses provided a new name: HIV (Human Immunodeficiency Virus).

Research into AIDS and HIV diversified in many directions. Epidemiologists traced the epidemic back to an African origin of indeterminate date. Pharmaceutical firms and medical science institutes sought vaccines and drug treatments. Burroughs Wellcome searched through catalogues of chemical substances, coming across azidothymidine (AZT), a substance that had first been synthesized in the mid-1960s and briefly trialled then as an anti-cancer agent. Burroughs Wellcome reported in 1986 that AZT slowed down the onset of AIDS. A clinical trial indeed showed great promise: one death among the group receiving AZT versus nineteen in the control group, which had received a placebo. The trial was so promising that it was stopped so that all patients could benefit. AZT was granted FDA approval in March 1987. However, access to the expensive drug was limited, nor was the drug universally or completely effective by any means.

This crisis developed in the context of social movements showing greater readiness to be critical of authority. In particular, members of the gay community in the United States were articulate, often well educated and relatively well resourced. In these conditions, the gay community was able to challenge the conventional restriction of scientific decision-making to those with credentialed expertise. 'The sheer complexity of AIDS from a scientific standpoint and the profound and differentiated impact of the epidemic have ensured the participation of scientists from a range of disciplines, all of them bringing their particular, often competing claims to credibility', writes Epstein of the immunologists, virologists, molecular biologists, epidemiologists, chemists and physicians involved. Yet 'the striking fact about AIDS is that politicization of the epidemic has brought about a further multiplication of the successful pathways to the establishment of credibility.'[54] In particular, 'the AIDS movement is . . . the first social movement . . . to accomplish the mass conversion of disease "victims" into activist-experts.'[55] AIDS groups, such as ACT UP, not only campaigned, protested and raised money for research, but also learned the science, intervened in debates over methodology,

489

and ended up contributing to the science as 'lay experts' and changing clinical practice. For example, their involvement tipped the debate over the design of clinical trials towards a more 'pragmatic' approach and away from a 'fastidious' one.[56]

Nor was this phenomenon confined to the communities engaged in AIDS research. Epstein identifies similar participatory science, science that was the outcome of contributions from both the traditionally certified experts and the lay expertise of social movements, in other medical areas, from Lyme disease and breast cancer to chronic fatigue.[57] The potential of participatory science was not confined to the working world of human health and its biomedical sciences. In 1986, a nuclear reactor at Chernobyl, then in the Soviet Union, suffered the worst nuclear accident (so far) in world history. A radioactive plume blew first north over Finland and Scandinavia and then west over the United Kingdom. Where it rained, nuclear contaminants fell from the sky and into the soil. Cumbria, the rainy upland county of the Lake District in England, was particularly badly hit by fallout of caesium-137. Brian Wynne, who led a sociological project interviewing local sheep farmers as well as scientists, argued that the Cumbrian sheep farmers possessed lay expertise that would have been directly relevant to guiding the formation of the best policy for handling the crisis.[58] They knew local variations in soil, as well as the cycles of farming life, which the accredited experts, the scientists, did not. They were also neighbours of the Sellafield nuclear-processing factory and the site of the Windscale fire of 1957, Britain's worst home-grown nuclear incident.

Cases such as Epstein's AIDS activists and Wynne's Cumbrian sheep farmers helped make arguments that proved one factor in facilitating a change in how the relationship between science and the public was pictured. Previously, scientists were considered to be the possessors of all the relevant knowledge, while members of the public knew at most watered-down versions of the scientists' knowledge. Anxieties expressed by the public were therefore, it followed, best managed either by authoritative reassurance or by filling the deficit in knowledge through popularization. In the 1980s, programmes of public understanding of science that made this 'deficit model' assumption were under way in countries such as the United Kingdom. However, a series of crises, often in food, began to change consumer attitudes: growth hormones in beef (1970s), salmonella in eggs (1980s), Bovine Spongiform Encephalopathy (BSE) or 'mad cow disease' and its relationship to the human disease variant Creutzfeldt–Jakob disease (vCJD, late 1980s and 1990s), and genetically modified foods (1990s). Scientists such as Stanley Prusiner argued that BSE in cows, scrapie in sheep and vCJD and kuru (in Papua New Guinea) in humans were examples of diseases caused by 'prions', a new scientific disease-causing entity identified in the twentieth century.[59] The furore over genetically modified (GM) foods, on the other hand, was sparked by the ambitions of companies such as Monsanto to widen their markets, and in particular to sell products that followed the first GM food, the 'Flav Savr' tomato (engineered for a long shelf life, sold in the United States in 1994), such as foodstuffs containing the 'Roundup Ready' soybean (engineered for resistance to the herbicide Roundup) in Europe.

By the 1990s, then, there existed a political crisis around regulation of new technologies and a popular and vocal scepticism of both the politicians seeking a solution and the accredited experts advising them. This second factor made the arguments for the existence and relevance of lay expertise more compelling. A fashion for public engagement and public dialogue flourished. 'We do not currently possess institutions which allow us to monitor technological change', wrote Anthony Giddens, sociologist and advisor to the Blair government in Britain in 1998, adding, 'We might have prevented the BSE debacle if a public dialogue had been established about technological change and its problematic consequences.'[60] Public engagement over science, even 'upstream' where the science had yet to leave the laboratories, was advocated widely in the 2000s, not least by think tanks.[61]

For scholars such as Wynne and Epstein, or think tanks such as Demos, lay expertise needed to be recognized in the name of bottom-up democratic reform, a morally justified widening in participation in decision-making through dialogue. For politicians and leaders in the science-based industries and the scientific establishment, dialogue was a way of heading off and managing public disquiet that threatened to interrupt the flow of innovation. This latter stance can be seen in the policies adopted towards nanoscience and nanotechnology, newly made fields breathlessly hyped in the 1990s and 2000s.

Nanoscience

In 1959, Richard Feynman had given a public lecture at the Californian branch of the American Physical Society entitled 'There's plenty of room at the bottom'. The physicist had speculated about the implications of being able to write the contents of the *Encyclopaedia Britannica* on a pinhead. If information could be so compressed, then machines working with such information were imaginable, even perhaps a microscopic computer. Feynman's lively lecture on micro-machines was nearly entirely forgotten. His contribution then became part of a 'creation story' memorialized in the late twentieth century as a way of attaching a famous ancestor to a contemporary interdisci-plinary project that was being driven together by a variety of forces.[62]

There is no essential reason why 'nanoscience' or 'nanotechnology' should exist. They were called into being by an overlap of sciences of quite different working worlds – synthetic chemistry, biotechnology, colloid science – with electronic component manufacture and enter-tainment. Popular science writing has to satisfy an audience, rather than a scientific community, to thrive. Eric Drexler, an MIT student in the 1970s, published speculative ideas about micro-machines. In particular, in his 1986 book *The Engines of Creation* he wrote of 'assemblers', self-replicating machines, and of their potential for good (building other structures from the micro-scale upwards) and evil (running amok, creating a world of 'grey goo'). The biotech debates of the 1970s had demonstrated the bootstrapping effect of Manichaean innovation. Since it was a commonplace that new tech-nologies created new risks, the publicization of the potential for risk (which, after all, was more likely to attract press interest) powerfully drew attention, illogically, to the potential of the technology. Drexler-style machines dominated the imagery of nanotechnology but were never mainstream science. Most of the time Drexler has been ignored. Occasionally, scientists, such as Richard Smalley in a 2001 article in *Scientific American*, have taken time out to do some boundary work and explain why the assemblers would not work.

Smalley had won a Nobel Prize for his contribution to the discov-eries of new forms of carbon in the 1980s. Carbon atoms can bond to each other to form two familiar structures: repeated tetrahedrons (diamond) and sheets of repeated hexagons (graphite). Harold Kroto, a chemist at the University of Sussex, was exploring the carbon chemistry of old stars with a view to accounting for the 'dark matter' deficit. He had contacted Smalley's group because of their develop-ment of laser spectroscopy techniques. In the course of simulating the carbon chemistry of stars, their collaboration uncovered a new form of carbon made of sixty atoms arranged in a ball shape reminiscent of the futuristic domes built by the American architect Buckminster Fuller in the 1960s. They called C_{60} 'buckminsterfullerene', which the popular science press shortened to the nickname 'bucky-balls'.

Large novel molecules of carbon, the most prodigious of chemi-cal elements, surely promised usefulness in a wealth of applications.

492

However, 'nanotechnology', as it coalesced in early 1990s, found even greater potential in other forms of carbon, in particular, rolls of graphite-type sheets known as 'nanotubes'. Evidence of filamentous carbon structures had gathered over the twentieth century and began to be imaged by Soviet scientists using electron microscopes in the 1950s. The observation had been repeated several times before the 1990s, generating little interest. This response was due to the fact that such filaments were regarded as irritants within the working worlds that first made sense of them: 'investigated by material scientists, whose main goal was to understand the growth mechanisms so that they could prevent their formation in coal and steel industry processing and in the coolant channels of nuclear reactors'.[63]

In 1991, Sumio Iijima, a solid-state physicist working for the NEC Corporation in Japan, published an article in *Nature*, reviewing the evidence for the filaments. Iijima's paper has been cited as the 'discovery' of 'carbon nanotubes'; it wasn't, but the sudden need for a named 'discoverer' was evidence for the gathering importance of the nano-realm. The 'undoubted tremendous impact of the 1991 Iijima paper came from the right combination of a number of factors', write the guest editors of the journal *Carbon*, 'a high quality paper, a top-rank journal read by all kinds of scientists, including those involved in basic research and fundamental physics, a boost received from its relation to the earlier worldwide research hit (fullerenes), and a fully mature scientific audience ready to surf on the "nano" wave.'[64]

The third major force which pulled together nanoscience and nanotechnology was the development of new techniques within the working world of electronics manufacture. Some techniques were imaging devices. In particular, IBM's Zurich laboratory made the Scanning Tunneling Microscope, an exquisitely sensitive device capable of both detecting and moving individual atoms, completed by 1981. The well-publicized 1989 image of 'IBM' written out in neon atoms on a nickel surface was not only a technical achievement but also advertised the capabilities of manipulation at the atomic level by literally stamping IBM's name on the technology. Other techniques came from one of the central drivers of technological change in the last third of the twentieth century: the inscription of components at ever finer detail on chips of silicon. The first microprocessor chip, Intel's 4004 of 1971, had had a mere 2,300 transistors, while the company's 2003 Intel Xeon had 108 million. Chip manufacturers, such as Intel, had accumulated an extraordinary repertoire of techniques for manipulating matter at the nano-level. 'There is no doubt', noted the authors of a landmark Royal Society and Royal Academy of Engineering report of 2004, that the information and communications technology sector of the economy 'has effectively driven a large proportion of nanoscience'.[65]

The nano-level encompasses objects ranging from large molecules to small components of cells. Biotechnology, with its suite of tools for manipulating DNA, was also considered to be nanotechnology, which in turn was cast as a convergence of 'bottom-up' synthesis and 'top-down' manufacture. Indeed, the final factor that led to the condensation of the fields of nanoscience and nanotechnology was the import of many expectations, the entrepreneurialism and the bootstrapping hype from biotechnology to 'nano', now seen as a prime investment opportunity both for private speculators and for national governments intent on 'catching the next wave'.

It was in the late 1990s, notes historian of nanotechnology Patrick McCray, that 'utopian imaginings and policy rationales met fiscal realities'.[66] One of the key agencies driving condensation was the National Science Foundation, where Mihail C. Roco, a mechanical engineer who had arrived in the United States from Romania in 1980, promoted an ambitious programme to support nanotechnology. As McCray points out, Roco's project was launched at a 'promising time': the end of the Cold War had prompted a shift in justification for supporting basic science away from countering the Soviet threat towards a greater emphasis on increasing international competitiveness.[67] In 1991 Japan's Ministry of International Trade and Industry had pledged $225 million to nanotechnology, located in the working world of electronics manufacture. Both Congress and wider public stakeholders were lobbied, through, for example, a glossy brochure, *Nanotechnology: Shaping the World Atom by Atom*, which 'drew upon the nano-utopian rhetoric that was rapidly becoming de rigueur – the vision of Feynman's 1959 talk, the progress made in manipulating atoms via electron microscopy, the potential to fabricate nanoscale electronic devices, the now famous IBM logo writ in xenon atoms, and the vast commercial and societal benefits that investment . . . could open up'.[68] The National Nanotechnology Initiative of 2000 committed the United States government to $465 million of support.

Following the model of the human genome project, a percentage of nanotechnology funds was earmarked for investigation of social and ethical issues. The National Science Foundation's *Societal Implications of Nanoscience and Nanotechnology*, edited by Roco and William Sims Bainbridge, published in March 2001, contained many predictions in which an existing problematic trend was extended into the future and nano posited as a solution. So, for example, the problem of a growing 'proportion of the population at retirement age' could be aided by using nano to help provide 'assistive means to maintain physical independence and tools to support cognitive capability'.[69] More critically, the NSF report tended to raise ways that nano could manage existing problems, deflecting attention from tackling their root causes. So, for example,

> Medical sensors that can, for example, 'detect an array of medically relevant signals at high sensitivity and selectivity' [quoting from a 2000 policy report] . . . promise to aid diagnosis and treatment of disease, but also to develop predictive health profiles of individuals. Today, health and life insurance companies often use pre-existing conditions as a basis for denying or restricting coverage. The advent of nanodetection capabilities will considerably expand the information that insurance companies will want to use in making decisions about coverage. The generation of new information might thus destabilize the risk-spreading approach that allows equitable delivery of social benefits to broad populations. How will society respond?[70]

Society, here, was not being asked to consider the root causes of inequalities in access to medical care. Likewise, the report correctly noted that the military would continue to be an eager customer of novel science and technology, but the root causes of conflict were not up for discussion:

> Nanotechnology offers a dizzying range of potential benefits for military application. Recent history suggests that some of the earliest applications of nanotechnology will come in the military realm, where specific needs are well articulated, and a customer – the Department of Defense – already exists. One area of desired nano-innovation lies in the 'increased use of enhanced automation and robotics to offset reductions in military manpower, reduce risks to troops, and improve vehicle performance' [quoting the same 2000 report as above] . . . How might progress in this realm interact with the current trend toward rising civilian casualties (in absolute terms and relative to military personnel) in armed conflict worldwide? As increased robotic capability is realized in warfare, will we enter an era when it is safer to be a soldier in wartime than a civilian?[71]

Even if society was to benefit from nanoscience and nanotechnology, several severe obstacles were apparent in the first decade of the twenty-first century. First, it was extraordinarily difficult to make comparable measurements in different laboratories. Similar 'nanometers' have not been exported, at least in a robust way, around the world. The Physikalisch-Technische Bundesanstalt (the post-war

incarnation of the Reichsanstalt) found that 'even apparently sophisticated users of atomic force microscopes can produce large variations in their measurements of the same artefacts', from which Royal
Society and the Royal Society of Engineering drew the conclusion that

> Without agreed standards, tools or machines cannot be calibrated at
> the nanometre scale. It is therefore not yet possible for laboratories and
> manufacturing plants to exchange or share data or physical compo-
> nents. Also health and safety standards cannot be set for legal require-
> ments.[72]

Even if and when the metrological issues were resolved, the future
public response to nano-products was unknown. In Britain, the
authorities adopted the strategy of 'upstream engagement', discussed
above, on the rationale that early discussion of nano issues would
forestall later surprises.

Nevertheless, as the philosopher of the politics of technology
Langdon Winner told the Committee on Science of the United States
House of Representatives in 2003: 'In writings on nanotechnology,
there seems little willingness to ask: What are society's basic needs at
present? . . . What we find instead is a kind of opportunistic means-
to-end logic.' While correct, Winner's complaint also misses the point
somewhat: it is not the case that needs are not being responded to.
Rather, as has been the case for science in the twentieth century and
beyond, more generally, the needs are those of particular working
worlds, which are not the same as those of society in total.

Part V

Conclusions

— 20 —

SCIENCE IN THE TWENTIETH CENTURY AND BEYOND

In *Science in the Twentieth Century and Beyond* I have told a largely chronological story. We have seen how the first decades of the twentieth century saw new developments in the physical, life and human sciences against a background of continuities from nineteenth-century science. Conflicts which tore and transformed twentieth-century lives also reshaped twentieth-century science. The final third of the twentieth century witnessed further dramatic events and trends. I have argued that 'working worlds', arenas of human projects that generate problems, are the settings in which much of the history of twentieth-century science makes sense. Much good science was done by measuring, comparing and contesting simplified and abstracted representative models of working world situations.

In conclusion I step back and provide two surveys. The second, looking forward, is an impression of twenty-first century science. The first, however, looks backward and is a deeper analysis of the four major cross-cutting themes that I notice when I read again the history I have told: the extraordinary and unambiguous importance of the working world of warfare in shaping the sciences, the emergence of the United States as the leading scientific power, the missing stories, and the swing from physical to life sciences in the second half of the twentieth century

On war

Science and inventiveness had been mobilized for war long before the twentieth century. But during the twentieth century the relationship was deepened. Several factors encouraged this trend. The duration of wars – the confounded expectations of a quick conflict in the First World War, the scale of the Second World War, and the frozen conditions of the Cold War encouraged permanent mobilization. So did the increased emphasis on speed, which in turn was both a product and a cause of scientific and technological change. Speed is an exemplary problem thrown up by the working world of warfare. Scientists responded to mobilization for a variety of motives: a sense of patriotic duty, an enjoyment in the successful provision of technically sweet solutions, or a desire to protect home institutions, among many others. Wars are also social ratchets: making the unthinkable thinkable, the socially or politically possible out of the impossible. Whole scientific disciplines, such as psychology in the United States in the First World War, could be lifted and permanently transformed by war service.

Furthermore, not only was the working world of warfare of importance for the changing shape of the sciences in the twentieth century, but its changing relationships to other working worlds were significant too. 'As Western nations were distancing themselves from the collective catastrophes of epidemics', notes Jeanne Guillemin of the policies of public health and the science of bacteriology born from the working world of human disease, 'some of their governments invented biological weapons as a means of achieving advantage in warfare.'[1] Control of disease at home was a condition, even a prompt, for the development of weapons designed for use abroad.

On America

Judgements of national leadership in the sciences, including judgements of national advance or decline, are so fraught with problems that most historians of science now avoid them. While 'science is not so like art and literature that judgments of relative national achievements often are simply impossible', notes historian Mary Jo Nye, it is 'enough like art and literature that very fine distinctions of national fluctuations in scientific achievement may be burdened by cultural predilections and distracting evaluations of national rank in the international hierarchy of economic and military power'.[2] She had in mind, in particular, assessments of French decline in the sciences in the nineteenth century. She showed that both qualitative and quantitative assessments, whether they were by historical actors or by later analysts (including the pioneer sociologist of science Joseph Ben-David and sophisticated historians of science such as Paul Forman,

John Heilbron, Spencer Weart and Terry Shinn), have not produced reliable indicators. Analysts in the twentieth century have counted scientists and quantified both their productivity (such as the numbers of research papers produced per head) and their success in prizes. Each indicator becomes questionable as its assumptions have been probed.

Furthermore, this sceptical position holds despite the working worlds of warfare and administration directing considerable investment in the production of indicators of science in the twentieth century. Examples are the launch of Naukovedenie, immensely ambitious quantitative studies of science in the Soviet Union in the 1920s, and the resurgence since the 1960s of scientometrics, including the international collation of comparative statistics by the Organization for Economic Cooperation and Development (OECD), other Western projects by individual authors such as Derek de Solla Price, and the Soviet revisiting of Naukovedenie.[3] Indeed, the post-war science of scientometrics was itself a response to a working world need: permanently mobilized science, lavishly supported because of military expectations, required administration and the solving of managerial problems – not the least of which in the 1960s were the pathological consequences of 'Big Science'.[4] Their chief discovery was growth, whether of the number of scientists, of their products, such as papers and patents, or of the scale of their funding.[5] Comparative international indicators have continued to be the essential tools of twenty-first-century science policy interventions, as illustrated, for example, by even a cursory glance through the 2007 report *Rising Above the Gathering Storm*, which was written by scientists in the United States to warn politicians of the coming competition with China and India.[6]

Yet national predominance, while problematic, can be acknowledged, while the need for explanation and critical probing of accounts is also recognized. American leadership in the sciences, especially in the second half of the twentieth century, has occasionally been explicitly addressed and more often implicitly conceded by historical actors and analysts. 'When we first met in 1929, American physics was not really very much, certainly not consonant with the great size and wealth of the country. We were very much concerned with raising the level of American physics. We were sick and tired of going to Europe as learners. We wanted to be independent', recalled the physicist Isidor I. Rabi, testifying at the Oppenheimer trial hearings in the 1950s, adding, significantly: 'I must say I think that our generation . . . did that job, and that ten years later we were at the top of the heap.'[7] Other disciplines followed physics, not least biology.[8]

In the secondary literature there is a paradox. On the one hand, the problematic of finding a sound explanation of American predominance (if the phenomenon exists) is rarely, if ever, explicitly stated. On the other hand, there is a wealth of candidate explanations that can be extracted. Some of these accounts are variants of arguments for American exceptionalism. Some are not, or at least are not in a strong way: for example, to argue that science is an expensive and resource-intensive activity and to note that the United States was in the twentieth century both wealthy and populous, and therefore a place where science could flourish, is a weak form of the argument. Many of the accounts are intended by their authors as explanations of specific features of American science that become candidate explanations for the predominance of the United States only if these features, singly or, more plausibly, in combination, can be shown to have a more general traction. Nevertheless, seven different arguments can be identified, premised on the particular character of: (1) the corporate and inventive response to markets in place by the late nineteenth century, (2) a version of the frontier thesis, (3) the relative institutional heterogeneity of the United States compared to European rivals, also in place by the late nineteenth century, (4) pragmatism, (5) the peculiar relationship between science and political structures, a changing constitutional dynamic since the eighteenth century, (6) wealth, and (7) military might. Some of these arguments support each other, some pull in different directions.

Historian of technology Thomas Hughes provides examples of type (1) arguments, for example in *American Genesis* (1989). One of the consequences of the American project of political and economic democracy was the development of 'a new and unprecedentedly large market for mass-produced goods and services for masses of the population'.[9] While Europeans expected high unit profits on a small turnover from the sale of luxury goods, entrepreneurs and inventors in the United States aimed for large turnovers based on the sale of cheaper goods. More specifically, American systems builders, including Edison and Ford and those who extended systems strategies, such as Samuel Insull and Frederick Winslow Taylor, focused inventiveness on the achievement of continuous flow of production.[10] Corporations, argues historian Alfred Chandler, formed to protect profitability from the demands of a large-scale market. They sought to control unpredictability – including unpredictabilities of market swings and independent invention – by bringing inventiveness in-house, a vertical integration into knowledge-production that created the corporate laboratory.[11] The result was a world in motion,

502

in which corporate inventiveness, including but not restricted to the work of corporate laboratories, had critical roles. As the industries of petroleum refining and use, long-distance telecommunications, automobility, electrical light and power, and many others provided the infrastructures of the twentieth century, so these working worlds called forth twentieth-century sciences.

The scale of the continent of North America also underpinned historian Frederick Jackson Turner's frontier thesis, a notorious explanation of American character and development articulated in the 1890s. The challenge of vast expanses of uncultivated land encouraged a ruggedness and self-reliance among pioneers. But, argued Turner, the 1890 census revealed that the frontier was now closed – with implications for the sustenance of distinctive American values. The Turner thesis has been accepted, revised, rejected and revisited in the century since its proposal. Frontier historiography in the history of science has been implicit rather than explicit. (Although it is interesting that American corporations tunnelled inwards, vertically integrating inventive knowledge, a source of raw materials, broadly after horizontal market integration slowed with the closing of the frontier.) Nevertheless, in Vannevar Bush's *Science: The Endless Frontier* (1945), which reported to the president of the United States on the need to learn from wartime experience and to harness science for the post-war world, there is a strong echo, even, for its educated readership, an invocation, of Turner's argument. American character and values would be kept vigorous by the challenge of working at the frontier of science.

Historian Terry Shinn argues that American exceptional achievement in the sciences stems from the extraordinarily diverse range of relationships tying together, sometimes loosely, education, industry and research.[12] From the nineteenth century, France and Germany offered contrasting models of these relationships: the model museum, model hospital and model research university. Germany provided a 'paradigm of heterogeneity', with diverse links between types of technical education (including but not restricted to the celebrated Technische Hochschulen), state-level support of university and industry, and national institutions such as the pioneering standards and testing laboratory, the Physikalisch-Technische Reichsanstalt.[13] France on the other hand was a 'paradigm of homogeneity', with structural rigidity and hierarchy marking educational institutions and firms.[14] The United States followed Germany, by accident as well as by design, creating an institutional 'polymorphism', in which educational establishments enjoyed both beneficial linkages and, when

503

required, autonomy from the demands of enterprise.[15] (MIT and Caltech have been the champions of this balancing act.)

In the United States there was an entrepreneurial dedication in the contribution research could make to solving working world problems, but also the diversity of funding (for example through philanthropic foundations) and a robustness of professional bodies to protect flexibility in research policy.[16] Following the conclusions of economic historians Nathan Rosenberg and Richard R. Nelson, Shinn also notes a temporal stratification ('polychronism', perhaps?) of mutual benefit: in the United States can be found a division of labour not between applied versus anti-applied learning but between short-term and long-term problem-solving, with academia gravitating to longer-term research and entrepreneurial firms specializing in shorter-term research, while both agreed on the importance of 'utility' as a guiding value.[17]

Appeals to pragmatism are the philosophical variant of this institutional argument for radical heterogeneity as the root of American distinctiveness. The extent to which informal appeals to pragmatism found in historical writing on science in the United States are indebted to the formal philosophical movement led by Charles Sanders Peirce, William James and John Dewey is an interesting question that can only be answered by further research.[18] What I note here are the frequent gestures towards a preference, for example, for rough working approximations over precise calculation, an engineering feel for scientific tractability over derivation from first principles, or a predilection, such as that noted in a contrast of American and European physicists when confronted by quantum theory in the first decades of the twentieth century, for 'practical applications' rather than 'endless discussions and heated debates about the philosophical implications'.[19]

Don K. Price's *Government and Science*, published in 1954, the same year that Price, a Washington insider, became vice president of the Ford Foundation, provides a strong example of an argument that identifies the peculiar relationship between science and political structures in the United States as the principal factor behind American distinctiveness. It is worth examining Price's argument in some detail. Following the revolution, American political institutions were complex, but that complexity created dynamism. 'It is dynamic', writes Price, in words in which the echo of the Trinity test reverberate, 'because science, which has been the most explosive force in modern society, has profoundly influenced the development of the American government.'[20] Specifically, Price emphasizes two stages. In the first, the weakness of divided authority pushed science to be the problem-solving agency of the working world of the American government.[21] Second, federal support for science, from first beginnings in the nineteenth century, to the turn of the century vogue for technocratic conservationism, to its dramatic confirmation in the mid-twentieth century, has not shaken a dynamic that rests, counter-intuitively, on a democratic distrust of authority. (This counter-intuitive link between weakness of political authority and support for quantitative objectivity has been supported by other historical studies.)[22] According to Price:

> In the American political system the pragmatic and experimental method prevailed. This was the method in which each issue was dealt with on an experimental basis, with the views of the interested technical or professional groups having more weight than party platforms or political theories. It was a method that gave far more weight to research and to scientists, and created a more dynamic economic and political system than could have prevailed under a more orderly and authoritative approach . . . Thus the alliance between science and the republican revolution first destroyed, and then rebuilt on a different pattern, the forms of organization and the systems of personnel that determine the practical working authority in the modern state.[23]

While these forces could be found 'at work in the whole Western world', they were 'most influential in America'. As American political power has grown, so has this dynamism become the leading force in world science. As is clear, this type (5) argument has clear connections with the institutional diversity argument of type (3) and the pragmatism argument of type (4).

Price was writing when the military (federal) sponsorship of science raised for many the issue of the autonomy of science. His argument is a defence, or rather a plea for a correct understanding of the nature, of the status quo. 'As science becomes an active ally of military power', writes the man who helped write legislation for the Atomic Energy Commission and the National Science Foundation and was Deputy Chairman of Research and Development Board of the Department of Defense, 'we shall do well to understand the principles that have guided its relation to government in America.'[24]

There are, of course, simpler arguments relating wealth, military might and the direction of the sciences in the twentieth century. The increased funding for science during the Second World War, and, crucially, sustained in the Cold War, was of such a scale as to appeal as the most important factor shaping the sciences. Many historians and other commentators have taken this view, although perhaps historian of science Paul Forman provides an exemplar. Forman argues

505

that not only did Cold War patrons fund a broad range of physical science, so-called pure and fundamental as well as applied, but that research directions were skewed, and that the very content of the scientific knowledge produced carries the stamp of Cold War interests.[25]

Forman was writing at the height of the second phase, Reagan-era Cold War. Post-Cold War analyses have continued the broad argument, while beginning to downplay the specific military interest. To take just one example, the 2007 report *Rising Above the Gathering Storm*, commissioned by congressmen and written by scientists with deep experience of Washington, not only related the success of science in the United States to its support by prosperous patrons, but, crucially and conventionally, completes the circle:

> Since World War II, the United States has led the world in science and technology, and our significant investment in research and education has translated into benefits from security to healthcare and from economic competitiveness to the creation of jobs . . . Central to prosperity over the last 50 years has been our massive investment in science and technology.[26]

According to the authors of this report, it was the circular relationship between wealth and investment in science and technology that made the 'American Century'. When the century started is a good question. But it certainly has not ended, despite anxieties discussed below. Post-Second World War success in Nobel prizes provides a crude, albeit problematic, demonstration of American achievement.[27] According to the OECD's *Main Science and Technology Indicators*, the dominance in investment was also clearly evident in the gross domestic expenditures on research and development. In 2008, the figures were $368 billion in the United States, compared with $147 billion in Japan, $102 billion in China, $71 billion in Germany, $43 billion in France, $41 billion in Korea and $38 billion in the United Kingdom. The report's authors could still warn: 'As we enter the 21st century, however, our leadership is being challenged.'

My general point is to note that we have available a range of possible answers for the question of how we might account for the rise of the United States as a dominant scientific power in the twentieth century. Arguments of types (1) to (7) are clearly interdependent and need further analysis. I have provided only a few indicative authors above, when there are in fact a host of commentators with nuanced and specific arguments. Additional types of account should be sought and also subjected to scrutiny. So far, many of these accounts have been implicit or have been offered as explanations of other historical or contemporary phenomena. But the question needs to be posed explicitly. Only when history of science paints on a broad canvas – landscapes rather than miniatures if you will – will the posing of such questions become unavoidable. I offer the working worlds approach as a way into the exploration of possible answers.

On missing stories

The case studies style of historiography of science leaves many gaps. Even when topics are addressed, casuist historiography can encourage their segregation. *Science in the Twentieth Century and Beyond*, so dependent on secondary literature, is guilty too. I am aware of many connections that I have endeavoured but failed to build. First, there is an emphasis on the ideas, theories, practices and instruments which work, which survive and which flourish. This preference has persisted even though historians of science have rightly taken to heart the methodological counsel to be 'impartial with respect to truth and falsity, rationality or irrationality, success or failure' and to be symmetrical in the sense of giving similar kinds of explanation for claims later judged to be on either side of these dichotomies.[28] It would be provocative to compile a list of ideas that died in the twentieth century and ask what sort of alternative synthetic history of science could be made from their study. Historian of biology Frederick B. Churchill offered the disputes between epigenesis and preformation and between vitalism and mechanism as two 'dead issues' by 1910.[29] Yet both, it could be argued, lurched back in later years: think, perhaps, of cybernetics. Is history of science a history of the undead?

Nationalities and national research systems can be missing too. Japanese scientists appear frequently as independent co-discoverers or as part of research programmes that parallel Western ones. Examples include Shoichi Sakata's proposal of the two meson theory that anticipated the work of Robert Marshak in fundamental physics, Kiyoo Wadati's account of sloping weak surfaces in seismology, or, in the emerging science of endocrinology, the ways that German and Chinese scientists isolated ephedrine in the 1920s reproducing Japanese research of the 1880s.[30] Despite the existence of some good history of Japanese science, the remarkable results of investment in the final third of the twentieth century are not yet well understood.[31]

A third sense of 'missing stories' comes from secrecy. The working worlds that call forth sciences to solve their problems can redact the knowledge, so that what's missing can only be guessed from the shape that has been left cut out. The post-war secret world of intelligence, security services and eavesdropping undoubtedly spurred the construction of immense technological systems, which in turn acted as a working world for new research. If all the documents were released it is certain, I feel, that the histories of computer science, linguistics, electronics, artificial intelligence, medical scanning instrumentation, information theory, nuclear science and many others would read significantly differently. As just one indicator, historian of physics Peter Galison estimates that the 'classified universe . . . is certainly not smaller and very probably is much larger than . . . [the] unclassified one'. He warns:

> The closed world is not a small strongbox in the corner of our collective house of codified and stored knowledge. It is we in the open world – we who study the world lodged in our libraries, from aardvarks to zymurgy, we who are living in a modest information booth facing outwards, our unseeing backs to a vast and classified empire we barely know.[32]

On the swing from the physical to the life sciences

My fourth and final headline finding concerns the apparent swing from the physical to the life sciences. The last decades of the twentieth century created the impression of a resurgent wave of work in the life sciences, with molecular biology and biotechnology at its crest. At the same time, the end of the Cold War brought into question whether a contract between the physical sciences and its military patrons would be continued in the same or a different form. In a separate development, many academic chemistry departments struggled to recruit, despite the continuing leadership of chemistry as a vocation for many scientists. Some departments closed, many others repositioned research according to more interdisciplinary agendas, including materials science, biochemistry and nanotechnology. Funding patterns and the step up in scale of some life science projects, not least genomics, also seem to provide evidence for a swing. But the nature and explanation of the swing are still in doubt and require further historical work. An alternative reading of the last third of the twentieth century, and beyond, would be to notice a resurgence in human sciences which has reshaped traditional biological and physical disciplinary boundaries and identities. Fortunately, the history of human sciences has attracted scholarly attention since the 1990s, and the tools, examples and arguments for rethinking late twentieth-century science are in place.

Beyond: science in the twenty-first century

It is difficult to judge the historical significance of recent events. This is as true for history of science as it is for political history. For example, in 2004, archaeologists working a cave site on the Indonesian island of Flores discovered remains that almost certainly point to the existence of a new human-like species, now named *Homo floresiensis*. It may be the most dramatic discovery in human evolution since the nineteenth century. But its status is still being contested and revised, and its place in history uncertain.

However, the usefulness of history to provide a fresh perspective on contemporary science is an entirely different question. One of the most compelling reasons for understanding the history of modern science is that we can know why the scientific world around us is as it is, as a survey of this early twenty-first-century world will illustrate.

The global picture

Nature in January 2008 published a world map of 'How the World Invests in R&D'.[33] It drew on data from the United States National Science Board and gives a big picture of the global funding of science in the first decade of the twenty-first century. It's a familiar North–South picture. The big spenders, as a proportion of GDP, are in North America, Europe, the Far East, plus Australia, and at the top of the league is Israel (4.71 per cent). Five countries alone were responsible for 59 per cent of all spending on science in the world.[34] The United States has dropped slightly in proportion to other countries: compared to 2006, 'both South Korea and Switzerland have leaped ahead' in terms of proportion of GDP.

But size matters: the United States was still comfortably the biggest spender ($340 billion invested), while China now came third in overall funding ($115 billion), not far behind Japan. Within this picture, international scientific collaboration is increasing and is proportionately more likely to generate good science.[35] 'Global science is increasing', noted a Royal Society report in 2011, adding, 'but it is also nothing new'.[36] Indeed, to take one example: an International Polar Year took place in 2007–8. The large-scale organization of science is old enough to have its established cycles. An International Polar Year had taken place in 1932–3, fifty years after the first. The International Geophysical Year of 1957–8, discussed in chapter 14, had been opportunistically shoe-horned in twenty-five years later, claiming space and the Antarctic for science during the Cold War freeze. The third IPY consisted of 170 projects, involving more than sixty nations, at a cost of $1.2 billion. In the North, scientists measured melting permafrost. In the South, China established a new Antarctic base. Global cooperation, however, ran parallel to national worries.

American anxieties

In mid-2005, a wave of anxiety spread among the shapers of science policy in the United States. It was a fear that, in a globalized world, other countries would soon overtake the United States as leaders in science and consequently, so the argument went, as leaders in innovation and, ultimately, in economic might. The United States moved from being a net exporter of high-technology goods ($45 billion in 1990) to being a net importer ($50 billion in 2001). Furthermore, because of the visa restrictions introduced after 9/11, the number of scientists migrating to the United States was dropping. At school and university, students were dropping science and engineering. The United States was becoming a less attractive place to be a scientist.

Politicians, further alarmed by a poll that recorded that 60 per cent of scientists felt that science in the United States was in decline, asked the National Academies (the National Academy of Sciences, the National Academy of Engineering and the Institute of Medicine) to suggest, urgently, 'ten actions ... that federal policy-makers could take to enhance the science and technology enterprise so that the United States can successfully compete, prosper, and be secure in the global community of the 21st century'.[37] They got back more than they asked for. The report, brainstormed by the elite of American science, *Rising Above the Gathering Storm: Energizing and Employing America for a Brighter Economic Future* (2007), mentioned above, ran to over 500 pages and urged twenty actions. These included training more teachers of science, increasing federal investment in basic research by 10 per cent a year for seven years, offering attractive grants to outstanding early-career scientists, setting up ARPA-E, funding lots of graduate fellowships, tweaking the tax incentives for research, and making sure everyone has access to broadband internet. Some of these actions were indeed implemented.

What is more significant for this 'history of science in the twentieth century and beyond' is the set of assumptions behind the panic. First, there was an argument that relates economic power to basic research. The authors were in no doubt that 'central to [American] prosperity over the last 50 years has been our massive investment in science and technology'.[38] Second, the competition that most worried the politicians was the gathering forces of China and, to a lesser extent, the European Union countries, South Korea and India. The huge populations of China and India mean that only a small percentage need be trained scientists or engineers to more than match the aggregate number in the United States.

Nevertheless, the evidence supporting a claim that the United States is losing its leadership in the sciences is slim. What is certainly happening is a globalization of research and development. Global companies are now very likely to tap the skills and cheap salary costs of China, India and other fast-developing nations. More research and development is conducted off-shore, and the trend will continue. But this does not mean a lessening of American predominance. Indeed, the United States (and the United Kingdom) is a beneficiary of this trend: foreign-funded research and development has rocketed, and in the 2000s more corporate research and development investment flowed into the United States than was sent out. Nevertheless, India and China are fascinating areas of growth.

India

Independent India has long had a technocratic streak. The relationship between Nehru and elite physicists (such as theoretical astrophysicist Meghnad Saha, but also including select Westerners such as Patrick Blackett) was strong. Support was given to research at sites such as the Saha Institute of Nuclear Physics (inaugurated in 1950) and the Tata Institute of Fundamental Research (1946). One result was a nuclear programme that culminated in the first Indian nuclear test in 1974. In 1958, Nehru had asked the father of the Indian bomb, physicist and founder of the Tata Institute Homi Jehangir Bhabha, to draft a new science policy. 'The gap between the advanced and backward countries has widened more and more', stated the motion that passed the Indian Parliament. 'It is only by adopting the most

511

vigorous measures and by putting forward our utmost effort into the development of science that we can bridge the gap.'[39] An outcome was another techno-scientific project whose primary output was, at least initially, national prestige: a space programme. An Indian satellite was launched from Kerala in 1980.

Following economic liberalization in 1991, India has been sending more and more students to train abroad, funded since 2001 by government loans on easy terms. In 2002, 'India surpassed China as the largest exporter of graduate students to the United States.'[40] Attractive policies encouraged the students to return, bringing skills home.

Furthermore, this accumulation of skills has encouraged science-based corporations to set up laboratories. India's science policy leaders were bullishly optimistic. 'More than 100 global companies . . . have established R&D centres in India in the past 5 years, and more are coming', noted Raghunath Mashelkar, director general of India's Council for Scientific and Industrial Research, speaking in 2005. 'As I see it from my perch in India's science and technology leadership, if India plays its cards right, it can become by 2020 the world's number-one knowledge production center.'[41]

And the prestige projects have benefited too. Between 1999 and 2003, under the leadership of Krishnaswamy Kasturirangan, a coalition of elite scientists and politicians drew up proposals for Indian moon missions. 'Their stated goals', notes Subhadra Menon, 'were to expand human knowledge, and to challenge India to go beyond geostationary orbit, thereby potentially attracting young talent to the space sciences and into the country's space programme.'[42] The moon probe Chandrayaan-I, carrying eleven payloads (five Indian, two from NASA, three from the European Space Agency, and one from Bulgaria) observed the moon in 2008 and 2009. In September 2009, just before expiring, it reported the presence of minute, but significant, quantities of water.

Indian bureaucracy is also becoming far more protective of the country's genetic riches. For example, in 2008 a collaboration between the Indian Ashoka Trust for Research in Ecology and the Environment and the American Illinois Natural History Survey stalled because of official complaints about biopiracy.[43] A researcher for the Askoka Trust said: 'We have to send the specimens abroad for identification as we do not have the expertise at home', while the senior official at the National Biodiversity Authority responded: 'exporting 200,000 specimens is not permissible'. Exporting photographs would be fine, but not gene-rich specimens.

China

Who will be a typical scientist in the twenty-first century? A case can be made for its being a technician working in a contract research organization in China. The numbers are certainly impressive. In 1949, China had 50,000 people who could broadly be categorized as working in science and technology in a total population of half a billion; by 1985, during Deng Xiaoping's relaunching of Chinese science (discussed in chapter 14), the figures were 10 million in a billion.[44] In 2006 China devoted 1.6 per cent of its GDP to research and development, and its leaders announced a target of raising this figure to 2.5 per cent.[45] (Reaching this figure would mean that China was spending proportionately more on research and development than any European nation, would be spending roughly to the same amount as the United States, and was approaching the scale of Japan's investment.) Furthermore, with the increase in funds, the output of Chinese science was also increasing: the Royal Society predicted that China would overtake the United States in annual number of scientific publications by 2013.[46] Like India, China has launched high-profile space missions, including a spacewalk in 2008.

Such growth is revealing some interesting issues and tensions. Starting with journals, large numbers of papers are not the same as large numbers of good papers. 'The average impact' of Chinese articles, noted journalist David Cyranoski in *Nature*, 'was below average even in China's strongest fields of materials science, physics and chemistry.'[47] Furthermore, researchers at the best institutions are encouraged to publish in journals listed in the *Science Citation Index*. These are predominantly English-language journals. An 'unintended consequence', notes science policy analyst Lan Xue, has been to 'threaten to obviate the roughly 8,000 national scientific journals published in Chinese'.[48]

Indeed, there is an unresolved tension between Western and native sciences. Lan Xue cites the case of earthquake prediction. In the 1960s and 1970s 'China set up a network of popular earthquake-prediction stations, using simple instruments and local knowledge.' This network was decommissioned in favour of a high-tech system. But when the new system failed to predict the Sichuan earthquake of 2008, it was claimed that the indigenous stations would have. Lan Xue argues that a 'one science fits all' approach should be rejected: 'One should tolerate or even encourage such indigenous research efforts in developing countries even if they do not fit the recognized international scientific paradigm.'

The tension is not just with indigenous traditional but also with the brute political fact that China has a slow-moving one-party system of rule overseeing a fast-moving economy, including an expanding science sector. In 2009, the editors of *Nature*, reflecting on the obstacles in the way of setting up a national stem-cell science society, urged the political authorities to ease the severe restrictions on forming groups.[49] The problem was that the Chinese authorities, in the absence of democratic legitimation, had an 'aversion to congregations', seeing groups, not least Falun Gong, as the seeds of political challenge. Furthermore, the established scientific organizations are concerned more about the maintenance of hierarchy and status than the encouragement of critical debate. 'Most of the current learned societies do not function well', ran the *Nature* editorial. 'Annual meetings are often a matter of pomp, with elite researchers showing up to swagger about and form cliques based on pedigree rather than scientific views . . . Constructive criticism is more likely [to] be taken as grounds for breaking off relations than as insightful advice.'

Political sensitivities also shape international relations. Environmental research, especially in Beijing's Olympic year of 2008, drew the watchful eye of political authorities. Two new laws, Measures for the Administration of Foreign-Related Meteorological Sounding and Information (January 2008) and Measures Governing the Survey and Mapping in China by Foreign Organizations and Individuals (March 2008) were introduced to clamp down on unauthorized release of data.[50] Collaborative projects have had their field stations in Yunnan dismantled; geologists working in the sensitive Tibet and Xinjiang regions objected to the use of GPS. Projects, to be approved, had to be seen to be 'in China's best interest'. But these laws were also about making sure that Chinese science benefited from scientific research conducted within Chinese borders: hence the insistence on a demonstrable 'equitable partnership' before approval.

Nor, of course, is China immune from the forces of globalization. It trades with the world, manufacturing much of its goods. Its people travel. China is therefore, by necessity, part of the international science-based networks of organization that are the infrastructure that make the global world of travel and trade. For example, the network of international disease notification and control has tested the Chinese authorities several times in recent years. The emergence in 2003 of a new viral disease, Severe Acute Respiratory Syndrome (SARS), revealed problems with Chinese healthcare and how it communicated information to the rest of the world. The outbreak eventually stung the authorities into action.[51]

The Chinese government has a stated aim of making China a place of predominantly home-grown innovation. (Only a minuscule proportion of new drugs approved in China, for example, are discovered there.)[52] China aims to be among the top five nations in areas such as producing new patents. 'These are uncomfortable goals for China', argues David Cyranoski, noting the industrial non-governmental source of most research and development funds, 'because, unlike space research, they are more difficult to mandate from the top down.'[53] Indeed it is science deriving from the working world of Chinese and multinational industries that has taken off, to such an extent that complaints such as the following (from a German academic in 2008 with extensive knowledge of working in Chinese science): 'Making money has become the major attraction in China and this has severe consequences at the university level: basic research is not considered as important and attractive as it had been.'[54] Let's look in more detail at the multinationals' activities in China. Motorola set up 'the first major foreign corporate research and development centre on Chinese soil' only in 1993.[55] In 1997, China had fewer than fifty research centres that were managed by multinational corporations; by mid-2004, there were more than 600.[56]

What are they doing there and why? First, there are companies, such as Pfizer, Roche and Eli Lilly, which set up offices in the Zhangjiang Hi-Tech Park in Shanghai, and are attracting the burgeoning number of contract research organizations 'selling almost every service a pharmaceutical or biotech company could want, including the production of active ingredients, genomics, analytical and combinatorial chemistry, and preclinical toxicology testing'.[57] Cheap clinical trials are a particular draw, costing between one-fifth and one-half the cost of running one in the United States. Eli Lilly's Shanghai office did no science itself, but put together chains of contract research organizations. The cost and flexibility were the overriding interests. Second, companies have an eye on China itself as a growing and lucrative market, especially for drugs and therapies. Locating a research and development centre in China, then, is done for tactical purposes rather than in any expectation of new worthwhile discoveries. A research branch, according to one analysis, 'helps build connections with regulatory agencies' and facilitates 'access to a market that some analysts predict will be the world's second largest after the United States by 2020'.[58] Third, some companies are intent on tapping local resources. These resources can be human knowledge and skills: GlaxoSmithKline, for example, have chosen to locate their 'R&D China' subsidiary next to the Shanghai Institutes for Biological

Sciences rather than across the city in the Zhangjiang Hi-Tech Park because the company wanted 'hard-core science', not 'services'.[59] Nevertheless, a secondary justification, given by Jingwu Zang, the leader of GSK's R&D China, was 'the benefits of being able to transition drugs to a growing Chinese market more easily'.

Of course, expensive Western-style healthcare is not an unmitigated blessing. Parts of China are booming, but other parts suffer desperate poverty. The uneven development within the country is directly mirrored in a map of Chinese science: overwhelmingly located in Eastern hotspots – Beijing, Shanghai and Guangdong. This final tension shows no signs of declining in force. A nation (and China is not alone) that can aim to 'send a satellite to Mars but not solve the most basic problems that threaten millions of lives in the developing world' can expect uncomfortable questions.[60]

Europe

Many twenty-first-century news stories concerning the sciences in European countries reveal continuities in theme that stretch back far into the twentieth century. The Russian Academy of Sciences, for example, discussed in chapter 9, has continued to have a close but difficult relationship with the supreme political powers of the nation. The appointment of Vladimir Putin's favourite for president of the academy, Mikhail Kovalchuk, controller of the purse-strings of a $7 billion nanotechnology initiative, was held up by academicians in 2008.[61] They nevertheless chose to retain Yuri Osipov, a 71-year-old mathematician, a probable placeholder for Kovalchuk, as president, rather than elect Vladimir Fortov, a modernizing physicist who did not have Putin's support. Fortov promised the introduction of international peer review and open competition for funding. Putin, on the other hand, doubled academicians' salaries.[62]

In France, too, the tensions were between the centralized institutions of science and the movements for political reform. President Nicolas Sarkozy pushed to make universities more autonomous, able to set their own budgets and pursue their own research programmes.[63] In a closely related move, Sarkozy also wanted to reform the CNRS, the body in France that both funds and performs, in its constituent laboratories, most French research. The idea was that CNRS would become more like a research council, a funder rather than a performer of science. In effect, these reforms would make the French system much more like the British system. This direction was, for Sarkozy's critics on the left, enough evidence that an Anglo-American liberalization, perhaps privatization, agenda was being followed. University staff went on strike in 2009, while CNRS scientists, in a campaign coordinated by Claire Lemercier, invaded CNRS's Paris headquarters in June 2008 in protest against the plan to break up CNRS into six parts.[64] To some this was reminiscent of the 1968 events I discussed in chapter 17.

Meanwhile, the European Research Council was set up in February 2007. The European Research Council is rather like a National Science Foundation for the European Union (plus Israel). It distributes about €1 billion a year – not new money, but rather old science funding money now allocated at a European rather than at a national level. The council is an agency, separate from, but not entirely legally independent of, the European Commission. (*Nature*, in July 2009, backed arguments calling for the European Research Council to be fully autonomous.)[65]

The difference, at least from the perspective of scientists in some European Union nations, was the commitment to award grants on the basis of international peer review rather than according to political choices. The selection process was designed to be blind to the national origin of the research proposals. Nevertheless, in the first round of 300 grants, in 2008, the allocations followed traditional – that is to say, historically patterned – geographies of scientific strengths, with the United Kingdom doing particularly well (58) followed by France (39), Germany (33), Italy (26), the Netherlands (26), Spain (24) and Israel (24).[66] In comparison Bulgaria and the Czech Republic received only one grant each.

Post-Cold War tensions?

In chapters 13 to 16 I traced many linkages between the Cold War and science. In chapter 19 I asked what consequences the end of the Cold War had had, particularly for the physical sciences. Following the devastating attacks on the World Trade Center in New York on 11 September 2001, many Cold War features – the simplification of the world into two polar enemies, a 'War on Terror' against hidden foes, home and abroad – were resurrected. However, while there was certainly some redirection of research programmes, the 'War on Terror' does not seem to have had the profound effect in science that the Cold War had. It is, alternatively, another of the 'missing stories' that history of science will have to work to tell.

Other apparent echoes of the Cold War have been heard, on occasion, in twenty-first-century science. 'The Arctic is Russian', explained Artur Chilingarov, leader of an expedition that planted a Russian flag on the sea floor at the North Pole in August 2007. 'We must prove the North Pole is an extension of the Russian coastal shelf.' As Daniel Cressey explains, under the 1982 United Nations Convention on the Law of the Sea, countries could lay claim to areas beyond the standard 200 nautical mile limit, so long as there was a 'natural prolongation' of its continental shelf.[67] The Russians argued that the Lomonosov shelf, which stretches undersea across the polar region, is connected to the motherland. Canada and Denmark (Greenland) disputed the claim. The dispute would revolve around interpretations of geophysical and geological data, snatched during explorations in the brief Arctic summer. At stake were unknown oil, gas and mineral wealth.

While reminiscent of Antarctic politics, in which scientific activity was a marker for future possible claims (discussed in chapter 14), and despite the Russian presence, this episode is not best understood as a re-emergence of older Cold War tensions. Rather it is a vigorous expression of nationalistic interests. There are comparable clashes between Western nations – between France, Spain, Britain and Ireland over the Bay of Biscay, for example.

Nevertheless, the two decades since the end of the Cold War have seen some of the organizations most associated, even defined, by the conflict struggle for new rationales and identities. DARPA (the Defense Advanced Research Project Agency, known as ARPA until 1972) provides a good example. Recall from chapter 14 that ARPA was designed as a lean, fast-moving agency to fund defence research projects to enable the United States to overtake the Soviet Union following Sputnik. Its successes included the ARPANET, Project Vela (detecting nuclear detonations), the anti-ballistic missile system Defender, and stealth aircraft (first flying in 1977). On the one hand, DARPA in the post-Cold War years attracted criticism. Sharon Weinberger notes insider views that DARPA conferences were in the twenty-first century 'light in substance' and too close to Hollywood gimmickry.[68] Furthermore, in the conflict in Iraq, when innovative military solutions were needed, the Pentagon turned to agencies such as the Joint Improvised Explosive Device Defeat Organization, and not DARPA. On the other hand, the old DARPA was regarded as a model of an innovative organization and as a model to copy. Examples include IARPA (for Intelligence), HSARPA (for Homeland Security) and ARPA-E (for

518

energy, with a particular emphasis on energy security and, latterly, climate change).

Stem cells and politics

President George W. Bush was frequently criticized for his policies affecting science. The values of the conservative Christian right had guided Bush's 2001 decision to stop federal funding of research on new embryonic stem-cell-lines. Since then, as I discuss below, the United States had been overtaken by countries that were more permissive in this specific biomedical field: Singapore, the United Kingdom, Israel, China and Australia. Bush's administration was accused of conducting a 'war on science', with allegations of interference and the rewriting of evidence, and of a 'failure to deal with the risks of nuclear proliferation' – walking away, for example, from the Nuclear Non-Proliferation Treaty.[69]

'Many researchers, of all political stripes, are deeply troubled by what they regard as the dysfunctional relationship between science and the outgoing Bush administration', noted a *Nature* editorial in January 2008.[70] The Democrat candidate, Barack Obama, was comfortable supporting investment in science and unambiguously backing the teaching of evolution by natural selection. The Republican ticket, John McCain and his running mate Sarah Palin, was split, with the two representing wildly different constituencies. Palin's comments during the campaign – hedging her views on whether climate change is being caused by humans, strongly opposing embryonic stem-cell research, promising to cut funding, as she said in debate, for 'fruit fly research in Paris, France. Can you believe that?' – threatened to overshadow McCain's stance.

However, a science debate, pushed for by a campaign given coverage by *Science* and *Nature*, never materialized. Science, in the end, was not a crucial issue for the 2008 election. On winning the election, president-elect Obama began appointing key officials. In December, the post of secretary of the Department of Energy, with a seat at the cabinet, was given to a scientist, the physicist Steven Chu, who was director of the Lawrence Berkeley National Laboratory and a Nobel Prize winner. In January 2009, Obama appointed his science adviser. John Holdren's career was varied in all the right places: a physicist, with experience as an engineer at Lockheed Missiles and Space Company, fusion research at the Lawrence Livermore National Laboratory, anti-nuclear proliferation work with the Pugwash organization, and now a Harvard environmental policy expert in charge of the Woods Hole Research Center. Other Obama picks included Harold Varmus (whom we met in chapter 18), who had directed the National Institutes of Health, and Eric Lander, one of the leaders of the Human Genome Project. Early in his presidency, in March 2009, Obama used an executive order to overturn Bush's restrictions on federally funded stem-cell research, 'issued a memo directing [John Holdren] to ensure scientific integrity in government decision-making', and budgeted for big increases in science funding.[71]

Stem cells have the potential to turn into any kind of cell in the body. It is this general-purpose nature that makes them such an extraordinarily attractive object to medical scientists developing new therapies and new tests. But, until 2007, human stem cells were very hard – physically and politically – to isolate, culture, manipulate and use. Previously, the only source of stem cells was the embryo. In 1981, embryonic stem cells had been isolated in the mouse. It took seventeen years before human embryonic stem cells were found. The difficulty was not just in finding the tiny cells but also in discovering the methods needed to grow and culture them – that is to say, to turn stem cells into stem-cell-lines.

The techniques were immediately controversial, especially, given the embryonic source of the cells, within the worldwide Catholic Church and the conservative, 'pro-life', anti-abortion movement in the United States. President George W. Bush banned the use of federal funds to pursue research on any other than stem-cell-lines derived before 9 August 2001. In Germany, church leaders, such as Roman Catholic Cardinal Karl Lehmann, Green politicians, a legacy of law written in the wake of Nazi atrocities, and intense debate over two years led to a similar restriction of researchers to work only on old stem-cell-lines. Only the strong support of Chancellor Gerhard Schroeder and the minister for research, Edelgard Bulmahn, secured the continuation of limited imports of stem-cell-lines into Germany.[72] Schroeder and Bulmahn argued that, if Germany did not do some stem-cell research, it would be done elsewhere, with no ethical restrictions and to others' economic gain.

Embryonic stem-cell research did indeed flourish in areas of the world with fewer restrictions: Japan, Britain, and individual American states such as California ($3 billion),[73] Massachusetts ($1 billion) and Wisconsin ($750 million), which have made up the federal shortfall in funding. (Some states, for example Louisiana and North Dakota, criminalized such science; others, such as New Jersey, found that an attempt to fund research was rejected at the polls.)[74]

This savage pulling of research in two directions – on one hand towards visions of making the blind see and the crippled walk again, and on the other towards criminalization and dogmatic abhorrence – explains why the announcement of a new, non-embryonic source of stem cells was met with such enthusiasm. In 2006, at Kyoto University, Shinya Yamanaka had taken ordinary mouse cells and, by making four, relatively simple genetic changes, had caused some of them to revert to stem-cell status. Within the year, working with his postdoc student Kazutoshi Takahashi in secrecy, Yamanaka had achieved the same with human cells. (He used viral vectors to introduce clones of four genes – Oct3/4, Sox2, c-Myc and Klf4 – to the human cells.) Yamanaka published his results on what he called induced pluripotent stem (iPS) cells in November 2007. Remarkably, on the same day, a Wisconsin pioneer of stem-cell methods, James Thomson of the Univesity of Wisconsin in Madison, one of the original discoverers of human embryonic stell cells, independently announced the same iPS technique.

iPS cells became a goldrush in 2008 and 2009. Universities such as Harvard, Toronto and Kyoto speedily established new facilities; scientists switched research fields; Addgene, a Massachusetts company that sold Yamanaka's four reprogramming vectors, received 6,000 requests from 1,000 laboratories since the original announcement of the mouse technique. In Japan, where Kyoto University had delayed applying for patents, the simultaneous publication sparked a national debate on whether the nation was losing its scientific lead.[75] Billions of yen were made available for iPS science, and a Japanese patent was rushed through the system by September 2008.[76]

iPS cell research has ethical issues of its own.[77] iPS cells might be used to derive gametes from any source of cells (a celebrity's hair?) or to clone humans (it has been done with mice), or they might induce cancers rather than cures.

Translational research

There's a paradoxical relationship between the flourishing of biotechnology and the commercialization of science. On one hand (as discussed in chapter 18) the received narrative says that the patenting of recombinant DNA techniques led to the launch of the biotech companies, the celebration of the professor-entrepreneur, and the pressure further to commercialize academic research. So the 1980 Bayh–Dole Act, which granted intellectual property rights on publicly funded research to the universities, prompted the growth in technology transfer offices (TTOs).[78] By 2008 there were TTOs at 230 universities in the United States; data from two years earlier records 16,000 patent applications, 697 licences for products and 553 start-up companies. Some generated immense income streams, for example the Gatorade sports drink (shared between faculty inventor Robert Cade and the University of Florida), taxol (discussed earlier) and cisplatin (a platinum compound, found to have anti-cancer properties at Michigan State University). Big funds such as Royalty Pharma specialize in acquiring biomedical patents and licensing them to manufacturers: Royalty moved the anti-convulsant pregabalin from Northwestern University to Pfizer to make Lyrica in a $700 million deal; and Royalty, again, moved filgrastim substances, which stimulated white blood-cell production, from the Memorial Sloan-Kettering Cancer Centre to Amgen to make Neupogen/Neulasta in a $263 million deal.

TTOs were a model to copy. A wave of TTOs was set up in European countries in the 1990s and 2000s. In other countries, institutes with special briefs to transfer technology have been established, for example the A*STAR institutes of Singapore. 'A multi-continental chorus of academic researchers', notes Meredith Wadman in *Nature*, complain that the 'plethora of TTOs that have sprung up ... are at best a mixed blessing'.[79] TTOs, they say, overvalue intellectual property, hoard inventions, have small stretched staff, of which the talented are poached by industry and venture capital firms, and drive the overcommercialization of academic science. Nevertheless, the assumption is that biotech is the cause of this movement close to market.

On the other hand, an examination of biomedical science in the thirty years since the late 1970s, precisely the same period that saw the growth of the TTOs, shows that basic biomedical science has grown away from clinical application and product, and the cause of the drift was biotech. In the 1950s and 1960s, notes Declan Butler, a typical medical researcher was a physician-scientist.[80] But, with the growth of molecular biology, clinical and biomedical research began to separate. Basic biomedical researchers looked to top academic journals for credit rather than their contribution to medicine. Basic scientists also regarded unfamiliar regulation and patenting with trepidation. Clinicians, meanwhile, 'who treat patients – and earn fees for doing so – have little time or inclination to keep up with an increasingly complex literature, let alone do research.' Furthermore, in parallel, genomics, proteomics and so on generated so many possible drug targets, on average more expensive to develop than older therapies, that the pharmaceutical firms felt overwhelmed.

In this second account, then, biotech has had the effect of driving the laboratory bench further from medical application. To counter the trend, key funding agencies have promoted offices of 'translational research', a term that first appeared in 1993 as part of the *BRCA2* controversy. In the United States, the National Institutes of Health began funding, from 2003, Clinical and Translational Science Centers, which encourage multidisciplinary teamwork and business-style assessment. In Britain the Wellcome Trust and the Medical Research Council followed suit. Europe was in the course of setting up, in 2008–9, its new European Advanced Translational Infrastructure in Medicine, linking translational research in Denmark, Finland, France, Greece, Germany, Norway, the Netherlands, Italy, Spain, Sweden and the United Kingdom. So biotech led both to and away from the market and application.

Reverse engineering the human

Molecular biology has been accused of offering a reductive science of life. The old dogma – DNA is transcribed as RNA, RNA is translated into proteins – seems simple enough. But the discoveries, in the 1990s and 2000s, of many different kinds of snippets of nucleic acids, performing all sorts of functions in the cell, painted a picture of bewildering complexity. The first reports of non-protein-coding RNA molecules were rejected as artefacts, since such RNA was expected to be rapidly broken down. But genomics techniques are now so fast that the signatures of these ncRNAs suggest a reality of hosts of these strands of nucleic acid rather than artefacts.[81] (Notice how dependent on the state of techniques what counts as 'really' in the cell is!)

The trend towards large-scale bioinformatics, combined with fine-scale biochemical investigation, is what leads to the analysis of this complexity. In particular, cancer programmes, deploying these techniques, have focused on gene regulation, and many of types of small RNA molecules have been discovered in the course of such research.

Micro RNAs (miRNAs) are really small – roughly 23 nucleotides long – and, by either breaking down messenger RNA or interfering with how messenger RNA is translated, help fine-tune the expression of proteins.[82] Discovered in the 1990s, miRNAs were being developed into drug therapies in the 2000s, an example being GlaxoSmithKline's deal with Regulus Therapeutics in 2008.[83]

But, in general, pharmaceutical companies had hoped that the great genomics boom of the 1990s and early 2000s would lead to lots of promising drug targets. What they got instead was complexity. As Alison Abbott reported in *Nature* in 2008:

> the more that geneticists and molecular biologists have discovered, the more complicated most diseases have become. As individual genes have fallen out of favour, 'systems' – multitudes of genes, proteins and other molecules interacting in an almost infinite number of ways – have come into vogue.[84]

The great hope of the late 2000s was this new science, 'systems biology', which tackled the complexity of the cell in ways inspired by the manner that an electrical engineer analyses a complex, black-boxed piece of electronic kit. Electrical engineers might record the responses to inputs of many different frequencies and then make deductions about the wiring inside. Likewise, for example, systems biologists, using new technologies (microfluidics to vary osmotic pressure, fluorescent cell reporters to track changes), subjected yeast cells to varying conditions and then made deductions about bio-chemical pathways.[85] *Nature* called it 'reverse engineering the cell'. It is computationally very demanding. Some of the modelling 'requires a scale of computing effort analogous to that required to predict weather and understand global warming.'[86] There are even calls in Japan for a three-decade programme to 'create a virtual representa-tion of the physiology of the entire human': model and deduce every biochemical pathway and all its variants – reverse engineer the human body. The cost in computing bills alone is eye-watering.

Systems biology received plenty of hype and attention as a potential 'saviour of the failing research and development pipelines' of a 'crisis-ridden pharmaceutical industry', as Abbott puts it. Giant companies such as Pfizer, AstraZeneca, Merck and Roche have all bought into the project, albeit in a 'suck it and see' manner. Genomics of the first kind – the human genome project kind – has not delivered new drugs. The companies hope that systems biology might. Either way, new medicine will not come cheap.

Genomics: cheap and personal?

Nevertheless, sequencing genomes has become remarkably cheaper and quicker. The Human Genome Project took thirteen years, the involvement of nearly 3,000 scientists across sixteen institutions in six countries, and a small matter of $2.7 billion to produce a sequence in 2003. In 2008, a handful of scientists took just over four months, spending less than $1.5 million, to provide a sequence of the DNA of James Watson.[87] The trend is expected to continue, perhaps reaching rates of a billion kilobases per day per machine in the 2010s. The X Prize Foundation has offered a $10 million prize for the first successful attempt to sequence 100 human genomes in ten days at under $10,000 per genome.

After his sequence had been read, James Watson was counselled by a team of experts, who explained what the implications were of the twenty detected mutations associated with increased disease risk. 'It was so profound', one expert told *Nature*, in a plea for further research, 'how little we were actually able to say,'

That doesn't bode well for personal genomics. As sequencing becomes cheaper, we may all become Jim Watsons. Indeed, direct-to-consumer whole-genome testing went on sale in 2007. Three companies were pioneers: 23andMe and Navigenics in California, and deCode in Iceland. On payment, these companies would cross-check your DNA against a million single-point genetic polymorphisms. Critics complained that the clinical usefulness of this information was unclear and led to unnecessarily frightened customers, and demanded regulation or restriction. Furthermore, the information meant very little unless it was placed in the context of family histories and other facts. The companies replied that individuals had the right to their own genetic knowledge. Furthermore, 23andMe's technological and financial links to Google, which was launching its Google Health, a facility for recording personal medical histories, points to how genetic information might be interpreted in the future.[88]

Cheaper sequencing has also aided the phylogenetic investigations of the tree of life. Draft sequences for more and more organisms were published in the 2000s. In addition to a host of micro-organisms, notable sequences published (reading like an alternative Chinese calendar) were the model fly *Drosophila melanogaster* (2000), the model plant *Arabidopsis thaliana* (2000), rice (2002), the mouse (2002), the malaria mosquito *Anopheles gambiae* (2002), the model organism *Neurospora crassa* (2003), the silk worm (2004), the rat (2004), the chicken (2004), the chimpanzee (2005), the dog (2005), a sea urchin (2006), the cat (2007), the horse (2007), the grape vine (2007), the platypus (2008), corn (2008), sorghum (2009) and the cow (2009). Nor did organisms have to be living to give up genetic secrets. Almost complete sequences of the mammoth (2008) and the Neanderthal (2009) were salvaged from fossils.

Big pharma and publishing science

Publishing on science opened up in the 2000s, with new open access journals and the blog phenomenon. However, big vested interests still clashed. When big pharma appears in the courts, the effects can both open up and close down the publishing processes in science. In May 2007, for example, Pfizer found itself in a court case related to its painkillers Celebrex (celecoxib) and Bextra (valdecoxib) and demanded that the *New England Journal of Medicine* hand over the relevant peer reviews, along with the names of reviewers and any documents of internal editorial deliberation.[89] In November 2007, the journal handed over some documents, but not all that the company wanted, and in 2008 it dug its heels in. Pfizer's lawyers argued that among the papers might be exonerating data vital to Pfizer's defence. *The New England Journal of Medicine*'s editors said that the move to strip reviewers of their anonymity would damage peer review. Other editors, such as Donald Kennedy, editor-in-chief at *Science*, agreed.

Elsewhere, one of the unanticipated consequences of the Vioxx (refecoxib) trial, in which accusations that the drug had serious side-effects hidden from users were rejected by the Merck company, has been to open up documents previously closed and reveal how trials were published. In 2008, analysts of these documents, writing in the *Journal of the American Medical Association*, found strong suggestions of ghost writing: 'one of the Merck-held documents lists a number of clinical trials in which a Merck employee is to be author of the first draft of a manuscript', but in sixteen out of twenty cases the name on the finally published article was of an external academic.[90]

The Large Hadron Collider in global news

While CERN started as a European project, the sheer expense of the Large Hadron Collider, alongside the cancellation of its American rival, the Superconducting Supercollider, made it, noted science journalist Geoff Brumfiel, 'the first truly global experimental undertaking'.[91] In 1994, Britain and Germany had considered pulling out of the project. To rescue it, some of CERN's leading scientists, including accelerator designer Lyn Evans and the director-general Christopher Llewellyn Smith, had persuaded CERN's European council to accept a reduced design and, in 1995, had prevailed upon Japan and Russia to fill the funding gap. India and the United States joined too.

So, 10 September 2008, 'beam day', the first time protons had been coaxed around the full LHC ring, was a global news event. The *Times of India* reported on how scientists had gathered in great excitement at the Tata Institute of Fundamental Research (before a data link crashed, leaving them in darkness), while the *Economic Times* of India noted with pride how 200 Indian scientists were involved, as well as a lucrative concrete contract. Israeli press reported the words of Prime Minister Ehud Olmert: 'I'm very proud of the contribution made by the Israeli scientists and Israeli technology to an experiment of such magnitude, one that has the power to impact all of humanity.' 'Scottish firm expands on Cern successes' was a headline in *The Scotsman*. And so on.

So there was a measure of global embarrassment when, nine days later, a cable linking two of the LHC's magnets lost superconductivity, melting almost instantaneously. Liquid helium then leaked into the LHC tunnels. The whole $4 billion project was suspended for a year while the fault was fixed. Racks of computers, poised to store petabytes of information, sat unused. With no data, scientists across the world, including students waiting to finish PhDs, had no choice but to wait.

The LHC was designed to collide protons with antiprotons at energies seven times that of its nearest competitor, Fermilab's Tevatron, and to search among the remnants for clues of physics beyond the standard model. With the LHC out of action, scientists with alternative schemes of generating such physics suddenly had a window of opportunity. The Tevatron was cranked up as high as it would go, with each new collision added to a slowly building statistical picture, hinting at anomalies (such as in the decay of the strange B meson) that might not be explained by the standard model. However, by 2011, the LHC was up and running and, perhaps, closing in on the Higgs particle.

A second approach was to investigate the ghostly neutrino particles. In 1998, Japanese scientists, using the Super-Kamiokande experiment in Hida, had shown that neutrinos switched between different types. This switching was only possible in neutrinos that possessed mass, albeit a tiny one. (This finding had also resolved a long-standing and increasingly worrisome observation of a shortage of solar neutrinos.) Massive neutrinos were not part of the standard model. New, even bigger neutrino experiments, such as IceCube, an array of detectors under the Antarctic ice, it was hoped in 2008, might reveal new phenomena.

Finally, while the standard model is a theory of the fundamental forces of the universe, and usually probed at the microphysics level, it is also central to cosmological theory. Cosmologists had already noted a deficit between the observable matter in the universe and the deduced mass necessary for cosmological theory to work. This deficit, labelled dark matter, was of an unknown nature in the first decade of the twenty-first century. A candidate material had not been observed, nor did the standard model have room for it. Yet dark matter was necessary to make sense of the speed of rotation of galaxies and makes up some 85 per cent of matter in the universe.

Dark matter was the target of several large-scale scientific projects. In 2008, scientists working with data from a satellite called PAMELA (Payload for Antimatter Matter Exploration and Light-nuclei Astrophysics, an Italian–Russian–German–Swedish collaboration) reported an unexpected number of antielectrons of certain energies. Did they come from colliding exotic dark matter particles, such as the 'neutralino'? Also in 2008, an underground dark matter detector at the Italian Gran Sasso National Laboratory reported a signal. Yet other detectors, including ironically another one under the same Italian mountain, have seen nothing.

These non-particle accelerator routes to new physics may well become typical. The LHC was expected to last twenty years. The machine that the global fundamental physics community regards as the successor to the LHC – an electron-positron collider called the International Linear Collider – barely limped on as design money was cut and partner countries, such as the United Kingdom in 2008, pulled out. In this book I have argued that national particle physics centres were largely justified, ultimately, by Cold War security concerns, while CERN was built partly to sustain and exemplify the European project. There are no international visions, as yet, that are powerful enough to support the ILC.

In contrast to stalling and uncertain experiments, theory scored an impressive goal in 2008 in a highly precise demonstration, published by Stephan Dürr and his colleagues in *Science*, that the predictions of mass made by the theory of quantum chromodynamics (QCD) matched observed masses of hadrons very closely.[92] The journal's rival, *Nature*, praised the achievement, both for its practical implications (understanding, for example, supernovae) and as a triumphant vindication of the Pythagorean credo that 'all things are number'.[93] 'The accurate, controlled calculation of hadron masses is a notable milestone', the theoretical physicist wrote. 'But the fact that it has taken decades to reach this milestone, and that even today it marks the frontier of ingenuity and computer power, emphasizes the limitations of existing methodology and challenges us to develop more powerful techniques.' An alternative way of putting this would be that theory, too, had trouble scaling up.

Another green world (and another, and another . . .)

There are five classical planets. The eighteenth, nineteenth and twentieth centuries add one each. The detection of hordes of fainter planets in the 1990s and 2000s has had some dramatic consequences. First, the discovery, many by Caltech astronomer Michael Brown, of a string of 'trans-Neptunian objects' has forced scientists to rethink what a 'planet' is. The International Astronomical Union created in 2006 a new category of 'dwarf planet', placing Pluto and Ceres with some of these new objects, now given names: Eris, Sedna (possibly), Haumea and Makemake. The demotion of Pluto, discovered in 1930 by Clyde Tombaugh, is a reminder of the revisable character of even the most apparently basic or venerable of scientific categories.

Second, beginning in 1995, planets around other ordinary stars have been detected. These 'exoplanets' have typically revealed themselves through the tiny wobbles, detectable with fine spectroscopic measurement, induced in the movement of their home stars. By 2008, more than 300 exoplanets had been identified, including one, HD 189733b, with a trace of water vapour in its spectrum. This discovery prompted serious speculation about 'biomarkers' to look for in exoplanets.

Trouble on our own planet

The political response in the 2000s to the scientific consensus of the existence of global warming was both slow and complex, with, for example, American states such as California moving in a different direction and speed to the national government. Climate change was a key issue in the election of a new prime minister, Kevin Rudd, in Australia in 2007. In Britain, climate change and energy security were used (not least by the then chief scientific advisor to the government, David King) to justify the start of a new wave of nuclear power station construction. The strange amalgam of science and politics that marks climate change action was honoured by the award of the 2007 Nobel Prize for Peace jointly to the Intergovernmental Panel on Climate Change and Al Gore, the ex-vice president who had turned his PowerPoint presentation into a 2006 feature film, *An Inconvenient Truth*.

Nevertheless, both the political will to act on climate change and the feasibility of large-scale technological solutions touted were not yet strong enough in the 2000s. Carbon capture and storage (CC&S), the plan to bury carbon dioxide, a leading candidate technology, even struggled to receive funding, as the case of FutureGen in Illinois illustrated. The fourth IPCC *Assessment Report*, released in 2007, was incrementally more forthright in its insistence on the reality of anthropogenic climate change than its predecessor, and even then fell short of what a mainstream scientific consensus might be. The international meeting at Copenhagen in 2009, expected to consider a replacement for the Kyoto Protocol, was widely regarded as a failure.

Attempts to use the IPCC as a model for other areas where global-scale problems needed guidance from complex and disputed scientific advice ran into trouble, too, in 2008. The International Assessment of Agricultural Science and Technology, a four-year $10 million project to 'do for hunger and poverty what the Intergovernmental Panel on Climate Change has done for another global challenge' was threatened by the withdrawal of Monsanto and Syngenta.[94] At issue was the contribution biotechnology – genetically modified organisms, specifically – could make to the mission. The conflict between the working worlds of energy-intensive societies, international environmental governance and global commerce look set to shape the twenty-first-century sciences.

NOTES

1 INTRODUCTION

1 Proctor, Robert N. (1988), *Racial Hygiene: Medicine under the Nazis*, Cambridge, MA: Harvard University Press, p. 9.
2 Gieryn, Thomas F. (1983), 'Boundary-work and the demarcation of science from non-science: strains and interests in professional ideologies of scientists', *American Sociological Review* 48, pp. 781–95.
3 Orwell, George (1946), 'Politics and the English language', *Horizon* 76 (April), repr. in *Inside the Whale and Other Essays*, London: Penguin, 1962.
4 For a parallel discussion, see Greenhalgh, Susan (2008), *Just One Child: Science and Policy in Deng's China*, Berkeley: University of California Press, p. 9.
5 Gaudillière, Jean-Paul (2009), 'Cancer', in Peter J. Bowler and John V. Pickstone (eds), *The Cambridge History of Science*, Vol. 6: *The Modern Biological and Earth Sciences*, Cambridge: Cambridge University Press, pp. 486–503; Anderson, Benedict (1983), *Imagined Communities: Reflections on the Origin and Spread of Nationalism*, London: Verso; Edwards, Paul N. (2010), *A Vast Machine: Computer Models, Climate Data, and the Politics of Global Warming*, Cambridge, MA: MIT Press; Latour, Bruno (1999), *Pandora's Hope: Essays on the Reality of Science Studies*, Cambridge, MA: Harvard University Press.
6 Latour, Bruno (1983), 'Give me a laboratory and I will raise the world', in Karin D. Knorr-Cetina and Michael J. Mulkay (eds), *Science Observed*, Beverly Hills, CA: Sage, pp. 141–70.
7 Reich, Leonard (1983), 'Irving Langmuir and the pursuit of science and technology in the corporate environment', *Technology and Culture* 24, pp. 199–221; Reich, Leonard (1985), *The Making of American Industrial Research: Science and Business at GE and Bell, 1876–1926*, Cambridge: Cambridge University Press; Dennis, Michael Aaron (1987), 'Accounting for research: new histories of corporate laboratories and the social history of American science', *Social Studies of Science* 17, pp. 479–518.
8 Rouse, Joseph (1987), *Knowledge and Power: Towards a Political*

Philosophy of Science, Ithaca, NY: Cornell University Press, p. 101; Golinski, Jan (1998), *Making Natural Knowledge: Constructivism and the History of Science*, Cambridge: Cambridge University Press, p. 35.

9 Latour, Bruno, and Steve Woolgar (1979), *Laboratory Life: The Social Construction of Scientific Facts*, London: Sage.

10 Quoted in Todes, Daniel P. (1997), 'Pavlov's physiological factory', *Isis* 88, pp. 205–46, at p. 205.

11 Latour, Bruno (1987), *Science in Action: How to Follow Scientists and Engineers through Society*, Milton Keynes: Open University Press.

12 Rudolph, John L. (2002), 'Portraying epistemology: school science in historical context', *Science Education* 87, pp. 64–79, at p. 70.

13 Gibbons, Michael, Camille Limoges, Helga Nowotny, Simon Schwartzman, Peter Scott and Martin Trow (1994), *The New Production of Knowledge: The Dynamics of Science and Research in Contemporary Societies*, London: Sage; Shinn, Terry (2002), 'The triple helix and new production of knowledge: prepackaged thinking on science and technology', *Social Studies of Science* 32, pp. 599–614; Godin, Benoît (1998), 'Writing performative history: the new New Atlantis?', *Social Studies of Science* 28, pp. 465–83.

14 Stokes, Donald E. (1997), *Pasteur's Quadrant: Basic Science and Technological Innovation*, Washington, DC: Brookings Institution Press.

15 Price, Don K. (1954), *Government and Science: Their Dynamic Relation in American Democracy*, New York: New York University Press, p. 5.

16 Quoted in Keller, Evelyn Fox (1996), 'The dilemma of scientific subjectivity in postvital culture', in Peter Galison and David J. Stump (eds), *The Disunity of Science: Boundaries, Contexts and Power*, Stanford, CA: Stanford University Press, pp. 417–27, at p. 422.

17 Nazi physician Hermann Boehm, quoted in Proctor, (1988), p. 84.

2 NEW PHYSICS

1 Hunt, Bruce J. (2003), 'Electrical theory and practice in the nineteenth century', in Mary Jo Nye (ed.), *The Cambridge History of Science*, Vol. 5: *The Modern Physical and Mathematical Sciences*, Cambridge: Cambridge University Press, pp. 311–27, at p. 317.

2 Quoted ibid., p. 319.

3 Hunt, Bruce J. (1991), *The Maxwellians*, Ithaca, NY: Cornell University Press.

4 Quoted in Hunt (2003), p. 319.

5 Buchwald, Jed Z. (1994), *The Creation of Scientific Effects: Heinrich Hertz and Electric Waves*, Chicago: University of Chicago Press.

6 Sungook Hong (2003), 'Theories and experiments on radiation from Thomas Young to X-rays', in Nye, *The Cambridge History of Science*, Vol. 5: *The Modern Physical and Mathematical Sciences*, p. 288.

7 Hunt (2003), p. 323.

8 Seliger, Howard H. (1995), 'Wilhelm Conrad Röntgen and the glimmer of light', *Physics Today* (November), pp. 25–31, at p. 26.

9 Quoted ibid., p. 27.

10 James, Frank A. J. L. (1983), 'The conservation of energy, theories of absorption and resonating molecules, 1851–1854: G. G. Stokes, A. J.

Angström and W. Thomson', *Notes and Records of the Royal Society of London* 38, pp. 79–107.

11 Seliger (1995), pp. 27–8.
12 Ibid., p. 29.
13 Ibid., p. 30.
14 Hong (2003), p. 287.
15 Electrical Engineer advertisement, in Seliger (1995), p. 30.
16 Kevles, Bettyann Holtzmann (2003), 'The physical sciences and the physician's eye', in Nye, *The Cambridge History of Science*, Vol. 5: *The Modern Physical and Mathematical Sciences*, pp. 615–33, at p. 616.
17 Laue, Max von (1915), 'Concerning the detection of X-ray interferences', Nobel Lecture, www.nobelprize.org/nobel_prizes/physics/laureates/1914/laue-lecture.pdf, pp. 351–2.
18 Hunt (2003), p. 325.
19 Ibid., p. 324.
20 Davis, Edward A., and Isobel Falconer (1997), *J. J. Thomson and the Discovery of the Electron*, London: Taylor & Francis; Hunt (2003), p. 325.
21 Hughes, Jeff (2003), 'Radioactivity and nuclear physics', in Nye, *The Cambridge History of Science*, Vol. 5: *The Modern Physical and Mathematical Sciences*, pp. 350–74, at p. 352.
22 Ibid., p. 353.
23 Ibid.
24 Ibid., p. 355.
25 Mazeron, Jean-Jacques, and Alain Gerbaulet (1998), 'The centenary of discovery of radium', *Radiotherapy and Oncology* 49, pp. 205–16, at p. 207.
26 Ibid., p. 209.
27 Boudia, Soraya (1997), 'The Curie laboratory: radioactivity and metrology', *History and Technology* 13, pp. 249–65.
28 Mazeron and Gerbaulet (1998), p. 209.
29 Ibid., p. 210.
30 Boudia (1997); Gaudillière, Jean-Paul (2009), 'Cancer', in Peter J. Bowler and John V. Pickstone (eds), *The Cambridge History of Science*, Vol. 6: *The Modern Biological and Earth Sciences*, Cambridge: Cambridge University Press, pp. 486–503, at p. 490.
31 Hughes (2003), p. 356.
32 Ibid., p. 357.
33 Ibid., p. 356.
34 Quoted in Pais, Abraham (1986), *Inward Bound: Of Matter and Forces in the Physical World*, Oxford: Clarendon Press, p. 189.
35 Hughes (2003), p. 357.
36 Cahan, David (1989), *An Institute for an Empire: The Physikalisch-Technische Reichsanstalt, 1871–1918*, Cambridge: Cambridge University Press, p. 146.
37 Quoted ibid., pp. 147–8.
38 Ibid., p. 154.
39 Darrigol, Olivier (2003), 'Quantum theory and atomic structure, 1900–1927', in Nye, *The Cambridge History of Science*, Vol. 5: *The Modern Physical and Mathematical Sciences*, pp. 331–49, p. 333.
40 Klein, Martin J. (1962), 'Max Planck and the beginning of quantum theory', *Archive for the History of Exact Sciences* 1, pp. 459–79; Klein,

Martin J. (1963), 'Planck, entropy and quanta', *Natural Philosopher* 1, pp. 83–108.
41 Kuhn, Thomas S. (1978), *Black-body Theory and the Quantum Discontinuity, 1894–1912*, Oxford: Oxford University Press; Darrigol (2003), p. 332.
42 Cassidy, David (1995), *Einstein and our World*, Atlantic Highlands, NJ: Humanities Press.
43 Ibid., p. 23.
44 Ibid., p. 27.
45 Galison, Peter (2003), *Einstein's Clocks, Poincaré's Maps: Empires of Time*, London: Hodder & Stoughton.
46 Cassidy (1995), p. 31.
47 McCormmach, Russell (1970), 'H. A. Lorentz and the electromagnetic view of nature', *Isis* 61, pp. 459–97.
48 This example is from Cassidy (1995), pp. 38–9.
49 Ibid., p. 39.
50 Darrigol (2003), p. 334.
51 Einstein, Albert (1905), 'Über einen die Erzeugung und Verwandlung des Lichtes betreffenden heuristischen Gesichtspunkt', *Annalen der Physik* 17, pp. 132–48; trans. as 'On a heuristic point of view about the creation and conversion of light', in D. ter Haar (ed.), *The Old Quantum Theory*, Oxford: Pergamon Press, 1967, pp. 91–107.
52 Darrigol (2003), p. 335.
53 Miller, Arthur I. (2003), 'Imagery and representation in twentieth-century physics', in Nye, *The Cambridge History of Science*, Vol. 5: *The Modern Physical and Mathematical Sciences*, pp. 191–215, at p. 199.
54 Bohr, Niels (1913), 'On the constitution of atoms and molecules', *Philosophical Magazine*, series 6, 26, pp. 1–25.
55 Ibid., pp. 3–4.
56 Wilson, William (1956), 'John William Nicholson, 1881–1955', *Biographical Memoirs of Fellows of the Royal Society* 2, pp. 209–14, at p. 210.
57 Quoted in Kevles, Daniel J. (1971), *The Physicists: The History of a Scientific Community in Modern America*, Cambridge, MA: Harvard University Press, p. 92.
58 Bohr (1913), p. 11.
59 Darrigol (2003), p. 337.
60 Hughes (2003), p. 360.
61 Cassidy (1995), p. 15.
62 Ibid., p. 17.
63 Klein, Martin J. (1971), 'Einstein', in Charles Coulston Gillispie (ed.), *Dictionary of Scientific Biography*, Vol. 4. New York: Charles Scribner's Sons.
64 Einstein, Albert (1905), 'Zur Elektrodynamik bewegter Körper', *Annalen der Physik* 17, pp. 891–921; trans. as 'On the electrodynamics of moving bodies', in *The Principle of Relativity*. London: Methuen, 1923.
65 Galison (2003), pp. 236–7.
66 Ibid., p. 37.
67 Ibid., p. 38.
68 Ibid., p. 30.

69 Ibid., pp. 243–4.
70 Ibid., pp. 246–8, 253.
71 McCormmach (1970); Jungnickel, Christa, and Russell McCormmach (1986), *Intellectual Mastery of Nature: Theoretical Physics from Ohm to Einstein*, 2 vols, Chicago: University of Chicago Press; Cassidy (1995), pp. 13–14.
72 Rigden, John S. (2005), *Einstein 1905: The Standard of Greatness*, Cambridge, MA: Harvard University Press.
73 Holton, Gerald (1969), 'Einstein, Michelson and the "crucial" experiment', *Isis* 60, pp. 132–97.
74 Ibid., p. 169.
75 Ibid., pp. 186–7.
76 Ibid., pp. 148–9.
77 Renn, Jürgen, and Tilman Sauer (2003), 'Errors and insights: reconstructing the genesis of general relativity from Einstein's Zurich notebook', in Frederic Lawrence Holmes, Jürgen Renn and Hans-Jörg Rheinberger (eds), *Reworking the Bench: Research Notebooks in the History of Science*, Berlin: Springer, pp. 253–68, at p. 264.
78 Einstein, Albert (1915), 'Die Feldgleichungen der Gravitation', *Sitzungsberichte der Preussischen Akademie der Wissenschaften zu Berlin*, pp. 844–7, at p. 843.
79 Rowe, David E. (2003), 'Mathematical schools, communities and networks', in Nye, *The Cambridge History of Science*, Vol. 5: *The Modern Physical and Mathematical Sciences*, pp. 113–32, at p. 125.
80 Barkan, Diana Kormos (1993), 'The witches' Sabbath: the First International Solvay Congress in Physics', *Science in Context* 6, pp. 59–82, at p. 61.
81 Ibid., p. 60; Hughes (2003), p. 357.
82 Kuhn (1978).
83 Quoted in Barkan (1993), p. 66.
84 Kevles (1971), pp. 93–4.
85 Quoted ibid., p. 94.
86 Cahan (1989).
87 Quoted in Kevles (1971), p. 98.
88 Hughes, Thomas P. (1989), *American Genesis: A Century of Invention and Technological Enthusiasm, 1870–1970*, New York: Penguin.
89 Kevles (1971), pp. 100–1.

3 NEW SCIENCES OF LIFE

1 Bowler, Peter J. (1983), *The Eclipse of Darwinism: Anti-Darwinian Evolution Theories in the Decades around 1900*, Baltimore: Johns Hopkins University Press.
2 Provine, William B. (1971), *The Origins of Theoretical Population Genetics*, Chicago: University of Chicago Press.
3 Brannigan, Augustine (1981), *The Social Basis of Scientific Discoveries*, Cambridge: Cambridge University Press.
4 Bowler, Peter J. (1989), *The Mendelian Revolution: The Emergence of Hereditary Concepts in Modern Science and Society*, London: Athlone Press.

5 Quoted in Brannigan (1981), p. 101.

6 Bowler (1989).

7 Ibid., p. 103.

8 Brannigan (1981), p. 94.

9 Agar, Nigel E. (2005), *Behind the Plough: Agrarian Society in Nineteenth-Century Hertfordshire*, Hatfield: University of Hertfordshire Press, p. 97; Rothamsted Research (2006), *Guide to the Classical and other Long-Term Experiments, Datasets and Sample Archive*. Harpenden: Rothamsted Research.

10 Rosenberg, Charles E. (1976), *No Other Gods: On Science and American Social Thought*, Baltimore: Johns Hopkins University Press.

11 Kimmelman, Barbara A. (1983), 'The American Breeders' Association: genetics and eugenics in an agricultural context, 1903–13', *Social Studies of Science* 13, pp. 163–204, at pp. 164, 168.

12 Ibid., pp. 170–1.

13 Gayon, Jean, and Doris T. Zallen (1998), 'The role of the Vilmorin company in the promotion and diffusion of the experimental science of heredity in France, 1840–1920', *Journal of the History of Biology* 31, pp. 241–62.

14 Kimmelman (1983), p. 172.

15 Porter, Theodore M. (1994), 'The English biometric tradition', in Ivor Grattan-Guinness (ed.), *Companion Encyclopedia of the History and Philosophy of the Mathematical Sciences*, 2 vols, London: Routledge, pp. 1335–40, at p. 1335.

16 Ibid., p. 1336.

17 Kevles, Daniel J. (1992), 'Out of eugenics: the historical politics of the human genome', in Kevles and Hood (eds), *Code of Codes: Scientific and Social Issues in the Human Genome Project*, Cambridge, MA: Harvard University Press, pp. 3–36.

18 Quoted in Turner, Frank M. (1980), 'Public science in Britain, 1880–1919', *Isis* 71, pp. 589–608, at p. 598.

19 Haldane, John Burdon Sanderson (1923), *Daedalus, or Science and the Future*, London: Kegan Paul.

20 Kevles (1992).

21 MacKenzie, Donald (1976), 'Eugenics in Britain', *Social Studies of Science* 6, pp. 499–532; see also Turner (1980).

22 Paul, Diane B. (1995), *Controlling Human Heredity: 1865 to the Present*, Atlantic Highlands, NJ: Humanities Press, p. 74.

23 Kevles (1992), p. 6.

24 Kimmelman (1983), pp. 184–6.

25 Kevles (1992), p. 7.

26 Rafter, Nicole Hahn (ed.) (1988), *White Trash: The Eugenic Family Studies, 1877–1919*, Boston: Northeastern University Press.

27 Proctor, Robert N. (1988), *Racial Hygiene: Medicine under the Nazis*, Cambridge, MA: Harvard University Press, p. 97; Kevles, Daniel J. (1995); *In the Name of Eugenics: Genetics and the Uses of Human Heredity*, 2nd edn, Cambridge, MA: Harvard University Press.

28 Procter (1988), p. 97.

29 MacKenzie (1976); Magnello, M. Eileen (1999), 'The non-correlation of biometrics and eugenics: rival forms of laboratory work in Karl Pearson's

career at University College London, Part 1', *History of Science* 37, pp. 45–78.

30 Porter (1994), p. 1337.

31 Edwards, A. W. F. (1994), 'Probability and statistics in genetics', in Grattan-Guinness, *Companion Encyclopedia of the History and Philosophy of the Mathematical Sciences*, pp. 1357–62, at p. 1360.

32 Kimmelman (1983), p. 176.

33 Kohler, Robert E. (1994), *Lords of the Fly: Drosophila Genetics and the Experimental Life*, Chicago: University of Chicago Press.

34 Judson, Horace Freeland (1992), 'A history of gene mapping and sequencing', in Daviel J. Kevles and Leroy Hood (eds), *The Code of Codes: Scientific and Social Issues in the Human Genome Project*, Cambridge, MA: Harvard University Press, pp. 37–80, at p. 48.

35 Kohler (1994).

36 Morgan, Thomas Hunt, et al. (1915), *The Mechanism of Mendelian Heredity*, rev. edn (1923), New York: Holt, p. 281; quoted in Burian, Richard M., and Doris T. Zallen (2009), 'Genes', in Peter J. Bowler and John V. Pickstone (eds), *The Cambridge History of Science*, Vol. 6: *The Modern Biological and Earth Sciences*, Cambridge: Cambridge University Press, pp. 431–50, at p. 437.

37 Basalla, George (1967), 'The spread of Western science', *Science* 156, pp. 611–22, at p. 620.

38 Burian and Zallen (2009), p. 436.

39 Harwood, Jonathan (2009), 'Universities', in Bowler and Pickstone, *The Cambridge History of Science*, Vol. 6: *The Modern Biological and Earth Sciences*, pp. 90–107.

40 Fitzgerald, Deborah Kay (1990), *The Business of Breeding: Hybrid Corn in Illinois, 1890–1940*. Ithaca, NY: Cornell University Press.

41 Cittadino, Eugene (2009), 'Botany', in Bowler and Pickstone, *The Cambridge History of Science*, Vol. 6: *The Modern Biological and Earth Sciences*, pp. 225–42, at p. 239.

42 Harwood (2009), p. 99.

43 Ibid., p. 98.

44 Kohler, Robert E. (1982), *From Medical Chemistry to Biochemistry: The Making of a Biomedical Discipline*, Cambridge: Cambridge University Press; Harwood (2009), p. 102.

45 Goujon, Philippe (2001), *From Biotechnology to Genomes: The Meaning of the Double Helix*, Singapore: World Scientific, p. 14.

46 Amsterdamska, Olga (2009), 'Microbiology', in Bowler and Pickstone, *The Cambridge History of Science*, Vol. 6: *The Modern Biological and Earth Sciences*, pp. 316–41, at p. 335.

47 Bud, Robert (1993), *The Uses of Life: A History of Biotechnology*, Cambridge: Cambridge University Press.

48 Goujon (2001), p. 15.

49 Cittadino (2009), p. 237.

50 Amsterdamska (2009), p. 333.

4 NEW SCIENCES OF THE SELF

1 Sulloway, Frank J. (1979), *Freud, Biologist of the Mind: Beyond the Psychoanalytic Legend*, London: Burnett Books.
2 Hopwood, Nick (2009), 'Embryology', in Peter J. Bowler and John V. Pickstone (eds), *The Cambridge History of Science*, Vol. 6: *The Modern Biological and Earth Sciences*, Cambridge: Cambridge University Press, pp. 285–315.
3 Sulloway (1979), p. 15.
4 Ibid., p. 29.
5 Ibid., p. 59.
6 Ibid., p. 62.
7 Quoted in Ash, Mitchell (1995), *Gestalt Psychology in German Culture, 1890–1967: Holism and the Quest for Objectivity*, Cambridge: Cambridge University Press, p. 71.
8 Hughes, Thomas P. (1983), *Networks of Power: Electrification in Western Society, 1880–1930*. Baltimore: Johns Hopkins University Press.
9 Hughes, Thomas P. (1989), *American Genesis: A Century of Invention and Technological Enthusiasm, 1870–1970*, New York: Penguin, p. 233.
10 Hughes (1983), p. 50.
11 Sulloway (1979), pp. 89–93.
12 Ibid., p. 98.
13 Ibid., pp. 248–9, 251.
14 Hopwood (2009), p. 296.
15 Quoted in Sulloway (1979), p. 272.
16 See ibid., p. 362.
17 Quoted ibid., p. 353.
18 Ibid., pp. 419–44.
19 Smith, Roger (1997), *The Fontana History of the Human Sciences*, London: HarperCollins, p. 711.
20 Ibid., p. 493.
21 Ibid., p. 497.
22 Ibid., p. 590.
23 Ibid., p. 592.
24 Ibid., p. 591.
25 Ibid., pp. 583–5.
26 Ibid., p. 586.
27 Ibid., p. 593.
28 Ibid., p. 595.
29 Harrington, Anne (2009), 'The brain and behavioural sciences', in Bowler and Pickstone, *The Cambridge History of Science*, Vol. 6: *The Modern Biological and Earth Sciences*, pp. 504–23, at p. 513.
30 Liddell, E. G. T. (1952), 'Charles Scott Sherrington, 1857–1952', *Obituary Notices of Fellows of the Royal Society* 8, pp. 241–70, at p. 245.
31 Smith (1997), p. 645.
32 Sherrington, Charles Scott (1940), *Man on his Nature: The Gifford Lectures, Edinburgh 1937–38*, Cambridge: Cambridge University Press; quoted in Liddell (1952), p. 258.
33 Todes, Daniel P. (1997), 'Pavlov's physiological factory', *Isis* 88, pp. 205–46, at p. 209.

34 Ibid., pp. 217–18.
35 Ibid., pp. 241, 211.
36 Smith (1997), p. 647.
37 Ibid., p. 649.
38 Harrington (2009), p. 515.
39 Smith (1997), p. 650.
40 Watson, John Broadus (1913), 'Psychology as the behaviorist views it', *Psychological Review* 20, pp. 158–77, at p. 158.
41 Smith (1997), p. 654.
42 Watson (1913).
43 Ibid.
44 Watson, quoted in Smith (1997), pp. 657, 656.
45 Ibid., p. 656.
46 Ibid., pp. 660–72.
47 Tolman, quoted ibid., p. 662.
48 Watson (1913).
49 Ibid.
50 Ibid.
51 Harrington (2009), p. 520.
52 Smith (1997), p. 819.
53 Schindler, Lydia Woods (1988), *Understanding the Immune System*, Washington, DC: US Department of Health and Human Services; quoted in Martin, Emily (1990), 'Toward an anthropology of immunology: the body as nation state', *Medical Anthropology Quarterly*, new series 4, pp. 410–26, at p. 411.
54 Nilsson, Lennart (1987), *The Body Victorious: The Illustrated Story of our Immune System and Other Defences of the Human Body*, New York: Delacorte Press; quoted in Martin (1990), p. 411.
55 Söderqvist, Thomas, Craig Stillwell and Mark Jackson (2009), 'Immunology', in Bowler and Pickstone, *The Cambridge History of Science*, Vol. 6: *The Modern Biological and Earth Sciences*, pp. 467–85.
56 Nobel Foundation (1967), *Nobel Lectures including Presentation Speeches and Laureates' Biographies: Physiology or Medicine 1901–1921*, Amsterdam: Elsevier.
57 Ibid.
58 Ibid.
59 Brandt, Allan M. (1985), *No Magic Bullet: A Social History of Venereal Disease in the United States since 1880*, Oxford: Oxford University Press.
60 Ibid., p. 40.
61 Ehrlich, Paul (1899–1900), 'Croonian Lecture: On immunity with special reference to cell life', *Proceedings of the Royal Society of London* 66, pp. 424–48, at p. 426.
62 Ibid., p. 428.
63 Ibid., pp. 433–4. Söderqvist et al. (2009).
64 Söderqvist et al. (2009), p. 470.
65 Ibid., p. 472.

5 SCIENCE AND THE FIRST WORLD WAR

1 Heilbron, John L. (1974), *H. G. J. Moseley: The Life and Letters of an English Physicist, 1887–1915*, Berkeley: University of California Press.
2 Sarton, George (1927), 'Moseley: the numbering of the elements', *Isis* 9, pp. 96–111.
3 Schütt, Hans-Werner (2003), 'Chemical atomism and chemical classification', in Mary Jo Nye (ed.), *The Cambridge History of Science*, Vol. 5: *The Modern Physical and Mathematical Sciences*, Cambridge: Cambridge University Press, pp. 237–54, at p. 253.
4 North, John David (1989), *The Universal Frame*, London: Hambledon, p. 347.
5 Kevles, Daniel J. (1971), *The Physicists: The History of a Scientific Community in Modern America*, Cambridge, MA: Harvard University Press, p. 113.
6 Sarton (1927), pp. 95–7.
7 Ibid., p. 97.
8 McNeill, William H. (1982), *The Pursuit of Power: Technology, Armed Force, and Society since A.D. 1000*, Chicago: University of Chicago Press, p. 318.
9 Swann, John P. (2009), 'The pharmaceutical industries', in Peter J. Bowler and John V. Pickstone (eds), *The Cambridge History of Science*, Vol. 6: *The Modern Biological and Earth Sciences*, Cambridge: Cambridge University Press, pp. 126–40, at p. 134.
10 Wiebe, Robert H. (1967), *The Search for Order, 1877–1920*, New York: Hill & Wang; Galambos, Louis (1970), 'The emerging organizational synthesis in modern American history', *Business History Review* 44, pp. 279–90; McNeill (1982), p. 317.
11 Turner, Frank M. (1980), 'Public science in Britain, 1880–1919', *Isis* 71, pp. 589–608, at pp. 589, 592.
12 Brock, William H. (1976), 'The spectrum of scientific patronage', in Gerard l'Estrange Turner (ed.), *The Patronage of Science in the Nineteenth Century*, Leyden: Noordhoff, pp. 173–206.
13 Quoted in Turner (1980), 601–2.
14 Quoted ibid., p. 602.
15 Quoted in MacLeod, Roy M., and E. Kay Andrews (1971), 'Scientific advice in the war at sea, 1915–1917: the Board of Invention and Research', *Journal of Contemporary History* 6, pp. 3–40, at p. 5.
16 Quoted ibid., p. 6.
17 For the quotation, see ibid., p. 11.
18 Ibid., p. 13.
19 Wilson, David A. H. (2001), 'Sea lions, greasepaint and the U-boat threat: Admiralty scientists turn to the music hall in 1916', *Notes and Records of the Royal Society of London* 55, pp. 425–55.
20 Hackmann, Willem (1984), *Seek and Strike: Sonar, Anti-Submarine Warfare and the Royal Navy, 1914–54*. London: HMSO.
21 Ibid.
22 MacLeod and Andrews (1971), p. 34.
23 Pattison, Michael (1983), 'Scientists, inventors and the military in Britain,

1915–19: the Munitions Inventions Department', *Social Studies of Science* 13, pp. 521–68.
24 Kevles (1971).
25 Ibid., p. 103.
26 Hughes, Thomas P. (1989), *American Genesis: A Century of Invention and Technological Enthusiasm, 1870–1970*, New York: Penguin, pp. 109, 137.
27 Quoted ibid., p. 119.
28 Quoted in Kevles (1971), p. 109; emphasis in the original.
29 Ibid., p. 110.
30 Ibid., p. 112.
31 Ibid., p. 120.
32 Ibid., p. 124.
33 Ibid., pp. 127–30.
34 Ibid., p. 138.
35 Ash, Mitchell (1995), *Gestalt Psychology in German Culture, 1890–1967: Holism and the Quest for Objectivity*, Cambridge: Cambridge University Press, p. 188.
36 Quoted in Charles, Daniel (2005), *Master Mind: The Rise and Fall of Fritz Haber, the Nobel Laureate who Launched the Age of Chemical Warfare*, New York: Ecco, p. 62.
37 Although, for an interesting rebuttal, see Loeb, Jacques (1917), 'Biology and war', *Science* 45 (26 January), pp. 73–6.
38 Charles (2005), p. 82.
39 Ibid., p. 11.
40 Ibid., p. 31.
41 Ibid., p. 92.
42 Quoted ibid., pp. 92–3.
43 Smil, Vaclav (2000), *Enriching the Earth: Fritz Haber, Carl Bosch, and the Transformation of World Food Production*, Cambridge, MA: MIT Press.
44 Ibid., pp. 155–60. Charles (2005), p. 103.
45 Macrakis, Kristie I. (1993), *Surviving the Swastika: Scientific Research in Nazi Germany*, Oxford: Oxford University Press, p. 17.
46 Quoted in Charles (2005), p. 118.
47 Ibid., p. 120.
48 Coleman, Kim (2005), *A History of Chemical Warfare*, Basingstoke: Palgrave Macmillan, p. 14; Spiers, Edward M. (1986), *Chemical Warfare*, London: Macmillan, p. 14.
49 Charles (2005), p. 156.
50 Ibid., p. 157.
51 Spiers (1986), p. 15.
52 Ibid., p. 16.
53 Nye, Mary Jo (1996), *Before Big Science: The Pursuit of Modern Chemistry and Physics, 1800–1940*, Cambridge, MA: Harvard University Press, p. 193.
54 Charles (2005), pp. 1, 164.
55 Slotten, Hugh R. (1990), 'Humane chemistry or scientific barbarism? American responses to World War I poison gas, 1915–1930', *Journal of American History* 77, pp. 476–98, at p. 486.
56 Charles (2005), p. 169.

57 Russell, Edmund (2001), *War and Nature: Fighting Humans and Insects with Chemicals from World War I to Silent Spring*, Cambridge: Cambridge University Press.
58 Eckert, Michael (2003), 'Plasmas and solid-state science', in Nye, *The Cambridge History of Science*, Vol. 5: *The Modern Physical and Mathematical Sciences*, pp. 413–28, at p. 415.
59 McNeill (1982), p. 334.
60 Ibid., p. 336.
61 Kevles, Daniel J. (1968), 'Testing the army's intelligence: psychologists and the military in World War I', *Journal of American History* 55, pp. 565–81, at p. 566.
62 Quoted ibid., pp. 566–7.
63 Quoted ibid., p. 570.
64 Ibid., p. 572.
65 Quoted in Carson, John (1993), 'Army alpha, army brass, and the search for army intelligence', *Isis* 84, pp. 278–309, at pp. 282–3.
66 Kevles (1968), p. 574.
67 Carson (1993), p. 278.
68 Kevles (1968), p. 576.
69 The full test is reprinted in Carson (1993), p. 303.
70 Samelson, Franz (1977), 'World War I intelligence testing and the development of psychology', *Journal of the History of the Behavioral Sciences* 13, pp. 274–82, at p. 279.
71 Quoted ibid., p. 280.
72 Ibid., p. 277.
73 Quoted ibid., p. 275.
74 Quoted in Kevles (1968), p. 566.
75 Carson (1993), p. 307.
76 Samelson (1977), p. 278.
77 Smith, Roger (1997), *The Fontana History of the Human Sciences*, London: HarperCollins, p. 641.
78 Bogacz, Ted (1989), 'War neurosis and cultural change in England, 1914–22: the work of the War Office Committee of Enquiry into "shell-shock"', *Journal of Contemporary History* 24, pp. 227–56, at p. 232.
79 Rivers, W. H. R. (1919), 'Psychiatry and the war', *Science* 49 (18 April), pp. 367–9, at p. 367.
80 Jones, Ernest (1921), 'War shock and Freud's theory of the neuroses', *International Psycho-Analytical Library* 2, pp. 44–59, at p. 47.
81 Bud, Robert (2009), 'History of biotechnology', in Bowler and Pickstone, *The Cambridge History of Science*, Vol. 6: *The Modern Biological and Earth Sciences*, pp. 524–38.
82 Bud, Robert (1993), *The Uses of Life: A History of Biotechnology*, Cambridge: Cambridge University Press.
83 Bud (2009), p. 531.
84 Higgs, Robert (1987), *Crisis and Leviathan: Critical Episodes in the Growth of American Government*, Oxford: Oxford University Press.
85 Rossiter, Margaret W. (2003), 'A twisted tale: women in the physical sciences in the nineteenth and twentieth centuries', in Nye, *The Cambridge History of Science*, Vol. 5: *The Modern Physical and Mathematical Sciences*, pp. 54–71, at p. 65.

86 Kevles (1971), p. 139.
87 Quoted ibid., p. 141.
88 Tom Wilkie (1991), *British Science and Politics since 1945*, Oxford: Blackwell, p. 30.
89 Roland, Alex (2003), 'Science, technology, and war', in Nye, *The Cambridge History of Science*, Vol. 5: *The Modern Physical and Mathematical Sciences*, pp. 561–78, at pp. 563–4; Gruber, Carol S. (1975), *Mars and Minerva: World War I and the Uses of the Higher Learning in America*, Baton Rouge: Louisiana State University Press.
90 Kevles (1971), p. 141.
91 Ibid., p. 143.
92 Quoted ibid., p. 153.
93 Quoted in Rotblat, Joseph (1979), 'Einstein the pacifist warrior', *Bulletin of the Atomic Scientists*, March, pp. 21–6, at p. 22.
94 Quoted in Stanley, Matthew (2003), '"An expedition to heal the wounds of war": the 1919 eclipse and Eddington as Quaker adventurer', *Isis* 94, pp. 57–89, at p. 58.
95 Ibid., p. 89.
96 Nye (1996), p. 195.
97 Rowe, David E. (2003), 'Mathematical schools, communities and networks', in Nye *The Cambridge History of Science*, Vol. 5: *The Modern Physical and Mathematical Sciences*, pp. 113–32, at p. 128.
98 Hughes, Jeff (2003), 'Radioactivity and nuclear physics', in Nye *The Cambridge History of Science*, Vol. 5: *The Modern Physical and Mathematical Sciences*, pp. 350–74, at p. 360.
99 Rudolph, John L. (2002), 'Portraying epistemology: school science in historical context', *Science Education* 87, pp. 64–79, at p. 70.
100 Dewey in 1916, quoted ibid., p. 69.
101 Wells, F. Lyman (1916), 'Science and war', *Science* 44 (25 August), pp. 275–6.

6 CRISIS

1 Ringer, Fritz K. (1969), *The Decline of the German Mandarins: The German Academic Community, 1890–1933*, Cambridge, MA: Harvard University Press; Forman, Paul (1971), 'Weimar culture, causality, and quantum theory, 1918–1927: adaptation by German physicists and mathematicians to a hostile intellectual environment', *Historical Studies in the Physical Sciences* 3, pp. 1–116, at pp. 26–7.
2 Gay, Peter (1968), *Weimar Culture: The Outsider as Insider*, New York: Harper & Row.
3 Mehra, Jagdish, and Helmut Rechenberg (1982–2001), *The Historical Development of Quantum Theory*, 6 vols, New York: Springer.
4 Darrigol, Olivier (2003), 'Quantum theory and atomic structure, 1900–1927', in Mary Jo Nye (ed.), *The Cambridge History of Science*, Vol. 5: *The Modern Physical and Mathematical Sciences*, Cambridge: Cambridge University Press, pp. 331–49, at pp. 334–5.
5 Ibid., p. 339.
6 Bohr, Niels (1923), 'The structure of the atom', Nobel Prize for Physics

Lecture 1922, *Nature* 112, pp. 29–44, available at www.nobelprize.org/nobel_prizes/physics/laureates/1922/bohr-lecture.pdf, p. 27.

7 Ibid., p. 28.
8 Darrigol (2003), p. 341.
9 Ibid., p. 341.
10 Quoted in Massimi, Michela (2005), *Pauli's Exclusion Principle: The Origin and Validation of a Scientific Principle*, Cambridge: Cambridge University Press, p. 73.
11 Cassidy, David (1992), *Uncertainty: The Life and Science of Werner Heisenberg*, New York: W. H. Freeman.
12 All quotations from Moore, Walter (1994), *A Life of Erwin Schrödinger*, Cambridge: Cambridge University Press, pp. 148–9.
13 Forman (1971), p. 3.
14 Ibid., p. 4.
15 Ibid., pp. 12–13.
16 Ibid., p. 7.
17 Ibid., pp. 58–9.
18 Heisenberg, *The Physical Properties of the Quantum Theory*, 1930, quoted ibid., p. 65.
19 Weyl, *Space–Time–Matter*, 4th edn, 1921, quoted ibid., p. 79.
20 Ibid., p. 99.
21 Ibid., pp. 86, 90.
22 Ibid., p. 94.
23 Ibid., pp. 109–110.
24 Kraft, P., and P. Kroes (1984), 'Adaptation of scientific knowledge to an intellectual environment. Paul Forman's "Weimar culture, causality, and quantum theory, 1918–1927": analysis and criticism', *Centaurus* 27, pp. 76–99, at p. 89.
25 Hendry, John (1980), 'Weimar culture and quantum causality', in Colin Chant and John Fauvel (eds), *Darwin to Einstein: Historical Studies on Science and Belief*, Harlow: Longman, in association with the Open University, pp. 303–26, at p. 303.
26 Ibid., p. 316.
27 Kraft and Kroes (1984), p. 97.
28 Hendry (1980), p. 307.
29 Darrigol (2003), p. 332.
30 Forman (1971), pp. 108–9.
31 See Miller, Arthur I. (2003), 'Imagery and representation in twentieth-century physics', in Nye, *The Cambridge History of Science*, Vol. 5: *The Modern Physical and Mathematical Sciences*, pp. 191–215.
32 Ibid., pp. 200–2.
33 Hendry (1980), p. 316.
34 Faye, Jan (2008), 'Copenhagen interpretation of quantum mechanics', *Stanford Encyclopedia of Philosophy*, at http://plato.stanford.edu/entries/qm-copenhagen/.
35 Einstein, Albert, Boris Podolsky and Nathan Rosen (1935), 'Can quantum-mechanical description of physical reality be considered complete?', *Physical Review* 77, pp. 777–80.
36 Cushing, James (1994), *Quantum Mechanics, Historical Contingency, and the Copenhagen Hegemony*, Chicago: University of Chicago Press.

NOTES TO PP. 131–140

37 Beller, Mara (1999), *Quantum Dialogue: The Making of a Revolution*, Chicago: University of Chicago Press.

38 Grattan-Guinness, Ivor (1997), *The Fontana History of the Mathematical Sciences: The Rainbow of Mathematics*, London: Fontana, p. 567.

39 Ibid., p. 666.

40 Hilbert, quoted in Shapiro, Stewart (ed.) (2005), *The Oxford Handbook of Philosophy of Mathematics and Logic*, New York: Oxford University Press, p. 278.

41 Ash, Mitchell (1995), *Gestalt Psychology in German Culture, 1890–1967: Holism and the Quest for Objectivity*, Cambridge: Cambridge University Press, p. 286.

42 Van Dalen, Dirk (1990), 'The war of the frogs and the mice, or the crisis of the *Mathematische Annalen*', *Mathematical Intelligencer* 12, pp. 17–31.

43 Agar, Jon (2003), *The Government Machine: A Revolutionary History of the Computer*, Cambridge, MA: MIT Press, pp. 69–74.

44 O'Connor. J. J., and E. F. Robertson (2005), 'Bourbaki: the pre-war years', www-history.mcs.st-andrews.ac.uk/HistTopics/Bourbaki_1.html.

45 Galison, Peter (1990), 'Aufbau/Bauhaus: logical positivism and architectural modernism', *Critical Inquiry* 16, pp. 709–52, at p. 725.

46 Quoted ibid., pp. 731–2, emphases removed.

47 Ibid., p. 733.

48 Ibid., pp. 710–11.

49 Smith, Roger (1997), *The Fontana History of the Human Sciences*, London: HarperCollins, pp. 619, 662.

50 Ibid., p. 663.

51 Ash (1995), p. 44.

52 Ibid., pp. 49–50.

53 Ibid., p. 98.

54 Ibid., pp. 106–8.

55 Ibid., p. 148.

56 Ibid., pp. 167, 169.

57 Quoted in Harrington, Anne (1996), *Reenchanted Science: Holism in German Culture from Wilhelm II to Hitler*, Princeton, NJ: Princeton University Press, p. 119.

58 Ash (1995), pp. 294–5.

59 Maulitz, Russell C. (2009), 'Pathology', in Peter J. Bowler and John V. Pickstone (eds), *The Cambridge History of Science*, Vol. 6: *The Modern Biological and Earth Sciences*, Cambridge: Cambridge University Press, pp. 367–81, at p. 376.

60 Quoted in Harrington (1996), p. 59.

61 Quoted in Ash (1995), p. 287.

62 Harrington (1996), p. 60.

63 Harwood, Jonathan (1993), *Styles of Scientific Thought: The German Genetics Community, 1900–1933*, Chicago: University of Chicago Press.

64 Hopwood, Nick (2009), 'Embryology', in Bowler and Pickstone, *The Cambridge History of Science*, Vol. 6: *The Modern Biological and Earth Sciences*, pp. 285–315, at p. 306.

7 SCIENCE AND IMPERIAL ORDER

1 Scott, James C. (1998), *Seeing Like a State: How Certain Schemes to Improve the Human Condition Have Failed*, New Haven, CT: Yale University Press.
2 Amsterdamska, Olga (2009), 'Microbiology', in Peter J. Bowler and John V. Pickstone (eds), *The Cambridge History of Science*, Vol. 6: *The Modern Biological and Earth Sciences*, Cambridge: Cambridge University Press, pp. 316–41, at p. 332.
3 Worboys, Michael (2009), 'Public and environmental health', ibid., pp. 141–63, at p. 154.
4 Ibid., p. 154.
5 Worster, Donald (1985), *Nature's Economy: A History of Ecological Ideas*, Cambridge: Cambridge University Press.
6 Acot, Pascal (2009), 'Ecosystems', in Bowler and Pickstone, *The Cambridge History of Science*, Vol. 6: *The Modern Biological and Earth Sciences*, pp. 451–66, at p. 455.
7 Quoted in Hagen, Joel B. (1992), *An Entangled Bank: The Origins of Ecosystem Ecology*, New Brunswick, NJ: Rutgers University Press, p. 9.
8 Ibid., p. 17.
9 Ibid., p. 21.
10 Worster, Donald (1997), 'The ecology of order and chaos', in Char Miller and Hal Rothman (eds), *Out of the Woods*, Pittsburgh: University of Pittsburgh Press, pp. 3–17, at p. 5.
11 Quoted in Acot (2009), p. 459.
12 Hagen (1992), p. 25.
13 Anker, Peder (2001), *Imperial Ecology: Environmental Order in the British Empire, 1895–1945*. Cambridge, MA: Harvard University Press.
14 Ibid., p. 17.
15 Ibid., p. 35.
16 Quoted ibid., pp. 70–1.
17 Ibid., pp. 60–70.
18 Ibid., p. 126.
19 Quoted ibid., p. 147
20 Ibid., p. 154; Acot (2009), p. 460.
21 Tansley, Arthur G. (1935), 'The use and abuse of vegetational concepts and terms', *Ecology* 16, pp. 284–307, at p. 297, his emphasis.
22 Ibid., p. 303.
23 Anker (2001), p. 89.
24 Ibid., p. 95.
25 Quoted ibid., p. 97.
26 Ibid., p. 77.
27 Quoted in Hagen (1992), p. 52.
28 Anker (2001), p. 102.
29 Worster (1997), pp. 3–4.
30 Schank, Jeffrey C., and Charles Twardy (2009), 'Mathematical models', in Bowler and Pickstone, *The Cambridge History of Science*, Vol. 6: *The Modern Biological and Earth Sciences*, pp. 416–31, at pp. 423–4.
31 Ibid., p. 424; Acot (2009), p. 461.
32 Worboys (2009), p. 153.

33 Packard, Randall (1993), 'The invention of the "tropical worker": medical research and the quest for Central African labor in the South African gold mines, 1903–36', *Journal of African History* 34, pp. 271–92, at p. 273.
34 Worboys (2009), p. 156.
35 Gilfoyle, Daniel (2006), 'Veterinary immunology as colonial science: method and quantification in the investigation of horsesickness in South Africa, c.1905–1945', *Journal of the History of Medicine and Allied Sciences* 61, pp. 26–65.
36 Carpenter, Kenneth J. (2000), *Beriberi, White Rice and Vitamin B: A Disease, a Cause, and a Cure*, Berkeley: University of California Press.
37 Grove, Richard (1995), *Green Imperialism: Colonial Expansion, Tropical Island Edens and the Origins of Environmentalism*, Cambridge: Cambridge University Press.
38 Beinart, William (1989), 'The politics of colonial conservation', *Journal of Southern African Studies* 15, pp. 143–62, at p. 147.
39 Harwood, Jonathan (2009), 'Universities', in Bowler and Pickstone, *The Cambridge History of Science*, Vol. 6: *The Modern Biological and Earth Sciences*, pp. 90–107, at p. 100.
40 Basalla, George (1967), 'The spread of Western science', *Science* 156 (5 May), pp. 611–22.
41 Ibid., p. 616.
42 Ibid., p. 617.
43 'China: in their words', *Nature* 454 (24 July 2008), pp. 399–401.
44 MacLeod, Roy (1987), 'On visiting the "moving metropolis": reflections on the architecture of imperial science', in Nathan Reingold and Marc Rothenberg (eds), *Scientific Colonialism: A Cross-Cultural Comparison*, Washington, DC: Smithsonian Institution Press, pp. 217–49, at p. 225 (orig. pubd 1982 in *Historical Records of Australian Science*).
45 Basalla (1967), p. 620.
46 MacLeod (1987), p. 226.
47 Inkster, Ian (1985), 'Scientific enterprise and the colonial "model": observations on Australian experience in historical context', *Social Studies of Science* 15, pp. 677–704, at p. 688.
48 Pyenson, Lewis (1985), *Cultural Imperialism and Exact Sciences: German Expansion Overseas, 1900–1930*, New York: Peter Lang; Pyenson, Lewis (1989), *Empire of Reason: Exact Sciences in Indonesia, 1840–1940*, Leiden: E. J. Brill; Pyenson, Lewis (1993), *Civilizing Mission: Exact Sciences and French Overseas Expansion, 1830–1940*, Baltimore: Johns Hopkins University Press.
49 Palladino, Paolo, and Michael Worboys (1993), 'Science and imperialism', *Isis* 84, pp. 91–102, at p. 95.
50 MacLeod (1987), p. 240.
51 Quoted ibid.
52 Kumar, Deepak (1997), *Science and the Raj, 1857–1905*, Oxford: Oxford University Press; Arnold, David (2000), *Science, Technology and Medicine in Colonial India*, Cambridge: Cambridge University Press; McCook, Stuart (2002), *States of Nature: Science, Agriculture, and Environment in the Spanish Caribbean, 1760–1940*, Austin: University of Texas Press.
53 MacLeod (1987), p. 245.

8 EXPANDING UNIVERSES

1 Rainger, Ronald (2009), 'Paleontology', in Peter J. Bowler and John V. Pickstone (eds), *The Cambridge History of Science*, Vol. 6: *The Modern Biological and Earth Sciences*, Cambridge: Cambridge University Press, pp. 185–204, at pp. 195–6.

2 Yergin, Daniel (1991), *The Prize: The Epic Quest for Oil, Money, and Power*, New York: Simon & Schuster, p. 36.

3 Quoted in Galbraith, John Kenneth (1958), *The Affluent Society*, Boston: Houghton Mifflin, p. 51.

4 Hughes, Thomas P. (1989), *American Genesis: A Century of Invention and Technological Enthusiasm, 1870–1970*, New York: Penguin, pp. 222, 232.

5 Yergin (1991), p. 49.

6 Ibid.

7 Dennis, Michael Aaron (1987), 'Accounting for research: new histories of corporate laboratories and the social history of American science', *Social Studies of Science 17*, pp. 479–518, at p. 502.

8 Hufbauer, Karl, and John Lankford (1997), 'Hale, George Ellery (1868–1938)', in John Lankford (ed.), *History of Astronomy: An Encyclopedia*, New York: Garland, pp. 249–50, at p. 250.

9 Smith, Robert W. (1982), *The Expanding Universe: Astronomy's 'Great Debate', 1900–1931*, Cambridge: Cambridge University Press, p. 164.

10 Ibid., p. x.

11 Ibid., p. 11.

12 Ibid., pp. 71–4.

13 Ibid., p. 40.

14 Ibid., p. 124.

15 Ibid., p. 17.

16 Ibid., pp. 41–2.

17 Kerszberg, Pierre (1989), *The Invented Universe: The Einstein–De Sitter Controversy (1916–17) and the Rise of Relativistic Cosmology*, Oxford: Clarendon Press.

18 Kragh, Helge (2003), 'Cosmologies and cosmogonies of space and time', in Mary Jo Nye (ed.), *The Cambridge History of Science*, Vol. 5: *The Modern Physical and Mathematical Sciences*, Cambridge: Cambridge University Press, pp. 522–37, at p. 527.

19 Ibid., p. 528.

20 Kay, Lily E. (1993), *The Molecular Vision of Life: Caltech, the Rockefeller Foundation, and the Rise of the New Biology*, Oxford: Oxford University Press, p. 65.

21 Kevles, Daniel J. (1971), *The Physicists: The History of a Scientific Community in Modern America*, Cambridge, MA: Harvard University Press, p. 117.

22 Quoted in Kay (1993), p. 65.

23 Owens, Larry (1997), 'Science in the United States', in John Krige and Dominique Pestre (eds), *Science in the Twentieth Century*, Amsterdam: Harwood Academic Press, pp. 821–37, at p. 832.

24 Kay (1993).

25 Ibid., p. 6.

26 Ibid., p. 29.

27 Ibid., p. 29.

28 Chandler, Alfred D. Jr. (1977), *The Visible Hand: The Managerial Revolution in American Business*, Cambridge, MA: Harvard University Press.
29 Kay (1993), p. 7.
30 Ibid., pp. 23, 220.
31 Quoted ibid., p. 26.
32 Quoted ibid., p. 33.
33 Servos, John W. (1976), 'The knowledge corporation: A. A. Noyes and chemistry at Cal-Tech, 1915–1930', *Ambix* 23, pp. 175–86.
34 Harwood, Jonathan (2009), 'Universities', in Bowler and Pickstone, *The Cambridge History of Science*, Vol. 6: *The Modern Biological and Earth Sciences*, pp. 90–107, at p. 94.
35 Ibid., p. 102.
36 Lucier, Paul (2009), 'Geological industries', in Bowler and Pickstone, *The Cambridge History of Science*, Vol. 6: *The Modern Biological and Earth Sciences*, pp. 108–25, at p. 121.
37 Yergin (1991), p. 25.
38 Ibid., p. 53.
39 Ibid., pp. 58–9.
40 Ibid., p. 59.
41 Ibid., pp. 62–3.
42 Lucier (2009), p. 122.
43 Yergin (1991), p. 111.
44 Ibid.
45 Ibid., p. 194.
46 Ibid., p. 218.
47 Ibid.
48 Greene, Mott T. (2009), 'Geology', in Bowler and Pickstone, *The Cambridge History of Science*, Vol. 6: *The Modern Biological and Earth Sciences*, pp. 167–84, at p. 182.
49 Yergin (1991), p. 219.
50 Harrison, Chris (1996), 'Lucien J. B. LaCoste: portrait of a scientist-inventor', *Earth in Space* 8(9), pp. 12–13; Oldroyd, David (2009), 'Geophysics and geochemistry', in Bowler and Pickstone, *The Cambridge History of Science*, Vol. 6: *The Modern Biological and Earth Sciences*, pp. 395–415, at p. 402.
51 Lucier (2009), p. 122.
52 Ibid., pp. 122–3.
53 Rainger (2009), p. 198.
54 Lucier (2009), p. 123.
55 Marvin, Ursula B. (1973), *Continental Drift: The Evolution of a Concept*, Washington, DC: Smithsonian Institution Press.
56 Oreskes, Naomi, and Ronald E. Doel (2003), 'The physics and chemistry of the earth', in Nye, *The Cambridge History of Science*, Vol. 5: *The Modern Physical and Mathematical Sciences*, pp. 538–57, at p. 543.
57 Ibid., p. 533.
58 Ibid., p. 549.
59 Frankel, Henry (2009), 'Plate tectonics', in Bowler and Pickstone, *The Cambridge History of Science*, Vol. 6: *The Modern Biological and Earth Sciences*, pp. 385–94, at p. 388.

60 Ibid., p. 389; Oreskes and Doel (2003), p. 543.
61 Marvin (1973).
62 Greene (2009), p. 182.
63 Rabinbach, Anson (1992), *The Human Motor: Energy, Fatigue, and the Origins of Modernity*, Berkeley: University of California Press.
64 Kanigel, Robert (1997), *The One Best Way: Frederick Winslow Taylor and the Enigma of Efficiency*, London: Viking.
65 Hughes (1989), p. 190.
66 Ibid., p. 191.
67 Ibid., p. 197.
68 Braverman, Harry (1974), *Labor and Monopoly Capital: The Degradation of Work in the Twentieth Century*, New York: Monthly Review Press.
69 Quoted in Maier, Charles S. (1970), 'Between Taylorism and technocracy: European ideologies and the vision of industrial productivity in the 1920s', *Journal of Contemporary History* 5(2), pp. 27–61, at pp. 31, 33.
70 Ibid., p. 28.
71 Kay (1993), p. 34.
72 Pauly, Philip J. (1987), *Controlling Life: Jacques Loeb and the Engineering Ideal in Biology*, Oxford: Oxford University Press.
73 Kay (1993), pp. 33–4.
74 Smith, Roger (1997), *The Fontana History of the Human Sciences*, London: HarperCollins, p. 608.
75 Kremer, Richard L. (2009), 'Physiology', in Bowler and Pickstone, *The Cambridge History of Science*, Vol. 6: *The Modern Biological and Earth Sciences*, pp. 342–66, at p. 355.
76 Smith (1997), p. 610.
77 Gillespie, Richard (1991), *Manufacturing Knowledge: A History of the Hawthorne Experiments*, Cambridge: Cambridge University Press; Smith (1997), pp. 771–3.
78 Hughes (1989), p. 203.
79 Bocking, Stephen A. (2009), 'Environmentalism', in Bowler and Pickstone, *The Cambridge History of Science*, Vol. 6: *The Modern Biological and Earth Sciences*, pp. 602–21, at p. 609.
80 Sicherman, Barbara (1984), *Alice Hamilton: A Life in Letters*. Cambridge, MA: Harvard University Press.
81 Maier (1970), p. 28.
82 James Moore, 'Religion and science', in Bowler and Pickstone, *The Cambridge History of Science*, Vol. 6: *The Modern Biological and Earth Sciences*, pp. 541–62, at p. 559.
83 Ibid., p. 559.
84 LaFollette, Marcel Chotkowski (2008), *Reframing Scopes: Journalists, Scientists, and Lost Photographs from the Trial of the Century*, Lawrence: University Press of Kansas.

9 REVOLUTIONS AND MATERIALISM

1 Graham, Loren R. (1998), *What Have We Learned about Science and Technology from the Russian Experience?* Stanford, CA: Stanford University Press.

2 Graham, Loren R. (1966), *Science and Philosophy in the Soviet Union*, New York: Alfred A. Knopf; Graham, Loren R. (1993a), *Science in Russia and the Soviet Union*, Cambridge: Cambridge University Press; Graham (1998); Alexandrov, Daniel A. (1993), 'Communities of science and culture in Russian science studies', *Configurations* 1, pp. 323–33; Holloway, David (1994), *Stalin and the Bomb: The Soviet Union and Atomic Energy, 1939–1956*, New Haven, CT: Yale University Press; Krementsov, Nikolai (1997), *Stalinist Science*, Princeton, NJ: Princeton University Press.
3 Graham (1993a), p. 82.
4 Ibid., p. 80.
5 Ibid., pp. 84–6.
6 Ibid., p. 87.
7 Quoted ibid., pp. 89–90.
8 Andrews, James T. (2003), *Science for the Masses: The Bolshevik State, Public Science, and the Popular Imagination in Soviet Russia, 1917–1934*, College Station: Texas A&M University Press.
9 Graham, Loren R. (1993b), *The Ghost of the Executed Engineer: Technology and the Fall of the Soviet Union*, Cambridge, MA: Harvard University Press.
10 Graham (1993a), p. 92.
11 Ibid., p. 93.
12 Ibid., p. 100.
13 Ibid., p. 101.
14 Joravsky, David (1961), *Soviet Marxism and Natural Science, 1917–1931*, New York: Columbia University Press; Josephson, Paul R. (1996), *Totalitarian Science and Technology*, Atlantic Highlands, NJ: Humanities Press, p. 51.
15 Graham (1966), p. 42.
16 Graham (1993a), pp. 93, 112.
17 Krementsov (1997), p. 29.
18 Bailes, Kendall E. (1990), *Science and Russian Culture in an Age of Revolutions: V. I. Vernadsky and his Scientific School, 1863–1945*, Bloomington: Indiana University Press.
19 Graham (1993a), p. 87.
20 Quoted ibid., p. 230.
21 Weart, Spencer R. (2003), *The Discovery of Global Warming*, Cambridge, MA: Harvard University Press, p. 15.
22 Weiner, Douglas R. (1982), 'The historical origins of Soviet environmentalism', *Environmental Review* 6(2), pp. 42–62; Weiner, Douglas R. (1988); *Models of Nature: Ecology, Conservation, and Cultural Revolution in Soviet Russia*, Bloomington: Indiana University Press.
23 Smith, Roger (1997), *The Fontana History of the Human Sciences*, London: HarperCollins, p. 834.
24 Graham (1993a), p. 107.
25 Vygotsky, Lev Semenovich (1962), *Thought and Language*, Cambridge, MA: MIT Press, quoted in Graham (1966), pp. 370–1.
26 Quoted ibid., p. 372; see also Graham (1993a), p. 106.
27 Graham (1993a), p. 103.
28 Ruse, Michael (2000), *The Evolution Wars: A Guide to the Debates*, Oxford: ABC-CLIO, p. 158.

29 Ibid., p. 159.
30 Graham (1966), p. 263.
31 Werskey, Gary (1978), *The Visible College*, London: Allen Lane, p. 158.
32 Graham (1993a), pp. 110–11.
33 Haldane, John Burdon Sanderson (1923), *Daedalus, or Science and the Future*, London: Kegan Paul.
34 Gaudillière, Jean-Paul (2009), 'Cancer', in Peter J. Bowler and John V. Pickstone (eds), *The Cambridge History of Science*, Vol. 6: *The Modern Biological and Earth Sciences*, Cambridge: Cambridge University Press, pp. 486–503, at p. 496.
35 Werskey (1978), p. 86.
36 Squier, Susan Merrill (1995), *Babies in Bottles: Twentieth-Century Visions of Reproductive Technology*, New Brunswick, NJ: Rutgers University Press.
37 Haldane (1923).
38 Bowler, Peter J. (1983), *The Eclipse of Darwinism: Anti-Darwinian Evolution Theories in the Decades around 1900*, Baltimore: Johns Hopkins University Press.
39 Mayr, Ernst, and William B. Provine (eds) (1980), *The Evolutionary Synthesis: Perspectives on the Unification of Biology*, Cambridge, MA: Harvard University Press.
40 Krementsov (1997), pp. 21–2.
41 Graham (1993a), p. 127.
42 Cittadino, Eugene (2009), 'Botany', in Bowler and Pickstone, *The Cambridge History of Science*, Vol. 6: *The Modern Biological and Earth Sciences*, pp. 225–42, at p. 240.
43 Medvedev, Zhores (1979), *Soviet Science*, Oxford: Oxford University Press, p. 13.
44 Graham (1993b), p. 61.
45 Graham (1993a), p. 100.
46 Quoted in Krementsov (1997), pp. 29–30.
47 Josephson, Paul R. (2003), 'Science, ideology and the state: physics in the twentieth century', in Mary Jo Nye (ed.), *The Cambridge History of Science*, Vol. 5: *The Modern Physical and Mathematical Sciences*, Cambridge: Cambridge University Press, pp. 579–97, at p. 583.
48 Graham (1993a), pp. 147–9.
49 Ibid., p. 149.
50 Graham (1993b), p. 43.
51 Quoted ibid.
52 Graham (1993a), p. 107.
53 Wiener (1982), p. 51.
54 Joravsky, David (1970), *The Lysenko Affair*, Cambridge, MA: Harvard University Press.
55 Graham (1993a), pp. 125–6.
56 Quoted ibid., p. 128.
57 Joravsky (1970), p. 116; Graham (1993a), p. 130.
58 Adams, Mark B. (1980), 'Sergei Chetverikov, the Kol'tsov Institute and the evolutionary synthesis', in Mayr and Provine, *The Evolutionary Synthesis: Perspectives on the Unification of Biology*, pp. 242–78.
59 Ibid., p. 245.

60 Todes, Daniel P. (1989), *Darwin without Malthus: The Struggle for Existence in Russian Evolutionary Thought*, Oxford: Oxford University Press; Graham (1993a), p. 67.
61 Adams (1980), p. 246.
62 Ibid.
63 Ibid., p. 257; Kohler, Robert E. (1994), *Lords of the Fly: Drosophila Genetics and the Experimental Life*, Chicago: University of Chicago Press.
64 Adams (1980), p. 262.
65 Ibid., p. 270.
66 Ibid., pp. 265–6.
67 Ibid., pp. 267–8.
68 Dobzhansky, Theodosius (1980), 'The birth of the genetic theory of evolution in the Soviet Union in the 1920s', in Mayr and Provine, *The Evolutionary Synthesis: Perspectives on the Unification of Biology*, pp. 229–42, at p. 235.
69 Mayr, Ernst (1980), 'Prologue: some thoughts on the history of the Evolutionary Synthesis', ibid., pp. 1–48, at p. 41.
70 Ibid., pp. 6, 40.
71 Ibid., p. 8.
72 Ibid., p. 11.
73 Ibid., p. 27.
74 Ibid., p. 13; Rensch, Bernhard (1980), 'Historical development of the present synthetic neo-Darwinism in Germany', in Mayr and Provine, *The Evolutionary Synthesis: Perspectives on the Unification of Biology*, pp. 284–303, at p. 290.
75 Quoted in Mayr (1980), p. 25.
76 Adams, Mark B. (1994), *The Evolution of Theodosius Dobzhansky: Essays on his Life and Thought in Russia and America*, Princeton, NJ: Princeton University Press.
77 Mayr (1980), p. 30.
78 Hodge, Jonathan (2009), 'The evolutionary sciences', in Bowler and Pickstone, *The Cambridge History of Science*, Vol. 6: *The Modern Biological and Earth Sciences*, pp. 243–64, at p. 260.
79 Huxley, Julian (1942), *Evolution: The Modern Synthesis*. London: George Allen & Unwin, p. 13.
80 Ibid., p. 8.
81 Morrell, Jack B. (1997), *Science at Oxford, 1914–1939: Transforming an Arts University*, Oxford: Clarendon Press; Harwood, Jonathan (2009), 'Universities', in Bowler and Pickstone, *The Cambridge History of Science*, Vol. 6: *The Modern Biological and Earth Sciences*, pp. 90–107, at p. 106.
82 Olby, Robert (1992), 'Huxley's place in twentieth-century biology', in C. Kenneth Waters and Albert Van Helden (eds), *Julian Huxley: Biologist and Statesman of Science*, Houston: Rice University Press.
83 Allen, David E. (2009), 'Amateurs and professionals', in Bowler and Pickstone, *The Cambridge History of Science*, Vol. 6: *The Modern Biological and Earth Sciences*, pp. 15–33, at p. 32.
84 Huxley (1942), p. 565.
85 Mayr, Ernst (1942), *Systematics and the Origin of Species from the Viewpoint of a Zoologist*, New York: Columbia University Press.
86 Winsor, Mary P. (2009), 'Museums', in Bowler and Pickstone, *The*

Cambridge History of Science, Vol. 6: *The Modern Biological and Earth Sciences*, pp. 60–75, at p. 75.

87 Gould, Stephen Jay (1980), 'G. G. Simpson, paleontology and the Modern Synthesis', in Mayr and Provine, *The Evolutionary Synthesis: Perspectives on the Unification of Biology*, pp. 153–72, at p. 154.

88 Laporte, Léo (1991), 'George G. Simpson, paleontology, and the expansion of biology', in Keith R. Benson, Jane Maienschein and Ronald Rainger (eds), *The Expansion of American Biology*, New Brunswick, NJ: Rutgers University Press, pp. 80–106; Rainger, Ronald (2009), 'Paleontology', in Bowler and Pickstone, *The Cambridge History of Science*, Vol. 6: *The Modern Biological and Earth Sciences*, pp. 185–204, at p. 198.

89 Rainger (2009), p. 198.

90 Mayr (1980), p. 29.

91 Cain, Joe (2009), 'Rethinking the synthesis period in evolutionary studies', *Journal of the History of Biology* 42, pp. 621–48.

92 Benson, Keith R. (2009), 'Field stations and surveys', in Bowler and Pickstone, *The Cambridge History of Science*, Vol. 6: *The Modern Biological and Earth Sciences*, pp. 76–89, at p. 87.

93 Schank, Jeffrey C., and Charles Twardy (2009), 'Mathematical models in the life sciences', ibid., pp. 416–31, at p. 422.

94 Huxley (1942), p. 116; Mayr (1980), p. 30.

95 Schank and Twardy (2009), p. 423.

96 Mayr (1980), pp. 30–2.

97 Ibid., p. 41.

98 Harwood (2009), p. 106.

10 NAZI SCIENCE

1 Taton, René (ed.) (1966), *Science in the Twentieth Century*, London: Thames & Hudson, p. 571 (orig. pubd as *La Science contemporaine*, II: *Le XXe Siècle*, Paris: Presses Universitaires de France, 1964).

2 Stern, Fritz (1961), *The Politics of Cultural Despair: A Study in the Rise of the Germanic Ideology*, Berkeley: University of California Press.

3 Herf, Jeffrey (1984), *Reactionary Modernism: Technology, Culture and Politics in Weimar and the Third Reich*, Cambridge: Cambridge University Press.

4 Proctor, Robert N. (1988), *Racial Hygiene: Medicine under the Nazis*, Cambridge, MA: Harvard University Press; Proctor, Robert N. (1999), *The Nazi War on Cancer*, Princeton, NJ: Princeton University Press; Proctor, Robert N. (2000), 'Nazi science and Nazi medical ethics: some myths and misconceptions', *Perspectives in Biology and Medicine* 43(3), pp. 335–46.

5 Proctor (1999).

6 Proctor (2000), p. 336.

7 Junker, Thomas, and Uwe Hossfeld (2002), 'The architects of the Evolutionary Synthesis in National Socialist Germany: science and politics', *Biology and Philosophy* 17, pp. 223–49.

8 Beyerchen, Alan D. (1977), *Scientists under Hitler: Politics and the Physics Community in the Third Reich*, New Haven, CT: Yale University Press.

9 Ash, Mitchell (1995), *Gestalt Psychology in German Culture, 1890–1967*:

Holism and the Quest for Objectivity, Cambridge: Cambridge University Press, p. 327.

10 Ibid., p. 329.
11 Quoted ibid., p. 330.
12 Ibid., p. 345.
13 Quoted in Harrington, Anne (1996), *Reenchanted Science: Holism in German Culture from Wilhelm II to Hitler*, Princeton, NJ: Princeton University Press, p. 178.
14 Quoted in Macrakis, Kristie I. (1993), *Surviving the Swastika: Scientific Research in Nazi Germany*, Oxford: Oxford University Press, p. 57; Ash (1995), p. 331.
15 Josephson, Paul R. (1996), *Totalitarian Science and Technology*, Atlantic Highlands, NJ: Humanities Press, p. 56; see also Beyerchen (1977).
16 Josephson, Paul R. (2003), 'Science, ideology and the state: physics in the twentieth century', in Mary Jo Nye (ed.), *The Cambridge History of Science*, Vol. 5: *The Modern Physical and Mathematical Sciences*, Cambridge: Cambridge University Press, pp. 579–97, at p. 587.
17 Beyerchen (1977); Josephson (1996), p. 58.
18 Josephson (2003), p. 586.
19 Josephson (1996), p. 60.
20 Ibid., p. 93.
21 Josephson (2003), p. 590; Walker, Mark (1989b), 'National Socialism and German physics', *Journal of Contemporary History* 24(1), pp. 63–89.
22 Trischler, Helmuth (1994), 'Self-mobilization or resistance? Aeronautical research and National Socialism', in Monika Renneberg and Mark Walker (eds), *Science, Technology and National Socialism*, Cambridge: Cambridge University Press, pp. 72–87, at p. 75.
23 Trischler, Helmuth (2001), 'Aeronautical research under National Socialism: Big Science or small science?', in Margit Szöllösi-Janze (ed.), *Science in the Third Reich*, Oxford: Berg, pp. 79–110, at p. 88.
24 Ceruzzi, Paul (1983), *Reckoners: The Prehistory of the Digital Computer, from Relays to the Stored Program Concept, 1935–1945*, Westport, CT: Greenwood Press; Williams, Michael (1985), *A History of Computing Technology*, Englewood Cliffs, NJ: Prentice-Hall, 1985; Ceruzzi, Paul (1990), 'Relay calculators', in William Aspray (ed.), *Computing before Computers*, Ames: Iowa State University Press, pp. 200–22.
25 Quoted in Wise, M. Norton (1994), 'Pascual Jordan: quantum mechanics, psychology, National Socialism', in Renneberg and Walker, *Science, Technology and National Socialism*, pp. 224–54, at pp. 226, 234.
26 Wise (1994), p. 244.
27 Deichmann, Ute (1996), *Biologists under Hitler*, Cambridge, MA: Harvard University Press, p. 323.
28 Proctor (1988), p. 6.
29 Ibid., p. 14.
30 Ibid., p. 17.
31 Ibid., p. 18.
32 Paul, Diane B. (1995), *Controlling Human Heredity: 1865 to the Present*, Atlantic Highlands, NJ: Humanities Press.
33 Proctor (1988), p. 20.
34 Ibid., p. 26.

35 Ibid., p. 27.
36 Berez, Thomas M., and Sheila Faith Weiss (2004), 'The Nazi symbiosis: politics and human genetics at the Kaiser Wilhelm Institute', *Endeavour* 28(4), pp. 172–7.
37 Proctor (1988), p. 41.
38 Ibid., p. 42.
39 Ibid., pp. 50–5.
40 Ibid., pp. 54–5.
41 Ibid., p. 60.
42 Quoted ibid., p. 62.
43 Ibid., p. 30.
44 Quoted ibid., pp. 37–8.
45 Quoted ibid., p. 84.
46 Ibid., p. 85.
47 Ibid., p. 62.
48 Ibid., p. 144.
49 Ibid., pp. 65–6.
50 Ibid., pp. 75–6.
51 Quoted ibid., p. 78.
52 Ibid., p. 117.
53 Ibid., p. 106.
54 Ibid., pp. 108–9.
55 Ibid., p. 132.
56 Ibid., p. 193.
57 Ibid., pp. 204–5.
58 Ibid., p. 207.
59 Ibid., p. 3.
60 Hughes Thomas P. (1969), 'Technological momentum in history: hydrogenation in Germany, 1898–1933', *Past and Present* 44, pp. 106–32.
61 Proctor (1988), pp. 4–5.
62 Allen, William Sheridan (1966), *The Nazi Seizure of Power: The Experience of a Single German Town, 1930–1935*, London: Eyre & Spottiswoode.
63 Ash, Mitchell, and Alfons Söllner (eds) (1996), *Forced Migration and Scientific Change: Émigré German-Speaking Scientists and Scholars after 1933*, Cambridge: Cambridge University Press, p. 7.
64 Kevles, Daniel J. (1971), *The Physicists: The History of a Scientific Community in Modern America*, Cambridge, MA: Harvard University Press, p. 281.
65 Charles, Daniel (2005), *Master Mind: The Rise and Fall of Fritz Haber, the Nobel Laureate who Launched the Age of Chemical Warfare*, New York: Ecco, p. 223.
66 Ibid., p. 263.
67 Geuter, Ulfried (1992), *The Professionalization of Psychology in Nazi Germany*, Cambridge: Cambridge University Press; Ash and Söllner (1996), p. 9.
68 Deichmann (1996), p. 320.
69 Ibid., pp. 320, 324.
70 Walker, Mark (1989a), *German National Socialism and the Quest for Nuclear Power, 1939–1949*, Cambridge: Cambridge University Press; Deichmann (1996).

71 Ash and Söllner (1996), p. 4.
72 Hoch, Paul K. (1987), 'Migration and the generation of scientific ideas', *Minerva* 25, pp. 209–37.
73 Hoch, Paul K., and Jennifer Platt (1993), 'Migration and the denationalization of science', in Elizabeth Crawford, Terry Shinn and Sverker Sörlin (eds), *Denationalizing Science: The Contexts of International Scientific Practice*, Amsterdam: Kluwer Academic, pp. 133–52; Ash and Söllner (1996), p. 5.
74 Kevles (1971), p. 281.

CHAPTER 11 SCALING UP, SCALING DOWN

1 Hughes, Jeff (2003), 'Radioactivity and nuclear physics', in Mary Jo Nye (ed.), *The Cambridge History of Science*, Vol. 5: *The Modern Physical and Mathematical Sciences*, Cambridge: Cambridge University Press, pp. 350–74, at p. 364.
2 Kevles, Daniel J. (1971), *The Physicists: The History of a Scientific Community in Modern America*, Cambridge, MA: Harvard University Press, p. 282.
3 Ibid., pp. 180–2.
4 Ibid., p. 237.
5 Kuznick, Peter J. (1987), *Beyond the Laboratory: Scientists as Political Activists in 1930s America*, Chicago: University of Chicago Press; Josephson, Paul R. (2003), 'Science, ideology and the state: physics in the twentieth century', in Nye, *The Cambridge History of Science*, Vol. 5: *The Modern Physical and Mathematical Sciences*, pp. 579–97, at p. 590.
6 Layton, Edwin, T., Jr. (1971), *The Revolt of the Engineers: Social Responsibility and the American Engineering Profession*, Cleveland: Case Western Reserve University Press.
7 Kevles (1971), p. 378.
8 Hughes, Thomas P. (1989), *American Genesis: A Century of Invention and Technological Enthusiasm, 1870–1970*, New York: Penguin, p. 353.
9 Quoted ibid., p. 378.
10 Gaudillière, Jean-Paul (2009), 'Cancer', in Peter J. Bowler and John V. Pickstone (eds), *The Cambridge History of Science*, Vol. 6: *The Modern Biological and Earth Sciences*, Cambridge: Cambridge University Press, pp. 486–503, at p. 493.
11 Simões, Ana (2003), 'Chemical physics and quantum chemistry in the twentieth century', in Nye, *The Cambridge History of Science*, Vol. 5: *The Modern Physical and Mathematical Sciences*, pp. 394–412, at pp. 404–5.
12 Seidel, Robert (1992), 'The origins of the Lawrence Berkeley Laboratory', in Peter Galison and Bruce Hevly (eds), *Big Science: The Growth of Large-Scale Research*, Stanford, CA: Stanford University Press, pp. 21–45, at pp. 23–4.
13 Ibid., p. 25.
14 Ibid, pp. 26–7.
15 Quoted in AIP (American Institute of Physics) (n.d.), 'Ernest Lawrence and the Cyclotron', www.aip.org/history/lawrence/ (accessed 2009).
16 Seidel (1992), p. 27.

17 AIP (n.d.).
18 Seidel (1992), p. 28.
19 Krige, John, and Dominique Pestre (1992), 'Some thoughts on the early history of CERN', in Galison and Hevly, *Big Science: The Growth of Large-Scale Research*, pp. 78–99, at p. 93, their emphasis removed.
20 Ibid., p. 94.
21 Seidel (1992), pp. 33–4.
22 Nye, Mary Jo (2004), *Blackett: Physics, War, and Politics in the Twentieth Century*, Cambridge, MA: Harvard University Press, p. 44.
23 Ibid., p. 45.
24 Hughes, Jeff (2002), *The Manhattan Project: Big Science and the Atom Bomb*, Cambridge: Icon Books, p. 34.
25 Hughes (2003), p. 365.
26 AIP (n.d.).
27 Kauffman, George B. (2002), 'In memoriam Martin D. Kamen (1913–2002), nuclear scientist and biochemist', *Chemical Educator* 7, pp. 304–8, at p. 305.
28 Hughes (2003), p. 368.
29 Bird, Kai, and Martin J. Sherwin (2005), *American Prometheus: The Triumph and Tragedy of J. Robert Oppenheimer*, London: Atlantic Books, pp. 9–10.
30 Quoted ibid., p. 22.
31 Quoted ibid., p. 39.
32 Ibid., p. 46; Nye (2004), p. 44.
33 Bird and Sherwin (2005), p. 96.
34 Ibid., p. 87.
35 Ibid., p. 88.
36 Quoted in Schweber, Silvan S. (2003), 'Quantum field theory: from QED to the Standard Model', in Nye, *The Cambridge History of Science*, Vol. 5: *The Modern Physical and Mathematical Sciences*, pp. 375–93, at p. 379.
37 Ibid., p. 380.
38 Brown, Laurie M. (1986), 'Hideki Yukawa and the meson theory', *Physics Today* 39(12), pp. 55–62.
39 Rechenberg, Helmut, and Laurie M. Brown (1990), 'Yukawa's heavy quantum and the mesotron (1935–1937)', *Centaurus* 33, pp. 214–52, at p. 216.
40 Nye (2004), p. 29; Rechenberg and Brown (1990), p. 226.
41 Quoted in Rechenberg and Brown (1990), p. 228.
42 Ibid., p. 229.
43 Ibid., p. 236.
44 Hager, Thomas (1995), *Force of Nature: The Life of Linus Pauling*, New York: Simon & Schuster, p. 35.
45 Ibid., p. 106.
46 Quoted ibid., p. 133.
47 Ibid., pp. 135–6.
48 Ibid., p. 147.
49 Ibid., p. 169.
50 Nye, Mary Jo (1993), *From Chemical Philosophy to Theoretical Chemistry: Dynamics of Matter and Dynamics of Disciplines, 1800–1950*, Berkeley: University of California Press; Hager (1995), p. 164; Simões (2003), p. 397.

51 Quoted in Hager (1995), p. 169.
52 Quoted ibid., p. 217.
53 Furukawa, Yasu (2003), 'Macromolecules: their structures and functions', in Nye, *The Cambridge History of Science*, Vol. 5: *The Modern Physical and Mathematical Sciences*, pp. 429–45, at p. 431.
54 Ibid., p. 434.
55 Ibid., p. 433.
56 Elzen, Boelie (1986), 'Two ultracentrifuges: a comparative study of the social construction of artefacts', *Social Studies of Science* 16, pp. 621–62, at p. 630.
57 Ibid., p. 642.
58 Quoted in Furukawa (2003), p. 435.
59 Ibid., p. 440.
60 Quoted in Hager (1995), p. 183.
61 Kevles (1971), p. 120.
62 Kay, Lily E. (1993), *The Molecular Vision of Life: Caltech, the Rockefeller Foundation, and the Rise of the New Biology*, Oxford: Oxford University Press, p. 8.
63 Quoted ibid., p. 45.
64 Quoted ibid.
65 Ibid., p. 9.
66 Servos, John W. (1976), 'The knowledge corporation: A. A. Noyes and chemistry at Cal-Tech, 1915–1930', *Ambix* 23, pp. 175–86.
67 Kay (1993), p. 69.
68 Ibid., p. 82.
69 Quoted ibid., p. 106.
70 Ibid., p. 90.
71 Ibid., p. 111.
72 Ibid., p. 110.
73 Beadle, G. W., and E. L. Tatum (1941), 'Genetic control of biochemical reactions in *Neurospora*', *Proceedings of the National Academy of Sciences of the United States of America* 27(11), pp. 499–506, at p. 499.
74 Hager (1995), p. 277.
75 Furukawa (2003), p. 439.
76 Kay (1993), p. 3.
77 Fleming, Donald (1969), 'Émigré physicists and the biological revolution', in Donald Fleming and Bernard Bailyn (eds), *The Intellectual Migration: Europe and America, 1930–1960*, Cambridge, MA: Harvard University Press, pp. 152–89.
78 Judson, Horace Freeland (1979), *The Eighth Day of Creation: Makers of the Revolution in Biology*, London: Penguin; Schank, Jeffrey C., and Charles Twardy (2009), 'Mathematical models in the life sciences', in Bowler and Pickstone, *The Cambridge History of Science*, Vol. 6: *The Modern Biological and Earth Sciences*, pp. 416–31, at p. 426.
79 Quoted in Judson (1979), p. 49.
80 Fleming (1969), p. 179.
81 Kay (1993), pp. 4–5.
82 Gaudillière (2009), p. 496.
83 De Duve, Christian, and George E. Palade (1983), 'Obituary: Albert Claude, 1899–1983', *Nature* 304 (18 August), p. 588.

84 Peter Morris (ed.) (2002), *From Classical to Modern Chemistry: The Instrumental Revolution*, Cambridge: Royal Society of Chemistry.
85 Baird, Davis (2002), 'Analytical chemistry and the "big" scientific instrumentation revolution', repr. ibid., pp. 29–56, at p. 29.
86 Ibid., p. 40.
87 Gaudillière (2009), p. 490.
88 Kremer, Richard L. (2009), 'Physiology', in Bowler and Pickstone, *The Cambridge History of Science*, Vol. 6: *The Modern Biological and Earth Sciences*, pp. 342–66.
89 Lederer, Susan E. (2009), 'Experimentation and ethics', ibid., pp. 583–601, at p. 590.
90 Bliss, Michael (1992), *Banting: A Biography*, Toronto: University of Toronto Press, p. 96.
91 Hounshell, David A., and John Kenly Smith (1988), *Science and Corporate Strategy: Du Pont R&D, 1902–1980*, Cambridge: Cambridge University Press; Hounshell, David A. (1992), 'Du Pont and the management of large-scale research and development', in Galison and Hevly, *Big Science: The Growth of Large-Scale Research*, pp. 236–61, at p. 238; Meikle, Jeffrey L. (1995), *American Plastic: A Cultural History*, New Brunswick, NJ: Rutgers University Press.
92 Hughes (1989), pp. 175–6.
93 Ibid., p. 177.
94 Ibid.
95 Hounshell (1992), p. 239.
96 Ibid., p. 241.
97 Ibid., pp. 243–4.
98 Ibid., p. 244.
99 Quoted in Cogdell, Christina (2000), 'The Futurama recontextualized: Norman Bel Geddes's eugenic "World of Tomorrow"', *American Quarterly* 52, pp. 193–245, at p. 225.
100 Barbrook, Richard (2007), *Imaginary Futures: From Thinking Machines to the Global Village*, London: Pluto Press.
101 Shinn, Terry (2003), 'The industry, research, and education nexus', in Nye, *The Cambridge History of Science*, Vol. 5: *The Modern Physical and Mathematical Sciences*, pp. 133–53, at p. 150.
102 For an opposite view, see Schabas, Margaret (2003), 'British economic theory from Locke to Marshall', in Theodore M. Porter and Dorothy Ross (eds), *The Cambridge History of Science*, Vol. 7: *The Modern Social Sciences*, Cambridge: Cambridge University Press, pp. 171–82, at p. 172.
103 Brown, Andrew (2007), *J. D. Bernal: The Sage of Science*, Oxford: Oxford University Press.

CHAPTER 12 SCIENCE AND THE SECOND WORLD WAR

1 Russell, Edmund (2001), *War and Nature: Fighting Humans and Insects with Chemicals from World War I to Silent Spring*, Cambridge: Cambridge University Press, p. 142.
2 Roland, Alex (2003), 'Science, technology, and war', in Mary Jo Nye (ed.), *The Cambridge History of Science*, Vol. 5: *The Modern Physical and*

Mathematical Sciences, Cambridge: Cambridge University Press, pp. 561–78, at p. 565.

3 Dennis, Michael Aaron (1987), 'Accounting for research: new histories of corporate laboratories and the social history of American science', *Social Studies of Science* 17, pp. 479–518, at p. 511.

4 Macrakis, Kristie I. (1993), *Surviving the Swastika: Scientific Research in Nazi Germany*, Oxford: Oxford University Press, p. 90.

5 Shinn, Terry (2003), 'The industry, research, and education nexus', in Nye, *The Cambridge History of Science*, Vol. 5: *The Modern Physical and Mathematical Sciences*, pp. 133–53, at p. 143.

6 McGucken, William (1984), *Scientists, Society and the State: The Social Relations of Science Movement in Great Britain, 1931–1947*, Columbus: Ohio State University Press, p. 196.

7 Kevles, Daniel J. (1971), *The Physicists: The History of a Scientific Community in Modern America*, Cambridge, MA: Harvard University Press, p. 296.

8 Ibid., p. 293.

9 Quoted in Mindell, David A. (2002), *Between Human and Machine: Feedback, Control, and Computing before Cybernetics*, Baltimore: Johns Hopkins University Press, pp. 187–8.

10 Kevles (1971), p. 298.

11 Sherry, Michael S. (1987), *The Rise of American Air Power: The Creation of Armageddon*, New Haven, CT: Yale University Press, p. 188.

12 Dupree, A. Hunter (1970), 'The Great Instauration of 1940: the organization of scientific research for war', in Gerald Holton (ed.), *The Twentieth-Century Sciences*, New York: W. W. Norton, pp. 443–67.

13 Mindell (2002), p. 190.

14 Quoted in Kevles (1971), p. 301; Pursell, Carroll W. (1979), 'Science agencies in World War II: the OSRD and its challengers', in Nathan Reingold (ed.), *The Sciences in the American Context*, Washington, DC: Smithsonian Institution, pp. 287–301.

15 Hartcup, Guy (2000), *The Effect of Science on the Second World War*, Basingstoke: Palgrave Macmillan, p. 123.

16 Ibid., p. 124.

17 Bud, Robert (2007), *Penicillin: Triumph and Tragedy*, Oxford: Oxford University Press.

18 Swann, John P. (2009), 'The pharmaceutical industries', in Peter J. Bowler and John V. Pickstone (eds), *The Cambridge History of Science*, Vol. 6: *The Modern Biological and Earth Sciences*, Cambridge: Cambridge University Press, pp. 126–40, at p. 135.

19 Bud, Robert (2009), 'History of biotechnology', ibid., pp. 524–38, at p. 533.

20 Hartcup (2000), p. 125.

21 Parascandola, John (ed.) (1980), *The History of Antibiotics: A Symposium*, Madison: American Institute of the History of Pharmacy; Bud (2009).

22 Gall, Yasha M., and Mikhail B. Konashev (2001), 'The discovery of Gramicidin S: the intellectual transformation of G. F. Gause from biologist to researcher of antibiotics and on its meaning for the fate of Russian genetics', *History and Philosophy of the Life Sciences* 23, pp. 137–50.

23 Rowe, A. P. (1948), *One Story of Radar*, Cambridge: Cambridge University

Press; Calder, Angus (1969), *The People's War: Britain, 1939–45*, London: Jonathan Cape, pp. 457–77; Bowen, Edward G. (1987), *Radar Days*, Bristol: Adam Hilger; Buderi, Robert (1996), *The Invention that Changed the World: How a Small Group of Radar Pioneers Won the Second World War and Launched a Technological Revolution*, New York: Simon & Schuster.

24 Agar, Jon (2003), *The Government Machine: A Revolutionary History of the Computer*, Cambridge, MA: MIT Press, p. 210.
25 Ibid., p. 216.
26 Hartcup (2000), p. 9.
27 Ibid., p. 37.
28 Agar, Jon, and Jeff Hughes (1995), 'Between government and industry: academic scientists and the reconfiguration of research practice at TRE', paper for CHIDE, conference at Bournemouth University.
29 Rowe (1948), p. 179.
30 Hartcup (2000), p. 102.
31 Quoted in Kirby, Maurice W. (2003), *Operational Research in War and Peace: The British Experience from the 1930s to 1970*, London: Imperial College, p. 110.
32 Hartcup (2000), p. 105.
33 Ibid., pp. 107–9.
34 Fortun, Michael, and Silvan S. Schweber (1994), 'Scientists and the legacy of World War II: the case of operations research (OR)', *Social Studies of Science* 23, pp. 595–642.
35 Hartcup (2000), p. 109.
36 Edwards, Paul N. (1996), *The Closed World: Computers and the Politics of Discourse in Cold War America*, Cambridge, MA: MIT Press, p. 200.
37 Smart, Jeffery K. (1997), 'History of chemical and biological warfare: an American perspective', in Frederick R. Sidell, Ernest T. Takafuji and David R. Franz (eds), *Medical Aspects of Chemical and Biological Warfare*, Washington, DC: Office of the Surgeon General, pp. 9–87, at p. 19.
38 Quoted ibid., p. 25.
39 Guillemin, Jeanne (2005), *Biological Weapons: From the Invention of State-Sponsored Programs to Contemporary Bioterrorism*, New York: Columbia University Press, p. 4.
40 Smart (1997), p. 27; Russell (2001), p. 65.
41 Smart (1997), p. 38.
42 Tucker, Jonathan B. (2007), *War of Nerves: Chemical Warfare from World War I to Al-Qaeda*, New York: Anchor Books, p. 24.
43 Smart (1997), p. 36.
44 Guillemin (2005), p. 24.
45 Ibid., p. 26.
46 Ibid., p. 41.
47 Ibid., p. 50.
48 Quoted in Balmer, Brian (2001), *Britain and Biological Warfare: Expert Advice and Science Policy, 1930–65*, London: Palgrave Macmillan, p. 37.
49 Guillemin (2005), p. 40.
50 Ibid, pp. 54–6.
51 Quoted ibid., p. 66.
52 Quoted in Balmer, Brian (2002), 'Killing "without the distressing prelimi-

naries": scientists' own defence of the British biological warfare programme', *Minerva* 40, pp. 57–75.

53 Quoted in Balmer (2001), p. 48.
54 Guillemin (2005), p. 73.
55 Quoted in Smart (1997), p. 34.
56 Price, Richard M. (1997), *The Chemical Weapons Taboo*, Ithaca, NY: Cornell University Press.
57 Müller-Hill, Benno (1988), *Murderous Science: Elimination by Scientific Selection of Jews, Gypsies, and Others, Germany 1933–1945*, Oxford: Oxford University Press, p. 70.
58 Harris, Sheldon H. (1994), *Factories of Death: Japanese Biological Warfare 1932–45, and the American Cover-Up*, London: Routledge.
59 Guillemin (2005), p. 81.
60 Lederer, Susan E. (2009), 'Experimentation and ethics', in Bowler and Pickstone, *The Cambridge History of Science*, Vol. 6: *The Modern Biological and Earth Sciences*, pp. 583–601, at pp. 594–5.
61 Guillemin (2005), p. 85.
62 Quoted ibid., p. 47.
63 Kay, Lily E. (1993), *The Molecular Vision of Life: Caltech, the Rockefeller Foundation, and the Rise of the New Biology*, Oxford: Oxford University Press, p. 199.
64 Hewlett, Richard, and Oscar E. Anderson, Jr. (1962), *A History of the United States Atomic Energy Commission*, Vol. 1: *The New World*, Washington, DC: Government Printing Office; Rhodes, Richard (1986), *The Making of the Atomic Bomb*, New York: Simon & Schuster.
65 Gowing, Margaret (1964), *Britain and Atomic Energy, 1939–1945*, London: Macmillan.
66 Ibid., p. 109.
67 Bird, Kai, and Martin J. Sherwin (2005), *American Prometheus: The Triumph and Tragedy of J. Robert Oppenheimer*, London: Atlantic Books, p. 180.
68 Ibid., p. 183.
69 Quoted in White, Richard (1995), *The Organic Machine*, New York: Hill & Wang, p. 81.
70 Seidel, Robert (1992), 'The origins of the Lawrence Berkeley Laboratory', in Peter Galison and Bruce Hevly (eds), *Big Science: The Growth of Large-Scale Research*, Stanford, CA: Stanford University Press, pp. 21–45, at p. 37.
71 Hughes, Thomas P. (1989), *American Genesis: A Century of Invention and Technological Enthusiasm, 1870–1970*, New York: Penguin, p. 382.
72 Eckert, Michael (2003), 'Plasmas and solid-state science', in Nye, *The Cambridge History of Science*, Vol. 5: *The Modern Physical and Mathematical Sciences*, pp. 413–28, at p. 419.
73 Hughes (1989), p. 414.
74 Ibid., p. 391.
75 Ibid., p. 392; Hounshell, David A. (1992), 'Du Pont and the management of large-scale research and development', in Galison and Hevly, *Big Science: The Growth of Large-Scale Research*, pp. 236–61, at p. 248.
76 White (1995), pp. 82–3.
77 Hughes (1989), p. 401.

78 Hoddeson, Lillian, Paul W. Henriksen, Roger A. Meade and Catherine L. Westfall (1993), *Critical Assembly: A Technical History of Los Alamos during the Oppenheimer Years, 1943–1945*, Cambridge: Cambridge University Press.
79 Bird and Sherwin (2005), p. 206.
80 Rossiter, Margaret W. (2003), 'A twisted tale: women in the physical sciences in the nineteenth and twentieth centuries', in Nye, *The Cambridge History of Science*, Vol. 5: *The Modern Physical and Mathematical Sciences*, pp. 54–71, at p. 65.
81 Bird and Sherwin (2005), p. 256.
82 Rotblat's recollection, in an interview for Bird and Sherwin, ibid., p. 284.
83 Quoted in Thorpe, Charles (2006), *Oppenheimer: The Tragic Intellect*, Chicago: University of Chicago Press, p. 149.
84 Shapin, Steven (2000), 'Don't let that crybaby in here again', *London Review of Books*, 7 September.
85 Bird and Sherwin (2005), p. 297.
86 Alperovitz, Gar (1995), *The Decision to Use the Atomic Bomb: The Architecture of an American Myth*, New York: Knopf; Walker, J. Samuel (2007), 'The decision to use the bomb: a historiographical update', *Diplomatic History* 14, pp. 97–114.
87 Seidel (1992).
88 Hounshell (1992).
89 Groves, Leslie R. (1962), *Now It Can Be Told: The Story of the Manhattan Project*, New York: Harper & Row.
90 Hughes (1989).
91 Josephson, Paul R. (2003), 'Science, ideology and the state: physics in the twentieth century', in Nye, *The Cambridge History of Science*, Vol. 5: *The Modern Physical and Mathematical Sciences*, pp. 579–97, at p. 590.
92 Guillemin (2005), p. 29.
93 Overy, Richard (1995), *Why the Allies Won*, London: Jonathan Cape.
94 Sherry (1987), p. 175.
95 Ibid., p. 53.
96 Porter, Theodore M. (1995), *Trust in Numbers: The Pursuit of Objectivity in Science and Public Life*, Princeton, NJ: Princeton University Press.
97 Sherry (1987), p. 193.
98 Guillemin (2005), p. 60.
99 Overy (1995), p. 110.
100 Guillemin (2005), p. 24.
101 Ibid., p. 56.
102 Galison, Peter (2004), 'Removing knowledge', *Critical Inquiry* 31, pp. 229–43, at p. 234.
103 Cassidy, David (1992), *Uncertainty: The Life and Science of Werner Heisenberg*, New York: W. H. Freeman.
104 Rose, Paul Lawrence (1998), *Heisenberg and the Nazi Atomic Bomb Project: A Study in German Culture*, Berkeley: University of California Press.
105 Walker, Mark (1989), *German National Socialism and the Quest for Nuclear Power, 1939–1949*, Cambridge: Cambridge University Press; Walker, Mark (1995), *Nazi Science: Myth, Truth, and the German Atomic Bomb*, New York: Plenum Press.

106 Goudsmit, Samuel A. (1947), *Alsos: The Failure in German Science*, London: SIGMA Books.
107 Neufeld, Michael (1994), 'The guided missile and the Third Reich: Peenemünde and the forging of a technological revolution', in Monika Renneberg and Mark Walker (eds), *Science, Technology, and National Socialism*, Cambridge: Cambridge University Press, pp. 51–71, at p. 51.
108 Ibid., p. 59.
109 Ibid.; Neufeld, Michael (1995), *The Rocket and the Reich: Peenemünde and the Coming of the Ballistic Missile Era*, New York: Free Press.
110 Quoted in Dickson, Paul (2001), *Sputnik: The Shock of the Century*, New York: Walker, p. 57.
111 Lederer (2009), p. 592.
112 Ibid.
113 Ibid., pp. 592–3.
114 Müller-Hill (1988), pp. 70–4.
115 Lederer (2009), p. 593.
116 Guillemin (2005), pp. 76–80.
117 Ibid., pp. 86, 87.
118 Gimbel, John (1990), *Science, Technology, and Reparations: Exploitation and Plunder in Post-War Germany*, Stanford, CA: Stanford University Press; Heinemann-Grüder, Andreas (1994), 'Keinerlei Untergang: German armaments engineers during the Second World War and in the service of the victorious powers', in Renneberg and Walker, *Science, Technology, and National Socialism*, pp. 30–50.
119 Heinemann-Grüder (1994), p. 48.
120 Ibid., p. 45; Dickson (2001), p. 58.

CHAPTER 13 TRIALS OF SCIENCE IN THE ATOMIC AGE

1 Gaddis, John Lewis (2005), *The Cold War: A New History*, New York: Penguin, p. 9.
2 Ibid., pp. 25–6.
3 Quoted ibid., p. 29.
4 Roland, Alex (2003), 'Science, technology, and war', in Mary Jo Nye (ed.), *The Cambridge History of Science*, Vol. 5: *The Modern Physical and Mathematical Sciences*, Cambridge: Cambridge University Press, pp. 561–78, at p. 565; see also Price, Don K. (1965), *The Scientific Estate*, New York: Oxford University Press.
5 Roland (2003), p. 575.
6 Leslie, Stuart W. (1993), *The Cold War and American Science: The Military-Industrial-Academic Complex at MIT and Stanford*, New York: Columbia University Press, pp. 252–6.
7 Reingold, Nathan (1987), 'Vannevar Bush's new deal for research: or the triumph of the old order', *Historical Studies in the Physical Sciences* 17, pp. 299–344.
8 Bush, Vannevar (1945), *Science: The Endless Frontier*, Washington, DC: United States Government Printing Office.
9 Kevles, Daniel J. (1971), *The Physicists: The History of a Scientific*

Community in Modern America, Cambridge, MA: Harvard University Press, p. 344.

10 Ibid., p. 347.

11 Quoted in Hollinger, David A. (1996), *Science, Jews and Secular Culture: Studies in Mid-Twentieth Century American Intellectual History*, Princeton, NJ: Princeton University Press, p. 107.

12 Paul Forman (1987), 'Beyond quantum electronics: national security as basis for physical research in the United States', *Historical Studies in the Physical Sciences* 18, pp. 149–229, at p. 183.

13 Leslie (1993).

14 Ibid.

15 Owens, Larry (1997), 'Science in the United States', in John Krige and Dominique Pestre (eds), *Science in the Twentieth Century*, Amsterdam: Harwood Academic Press, pp. 821–37, at p. 822.

16 Forman (1987), p. 184.

17 Ibid., pp. 186–7.

18 Ibid., p. 194.

19 Ibid., pp. 198–9.

20 Leslie (1993).

21 Bromberg, Joan Lisa (1991), *The Laser in America, 1950–1970*, Cambridge, MA: MIT Press.

22 Forman (1987), p. 210.

23 Quoted ibid., p. 218.

24 Ibid., p. 223.

25 Hoch, Paul K. (1988), 'The crystallization of a strategic alliance: the American physics elite and the military in the 1940s', in Everett Mendelsohn, M. Roe Smith and Peter Weingart (eds), *Science, Technology, and the Military*, Boston: Kluwer Academic, pp. 87–116.

26 Kevles, Daniel J. (1990), 'Cold War and hot physics: science, security, and the American State, 1945–56', *Historical Studies in the Physical and Biological Sciences* 20, pp. 239–64.

27 Bird, Kai, and Martin J. Sherwin (2005), *American Prometheus: The Triumph and Tragedy of J. Robert Oppenheimer*, London: Atlantic Books, p. 273.

28 Ibid., pp. 288–9.

29 Quotations from ibid., pp. 293–5.

30 Smith, Alice Kimball (1965), *A Peril and a Hope: The Scientists' Movement in America, 1945–47*, Chicago: University of Chicago Press.

31 Bird and Sherwin (2005), p. 325.

32 Hughes, Thomas P. (1989), *American Genesis: A Century of Invention and Technological Enthusiasm, 1870–1970*, New York: Penguin, p. 422.

33 Gowing, Margaret, with Lorna Arnold (1974), *Independence and Deterrence: Britain and Atomic Energy, 1945–1952*, Vol. 1: *Policy Making*, London: Macmillan, pp. 104–12.

34 Bird and Sherwin (2005), p. 342.

35 Holloway, David (1994), *Stalin and the Bomb: The Soviet Union and Atomic Energy, 1939–1956*, New Haven, CT: Yale University Press.

36 Oldroyd, David (2009), 'Geophysics and geochemistry', in Peter J. Bowler and John V. Pickstone (eds), *The Cambridge History of Science*, Vol. 6: *The*

Modern Biological and Earth Sciences, Cambridge: Cambridge University Press, pp. 395–415, at p. 414.

37 Holloway (1994).

38 Ibid.

39 Rotter, Andrew J. (2008), *Hiroshima: The World's Bomb*, Oxford: Oxford University Press, p. 236.

40 Holloway (1994), p. 213.

41 Quoted in Pollock, Ethan (2001), 'Science under socialism in the USSR and beyond', *Contemporary European History* 10, pp. 523–35, at p. 525.

42 Quoted ibid., p. 523.

43 Quoted ibid., p. 533.

44 Josephson, Paul R. (2003), 'Science, ideology and the state: physics in the twentieth century', in Nye, *The Cambridge History of Science*, Vol. 5: *The Modern Physical and Mathematical Sciences*, Cambridge: Cambridge University Press, pp. 579–97, at p. 585.

45 Pollock, Ethan (2006), *Stalin and the Soviet Science Wars*, Princeton, NJ: Princeton University Press.

46 Quoted in Krementsov, Nikolai (1997), *Stalinist Science*, Princeton, NJ: Princeton University Press, p. 172.

47 Joravsky, David (1970), *The Lysenko Affair*, Cambridge, MA: Harvard University Press.

48 Krementsov (1997), p. 170.

49 Ibid., p. 159.

50 Pollock (2001), p. 527.

51 Pollock (2006), p. 2.

52 Ibid.

53 Holloway (1994), p. 211.

54 Pollock (2006), p. 92.

55 Josephson, Paul R. (1997), *New Atlantis Revisited: Akademgorodok, the Siberian City of Science*, Princeton, NJ: Princeton University Press.

56 Quoted in Hennessy, Peter (2002), *The Secret State: Whitehall and the Cold War*, London: Penguin, p. 48.

57 Ibid.; DeGroot, Gerard J. (2005), *The Bomb: A Life*, Cambridge, MA: Harvard University Press, p. 219.

58 Cathcart, Brian (1994), *Test of Greatness: Britain's Struggle for the Atom Bomb*, London: John Murray.

59 Badash, Lawrence (2000), 'Science and McCarthyism', *Minerva* 38, pp. 53–80, at p. 63.

60 Kauffman, George B. (2002), 'In memoriam Martin D. Kamen (1913–2002), nuclear scientist and biochemist', *Chemical Educator* 7, pp. 304–8.

61 Hager, Thomas (1995), *Force of Nature: The Life of Linus Pauling*, New York: Simon & Schuster, p. 401.

62 Josephson (2003), p. 590.

63 Quoted in Carl Sagan (1996), *The Demon-Haunted World: Science as a Candle in the Dark*, London: Headline, p. 248.

64 Badash (2000), p. 72.

65 Ibid., p. 73.

66 Bernstein, Barton J. (1990), 'The Oppenheimer loyalty-security case reconsidered', *Stanford Law Review* 42, pp. 1383–484; Bird and Sherwin (2005);

Thorpe, Charles (2006), *Oppenheimer: The Tragic Intellect*, Chicago: University of Chicago Press.
67 Quoted in Bernstein (1990), p. 1404.
68 Quoted ibid., p. 1405.
69 Bird and Sherwin (2005), p. 325.
70 Quoted ibid., p. 418.
71 Ibid., pp. 422–3.
72 Quoted ibid., p. 428.
73 Bernstein (1990), p. 1407.
74 Quoted in Bird and Sherwin (2005), p. 362.
75 Bernstein (1990), p. 1445.
76 Ibid.
77 Bird and Sherwin (2005), p. 533; Thorpe (2006), p. 241, has 'unfrock'.
78 Quoted in Thorpe (2006), p. 223.
79 Bernstein (1990), p. 1471.
80 Ibid., p. 1387.
81 Ibid., p. 1484.
82 Bird and Sherwin (2005), p. 549.
83 Thorpe (2006), p. 238.
84 Ibid., p. 202.
85 Krige, John (2006), 'Atoms for Peace, scientific internationalism, and scientific intelligence', *Osiris* 21, pp. 161–81, at pp. 162–5.
86 Ibid., p. 164.
87 Quoted ibid., p. 169.
88 Ibid., p. 175.
89 Eckert, Michael (2003), 'Plasmas and solid-state science', in Nye, *The Cambridge History of Science*, Vol. 5: *The Modern Physical and Mathematical Sciences*, pp. 413–28.
90 Bromberg, Joan Lisa (1982), *Fusion: Science, Politics, and the Invention of a New Energy Source*, Cambridge, MA: MIT Press, p. 13; Eckert (2003).
91 Gaddis (2005), p. 42.
92 Ibid., p. 59.
93 Ibid., p. 53.
94 Kevles, Daniel J. (1992), 'K1S2: Korea, science, and the state', in Peter Galison and Bruce Hevly (eds), *Big Science: The Growth of Large-Scale Research*, Stanford, CA: Stanford University Press, pp. 312–33.
95 Ibid., p. 319.
96 National Security Council (1950), *NSC-68: United States Objectives and Programs for National Security (April 14, 1950): a Report to the President pursuant of the President's Directive of January 31, 1950.*
97 Gaddis (2005).
98 Kevles (1992), p. 320.
99 Ibid., p. 327.

CHAPTER 14 COLD WAR SPACES

1 Capshew, James H., and Karen A. Rader (1992), 'Big Science: price to the present', *Osiris* 7, pp. 3–25, at p. 4.

2 Ibid., pp. 20–3.
3 Galison, Peter (1997), *Image and Logic: A Material Culture of Microphysics*, Chicago: University of Chicago Press.
4 Hollinger, David A. (1996), *Science, Jews and Secular Culture: Studies in Mid-Twentieth Century American Intellectual History*, Princeton, NJ: Princeton University Press, p. 8.
5 Rudolph, John L. (2002), 'Portraying epistemology: school science in historical context', *Science Education* 87, pp. 64–79.
6 Feynman, Richard P. (1985). *'Surely You're Joking, Mr. Feynman!': Adventures of a Curious Character*, London: Unwin.
7 Roland, Alex (2003), 'Science, technology, and war', in Mary Jo Nye (ed.), *The Cambridge History of Science*, Vol. 5: *The Modern Physical and Mathematical Sciences*, Cambridge: Cambridge University Press, pp. 561–78, at p. 571.
8 Bok, Sissela (1982), 'Secrecy and openness in science: ethical considerations', *Science, Technology and Human Values* 7, pp. 32–41, at p. 35.
9 Quoted in Paul Forman (1987), 'Beyond quantum electronics: national security as basis for physical research in the United States', *Historical Studies in the Physical Sciences* 18, pp. 149–229, at p. 185.
10 Galison, Peter (2004), 'Removing knowledge', *Critical Inquiry* 31, pp. 229–43.
11 [Gregory, Richard] (1941), 'The new charter of scientific fellowship: declaration of scientific principles', *Nature* 148, p. 393.
12 Merton, Robert K. (1973), 'The normative structure of science, 1942', in Robert K. Merton, *The Sociology of Science: Theoretical and Empirical Investigations*, Chicago: University of Chicago Press, pp. 267–78, at pp. 270–8.
13 Hollinger (1996), p. 84.
14 Ibid., p. 92.
15 Josephson, Paul R. (2003), 'Science, ideology and the state: physics in the twentieth century', in Nye, *The Cambridge History of Science*, Vol. 5: *The Modern Physical and Mathematical Sciences*, pp. 579–97, at p. 591.
16 Polanyi, Michael (1962), 'The republic of science: its political and economic theory', *Minerva* 1, pp. 54–74.
17 Saunders, Frances Stonor (1999), *Who Paid the Piper? The CIA and the Cultural Cold War*, London: Granta Books.
18 Hollinger (1996), pp. 99–101.
19 Ibid., p. 110.
20 Killian, James R., Jr. (1977), *Sputnik, Scientists, and Eisenhower: A Memoir of the First Special Assistant to the President for Science and Technology*, Cambridge, MA: MIT Press; Roland (2003), p. 574.
21 Lilienfeld, Robert (1978), *The Rise of Systems Theory: An Ideological Analysis*, New York: Wiley; Williams, Rosalind (1993), 'Cultural origins and environmental implications of large technological systems', *Science in Context* 6, pp. 75–101; Hughes, Thomas P. (1983), *Networks of Power: Electrification in Western Society, 1880–1930*, Baltimore: Johns Hopkins University Press.
22 Hughes, Agatha C., and Thomas P. Hughes (eds) (2000), *Systems, Experts, and Computers: The Systems Approach in Management and Engineering, World War II and After*, Cambridge, MA: MIT Press, p. 2.

23 Mindell, David A. (2000), 'Automation's finest hour: radar and system integration in World War II', ibid., pp. 27–56.
24 Agar, Jon (2003), *The Government Machine: A Revolutionary History of the Computer*, Cambridge, MA: MIT Press, p. 216.
25 Hounshell, David A. (1997), 'The Cold War, RAND, and the generation of knowledge, 1946–1962', *Historical Studies in the Physical and Biological Sciences* 27, pp. 237–67.
26 Ibid., p. 254.
27 Ibid., p. 245.
28 Ghamari-Tabrizi, Sharon (2005), *The Worlds of Herman Kahn: The Intuitive Science of Thermonuclear War*, Cambridge, MA: Harvard University Press.
29 Hounshell (1997), p. 265.
30 Nelson, Richard R. (1959), 'The simple economics of basic scientific research', *Journal of Political Economy* 67, pp. 297–306; Arrow, Kenneth (1962), 'Economic welfare and the allocation of resources for invention', in Richard R. Nelson (ed.), *The Rate and Direction of Inventive Activity: Economic and Social Factors*, Princeton, NJ: Princeton University Press, pp. 609–25; Hounshell (1997); Hounshell, David A. (2000), 'The medium is the message, or how context matters: the RAND Corporation builds an economics of innovation, 1946–1962', in Hughes and Hughes, *Systems, Experts, and Computers: The Systems Approach in Management and Engineering, World War II and After*, pp. 255–310; Dennis, Michael Aaron (1997), 'Historiography of science – an American perspective', in John Krige and Dominique Pestre (eds), *Science in the Twentieth Century*, Amsterdam: Harwood Academic Press, pp. 1–26, at p. 18.
31 Misa, Thomas (1985), 'Military needs, commercial realities, and the development of the transistor, 1948–1958', in Merritt Roe Smith (ed.), *Military Enterprise and Technological Change*, Cambridge, MA: MIT Press, pp. 253–87.
32 Riordan, Michael, and Lillian Hoddeson (1997), *Crystal Fire: The Birth of the Information Age*, New York: W. W. Norton, pp. 82–4.
33 Eckert, Michael (2003), 'Plasmas and solid-state science', in Nye, *The Cambridge History of Science*, Vol. 5: *The Modern Physical and Mathematical Sciences*, pp. 413–28, at p. 421.
34 Riordan and Hoddeson (1997), p. 95.
35 Ibid., p. 159.
36 Kevles (1992), p. 314.
37 Riordan and Hoddeson (1997), p. 196.
38 Ibid., p. 203.
39 Quoted ibid., p. 232.
40 Dickson, Paul (2001), *Sputnik: The Shock of the Century*, New York: Walker, p. 81.
41 McDougall, Walter A. (1985), *The Heavens and the Earth: A Political History of the Space Age*, New York: Basic Books; Hall, R. Cargill (1995), 'Origins of US space policy: Eisenhower, Open Skies, and freedom of space', in John M. Logsdon et al. (eds), *Exploring the Unknown: Selected Documents in the History of the US Civil Space Program*, Vol. 1: *Organizing for Exploration*, Washington, DC: NASA, pp. 213–29.
42 DeVorkin, David H. (1992), *Science with a Vengeance: How the Military*

Created the US Space Sciences after World War II, New York: Springer, pp. 323–39.

43 Quoted in Dickson (2001), p. 105.

44 Ibid., p. 117.

45 Bulkeley, Rip (1991), *The Sputniks Crisis and Early United States Space Policy: A Critique of the Historiography of Space*, London: Macmillan.

46 Agar, Jon (1998), *Science and Spectacle: The Work of Jodrell Bank in Post-War British Culture*, Amsterdam: Harwood Academic Press, p. 123.

47 Dickson (2001), p. 101.

48 Norberg, Arthur L., and Judy O'Neill (1996), *Transforming Computer Technology: Information Processing for the Pentagon, 1962–1986*, Baltimore: Johns Hopkins University Press, p. 5.

49 Ibid., pp. 18–19.

50 Dickson (2001), p. 214.

51 Smith, Robert W. (2003), 'Remaking astronomy: instruments and practice in the nineteenth and twentieth centuries', in Nye, *The Cambridge History of Science*, Vol. 5: *The Modern Physical and Mathematical Sciences*, pp.154–73, at p. 168.

52 Doel, Ronald E. (1996), *Solar System Astronomy in America: Communities, Patronage and Interdisciplinary Science, 1920–1960*, Cambridge: Cambridge University Press, p. 224.

53 Gaddis, John Lewis (2005), *The Cold War: A New History*, New York: Penguin, p. 73.

54 Quoted in Dodds, Klaus (2002), *Pink Ice: Britain and the South Atlantic Empire*, London: I. B. Tauris, p. 28.

55 Beck, Peter (1986), *The International Politics of Antarctica*, London: Croom Helm.

56 Worboys, Michael (2009), 'Public and environmental health', in Peter J. Bowler and John V. Pickstone (eds), *The Cambridge History of Science*, Vol. 6: *The Modern Biological and Earth Sciences*, Cambridge: Cambridge University Press, pp. 141–63, at p. 159.

57 Henderson, D. A. (2009), *Smallpox: The Death of a Disease*, New York: Prometheus Books.

58 Bocking, Stephen A. (2009), 'Environmentalism', in Bowler and Pickstone, *The Cambridge History of Science*, Vol. 6: *The Modern Biological and Earth Sciences*, pp. 602–21, at p. 613.

59 Weart, Spencer R. (2003), *The Discovery of Global Warming*, Cambridge, MA: Harvard University Press.

60 Perkins, John H. (1997), *Geopolitics and the Green Revolution: Wheat, Genes and the Cold War*, Oxford: Oxford University Press.

61 Ibid., p. 108.

62 Ibid., pp. 111–13.

63 Quoted ibid., p. 144.

64 Quoted ibid., p. 138.

65 Ibid., p. 119.

66 Quoted ibid., p. 138.

67 Shiva, Vandana (1991), *The Violence of the Green Revolution: Third World Agriculture, Ecology and Politics*, London: Zed Books.

68 Edwards, Paul N. (1996), *The Closed World: Computers and the Politics of Discourse in Cold War America*, Cambridge, MA: MIT Press.

69 Gaddis (2005), p. 75.
70 Ibid., p. 78.
71 Burr, William (ed.) (2004), 'The creation of SIOP-62: more evidence on the origins of overkill', National Security Archive Electronic Briefing Book no. 130, at www.gwu.edu/~nsarchiv/NSAEBB/NSAEBB130/.
72 Gaddis (2005), p. 80.
73 Cohen, Avner (1998), *Israel and the Bomb*, New York: Columbia University Press, p. 49.
74 Ibid.
75 Galison (2004), pp. 241–2.

CHAPTER 15 COLD WAR SCIENCES (1)

1 Beatty, John (1991), 'Genetics in the atomic age: the Atomic Bomb Casualty Commission, 1947–1956', in Keith R. Benson, Jane Maienschein and Ronald Rainger (eds), *The Expansion of American Biology*, New Brunswick, NJ: Rutgers University Press, pp. 284–324, at p. 285.
2 Lindee, Susan (1999), 'The repatriation of atomic bomb victim body parts to Japan: natural objects and diplomacy', *Osiris* 13, pp. 376–409.
3 An ABCC preliminary report, quoted in Beatty (1991), pp. 287–8.
4 Ibid., p. 304.
5 Kraft, Alison (2009), 'Manhattan transfer: lethal radiation, bone marrow transplantation, and the birth of stem cell biology', *Historical Studies in the Natural Sciences* 39, pp. 171–218.
6 Holmes, Frederic L. (1991), *Hans Krebs*, Vol. 1: *The Formation of a Scientific Life, 1900–1933*, Oxford: Oxford University Press; Holmes, Frederic L. (1993); *Hans Krebs*, Vol. 2: *Architect of Intermediary Metabolism, 1933–1937*, Oxford: Oxford University Press.
7 Hagen, Joel B. (1992), *An Entangled Bank: The Origins of Ecosystem Ecology*, New Brunswick, NJ: Rutgers University Press, p. 100.
8 Ibid., p. 106.
9 The Odums' report, quoted ibid., p. 102.
10 Ibid., p. 105.
11 Acot, Pascal (2009), 'Ecosystems', in Peter J. Bowler and John V. Pickstone (eds), *The Cambridge History of Science*, Vol. 6: *The Modern Biological and Earth Sciences*, Cambridge: Cambridge University Press, pp. 451–66.
12 Ibid., p. 463.
13 Oldroyd, David (2009), 'Geophysics and geochemistry', ibid., pp. 395–415, at p. 414.
14 Schweber, Silvan S. (2003), 'Quantum field theory: from QED to the standard model', in Mary Jo Nye (ed.), *The Cambridge History of Science*, Vol. 5: *The Modern Physical and Mathematical Sciences*, Cambridge: Cambridge University Press, pp. 375–93, at p. 381.
15 Ibid., p. 383.
16 Ibid., p. 384.
17 Kaiser, David (2005), *Drawing Theories Apart: The Dispersion of Feynman Diagrams in Postwar Physics*, Chicago: University of Chicago Press.
18 Schweber, Silvan S. (1994), *QED and the Men Who Made It: Dyson,*

Feynman, Schwinger, and Tomonaga, Princeton, NJ: Princeton University Press.

19 Schweber (2003), p. 381.
20 Ibid., p. 387.
21 Ibid.
22 Galison, Peter (1997), *Image and Logic: A Material Culture of Microphysics*, Chicago: University of Chicago Press.
23 Schweber (2003), pp. 389–90.
24 Agar, Jon (1998), *Science and Spectacle: The Work of Jodrell Bank in Post-War British Culture*, Amsterdam: Harwood Academic Press, p. 12.
25 Krige, John (1989), 'The installation of high-energy accelerators in Britain after the war: big equipment but not "Big Science"', in Michelangelo De Maria, Mario Grilli and Fabio Sebastiani (eds), *The Restructuring of Physical Sciences in Europe and the United States, 1945–1960*, Singapore: World Scientific, pp. 488–501.
26 John Krige, in Hermann, Armin, et al. (1987), *History of CERN*, Vol. 1: *Launching the European Organization for Nuclear Research*, Amsterdam: North-Holland, p. 435.
27 Schweber (2003), p. 393.
28 Kragh, Helge (1996), *Cosmology and Controversy: The Historical Development of Two Theories of the Universe*, Princeton, NJ: Princeton University Press, pp. 101–2.
29 Kragh, Helge (2003), 'Cosmologies and cosmogonies of space and time', in Nye, *The Cambridge History of Science*, Vol. 5: *The Modern Physical and Mathematical Sciences*, pp. 522–37.
30 Gregory, Jane (2005), *Fred Hoyle's Universe*, Oxford: Oxford University Press.
31 Kragh (2003), p. 532.

CHAPTER 16 COLD WAR SCIENCES (2)

1 Sullivan, Walter T., III (ed.) (1984), *The Early Years of Radio Astronomy: Reflections Fifty Years after Jansky's Discovery*, Cambridge: Cambridge University Press.
2 Edge, David O., and Michael J. Mulkay (1976), *Astronomy Transformed: The Emergence of Radio Astronomy in Britain*, London: John Wiley.
3 Lovell, A. C. Bernard (1968), *The Story of Jodrell Bank*, London: Oxford University Press; Edge and Mulkay (1976); Agar, Jon (1998), *Science and Spectacle: The Work of Jodrell Bank in Post-War British Culture*, Amsterdam: Harwood Academic Press.
4 Agar (1998), p. 51.
5 Ibid., p. 59.
6 Quoted in Clark, Ronald W. (1972), *A Biography of the Nuffield Foundation*, London: Longman, p. 103.
7 Van Keuren, David K. (2001), 'Cold War science in black and white', *Social Studies of Science* 31, pp. 207–52.
8 Smith, Robert W. (2003), 'Remaking astronomy: instruments and practice in the nineteenth and twentieth centuries', in Mary Jo Nye (ed.), *The Cambridge History of Science*, Vol. 5: *The Modern Physical and*

Mathematical Sciences, Cambridge: Cambridge University Press, pp. 154–73, at p. 169.

9 Mindell, David A. (2002), *Between Human and Machine: Feedback, Control, and Computing before Cybernetics*, Baltimore: Johns Hopkins University Press, p. 185.

10 Galison, Peter (1994), 'The ontology of the enemy: Norbert Wiener and the cybernetic vision', *Critical Inquiry* 21, pp. 228–66, at p. 233.

11 Quoted ibid., p. 236, emphasis removed.

12 Ibid., p. 238.

13 Quoted ibid., p. 240.

14 Quoted in Heims, Steve J. (1980), *John von Neumann and Norbert Wiener: From Mathematics to the Technologies of Life and Death*, Cambridge, MA: MIT Press, p. 186.

15 Bowker, Geof (1993), 'How to be universal: some cybernetic strategies, 1943–70', *Social Studies of Science* 23, pp. 107–27.

16 Quoted in Galison (1994), pp. 250–1.

17 Hayward, Rhodri (2001), 'The tortoise and the love machine: Grey Walter and the politics of electro-encephalography', *Science in Context* 14, pp. 615–42, at p. 621.

18 Quoted in Bowker (1993), p. 111.

19 Gerovitch, Slava (2002), *From Newspeak to Cyberspeak: A History of Soviet Cybernetics*, Cambridge, MA: MIT Press.

20 Edwards, Paul N. (1996), *The Closed World: Computers and the Politics of Discourse in Cold War America*, Cambridge, MA: MIT Press, p. 99.

21 Ibid., p. 96.

22 Ibid., pp. 1, 15.

23 Gerovitch (2002), p. 133.

24 Nebeker, Frederik (1995), *Calculating the Weather: Meteorology in the 20th Century*, San Diego: Academic Press.

25 Galison, Peter (1996), 'Computer simulations and the trading zone', in Peter Galison and David J. Stump (eds), *The Disunity of Science: Boundaries, Contexts and Power*, Stanford, CA: Stanford University Press, p. 122; Galison (1997), p. 44.

26 Agar, Jon (2006), 'What difference did computers make?', *Social Studies of Science* 36, pp. 869–907.

27 Ferry, Georgina (1998), *Dorothy Hodgkin: A Life*, London: Granta Books.

28 De Chadarevian, Soraya (2002), *Designs for Life: Molecular Biology after World War II*, Cambridge: Cambridge University Press.

29 Galison (1997).

30 Ibid., p. 532.

31 Robertson, Douglas S. (2003), *Phase Change: The Computer Revolution in Science and Mathematics*, Oxford: Oxford University Press.

32 Ibid., pp. 8–9.

33 Agar (2006), p. 872.

34 Francoeur, Eric (2002), 'Cyrus Levinthal, the Kluge and the origins of inter-active molecular graphics', *Endeavour* 26, pp. 127–31; Francoeur, Eric, and Jérôme Segal (2004), 'From model kits to interactive computer graphics', in Soraya de Chadarevian and Nick Hopwood (eds), *Models: The Third Dimension of Science*, Stanford, CA: Stanford University Press, pp. 402–29.

35 Kevles, Bettyann Holtzmann (1997), *Naked to the Bone: Medical Imaging in the Twentieth Century*, New Brunswick, NJ: Rutgers University Press.

36 Kevles, Bettyann Holtzmann (2003), 'The physical sciences and the physician's eye', in Nye, *The Cambridge History of Science*, Vol. 5: *The Modern Physical and Mathematical Sciences*, pp. 615–33, at pp. 626–7.

37 Galison (1996), pp. 156–7; Galison (1997), p. 777.

38 Edwards (1996), p. 188.

39 Quoted in Crevier, Daniel (1993), *AI: The Tumultuous History of the Search for Artificial Intelligence*, New York: Basic Books, p. 103.

40 Crowther-Heyck, Hunter (2005), *Herbert A. Simon: The Bounds of Reason in Modern America*, Baltimore: Johns Hopkins University Press.

41 Ibid., p. 227.

42 Simon, Herbert (1991), *Models of my Life*, New York: Basic Books, p. 190.

43 Quoted in Crowther-Heyck (2005), pp. 228–9.

44 Norberg, Arthur L., and Judy O'Neill (1996), *Transforming Computer Technology: Information Processing for the Pentagon, 1962–1986*, Baltimore: Johns Hopkins University Press, p. 201.

45 Quoted in Edwards (1996), p. 266.

46 Hounshell, David A. (1997), 'The Cold War, RAND, and the generation of knowledge, 1946–1962', *Historical Studies in the Physical and Biological Sciences* 27, pp. 237–67, at pp. 260–1.

47 Lederberg, Joshua (1963), 'An instrumentation crisis in biology', unpublished manuscript in Lederberg archive, National Institutes of Health.

48 Quoted in Crevier (1993), p. 33.

49 Edwards (1996), pp. 222–33.

50 Ibid., p. 233.

51 Aspray, William (2003), 'Computer science and the computer revolution', in Nye, *The Cambridge History of Science*, Vol. 5: *The Modern Physical and Mathematical Sciences*, pp. 598–614, at p. 608.

52 Quoted in Ceruzzi, Paul (1998), *A History of Modern Computing*, Cambridge, MA: MIT Press, p. 102.

53 Agar, Jon (2003), *The Government Machine: A Revolutionary History of the Computer*, Cambridge, MA: MIT Press.

54 Ceruzzi (1998), p. 101.

55 Ibid., pp. 90–2.

56 Burian, Richard M., and Doris T. Zallen (2009), 'Genes', in Peter J. Bowler and John V. Pickstone (eds), *The Cambridge History of Science*, Vol. 6: *The Modern Biological and Earth Sciences*, Cambridge: Cambridge University Press, pp. 432–50, at p. 438.

57 Kay, Lily E. (1993), *The Molecular Vision of Life: Caltech, the Rockefeller Foundation, and the Rise of the New Biology*, Oxford: Oxford University Press, p. 110.

58 Judson, Horace Freeland (1979), *The Eighth Day of Creation: Makers of the Revolution in Biology*, London: Penguin, p. 34.

59 Quoted ibid., p. 39.

60 Ibid., p. 34.

61 Quoted ibid., p. 54.

62 Quoted ibid., p. 57.

63 Ibid., p. 130.

64 Schrödinger, Erwin (1944), *What is Life? The Physical Aspect of the Living Cell*, Cambridge: Cambridge University Press, p. 91.

65 Olby, Robert (1971), 'Schrödinger's problem: *What is Life?*', *Journal of the History of Biology* 4, pp. 119–48; Yoxen, Edward (1979), 'Where does Schroedinger's *What is Life?* belong in the history of molecular biology?', *History of Science* 17, pp. 17–52; Keller, Evelyn Fox (1990), 'Physics and the emergence of molecular biology: a history of cognitive and political synergy', *Journal of the History of Biology* 23, pp. 389–409; Morange, Michel (1998), *A History of Molecular Biology*, trans. Matthew Cobb, Cambridge, MA: Harvard University Press.

66 Fleming, Donald (1969), 'Émigré physicists and the biological revolution', in Donald Fleming and Bernard Bailyn (eds), *The Intellectual Migration: Europe and America, 1930–1960*, Cambridge, MA: Harvard University Press, pp. 152–89.

67 Ibid., pp. 159–60.

68 Sayre, Anne (1975), *Rosalind Franklin and DNA*, New York: W. W. Norton; Maddox, Brenda (2002), *Rosalind Franklin: The Dark Lady of DNA*, London: HarperCollins.

69 Hager, Thomas (1995), *Force of Nature: The Life of Linus Pauling*, New York: Simon & Schuster, p. 396.

70 Watson, James (1968), *The Double Helix: A Personal Account of the Discovery of the Structure of DNA*, London: Weidenfeld & Nicolson.

71 Wilkins, Maurice (2003), *The Third Man of the Double Helix*, Oxford: Oxford University Press, pp. 123–4.

72 Ibid., p. 205.

73 Watson, James D., and Francis H. C. Crick (1953), 'A structure for deoxyribose nucleic acid', *Nature* 171 (25 April), pp. 737–8.

74 Kay, Lily E. (2000), *Who Wrote the Book of Life? A History of the Genetic Code*, Stanford, CA: Stanford University Press, p. 5.

75 Quoted ibid., p. 131.

76 1968 interview with Gamow, quoted ibid., p. 132.

77 Ibid., p. 138.

78 Ibid., p. 163.

79 Both quoted ibid., p. 174.

80 Ibid., p. 189.

81 Ibid., p. 8.

82 Judson, Horace Freeland (1992), 'A history of gene mapping and sequencing', in Daniel J. Kevles and Leroy Hood (eds), *The Code of Codes: Scientific and Social Issues in the Human Genome Project*, Cambridge, MA: Harvard University Press, pp. 37–80, at p. 52.

83 Kay (2000), p. 195.

84 Quoted ibid., p. 215.

85 Amsterdamska, Olga (2009), 'Microbiology', in Bowler and Pickstone, *The Cambridge History of Science*, Vol. 6: *The Modern Biological and Earth Sciences*, pp. 316–41, at p. 340.

86 See discussion in Kay (2000), pp. 16–17, 291.

87 Wright, Susan (1986), 'Recombinant DNA technology and its social transformation, 1972–1982', *Osiris* 2, pp. 303–60, at p. 305; Allen, Garland

(1975), *Life Science in the Twentieth Century*, Cambridge: Cambridge University Press.

88 Oreskes, Naomi, and Ronald E. Doel (2003), 'The physics and chemistry of the earth', in Nye, *The Cambridge History of Science*, Vol. 5: *The Modern Physical and Mathematical Sciences*, pp. 538–57, at p. 552.

89 Wood, Robert Muir (1985), *The Dark Side of the Earth*, London: Allen & Unwin.

90 Frankel, Henry (2009), 'Plate tectonics', in Bowler and Pickstone, *The Cambridge History of Science*, Vol. 6: *The Modern Biological and Earth Sciences*, pp. 385–94, at p. 391.

91 Oreskes and Doel (2003), p. 555.

92 Frankel (2009), p. 392.

93 Ibid., p. 393.

94 Oldroyd, David (2009), 'Geophysics and geochemistry', in Bowler and Pickstone, *The Cambridge History of Science*, Vol. 6: *The Modern Biological and Earth Sciences*, pp. 395–415, at p. 409.

95 Ibid.

96 Greene, Mott T. (2009), 'Geology', in Bowler and Pickstone, *The Cambridge History of Science*, Vol. 6: *The Modern Biological and Earth Sciences*, pp. 167–84, at p. 184.

97 Ibid.

98 Svante Arrhenius, *Worlds in the Making*, 1908, quoted in Fleming, James Rodger (2003), 'Global environmental change and the history of science', in Nye, *The Cambridge History of Science*, Vol. 5: *The Modern Physical and Mathematical Sciences*, pp. 634–50, at p. 642.

99 Fleming, James Rodger (1998), *Historical Perspectives on Climate Change*, Oxford: Oxford University Press; Fleming (2003), p. 646; Weart, Spencer R. (2003), *The Discovery of Global Warming*, Cambridge, MA: Harvard University Press, p. 18.

100 Weart (2003), p. 59.

101 Ibid., p. 24.

102 Quoted ibid., p. 30.

103 Ibid., pp. 30–1.

CHAPTER 17 TRANSITION

1 For a model of this process, see Balogh, Brian (1991), *Chain Reaction: Expert Debate and Public Participation in American Nuclear Power, 1945–1975*, Cambridge: Cambridge University Press.

2 Brick, Howard (1998), *Age of Contradiction: American Thought and Culture in the 1960s*, New York: Twayne; for a general discussion see Agar, Jon (2008), 'What happened in the sixties?', *British Journal for the History of Science* 41, pp. 567–600.

3 Mendelsohn, Everett (1994), 'The politics of pessimism: science and technology circa 1968', in Yaron Ezrahi, Everett Mendelsohn and Howard Segal (eds), *Technology, Pessimism and Postmodernism*, London: Kluwer Academic, pp. 151–73, at p. 159.

4 Winston, Andrew S. (1998), 'Science in the service of the far right: Henry E. Garrett, the IAAEE, and the liberty lobby', *Journal of Social Issues* 54, pp.

179–210; Jackson, John P., Jr. (1998), 'Creating a consensus: psychologists, the Supreme Court, and school desegregation, 1952–1955', *Journal of Social Issues* 54, pp. 143–77.

5 Ezrahi, Yaron (1974), 'The authority of science in politics', in Arnold Thackray and Everett Mendelsohn (eds), *Science and Values: Patterns of Tradition and Change*, New York: Humanities Press, pp. 215–51, at p. 232.

6 Wittner, Lawrence S. (1993), *The Struggle against the Bomb*, Vol. 1: *One World or None: A History of the World Nuclear Disarmament Movement through 1953*, Stanford, CA: Stanford University Press, p. 29.

7 Moore, Kelly (1996), 'Organizing integrity: American science and the creation of public interest organizations, 1955–1975', *American Journal of Sociology* 101, pp. 1592–627, at p. 1614.

8 Wittner, Lawrence S. (1997), *The Struggle against the Bomb*, Vol. 2: *Resisting the Bomb: A History of the World Nuclear Disarmament Movement, 1954–1970*, Stanford, CA: Stanford University Press, p. 39.

9 Lear, Linda (1997), *Rachel Carson: Witness for Nature*, New York: Henry Holt.

10 Jamison, Andrew and Ron Eyerman (1994), *Seeds of the Sixties*, Berkeley: University of California Press, pp. 99–100.

11 Russell, Edmund (2001), *War and Nature: Fighting Humans and Insects with Chemicals from World War I to Silent Spring*, Cambridge: Cambridge University Press, pp. 158–63.

12 Gottlieb, Robert (1993), *Forcing the Spring: The Transformation of the American Environmental Movement*, Washington, DC: Island Press.

13 Monsanto (1962), 'The desolate year', *Monsanto Magazine* 42(4), pp. 4–9.

14 Proctor, Robert N. (1995), *Cancer Wars: How Politics Shapes What We Know and Don't Know about Cancer*, New York: Basic Books; Gaudillière, Jean-Paul (2009), 'Cancer', in Peter J. Bowler and John V. Pickstone (eds), *The Cambridge History of Science*, Vol. 6: *The Modern Biological and Earth Sciences*, Cambridge: Cambridge University Press, pp. 486–503, at p. 494.

15 Rachel Carson, speech at National Book Award, quoted in Lear (1997), p. 219.

16 Gottlieb (1993), p. 84.

17 Quoted in Lear (1997), p. 409.

18 Commoner, Barry (1966), *Science and Survival*, London: Victor Gollancz, p. 106.

19 An alternative argument is that mobilization for war suppressed the sociology of science between the 1930s and 1960s; see Rose, Hilary, and Steven Rose (1976), *The Radicalisation of Science*, London: Macmillan.

20 Ravetz, Jerome R. (1971), *Scientific Knowledge and its Social Problems*, Oxford: Oxford University Press, pp. 423–4.

21 Wittner (1997), p. 355.

22 Feenberg, Andrew (1999), *Questioning Technology*, London: Routledge.

23 Ibid., p. 31.

24 Ibid., p. 43.

25 Ibid., p. 4.

26 Quoted in Brick (1998), pp. 24–5.

27 Wisnioski, Matt (2003), 'Inside "the system": engineers, scientists, and the

boundaries of social protest in the long 1960s', *History and Technology* 19, pp. 313–33, at p. 320.

28 Leslie, Stuart W. (1993), *The Cold War and American Science: The Military-Industrial-Academic Complex at MIT and Stanford*, New York: Columbia University Press, p. 242.

29 Ibid., pp. 242–4.

30 Ibid., p. 235.

31 Dennis, Michael Aaron (1994), '"Our first line of defense": two university laboratories in the postwar American state', *Isis* 85, pp. 427–55.

32 Wisnioski (2003), p. 323.

33 Leslie (1993), p. 233.

34 Ibid.; Allen, Jonathan (ed.) (1970), *March 4: Students, Scientists, and Society*, Cambridge, MA: MIT Press.

35 Leslie (1993), p. 250.

36 Rose and Rose (1976).

37 Wittner, Lawrence S. (2003), *The Struggle Against the Bomb*, Vol. 3: *Toward Nuclear Abolition: A History of the World Nuclear Disarmament Movement, 1971 to the Present*, Stanford: Stanford University Press, p. 11.

38 Wittner (1997), pp. 172–3.

39 Moore (1996), p. 1594.

40 Ibid., p. 1608.

41 See also Ravetz, Jerome R. (1990), 'Orthodoxies, critiques and alternatives', in Robert Olby et al. (eds), *The Companion to the History of Modern Science*, London: Routledge, pp. 898–908.

42 Hollinger, David A. (1995), 'Science as a weapon in Kulturkampfe in the United States during and after World War II', *Isis* 86, pp. 440–54.

43 Jacques Ellul (1964), *The Technological Society*, New York: Knopf, pp. 10, 312, 45.

44 Marwick, Arthur (1998), *The Sixties: Cultural Revolution in Britain, France, Italy, and the United States, c.1958–c.1974*, Oxford: Oxford University Press.

45 Anton, Ted (2000), *Bold Science: Seven Scientists who are Changing our World*, New York: W. H. Freeman, p. 11.

46 Wilkins, Maurice (2003), *The Third Man of the Double Helix*, Oxford: Oxford University Press.

47 Capshew, James H., and Karen Rader (1992), 'Big Science: Price to present', *Osiris* 7, pp. 3–25.

48 Chargaff, Erwin (1968), 'A quick climb up Mount Olympus' (review of *The Double Helix*), *Science* 159 (29 March), pp. 1448–9.

49 Brick (1998), p. 23.

50 Hollinger, David A. (1996), *Science, Jews and Secular Culture: Studies in Mid-Twentieth Century American Intellectual History*, Princeton, NJ: Princeton University Press.

51 Brick (1998), p. 23.

52 Bird, Alexander (2000), *Thomas Kuhn*, Chesham: Acumen; Fuller, Steve (2000), *Thomas Kuhn: A Philosophical History for our Times*, Chicago: University of Chicago Press.

53 Ceruzzi, Paul (1998), *A History of Modern Computing*, Cambridge, MA: MIT Press, p. 108.

54 Gouldner, Alvin W. (1970), *The Coming Crisis of Western Sociology*, New York: Basic Books.
55 Wilkins, Maurice (1971), 'Introduction', in Watson Fuller (ed.), *The Social Impact of Modern Biology*, London: Routledge & Kegan Paul, pp. 5–10.
56 Monod, Jacques (1971), 'On the logical relationship between knowledge and values', ibid., pp. 11–21.
57 Etzioni, Amitai, and Clyde Z. Nunn (1973), 'Public views of scientists', *Science* 181 (21 September), p. 1123.
58 Watkins, Elizabeth Siegel (1998), *On the Pill: A Social History of Oral Contraceptives, 1950–1970*, Baltimore: Johns Hopkins University Press.
59 Ibid., p. 61.
60 Ibid., p. 14.
61 Ibid., p. 20.
62 Ibid., p. 15.
63 Ibid., p. 21.
64 Ibid., p. 27.
65 Ibid., p. 32.
66 Clarke, Adele E., Janet K. Shim, Laura Mamo, Jennifer Ruth Fosket and Jennifer R. Fishman (2003), 'Biomedicalization: technoscientific transformations of health, illness, and U.S. biomedicine', *American Sociological Review* 68, pp. 161–94.
67 Quoted in Watkins (1998), p. 66.
68 Ibid., pp. 57, 2.
69 Alice Wolfson, quoted ibid., p. 109.
70 Ibid., p. 127.
71 Ibid., p. 131.
72 Healy, David (2002), *The Creation of Psychopharmacology*, Cambridge, MA: Harvard University Press.
73 Ibid., p. 81.
74 Ibid., p. 92.
75 Ibid., pp. 92–3, 85.
76 Ibid., p. 107.
77 Ibid., pp. 133, 135.
78 Ibid., p. 176.
79 Ibid., pp. 176–7.
80 Ibid., p. 176.
81 Ibid., p. 364.
82 Jones, James H. (1981), *Bad Blood: The Tuskegee Syphilis Experiment*, New York: Free Press.
83 Lederer, Susan E. (2009), 'Experimentation and ethics', in Bowler and Pickstone, *The Cambridge History of Science*, Vol. 6: *The Modern Biological and Earth Sciences*, pp. 583–601, at p. 597.
84 Healy (2002), p. 159.
85 Hill, Gerry B. (2000), 'Archie Cochrane and his legacy: an internal challenge to physicians' autonomy?', *Journal of Clinical Epidemiology* 53, pp. 1189–92, at p. 1191.
86 Edwards, Paul N. (2000), 'The world in a machine: origins and impacts of early computerized global systems models', in Agatha C. Hughes and Thomas P. Hughes (eds), *Systems, Experts, and Computers: The Systems*

Approach in Management and Engineering, World War II and After, Cambridge, MA: MIT Press, pp. 221–53, 242–5.

87 Hagen, Joel B. (1992), *An Entangled Bank: The Origins of Ecosystem Ecology*, New Brunswick, NJ: Rutgers University Press, p. 135.

88 Veldman, Meredith (1994), *Fantasy, the Bomb and the Greening of Britain: Romantic Protest, 1945–1980*, Cambridge: Cambridge University Press, pp. 206–93.

89 Ibid., p. 269.

90 Ibid., p. 232.

91 Greenhalgh, Susan (2008), *Just One Child: Science and Policy in Deng's China*, Berkeley: University of California Press.

92 Elman, Benjamin A. (2007), 'New directions in the history of modern science in China: global science and comparative history', *Isis* 98, pp. 517–23, at p. 523.

93 Greenhalgh (2008), p. 55.

94 Schmalzer, Sigrid (2008), *The People's Peking Man: Popular Science and Human Identity in Twentieth-Century China*, Chicago: University of Chicago Press.

95 Greenhalgh (2008), p. 58.

96 Elman (2007), p. 520.

97 Greenhalgh (2008), p. 128.

98 Ibid.

99 Ibid., pp. 77, 49.

100 Elman (2007), p. 520.

101 Greenhalgh (2008), p. 131.

102 Ibid., p. 136.

103 Ibid.

104 Ibid., pp. 1, 318.

105 Peter J. T. Morris (2002), '"Parts per trillion is a fairy tale": the development of the Electron Capture Detector and its impact on the monitoring of DDT', in Morris, *From Classical to Modern Chemistry: The Instrumental Revolution*, Cambridge: RSC, pp. 259–84.

106 Hounshell, David A. (1997), 'The Cold War, RAND, and the generation of knowledge, 1946–1962', *Historical Studies in the Physical and Biological Sciences* 27, pp. 237–67, at p. 257.

107 Mendelsohn (1994).

108 Russell (2001), pp. 222–3.

109 Mendelsohn (1994); Ravetz (1990).

110 Rainger, Ronald (2009), 'Paleontology', in Bowler and Pickstone, *The Cambridge History of Science*, Vol. 6: *The Modern Biological and Earth Sciences*, pp. 185–204, at p. 199.

111 For an engaging account, see Gould, Stephen Jay (1991), *Wonderful Life: The Burgess Shale and the Nature of History*, Harmondsworth: Penguin.

112 Segerstråle, Ullica (2000), *Defenders of the Truth: The Battle for Science in the Sociobiology Debate and Beyond*, Oxford: Oxford University Press.

113 Rainger (2009), p. 199.

114 Calder, Nigel (1969), *Violent Universe: An Eye-Witness Account of the Commotion in Astronomy, 1968–69*, London: British Broadcasting Corporation.

115 Ibid., p. 7.

116 Weart, Spencer R. (2003), *The Discovery of Global Warming*, Cambridge, MA: Harvard University Press, pp. 46–8.
117 Ibid., pp. 61–3.
118 Ibid., p. 92.
119 Ibid., p. 93.
120 Gaddis, John Lewis (2005), *The Cold War: A New History*, New York: Penguin, p. 81.
121 Ibid.
122 Guillemin, Jeanne (2005), *Biological Weapons: From the Invention of State-Sponsored Programs to Contemporary Bioterrorism*, New York: Columbia University Press, pp. 134–6.

CHAPTER 18 NETWORKS

1 Beck, Ulrich (1992), *Risk Society: Towards a New Modernity*, London: Sage.
2 Proctor, Robert N. (1995), *Cancer Wars: How Politics Shapes What We Know and Don't Know about Cancer*, New York: Basic Books.
3 Beck, Ulrich (1998), 'Politics of risk society', in Jane Franklin (ed.), *The Politics of Risk Society*, Cambridge: Polity, pp. 9–22, at p. 10.
4 Gibbons, Michael, Camille Limoges, Helga Nowotny, Simon Schwatzman, Peter Scott and Martin Trow (1994), *The New Production of Knowledge: The Dynamics of Science and Research in Contemporary Societies*, London: Sage.
5 Scott, Alan (2000), 'Risk society or angst society? Two views of risk, consciousness and community', in Barbara Adam, Ulrich Beck and Joost van Loon (eds), *The Risk Society and Beyond: Critical Issues for Social Theory*, London: Sage, pp. 33–46; Godin, Benoît (1998), 'Writing performative history: the new *New Atlantis?*', *Social Studies of Science* 28, pp. 465–83; Shinn, Terry (2002), 'The triple helix and new production of knowledge: prepackaged thinking on science and technology', *Social Studies of Science* 32, pp. 599–614.
6 Godin (1998), p. 467.
7 Ibid., p. 478.
8 Leslie, Stuart W. (1993), *The Cold War and American Science: The Military–Industrial–Academic Complex at MIT and Stanford*, New York: Columbia University Press, p. 3.
9 Rettig, Richard A. (1977), *Cancer Crusade: The Story of the National Cancer Act of 1971*, Princeton, NJ: Princeton University Press.
10 Wynne, Brian (1996), 'May the sheep safely graze? A reflexive view of the expert–lay knowledge divide', in Scott Lash, Bronislaw Szerszynski and Brian Wynne (eds), *Risk, Environment and Modernity: Towards a New Ecology*, London: Sage, pp. 44–83.
11 Kay, Lily E. (2000), *Who Wrote the Book of Life? A History of the Genetic Code*, Stanford, CA: Stanford University Press, p. 292.
12 Wright, Susan (1986), 'Recombinant DNA technology and its social transformation, 1972–1982', *Osiris* 2, pp. 303–60, at p. 307.
13 Ibid., p. 310.
14 Ibid., p. 312.

15 Quoted ibid., p. 315.
16 Quoted in Kenney, Martin (1986), *Biotechnology: The University–Industrial Complex*, New Haven, CT: Yale University Press, p. 21.
17 Quoted in Wade, Nicholas (2001), 'Reporting recombinant DNA', *Perspectives in Biology and Medicine* 44, pp. 192–8, at p. 195.
18 Wright (1986), p. 324.
19 Quoted ibid.
20 Ibid., pp. 332–3.
21 Keller, Evelyn Fox (1983), *A Feeling for the Organism: The Life and Work of Barbara McClintock*, San Francisco: W. H. Freeman.
22 Wright (1986), p. 360.
23 Ibid., p. 338.
24 Bud, Robert (2009), 'History of biotechnology', in Peter J. Bowler and John V. Pickstone (eds), *The Cambridge History of Science*, Vol. 6: *The Modern Biological and Earth Sciences*, Cambridge: Cambridge University Press, pp. 524–38, at p. 536.
25 Kenney (1986), p. 4.
26 Wright (1986), p. 352.
27 Ibid., pp. 336–7.
28 Ibid., pp. 335, 344.
29 Kohler, Robert E. (1994), *Lords of the Fly*: Drosophila *Genetics and the Experimental Life*, Chicago: University of Chicago Press.
30 Kenney (1986), p. 131; Wright (1986), p. 358.
31 Kevles, Daniel J. (1998), '*Diamond v. Chakrabarty* and beyond: the political economy of patenting life', in Arnold Thackray (ed.), *Private Science: Biotechnology and the Rise of the Molecular Sciences*, Philadelphia: University of Pennsylvania Press, pp. 65–79.
32 Kevles, Daniel J. (2002), 'Of mice & money: the story of the world's first animal patent', *Daedalus* 131, pp. 78–88.
33 Quoted in Wright (1986), p. 357.
34 Kenney (1986), pp. 5–6.
35 Ibid., pp. 98–100.
36 Bud (2009), p. 537.
37 Swann, John P. (2009), 'The pharmaceutical industries', in Bowler and Pickstone, *The Cambridge History of Science*, Vol. 6: *The Modern Biological and Earth Sciences*, pp. 126–40, at p. 140.
38 Healy, David (2002), *The Creation of Psychopharmacology*, Cambridge, MA: Harvard University Press, p. 366.
39 Ibid., p. 368.
40 Ibid., p. 48.
41 Ibid., p. 215.
42 Ibid., p. 304.
43 Bowker, Geoffrey C., and Susan Leigh Star (1999), *Sorting Things Out: Classification and its Consequences*, Cambridge, MA: MIT Press.
44 Healy (2002), p. 330.
45 Fujimura, Joan H. (1996), *Crafting Science: A Sociohistory of the Quest for the Genetics of Cancer*, Cambridge, MA: Harvard University Press.
46 Morange, Michel (1997), 'From the regulatory vision of cancer to the oncogene paradigm, 1975–1985', *Journal of the History of Biology* 30, pp. 1–29, at p. 3.

47 Gaudillière, Jean-Paul (2009), 'Cancer', in Bowler and Pickstone, *The Cambridge History of Science*, Vol. 6: *The Modern Biological and Earth Sciences*, pp. 486–503, at p. 495.

48 Soderqvist, Thomas, Craig Stillwell and Mark Jackson (2009), 'Immunology', ibid., pp. 467–85, at p. 477.

49 Winsor, Mary P. (2009), 'Museums', ibid., pp. 60–75, at p. 75.

50 Hull, David L. (1988), *Science as a Process: An Evolutionary Account of the Social and Conceptual Development of Science*, Chicago: University of Chicago Press.

51 Futuyma, Douglas J. (2004), 'The fruit of the tree of life: insights into evolution and ecology', in Joel Cracraft and Michael J. Donoghue (eds), *Assembling the Tree of Life*, Oxford: Oxford University Press, pp. 25–39, at p. 26.

52 Hager, Thomas (1995), *Force of Nature: The Life of Linus Pauling*, New York: Simon & Schuster, p. 541.

53 Sapp, Jan (2009), *The New Foundations of Evolution: On the Tree of Life*, Oxford: Oxford University Press, p. 213.

54 Woese, Carl R., and George E. Fox (1977), 'Phylogenetic structure of the prokaryotic domain: the primary kingdoms', *Proceedings of the National Academy of Sciences* 74, pp. 5088–90, at p. 5088.

55 Lawrence, Susan C. (2009), 'Anatomy, histology and cytology', in Bowler and Pickstone, *The Cambridge History of Science*, Vol. 6: *The Modern Biological and Earth Sciences*, pp. 265–84, at p. 279.

56 Ralph S. Wolfe (2006), 'The Archaea: a personal overview of the formative years', *Prokaryotes* 3, pp. 3–9, at p. 3.

57 Castells, Manuel (1996, 1997, 1998), *The Information Age: Economy, Society and Culture*, Vol. 1: *The Rise of the Network Society*, Vol. 2: *The Power of Identity*, Vol. 3: *End of Millennium*, Oxford: Blackwell; Barry, Andrew (2001), *Political Machines: Governing a Technological Society*, London: Athlone Press.

58 Campbell-Kelly, Martin, and William Aspray (1996), *Computer: A History of the Information Machine*, New York: Basic Books, p. 293.

59 Abbate, Janet (1999), *Inventing the Internet*, Cambridge, MA: MIT Press, p. 101.

60 Ibid., p. 103.

61 Ceruzzi, Paul (1998), *A History of Modern Computing*, Cambridge, MA: MIT Press, p. 299.

62 Ding, Waverly W., Sharon G. Levin, Paula E. Stephan and Anne E. Winkler (2009), *The Impact of Information Technology on Scientists' Productivity, Quality and Collaboration Patterns*, Cambridge, MA: National Bureau of Economic Research Working Paper no. 15285, p. 5.

63 Jones, Ben, Stefan Wuchty and Brian Uzzi (2008), 'Multi-university research teams: shifting impact, geography, and stratification in science', *Science* 322 (21 November), pp. 1259–62.

64 Ding et al. (2009), p. 6.

65 Rowe, David E. (2003), 'Mathematical schools, communities and networks', in Mary Jo Nye (ed.), *The Cambridge History of Science*, Vol. 5: *The Modern Physical and Mathematical Sciences*, Cambridge: Cambridge University Press, pp. 113–32, at p. 131.

66 Sulston, John, and Georgina Ferry (2002), *The Common Thread: A Story of*

Science, Politics, Ethics and the Human Genome, London: Bantam Press, p. 55.

67 Lederberg, Joshua (1963), 'An instrumentation crisis in biology', unpublished manuscript in Lederberg archive, National Institutes of Health, p. 2, http://profiles.nlm.nih.gov/ps/access/BBGCVS.pdf.

68 Hopwood, Nick (2009), 'Embryology', in Bowler and Pickstone, *The Cambridge History of Science*, Vol. 6: *The Modern Biological and Earth Sciences*, pp. 285–315, at p. 310.

69 Judson, Horace Freeland (1992), 'A history of gene mapping and sequencing', in Daniel J. Kevles and Leroy Hood (eds), *The Code of Codes: Scientific and Social Issues in the Human Genome Project*, Cambridge, MA: Harvard University Press, pp. 37–80, at p. 65.

70 Ibid., p. 66.

71 Kevles, Daniel J. (1992), 'Out of eugenics: the historical politics of the human genome', in Kevles and Hood, *The Code of Codes: Scientific and Social Issues in the Human Genome Project*, pp. 3–36, at p. 14.

72 Judson (1992), p. 68.

73 Sulston and Ferry (2002), p. 23.

74 Ibid., pp. 43–4.

75 Zallen, Doris T., and Richard M. Burian (1992), 'On the beginnings of somatic cell hybridization: Boris Ephrussi and chromosome transplantation', *Genetics* 132, pp. 1–8, at p. 6.

76 Judson (1992), p. 69.

77 Ibid., p. 71.

78 Rabinow, Paul (1996), *Making PCR: A Story of Biotechnology*, Chicago: University of Chicago Press.

79 Morange, Michel (1997), 'EMBO and EMBL', in John Krige and Luca Guzzetti (eds), *History of European Scientific and Technological Cooperation*, Luxembourg: Office for Official Publications of the European Communities, pp. 77–92, at p. 89.

80 Kevles (1992), p. 27.

81 Ibid., p. 18.

82 Sulston and Ferry (2002), p. 58.

83 Kevles (1992), pp. 18, 22; Kevles, Daniel J. (1997), 'Big Science and big politics in the United States: reflections on the death of the SSC and the life of the human genome project', *Historical Studies in the Physical and Biological Sciences* 27, pp. 269–97.

84 Kevles (1992), p. 19.

85 Quoted in Cook-Deegan, Robert (1994), *The Gene Wars: Science, Politics, and the Human Genome*, New York: W. W. Norton, p. 161.

86 Sulston and Ferry (2002), p. 77.

87 Ibid., p. 88.

88 Ibid., p. 111.

89 Ibid., p. 82.

90 Ibid., pp. 140–1.

91 Gaudillière (2009), p. 501.

92 Sulston and Ferry (2002), p. 261.

93 Nelson, Bryn (2009), 'Empty archives', *Nature* 461 (9 September), pp. 160–3.

94 Burian, Richard M., and Doris T. Zallen (2009), 'The gene', in Bowler and

Pickstone, *The Cambridge History of Science*, Vol. 6: *The Modern Biological and Earth Sciences*, pp. 432–50, at p. 448.

95 Sulston and Ferry (2002), p. 180.
96 From Kevin Toolis (2000), 'DNA: it's war', *The Guardian*, 6 May.
97 Sulston and Ferry (2002), p. 234.
98 International Human Genome Sequencing Consortium (2001), 'Initial sequencing and analysis of the human genome', *Nature* 409 (15 February), pp. 860–921; Venter, J. Craig et al. (2001), 'The sequence of the human genome', *Science* 291 (16 February), pp. 1304–51.
99 Sulston and Ferry (2002), p. 241.
100 Futuyma (2004), p. 27.
101 Chase, Mark W., et al. (1993) 'Phylogenetics of seed plants: an analysis of nucleotide sequences from the plastid gene rbcL', *Annals of the Missouri Botanical Garden* 80, pp. 528–48, 550–80.
102 Baldauf, S. L., et al. (2004) 'The tree of life: an overview', in Cracraft and Donoghue, *Assembling the Tree of Life*, pp. 43–75.

CHAPTER 19 CONNECTING ENDS

1 Gillies, James, and Robert Cailliau (2000), *How the Web was Born*, Oxford: Oxford University Press, pp. 86–7.
2 Ibid., p. 151.
3 Ceruzzi, Paul (1998), *A History of Modern Computing*, Cambridge, MA: MIT Press, p. 367.
4 Quoted in Gillies and Cailliau (2000), p. 182.
5 On the memorandum reproduced ibid., 181.
6 Ibid., p. 199.
7 Kolb, Adrienne, and Lillian Hoddeson (1993), 'The mirage of the "world accelerator for world peace" and the origins of the SSC, 1953–1983', *Historical Studies in the Physical and Biological Sciences* 24, pp. 101–24.
8 Riordan, Michael (2001), 'A tale of two cultures: building the Superconducting Super Collider, 1988–1993', *Historical Studies in the Physical and Biological Sciences* 32, pp. 125–44; Kevles, Daniel J. (1997), 'Big Science and big politics in the United States: Reflections on the death of the SSC and the life of the human genome project', *Historical Studies in the Physical and Biological Sciences* 27, pp. 269–97.
9 Riordan (2001); Kevles (1997).
10 Leslie, Stuart W. (1993), *The Cold War and American Science: The Military–Industrial–Academic Complex at MIT and Stanford*, New York: Columbia University Press, p. 3.
11 Gaddis, John Lewis (2005), *The Cold War: A New History*, New York: Penguin, p. 227.
12 Ibid., p. 228.
13 Garthoff, Raymond L. (1994), *The Great Transition: American–Soviet Relations and the End of the Cold War*, Washington, DC: Brookings Institution.
14 Kevles (1997).
15 CERN (2009), *LHC: The Guide*, CERN Communication Group, http://cdsmedia.cern.ch/img/CERN-Brochure-2008-001-Eng.pdf.

16 Smith, Robert W., et al. (1989) *The Space Telescope: A Study of NASA, Science, Technology, and Politics*, Cambridge: Cambridge University Press; Smith, Robert W. (1992), 'The biggest kind of big science: astronomers and the Space Telescope', in Peter Galison and Bruce Hevly (eds), *Big Science: The Growth of Large-Scale Research*, Stanford, CA: Stanford University Press, pp. 184–211, at p. 184.

17 Smith (1992), p. 191.

18 Ibid., p. 195.

19 Ibid., p. 187.

20 Kevles, Bettyann Holtzmann (2003), 'The physical sciences and the physician's eye', in Mary Jo Nye (ed.), *The Cambridge History of Science*, Vol. 5: *The Modern Physical and Mathematical Sciences*, Cambridge: Cambridge University Press, pp. 615–33, at p. 632.

21 Dunwoody, Sharon (1980), 'The science writing inner club: a communication link between science and the lay public', *Science, Technology & Human Values* 5, pp. 14–22, at p. 14.

22 Ibid., p. 15.

23 Ibid., p. 19.

24 Rainger, Ronald (2009), 'Paleontology', in Peter J. Bowler and John V. Pickstone (eds), *The Cambridge History of Science*, Vol. 6: *The Modern Biological and Earth Sciences*, Cambridge: Cambridge University Press, pp. 185–204, at p. 203.

25 Bowler, Peter J. (2009), 'Popular science', ibid., pp. 622–33, p. 631.

26 Toy, Jennifer (2002), 'The Ingelfinger rule: Franz Ingelfinger at the New England Journal of Medicine 1967–77', *Science Editor* 25, pp. 195–8.

27 Sismondo, Sergio (2004), 'Pharmaceutical maneuvers', *Social Studies of Science* 34, pp. 149–59; Sismondo, Sergio (2007), 'Ghost management: how much of the medical literature is shaped behind the scenes by the pharmaceutical industry?', *PLoS Medicine* 4(9), e286; Sismondo, Sergio (2008), 'Pharmaceutical company funding and its consequences: a qualitative systematic review', *Contemporary Clinical Trials* 29, pp. 109–13.

28 Close, Frank (1991), *Too Hot to Handle: The Race for Cold Fusion*, Princeton, NJ: Princeton University Press.

29 Gieryn, Thomas F. (1991), 'The events stemming from Utah', *Science* 252 (17 May), pp. 994–5, at p. 995.

30 See Mandelbrot, Benoît B. (1982), *The Fractal Geometry of Nature*, New York: W. H. Freeman.

31 Weart, Spencer R. (2003), *The Discovery of Global Warming*, Cambridge, MA: Harvard University Press, p. 89.

32 Van der Sluijs, Jeroen, Josee van Eijndhoven, Simon Shackley and Brian Wynne (1998), 'Anchoring devices in science for policy: the case of consensus around climate sensitivity', *Social Studies of Science* 28, pp. 291–323.

33 Weart (2003), p. 151.

34 Parson, Edward A. (2003), *Protecting the Ozone Layer: Science and Strategy*, Oxford: Oxford University Press.

35 Quoted in Demeritt, David (2001), 'The construction of global warming and the politics of science', *Annals of the Association of American Geographers* 91, pp. 307–37, at p. 308.

36 Ibid., p. 328.

37 Weart (2003), pp. 156, 159.

38 Colwell, Rita R. (2004), 'A tangled bank: reflections on the tree of life and human health', in Joel Cracraft and Michael J. Donoghue (eds), *Assembling the Tree of Life*, Oxford: Oxford University Press, pp. 18–24.
39 Wilson, Edward O. (1995), *Naturalist*, London: Penguin, p. 358.
40 Ibid., p. 359.
41 Merson, John (2000), 'Bio-prospecting or bio-piracy: intellectual property rights and biodiversity in a colonial and postcolonial context', *Osiris* 15, pp. 282–96; Parry, Bronwyn (2004), *Trading the Genome: Investigating the Commodification of Bio-Information*, New York: Columbia University Press.
42 Merson (2000), p. 283.
43 Ibid., p. 284.
44 Drahos, Peter, and John Braithwaite (2002), *Information Feudalism: Who Owns the Knowledge Economy?* London: Earthscan; Drahos, Peter, and John Braithwaite (2007), 'Intellectual property, corporate strategy, globalisation: TRIPS in context', in William T. Gallagher (ed.), *Intellectual Property*, London: Ashgate, pp. 233–62.
45 Goodman, Jordan, and Vivien Walsh (2001), *The Story of Taxol: Nature and Politics in the Pursuit of an Anti-Cancer Drug*, Cambridge: Cambridge University Press, pp. 170, 193.
46 Merson (2000), pp. 288–9.
47 Shiva, Vandana (1990), 'Biodiversity, biotechnology and profit: the need for a people's plan to protect biological diversity', *The Ecologist* 20, pp. 44–7; Merson (2000), p. 289.
48 Merson (2000), p. 290.
49 Parry (2004), p. 4.
50 Ibid., p. 5.
51 Merson (2000), p. 293.
52 Ibid., p. 294.
53 Epstein, Steven (1995), 'The construction of lay expertise: AIDS activism and the forging of credibility in the reform of clinical trials', *Science, Technology and Human Values* 20, pp. 408–37, at p. 408; Epstein, Steven (1996), *Impure Science: AIDS, Activism, and the Politics of Knowledge*, Berkeley: University of California Press.
54 Epstein (1995), p. 411.
55 Ibid., pp. 413–14; Epstein (1996).
56 Epstein (1995), p. 422.
57 Ibid., p. 428.
58 Wynne, Brian (1989), 'Sheepfarming after Chernobyl: a case study in communicating scientific information', *Environment* 31(2), pp. 10–15.
59 Kim, Kiheung (2007), *The Social Construction of Disease: From Scrapie to Prion*, London: Routledge.
60 Giddens, Anthony (1998), 'Risk society: the context of British politics', in Jane Franklin (ed.), *The Politics of Risk Society*, Cambridge: Polity, pp. 23–35, at p. 32.
61 Wilsdon, James, and Rebecca Willis, *See-Through Science: Why Public Engagement Needs to Move Upstream*, London: Demos.
62 McCray, W. Patrick (2005), 'Will small be beautiful? Making policies for our nanotech future', *History and Technology* 21, pp. 177–203, at p. 181.
63 Monthioux, Marc, and Vladimir L. Kuznetsov (2006), 'Who should be

given the credit for the discovery of carbon nanotubes?' *Carbon* 44, pp. 1621–3, at p. 1622.

64 Ibid., p. 1623.
65 Royal Society and Royal Academy of Engineering (2004), *Nanoscience and Nanotechnologies: Opportunities and Uncertainties*, London: Royal Society, p. 17.
66 McCray (2005), p. 180.
67 Ibid., p. 184.
68 Ibid., p. 187.
69 Roco, Mihail C., and William Sims Bainbridge (eds) (2001), *Societal Implications of Nanoscience and Nanotechnology*, Arlington, VA: NSET Workshop Report, p. 41, www.wtec.org/loyola/nano/NSET.Societal.Implications/nanosi.pdf.
70 Ibid., p. 50.
71 Ibid.
72 Royal Society and Royal Society of Engineering (2004), p. 13.

CHAPTER 20 SCIENCE IN THE TWENTIETH CENTURY AND BEYOND

1 Guillemin, Jeanne (2005), *Biological Weapons: From the Invention of State-Sponsored Programs to Contemporary Bioterrorism*, New York: Columbia University Press, p. 21.
2 Nye, Mary Jo (1984), 'Scientific decline: is quantitative evaluation enough?', *Isis* 75, pp. 697–708, at p. 708.
3 Graham, Loren R. (1993), *Science in Russia and the Soviet Union*, Cambridge: Cambridge University Press, pp. 151–3.
4 Capshew, James H., and Karen Rader (1992), 'Big Science: Price to present', *Osiris* 7, pp. 3–25.
5 For a survey, see Cozzens, Susan E. (1997), 'The discovery of growth: statistical glimpses of twentieth-century science', in John Krige and Dominique Pestre (eds), *Science in the Twentieth Century*, Amsterdam: Harwood Academic Press, pp. 127–42.
6 Committee on Prospering in the Global Economy of the 21st Century (2007), *Rising Above the Gathering Storm: Energizing and Employing America for a Brighter Economic Future*, Washington, DC: National Academies Press.
7 Basalla, George (1967), 'The spread of Western science', *Science* 156 (5 May), pp. 611–22, at p. 620.
8 Hodge, Jonathan (2009), 'Evolution', in Peter J. Bowler and John V. Pickstone (eds), *The Cambridge History of Science*, Vol. 6: *The Modern Biological and Earth Sciences*, Cambridge: Cambridge University Press, pp. 244–64.
9 Hughes, Thomas P. (1989), *American Genesis: A Century of Invention and Technological Enthusiasm, 1870–1970*, New York: Penguin, p. 187.
10 Ibid., pp. 222, 232.
11 Chandler, Alfred D., Jr. (1977), *The Visible Hand: The Managerial Revolution in American Business*, Cambridge, MA: Harvard University Press; Dennis, Michael Aaron (1987), 'Accounting for research: new histories of corporate laboratories and the social history of American science',

Social Studies of Science 17, pp. 479–518, at p. 487; Noble, David F. (1977), *America by Design: Science, Technology, and the Rise of Corporate Capitalism*, New York: Alfred A. Knopf.

12 Shinn, Terry (2003), 'The industry, research, and education nexus', in Mary Jo Nye (ed.), *The Cambridge History of Science*, Vol. 5: *The Modern Physical and Mathematical Sciences*, Cambridge: Cambridge University Press, pp. 133–53.

13 Ibid., p. 134.

14 Ibid., p. 138.

15 Ibid., p. 147.

16 Geiger, Roger (1986), *To Advance Knowledge: The Growth of American Research Universities, 1900–1940*, Oxford: Oxford University Press; Servos, John W. (1980), 'The industrial relations of science: chemical engineering at MIT, 1900–1939', *Isis* 71, pp. 531–49; Layton, Edwin T., Jr. (1971), *The Revolt of the Engineers: Social Responsibility and the American Engineering Profession*, Cleveland: Case Western Reserve University Press; Shinn (2003).

17 Rosenberg, Nathan, and Richard R. Nelson (1994), 'American universities and technical advance in industry', *Research Policy* 23, pp. 323–48. Shinn (2003), p.152.

18 Although see the Peirce family connections discussed in Kevles, Daniel J. (1971), *The Physicists: The History of a Scientific Community in Modern America*, Cambridge, MA: Harvard University Press, p. 48.

19 Riordan, Michael, and Lillian Hoddeson (1997), *Crystal Fire: The Birth of the Information Age*, New York: W.W. Norton, p. 74; Simões, Ana (2003), 'Chemical physics and quantum chemistry in the twentieth century', in Nye, *The Cambridge History of Science*, Vol. 5: *The Modern Physical and Mathematical Sciences*, pp. 394–412, at p. 402.

20 Price, Don K. (1954), *Government and Science: Their Dynamic Relation in American Democracy*, New York: New York University Press, p. 3.

21 Ibid., p. 9.

22 Porter, Theodore M. (1995), *Trust in Numbers: The Pursuit of Objectivity in Science and Public Life*, Princeton, NJ: Princeton University Press.

23 Price (1954), pp. 30–1.

24 Ibid., p. 190.

25 Paul Forman (1987), 'Beyond quantum electronics: national security as basis for physical research in the United States', *Historical Studies in the Physical Sciences* 18, pp. 149–229.

26 Committee on Prospering in the Global Economy of the 21st Century (2007), pp. 204–5.

27 See Nye (1984) for a discussion of some of the problems.

28 Bloor, David (1976), *Knowledge and Social Imagery*, London: Routledge & Kegan Paul.

29 Churchill, Frederick B. (1980), 'The modern evolutionary synthesis and biogenetic law', in Ernst Mayr and William B. Provine (eds), *The Evolutionary Synthesis: Perspectives on the Unification of Biology*, Cambridge, MA: Harvard University Press, pp. 112–22, at p. 113.

30 Schweber, Silvan S. (2003), 'Quantum field theory: from QED to the Standard Model', in Nye, *The Cambridge History of Science*, Vol. 5: *The Modern Physical and Mathematical Sciences*, pp. 375–93, at p. 383; Oldroyd,

David (2009), 'Geophysics and geochemistry', in Bowler and Pickstone, *The Cambridge History of Science*, Vol. 6: *The Modern Biological and Earth Sciences*, pp. 395–415; Cittadino, Eugene (2009), 'Botany', ibid., pp. 225–42.

31 For an introduction, see Bartholomew, James R. (1997), 'Science in twentieth-century Japan', in Krige and Pestre, *Science in the Twentieth Century*, pp. 879–96; see also Low, Morris, Shigeru Nakayama and Hitoshi Yoshioka (1999), *Science, Technology and Society in Contemporary Japan*, Cambridge: Cambridge University Press.

32 Galison, Peter (2004), 'Removing knowledge', *Critical Inquiry* 31, pp. 229–43, at pp. 229, 231, his emphasis removed.

33 Courtland, Rachel (2008), 'How the world invests in R&D', *Nature* 451 (23 January) p. 378, www.nature.com/news/2008/080123/full/451378a. html.

34 Royal Society (2011), *Knowledge, Networks and Nations: Global Scientific Collaboration in the 21st Century*, London: Royal Society, p. 16.

35 Ibid.

36 Ibid., p. 14.

37 Committee on Prospering in the Global Economy of the 21st Century (2007).

38 Ibid., p. 205.

39 Quoted in Menon, Subhadra (2008), 'India's rise to the moon', *Nature* 455 (16 October), pp. 874–5, www.nature.com/nature/journal/v455/n7215/ full/455874a.html.

40 Committee on Prospering in the Global Economy of the 21st Century (2007), p. 79.

41 Mashelkar, Raghunath (2005), 'India's R&D: reaching for the top', *Science* 307 (4 March), pp. 1415–17, www.sciencemag.org/content/307/5714/1415. full.

42 Menon (2008).

43 Jayaraman, K. S. (2008), 'Entomologists stifled by Indian bureaucracy', *Nature* 452 (5 March), www.nature.com/news/2008/080305/full/452007a. html.

44 Hassan, Mohamed (2008), 'Beijing 1987: China's coming-out party', *Nature* 455 (2 October), pp. 598–9, www.nature.com/nature/journal/v455/ n7213/full/455598a.html.

45 Cyranoski, David (2008a), 'China: visions of China', *Nature* 454 (24 July), pp. 384–7, www.nature.com/news/2008/080723/full/454384a.html.

46 Royal Society (2011), p. 43.

47 Cyranoski (2008a).

48 Xue, Lan (2008), 'China: the prizes and pitfalls of progress', *Nature* 454 (24 July), pp. 398–401, www.nature.com/nature/journal/v454/n7203/ full/454398a.html.

49 Editorial (2009a), 'Collective responsibilities', *Nature* 457 (19 February), p. 935, www.nature.com/nature/journal/v457/n7232/full/457935a.html.

50 Cyranoski, David (2008b), 'Check your GPS at the border', *Nature* 451 (20 February), p. 871, www.nature.com/news/2008/080220/full/451871a. html.

51 Cyranoski (2008a).

52 Ibid.

53 Ibid.

54 Quoted in 'China: in their words', *Nature* 454 (24 July 2008), pp. 399–401, www.nature.com/nature/journal/v454/n7203/full/454399a.html.

55 Butler, Declan (2008a), 'China: the great contender', *Nature* 454 (24 July), pp. 382–3, www.nature.com/news/2008/080723/full/454382a.html.

56 Committee on Prospering in the Global Economy of the 21st Century (2007), p. 210.

57 Cyranoski, David (2008c), 'Pharmaceutical futures: made in China?', *Nature* 455 (29 October), pp. 1168–70, www.nature.com/news/2008/081029/full/4551168a.html.

58 Cyranoski (2008c).

59 Ibid.

60 Xue (2008).

61 Schiermeier, Quirin (2008), 'Russian science academy rejects Putin ally', *Nature* 453 (4 June), pp. 702–3, www.nature.com/news/2008/080604/full/453702a.html.

62 'Presidential election disappoints reformists', *Nature* (4 June 2008), www.nature.com/news/2008/080604/full/453702a/box/1.html.

63 Butler, Declan (2008b), 'French scientists revolt against government reforms', *Nature* 457 (4 February), pp. 640–1, www.nature.com/news/2009/090204/full/457640b.html.

64 Butler, Declan (2008c), 'Researcher battles CNRS reforms', *Nature* 454 (9 July), p. 143, www.nature.com/news/2008/080709/full/454143a.html.

65 Editorial (2009b), 'Growing pains', *Nature* 460 (23 July), www.nature.com/nature/journal/v460/n7254/full/460435a.html.

66 Brumfiel, Geoff (2008a), 'Funding: the research revolution', *Nature* (18 June), pp. 975–6, www.nature.com/news/2008/180608/full/453975a.html.

67 Cressey, Daniel (2008a), 'Geology: the next land rush', *Nature* 451 (2 January), pp. 12–15, www.nature.com/news/2008/020108/full/451012a.html.

68 Weinberger, Sharon (2008), 'Defence research: still in the lead?', *Nature* 451 (23 January), pp. 390–9, www.nature.com/news/2008/080123/full/451390a.html.

69 Butler, Declan (2009), 'Bush's legacy: the wasted years', *Nature* 457 (14 January), pp. 250–1, www.nature.com/news/2009/090114/full/457250a.html.

70 Editorial (2008a), 'Election fireworks', *Nature* 451 (3 January), www.nature.com/nature/journal/v451/n7174/full/451001a.html.

71 Hayden, Erika Check (2009), 'Obama overturns stem-cell ban', *Nature* (9 March), www.nature.com/news/2009/090309/full/458130a.html.

72 'Germany authorises stem cell imports', BBC News (30 January 2002), http://news.bbc.co.uk/1/hi/world/europe/1791365.stm.

73 Hayden, Erika Check (2008a), 'Stem cells: the 3-billion-dollar question', *Nature* 453 (30 April), pp. 18–21, www.nature.com/news/2008/080430/full/453018a.html.

74 Wadman, Meredith (2008a), 'Stem cells: stuck in New Jersey', *Nature* 451 (6 February), pp. 622–6, www.nature.com/news/2008/080206/full/451622a.html.

75 Cyranoski, David (2008d), 'Japan ramps up patent effort to keep up iPS lead', *Nature* 453 (18 June), pp. 962–3, www.nature.com/news/2008/080618/full/453962a.html.

76 Cyranoski, David (2008e), 'Stem cells: a national project', *Nature* 451 (16 January), p. 229, www.nature.com/news/2008/080116/full/451229a.html.

77 Cyranoski, David (2008f), 'Stem cells: 5 things to know before jumping on the iPS bandwagon', *Nature* 452 (26 March), pp. 406–8, www.nature.com/news/2008/080326/full/452406a.html.

78 Wadman, Meredith (2008b), 'The winding road from ideas to income', *Nature* 453 (11 June), pp. 830–1, www.nature.com/news/2008/080611/full/453830a.html.

79 Ibid.

80 Butler, Declan (2008d), 'Translational research: crossing the valley of death', *Nature* 453 (11 June), pp. 840–2, www.nature.com/news/2008/080611/full/453840a.html.

81 Carninci, Piero (2009), 'Molecular biology: The long and short of RNAs', *Nature* 457 (19 February), pp. 974–5, www.nature.com/nature/journal/v457/n7232/full/457974b.html.

82 Hayden, Erika Check (2008b), 'Thousands of proteins affected by miRNAs', *Nature* 454 (30 July), p. 562, www.nature.com/news/2008/080730/full/454562b.html.

83 'GlaxoSmithKline does deal to develop microRNA drugs', *Nature* 452 (23 April 2008), p. 925, www.nature.com/news/2008/080423/full/452925d.html.

84 Abbott, Alison (2008), 'Pharmaceutical futures: a fiendish puzzle', *Nature* 455 (29 October), pp. 1164–7, www.nature.com/news/2008/081029/full/4551164a.html.

85 Ingolia, Nicholas T., and Jonathan S. Weissman (2008), 'Systems biology: reverse engineering the cell', *Nature* 454 (28 August), pp. 1059–62, www.nature.com/nature/journal/v454/n7208/full/4541059a.html.

86 Abbott (2008).

87 Wadman, Meredith (2008c), 'James Watson's genome sequenced at high speed', *Nature* 452 (16 April), p. 788, www.nature.com/news/2008/080416/full/452788b.html.

88 Ibid.

89 Wadman, Meredith (2008d), 'Crunch time for peer review in lawsuit', *Nature* 452 (5 March), pp. 6–7, www.nature.com/news/2008/080305/full/452006a.html.

90 Cressey, Daniel (2008b), 'Merck accused of disguising its role in research', *Nature* 453 (15 April), p. 791, www.nature.com/news/2008/080415/full/452791b.html.

91 Brumfiel, Geoff (2008b), 'Newsmaker of the year: the machine maker', *Nature* 456 (17 December), pp. 862–8, www.nature.com/news/2008/081217/full/456862a.html.

92 Stephan Dürr et al. (2008), 'Ab initio determination of light hadron masses', *Science* 322 (21 November), pp. 1224–7, www.sciencemag.org/content/322/5905/1224.abstract.

93 Wilczek, Frank (2008), 'Particle physics: mass by numbers', *Nature* 456 (27 November), pp. 449–50, www.nature.com/nature/journal/v456/n7221/full/456449a.html.

94 Editorial (2008b), 'Deserting the hungry', *Nature* 451 (17 January), pp. 223–4, www.nature.com/nature/journal/v451/n7176/full/451223b.html.

INDEX